Fundamentals of Infrastructure Engineering

Civil Engineering Systems

Civil and Environmental Engineering

A Series of Reference Books and Textbooks

1. **Preliminary Design of Bridges for Architects and Engineers**
   Michele Melaragno
2. **Concrete Formwork Systems**
   Awad S. Hanna
3. **Multilayered Aquifer Systems: Fundamentals and Applications**
   Alexander H.-D. Cheng
4. **Matrix Analysis of Structural Dynamics: Applications and Earthquake Engineering**
   Franklin Y. Cheng
5. **Hazardous Gases Underground: Applications to Tunnel Engineering**
   Barry R. Doyle
6. **Cold-Formed Steel Structures to the AISI Specification**
   Gregory J. Hancock, Thomas M. Murray, Duane S. Ellifritt
7. **Fundamentals of Infrastructure Engineering: Civil Engineering Systems: Second Edition, Revised and Expanded**
   Patrick H. McDonald

Additional Volumes in Production

Handbook of Pollution Control and Waste Minimization
Abbas Ghassemi

Introduction to Approximate Solution Techniques, Numerical Modeling, and Finite Element Methods
Victor N. Kaliakin

Fundamentals of Infrastructure Engineering

Civil Engineering Systems

Second Edition, Revised and Expanded

Patrick H. McDonald
North Carolina State University
Raleigh, North Carolina

MARCEL DEKKER, INC. NEW YORK • BASEL

The first edition was published as *Civil Engineering Systems* by Patrick McDonald, New York: McGraw-Hill, 1993.

ISBN: 0-8247-0612-9

This book is printed on acid-free paper.

Headquarters
Marcel Dekker, Inc.
270 Madison Avenue, New York, NY 10016
tel: 212-696-9000; fax: 212-685-4540

Eastern Hemisphere Distribution
Marcel Dekker AG
Hutgasse 4, Postfach 812, CH-4001 Basel, Switzerland
tel: 41-61-261-8482; fax: 41-61-261-8896

World Wide Web
http://www.dekker.com

The publisher offers discounts on this book when ordered in bulk quantities. For more information, write to Special Sales/Professional Marketing at the headquarters address above.

Current printing (last digit):
10 9 8 7 6 5 4 3 2 1

PRINTED IN THE UNITED STATES OF AMERICA

PREFACE

In 1987 the new required course CE 375 Civil Engineering Systems was introduced into the curriculum at North Carolina State University. In the prospectus it was described as:

> "A broad-based, systematic approach to Civil Engineering planning, analysis and design for large-scale projects in construction, structures, transportation, water resources, and other Civil Engineering areas."

The justification for the course was given as:

> "The Civil Engineer is intimately involved in the planning, design, analysis, construction and operation, management and maintenance of large-scale, unique (usually one of a kind), multimillion-dollar projects. These projects, over the years, have increased in complexity manyfold. The course is intended to introduce the Civil Engineering student to the use of the systems approach from planning to implementation of these large-scale projects with which the student will be involved during a professional career."

The course objectives were stated as:

> "The main objective of this course is to introduce the student to systems methodology for analyzing large-scale projects. The student will follow a large-scale Civil Engineering project from conception through maturation using the systems approach to analyze the project and determine and choose (along with the client) the best alternative and design, and the application of sound management techniques during the construction phase."

Taken together, these expressions describe an undergraduate curriculum unit which is itself probably as unique as the projects which it purports to treat. The main focus is stated expressly to be an exposition of systems methodology—that is, the theory of systems. This theory is frequently very mathematical in nature, and, indeed, can sometimes become quite esoteric. On the other hand, it is clearly the purpose to apply this focus to the central feature of pragmatic engineering design work; namely, optimization.

Some might believe that these central features are mutually exclusive, as disparate in character as they are often found to be in other circumstances. Others may say that it is unsound pedagogy to commingle such incommensurate elements in one course offering.

Teachers experienced in trying to do this difficult and very demanding yet highly rewarding exercise may well answer that it not only can be done effectively, but that a certain unity of fabric is woven in the attempt which results in a pleasing tartan. As a matter of fact, the full spectrum of engineering competence is invoked, and students who are adequately challenged by this broader outlook on their chosen careers are usually compensated with a high degree of satisfaction as well as the confirmation of their career choices.

On the face of it, the most modern methods available should be chosen to support this instructional program. Specifically, this means, of course, computer programs designed to provide real assistance in the analysis, optimization and simulation stages of project work. In post-baccalaureate career practice, the student will encounter routines for mainframes, networks and personal machines. Various softwares will be offered for these devices, and most of them will likely be very specialized for reasons of efficiency.

Scaled-down versions of such larger-scoped algorithms are used in teaching the course. These feature user-friendly interactive programs carefully honed to a state in which the student can see the outcomes without excessive interpretation. In their most recent embodiments, these supply the contents of a locker on the *eos* net, so that students can call them up at will at any workstation (or through a modem).

A prerequisite to the course is therefore a minimum working knowledge of computers, including one or more of the standard transportable languages, although for the most part the codes are compiled and ready to run. Differential equations is also at least a corequisite, and some exposure to matrix methods is highly desirable.

It would be preferable, naturally, to have the student come with some knowledge of structures, hydraulics, first courses in transportation, materials, and construction practices. But if the course waits in the curriculum until all of these are available, it will be too late to be of much benefit in the senior design work. As it is, either semester of the third year is acceptable, and if some of the jargon of the fields must be acquired "on-the-job," so be it! Possibly this may serve to increase the student's motivation in parallel or subsequent coursework.

A number of topics which are beyond the scope of the core-curriculum one-semester course previously described are of great interest and applicability in civil engineering professional practice. These higher-level systems topics find their most significant utility at the frontiers of research in the several fields of professional activity today. To accommodate the exposure of these subjects to students and others at a level commensurate with current backgrounds and preparations is the motive behind the addition of two chapters beyond those which can reasonably be expected to be covered in the required undergraduate course.

Chapters 15 and 16 are intended for the use of advanced undergraduates and first-year graduate students. The instructor may wish to plan a second semester for covering this work, or it may be used as the basis for independent study. Modeling of systems using state-space methods, and the analysis and optimization of their designs, is increasingly prevalent in the research literature. So also are the methods of autonomous systems. In fact, the principal topics in this category are likely to dominate their several fields in the near future.

In addition, Chapters 15 and 16 are designed to serve as an introduction to the topics contained therein for the benefit of practicing professionals who may find themselves caught up in the rapidly developing advances of these fields, and who might not have experienced coursework in these subjects during their collegiate careers. A number of sources for initiating self-study in such fields are cited.

The author asks of those who read and study this book the following:

- The student is asked to be patient—some of the more valuable aspects of the subject necessarily come later rather than earlier or at once, and usually only after a substantial investment in study and work.

- Other teachers are requested to forbear from criticizing the writer's more unconventional methods. The overall sweep of so broad a subject and the novel assimilation of disparate topics in a unified amalgam can be expected to give rise to some peculiarities, oddities, prejudices, and uneven tendencies.

- The practicing professional, if any, who reads this will please remember that the more delicious aspects of the subject are those for which one is paid handsomely in "real" life, and that generally rules out students (and faculty).

The writer hopes that some small nuggets of value may be found herein, and will be satisfied if the errors are also small. He expresses his special thanks to his colleagues John Baugh, Bill Galler, and Kerry Havner for helpful discussions.

The author particularly thanks CIVIL-COMP PRESS LTD of Edinburgh and its Editor Professor Barry H. V. Topping for their kind concurrence in his adaptation in Chapters 15 and 16 of material from OPTIMIZATION AND CONTROL IN CIVIL AND STRUCTURAL ENGINEERING and NOVEL DESIGN AND INFORMATION TECHNOLOGY APPLICATIONS FOR CIVIL AND STRUCTURAL ENGINEERING, journals containing papers presented at the *Seventh International Conference on Civil and Structural Engineering* and *The Fifth International Conference on the Applications of Artificial Intelligence to Civil and Structural Engineering* held at Oxford University in 1999. He heartily recommends these and their companion journals covering other papers at the same conferences as collateral reading directly appropriate to the current frontiers of research in Civil Engineering Systems.

P. H. McDonald

CONTENTS

CHAPTER 1

THE NATURE OF SYSTEMS

1.1 INTRODUCTION

The word *system* has one of the richer etymological developments in the lexicon. In large measure, this can be attributed to the fact that persons from many differing backgrounds employ it to carry a meaning which few if any other words can supply. In any event, the shadings and nuances of meanings together with the splendid variety of modifying words and phrases coupled to it in its multiple usages are what endow it with its widespread utility.

A recent unabridged dictionary lists thirteen major meanings of *system*, with several sub-meanings in many of these larger senses. Even a collegiate dictionary will mention not less than 8-to-10 usages.

Of course, some of these are appropriate to fields of knowledge far removed from engineering activities. In particular, professionals in ecclesiastic and philosophic fields make large and important use of the word.

The first and probably most significant meaning can be quoted as follows:

> "A complex unity formed of many diverse parts subject to a common plan or serving a common purpose: an aggregation or assemblage of objects joined in regular interaction or interdependence: a set of units combined by nature or art to form an integral, organic, or organized whole: an orderly working totality: a coherent unification: a group of devices or artificial objects forming a network or used for a common purpose."[1]

Alternatively:

> "An assemblage of objects united by some form of regular interaction or interdependence: an organic or organized whole."[2]

There are thus several generic components of systems. These include:

elements, relationships, boundaries, and totality.

---

[1] "Webster's Third New International Dictionary," Philip Babcock Gove, Editor-in-Chief, *G. & C. Merriam Company*, Springfield, Massachusetts, USA, 1981, p. 2322.

[2] "Webster's Collegiate Dictionary," Fifth Edition, *G. & C. Merriam Company*, Springfield, Massachusetts, USA, 1941, p. 1013.

An examination of all systems worthy of the name will disclose the existence of some things which function to fulfil the roles of each of the above participants in the systems.

1.2 ELEMENTS

What are the elements of a system? One way to begin to understand these fundamental components is to construct a list of synonyms commonly employed to describe their behavior in context. Such a list might be similar to that below.

SYNONYMS for ELEMENTS

PARTS	COMPONENTS	UNITS
BODIES	ORGANS	FORCES
BITS	IDEAS	LAWS
DOCTRINES	PRINCIPLES	PORTIONS
OBJECTS	SUBSTANCES	PIECES
FRAGMENTS	TRAITS	RUDIMENTS
CONSTITUENTS	FEATURES	INGREDIENTS
INDIVIDUALS	FRACTIONS	MEMBERS
SHARES	CHARACTERS	VOICES
THINGS	DIVISIONS	DEVICES
BITES	MORSELS	WHITS
QUANTITIES		LINEAMENTS

It will be noted immediately that this landscape of elements includes things of widely-differing measures. Some are hardwares, whereas others are fluffy concepts that can barely be defined even in their own terms. Some are visible, others are conceptual. Some are of rigid shape—permanent—at the same level with those which are ephemeral—continuously changing shape.

System elements are frequently systems themselves. Some are active, moving, while others remain stationary, static. The range can thus be from inanimate to autonomous hypercomplex.

In sharp contrast to the commonplace assumptions found in the analysis of many special systems, the more general case is that in which the elements are incommensurate with each other, rather than being all alike in some way. It is rather like a fruit salad, wherein the elements—apples, oranges, pears, bananas, strawberries, cherries, grapes, pineapples, peaches, etc.—all have distinct colors, shapes, textures, sizes, tastes, and additional properties.

The fundamental nature of the system element, is, therefore, to be itself with all the attributes that that entails. It stands in its own place and does its own thing without a worry about the other elements, its relationships to the other elements or the overall functioning of the system into which it is incorporated.

1.3 RELATIONSHIPS

If the system in question is, say, a machine or a structure, the elements are related to each other in geometric ways, either statically or through the geometry of motion—that is, by kinematics. This geometry may be enforced by the constraint of "rigid" steel pins or other metal surfaces. Of course, all students of mechanics realize that it is in reality a law of force which does the confining.

The general system is held together with constraints which may be "laws" binding the elements to a system dynamic which is proscribed thereby in some fashion. In other words, the elements are not allowed to exercise their natural workings to the fullest extent; at least not necessarily. The law of constraint in a given case may literally be enforced by mechanical force, or it may be a soft feedback path with only visual or other sensory appreciation connecting the elements in question. It may be a flexible (extensible) string or an unbending total sum of money, materials, time, property, or other substance.

Relationships are thus functional in character. They may be (indeed usually are) ultimately expressed mathematically. This allows for a degree of generality that transcends the elementary fasteners of conventional hardware analysis. It may even permit the system designer to include a microprocessor in the differential trajectory between elements. Whatever mentality can be expressed by logic is therefore fair game as a constraint.

A system thereby incorporates its elements for their individuality, but impresses upon them an overriding demand that they function for the benefit of the whole. The shackles are usually called *constraints*.

1.4 BOUNDARIES

It is necessary to bound the system to a delimited amount of space and time. The envelope which cloaks the system is then called its boundary. One can make a fetish of drawing the system boundary, as is done in thermodynamics and in mechanics where it is euphemistically called a *free-body-diagram*. Whether drawn or not drawn, the edge of the region or space in which a system moves is a concept necessary to its analysis.

For one thing, the boundary rules out elements, effects, constraints, and functionings that are not germain to the system under definition. The City Limits is a valuable construct because it proscribes the polity to that area within the contour and rules out adjacent municipalities as well as the rural hinterlands.

Sometimes, the region is defined by "infeasibilities" or other impracticalities or impossibilities. Negative values for physical quantities, or, worse, complex or imaginary outcomes for parameters known to be real (as in real estate) things may draw practical lines in the dust to estop the system's running.

Occasionally, a system boundary will be drawn to exclude regions of space which should not be penetrated by the system dynamic on ethical, moral or legal grounds. Sometimes a limit is placed on operations because of good or bad practices as established by industry standards or codes. Not infrequently it may be the case that a particular requirement may be considered as either a constraint or a boundary, depending upon the point of view.

1.5 TOTALITY

The totality of systems can be summarized in the overall names applied to them in specific cases. Such names amount to synonyms for the functions representing their behaviors. Some synonyms for system are:

SYNONYMS for SYSTEM

SCHEME	NETWORK	COMPLEX
ORGANISM	ECONOMY	AGGREGATE
GROUP	BODY	BEING
STRUCTURE	WHOLE	ARRANGEMENT
CLASSIFICATION	PLAN	ORGANIZATION
PRACTICE	HYPOTHESIS	TREATISE
SET	PATTERN	FORM
PHILOSOPHY	NOTATION	METHOD
DESIGN	SEQUENCE	SERIES
COLLECTION	ASSEMBLAGE	PROCEDURE
SOCIETY	RELIGION	UNION
GOVERNMENT		POLITICAL ORDER

A perusal of the words listed above quickly gives the intended impression that there is a unity, oneness, wholeness, or totality expressed by calling these names. The label itself connotes the functioning of the system named in most cases. In any case the cementation or bondage implied by the name carries with it the assumption that, let loose, the thing can be on its own. It can operate autonomously, or be free-standing, and is self-contained, logistically adequate, and independent.

Think of a sports enthusiast who carries a duffel bag to the site of an athletic event. The sporter can be transformed in a short time using the elements in the duffel bag from an ordinary traveler into a uniformed, armored, prepared athlete ready to participate in the game. There is a unification of the system elements, constraints and boundaries for the purpose of the competition.

The German word *einheit* is uniquely appropriate to express the concept of totality which is embodied in the meaning of a system. A close rendition into English may be the "one-ness" used above.

It may be appropriate to mention here the system-analyst's favorite expression concerning systems. It is sometimes said that

"a system is more than the sum of its parts."

What this means is that a system is a complex entity of higher order than any and all properties borne by its elements, the constraints, the boundaries, and the functional purpose. It has, so-to-speak, a "mind of its own," and will behave in manners oftentimes surprising to even the mind which created it in the first place. If this sounds a bit like animism, well, it can certainly be taken to imply that higher-order systems may in fact demonstrate reflections of the intellects of their creators. Adaptive behaviors are a commonplace in work with systems, and learning machine theory is a standard model in the theory of automata.

Expert systems and artificial intelligence are practical utilizations of the idea that systems can behave in approximation to at least some of the lower levels of ratiocinative processes in humans. It is not necessary to postulate fiendish, malevolent or capricious motives to systems or indulge in other forms of anthropomorphism in order to recognize the higher-order construct represented by them.

Self-ordering systems are indeed anti-entropic constructs, and self-regulating systems do in fact display mentality. Moreover, systems operative in human societies do surely exhibit collective human attributes, as is only to be expected.

As in all of life, the poets have the last—the best—words:

"Our little systems have their day;
They have their day and cease to be;
They are but broken lights of thee;
And thou, O Lord, art more than they."

Alfred Lord Tennyson
IN MEMORIAM
1850

1.6 EOS

eos

The system is quiet today,
A clean face on the video;
No white-on-black banners running,
To surcharge the log-in window.

It bucks up in double-quick time,
Neither messages nor e-mail;
Call for TED and another TED,
To work the main and sub as well.

Drag the exes to their places;
Change the codes to make it better;
Save and exit, click on window,
Call for pee-cee dash-oh, get it!

System.go and eos percent—
One hundred percent—it goes fast;
The program runs, talks back to me,
Returns again to prompt at last.

No lost pee-eye-dees, other junk;
The system is quiet today,
So eff-tee-pee to ell-pee-are,
"Is anyone out there, I say?"

P. H. McDonald
December 13, 1991

CHAPTER 2

SYSTEMS IN ENGINEERING

2.1 INTRODUCTION

Every field of engineering concerns itself with some forms of systems. Of course, some fields utilize systems science and theory to a greater extent than do others, but a close inspection of the curriculum in the several branches will show a number of courses which deal in whole or in part with the systems appropriate to that discipline.

The names of some such offerings at North Carolina State University include the following:

MAE 435	PRINCIPLES OF AUTOMATIC CONTROL
MAE 520	INDUSTRIAL ROBOTICS
NE 405	REACTOR SYSTEMS
NE 522	REACTOR DYNAMICS AND CONTROL
BAE 461	ANALYSIS OF AGRICULTURAL SYSTEMS
CHE 225	CHEMICAL PROCESS SYSTEMS
CHE 425	PROCESS SYSTEM ANALYSIS AND CONTROL
CE 375	CIVIL ENGINEERING SYSTEMS
CE 411	ENGINEERING CYBERNETICS
CE 501	TRANSPORTATION SYSTEMS ANALYSIS
CE 575	CIVIL ENGINEERING SYSTEMS
MA 585	GRAPH THEORY
MA 586	NETWORK FLOWS
OR 501	INTRODUCTION TO OPERATIONS RESEARCH
OR 527	OPTIMIZATION OF ENGINEERING PROCESSES
OR 531	DYNAMICAL SYSTEMS & MULTIVARIATE CONTROL
OR 561	QUEUES AND STOCHASTIC SERVICE SYSTEMS

CSC 202	OPERATING SYSTEMS CONCEPTS & FACILITIES
CSC 421,2	MANAGEMENT INFORMATION SYSTEMS
CSC 432	DATABASE MANAGEMENT SYSTEMS
CSC 501	DESIGN OF SYSTEMS PROGRAMS
CSC 506	DIGITAL SYSTEMS ARCHITECTURE
ECE 301	LINEAR SYSTEMS
ECE 305	ELECTRIC POWER SYSTEMS
ECE 342	DESIGN OF COMPLEX DIGITAL SYSTEMS
ECE 435	ELEMENTS OF CONTROL
ECE 446	VLSI SYSTEMS DESIGN
ECE 451	POWER SYSTEM ANALYSIS
ECE 516	SYSTEM CONTROL ENGINEERING
IE 308	CONTROL OF PRODUCTION & SERVICE SYSTEMS
IE 361	DETERMINISTIC MODELS IN IE (LP)
IE 401	STOCHASTIC MODELS IN IE
IE 421	INFORMATION AND CONTROL SYSTEMS
IE 441	INTRODUCTION TO SIMULATION
IE 505	LINEAR PROGRAMMING
IE 509	DYNAMIC PROGRAMMING
IE 521	MANAGEMENT DECISION AND CONTROL SYSTEMS

This list, although by no means exhaustive of the courses in which systems theory is expounded, should serve to illustrate the central role played by this topic in the engineering curriculum. In fact, there are those who feel that systems studies are as much a part of the core curriculum as are any of the other engineering sciences.

2.2 CIVIL ENGINEERING PROJECTS

To examine the role of systems analysis in the practice of civil engineering, it is desirable to describe the several features of civil engineering project work. There are four convenient classifications or phases into which these activities fall. These are *planning*, *design*, *construction*, and *operation*.

The initial activity relative to most projects is that which can best be described as planning. The primary objects of this procedure are investigations of the needs for the project, its scope, the impacts it will make, alternatives to it, comparisons with the alternatives, and the resources which can be committed to it. Goal setting is a primary objective of planning.

However, it must be pointed out immediately that too-early establishment of criteria is to be avoided at all costs, since it has been proven that this precludes the evolution of the more creative and the more effective solutions. The Poincaré hypothesis of creativity assumes that the easier solutions to a problem are those which are least valuable; i.e., those which "come up" first. The more valuable alternatives surface later, only after the less-preferable, easier, earlier ones have been dispatched. So it is imperative that criteria for a project's conduct are not allowed to jell before the best options are "out."

Some specific questions can be asked in the planning stage (and, hopefully, answered) which will focus the project and lead to the establishment of guidelines for the later stages. Among these are:

- What are the evidences of the project's need?
- What are the perceived benefits of the project?
- Are there alternatives to carrying out the project?
- What are comparative advantages and disadvantages of the alternatives and the proposed project?
- Where will it be placed (located)?
- How large is it to be?
- Who will pay for it?
- Can it be financed?
- What are its environmental impacts?

The planning stage involves creative discussions by and among project leaders. Activities which characterize this stage are free-form ideation (brainstorming), followed by hard-headed pragmatic evaluation and, ultimately, go-or-no go decision making. Although the quality of effort exerted in this phase is assumed to be at the highest level, the out-of-pocket expense is generally small compared to the heavier costs in later portions of the project. It is thus much better to refine the scope, goals, and features of the project at this time. Major alterations at later stages are considerably more costly.

The second phase of project work is called design. It is frequently claimed that design is the hall-mark of any activity characteristic of engineering. In 1980 the Accrediting Board for Engineering and Technology (ABET) issued a definition of engineering design intended to serve as a standard against which something alleged to be design might be tested.

> "Design is the process of devising a system, component, or process to meet desired needs. It is a decision-making process (often iterative), in which the basic sciences, mathematics and engineering sciences are applied to convert resources optimally to meet a stated objective. Among the fundamental elements of the design process are the establishment of objectives and criteria, analysis, synthesis, construction, testing, and evaluation."

Clearly, the design phase interfaces with planning at the point where objectives and criteria are established. A design normally begins where specifications can be furnished or wishes, desires, or intent of a client expressed. After that, the design engineer will move to make specific the particulars of the design or project. A question-and-answer procedure, with answers evolving steadily, is an appropriate *modus operandi*.

Some typical questions (probably biased along the lines of site layout work) might be:

- How many units should there be?
- What sizes ought there to be?
- Where should each module be placed?
- Should some parts be below grade?
- How high can module structures go?
- What are the infrastructural elements?
- How are modules placed relative to the infrastructure?
- What about access?
- What are the materials options?

It is especially in the design phase that optimization comes to the forefront. Formulas, laws, principles, codes, and guidelines have been reoriented to allow for examination of the "best" outcomes in terms of established criteria. The meaning of "best"—frequently interpreted as least costly—is certainly open to question and interpretation. It can be "most reliable" or "longest lived" or "least maintenance" or a host of other things. Indeed, optimum solutions are often those based upon a number of subjective considerations.

However adjudicated, a design is ultimately produced. It is committed to hard copy and reproduced in all those necessary forms for communication to clients, colleagues, consultants, collaborators, and construction representatives. Blueprints and specifications are the typical design product.

At the completion of the design phase, there is a bid-letting, a bidding procedure and a ceremony of bid opening, after which, assuming a normal situation, a contract will be signed. The project moves into the construction phase.

The construction phase of a civil engineering project is characterized by the mobilization of people, property, materials, and money in a frenzy of activity which results in a building, structure, development, works, terminal, or other installation or institution which is the physical embodiment of the project's purpose. Some questions which can be raised and answered to guide the development of the process are:

- How can the activities be ordered in time to achieve the optimum result?
- What is the estimate of each activity's duration?
- Are there special facilities or equipment needed?
- How much supply of labor will be required?
- Which specialties of craftspeople are called for?
- What costs are expected for each activity?
- Where can the required materials be obtained?
- What institutions can lend construction money?

- Is there a cash-flow schedule?

To be sure, there are design modifications and change orders during the construction stage. But the general expectation is that at the beginning of construction, the design is fairly-well fixed; therefore, the dominant feature of this phase is the logistic aspect.

It is entirely possible that the construction of a project may be carried out by the same organization which created the plans and designs for it, but the more usual case is that an organization specializing in construction will perform the task under a close-ordered contract. The contract will usually specify points in time for the beginning and completion of the work, together with suitable penalties for failure to meet this schedule. The project is thereby delimited, in at least the timewise respect, and, of course, time is at a high premium along with supplies of materiel, machines, money, and personnel.

Both general contractors and subcontractors consequently specialize in procedural systems for managing the work tasks efficiently. This is the essence of the phase called construction.

At the completion of the construction phase, a civil engineering project begins a new phase called operations. The juncture between these two stages is typically the so-called "turn-key" event. The contractor "turns over" the key to the newly-created facility to the owner-operator who, in turn, "turns the key" to open, occupy, use, and operate it.

At this stage a capital installation is freshly in-being, ready to function to fulfil the planned purpose for which it was conceived. In most instances, certainly in the case of the large-scale project, a long life has been the design assumption, so the need arises for an efficient and orderly scheme to keep it running smoothly out to the planning horizon. The emphases, procedures, and practices in such a phase are very different from those found in the others.

The positing of questions to sharpen the focus of an analysis about how to govern this phase can be as useful as were other questions in the previous cases. Some are:

- How can the finished project be run efficiently?
- What are efficient operating policies?
- How often shall the policies be reviewed to account for changed circumstances?
- What maintenance schedules should be put in place for the physical plant, vehicles and the service infrastructure?
- What long-term changes can be expected in supply and demand for materials and services?
- How can modernized equipment be evaluated?

It is clear that a dominant feature of this phase of the project's life is the economic aspect. Application of the principles of the time-value of money, projections of growth or decline in demand and services, optimized maintenance, and timely replacement of obsolete elements come to the fore.

2.3 HIERARCHIES

In reviewing the analysis of a civil engineering project as outlined above, it seems clear that different specializations are called for in each of the phases. It is also clear that systems and subsystems are involved in each, and the overall project itself can be considered a system, hence there are systems within systems within systems in a complex of elements, relationships, boundaries, and totality. When considering just how large a project can be, it seems impossible for the full scope and significance of it to be comprehended in a reasonable way.

Let us say that the project is a superhighway, for instance. One might imagine that it could lead from one metropolitan area to another, but the full temporal, spatial, logistic, technical, and fiscal aspects would be very difficult to incorporate into one entity at the outset. The question is, "Where to enter the business of conceiving, planning, designing, building, and operating this roadway." It would be patently foolish to design the signage first, for example. This item, as complex a subsystem as it is in its own right, is well down on the list of things to do in order.

There is, in systems, a hierarchy of features which puts the elements of a complex system in a rank-ordered structure with the unity, totality, or one-ness at the apex of a pyramidal shape, and lesser details at increasingly lower levels. Each decreasing level displays increasing fineness of scale, until, in the case of the roadway, each roadside sign, each culvert, each guard-rail, every center-line reflector, and every other individual element has its place and relationship to the other elements.

Hierarchy is another word with multiple and varied meanings, one of the more relevant of which for systems theory is the following:

> "The arrangement of objects, elements, or values in a graduated series: a series of objects, elements, or values so arranged."

A complex system can be conceptualized in a hierarchical manner, so that it can be discussed, planned, designed, built, operated, and evaluated in a rational and effective manner.

CHAPTER 3

THEORIES OF SYSTEMS

3.1 INTRODUCTION

The varied nature of systems has resulted in a number of different methods for modeling, analyzing, and optimizing these systems. It has especially evidenced itself in the variety of devices for depicting and visualizing systems.

There are six major systems analysis theories which are in common use in civil engineering applications. These are *The Black Box Model, Linked Components Methods, Linear Graphs and Networks, Decision Process Approach, State Space Theory*, and *Differential Equations*. Each of them requires an extensive development, together with examples, and some are obviously of greater significance (and therefore of considerably greater length in a text) than others. Whole chapters will later be spent in discussing a theory and its application areas, as is indicated by the circumstances.

The purpose of the present chapter is to give a brief preview of these coming specializations, so that a better approach can be prepared for them. It is desirable to provide a stage setting or frame of reference for the sequels in somewhat the same manner as an overture functions to sensitize the audience to coming themes in a major musical performance.

3.2 THE BLACK BOX MODEL

The name of this concept comes from the time when scientific and engineering equipment, especially electrical and radio laboratory devices, was housed in a chassis and its cover painted with black wrinkle finish. There were plug-in type inputs and outputs, and dials to set or read.

However, since the box was kept closed, by being screwed together, or with latching clamps, or possibly Dzus fasteners, a knowledge of its interior workings was limited to the builders and service persons, all of whom supposedly possessed the arcane secrets. Even the generic composition of the contents of the black box was unknown, not to mention the specifics of what went on in there. Nowadays, of course, the box will be beige (desert sand) or some other designer color, and its covers will be easy to open so that servicing the contents is readily possible.

The black box model for systems is one which recognizes the facts that not much, if anything, is known about what constitutes the system, and that details of the mechanisms and parameters and their relationships and workings is either scarce, nonexistent, or irrelevant.

The positive side of the black box model is that it emphasizes the *functioning* of the systems to which it is applied. A knob or dial is turned through some angle on the face of the box and something functionally proportional takes place on the output side. If the box is literally a radio, moving the dial from one setting to another produces a change from the rock-and-roll station to the one playing classical music. The graduations of the dial need not be linear, of course, as in the case of the radio they are more nearly logarithmic.

In general, the black box system is described in purely mathematical terms, as for example,

$$y = f(x) \tag{3.1}$$

where the output is a function of the input and the function is assumed to be provided as a proper mathematical function. It may be that an extended definition of "function" is implied in some cases, such as jumps, discontinuities, and some others referred to as pathological functions. A common occurrence is the hereditary integral functional which provides for memory of a past history or for relaxation of a response parameter over a significantly lengthy period. Backlash in the input and output mechanisms is also common.

Frequently the input, and, therefore, presumably also the output, will be of a spectral nature. That is, it will depend upon a parameter called the "frequency." The response of dynamical systems is generally assumed *a priori* to be of this character because of the nature of the excitation and of the substances composing the elements. The function $y = f(x)$ in this case will adopt the form of an hereditary spectral tensor functional; hereditary because past outputs are remembered and incorporated in the conditions, spectral because of the dependence upon frequency, and functional in the technical sense of the noun *functional* as used in mathematics.

A good example of the black box modeling of an important system in civil engineering practice is that of the *unit hydrograph*. The branch of civil engineering from which this example is drawn is called Hydrology, or, more likely, Water Resources Engineering. In the example a geographic area of some considerable extent is drained by rivulets, branches, creeks, rivers and other streams, so that when a rainstorm passes over the area each of these tributaries collects water that runs off from the land and passes it downstream until the cumulative discharge flows through a major river to the sea.

The total land area drained by a particular stream is called the *watershed*. Naturally, the topography of this area—its slopes, contours, vegetation, detention basins, and urban developments—will determine the speed with which rainfall runoff occurs. Say that it has been some time since the last rain, so that the streams in the watershed are at their normal levels (i.e., mainly fed by subsurface flows), the upper level of the soil is fairly dry as is vegetation, and detention basins are empty.

When the rains come, it will be some time before runoff takes place, because all of those natural sources of storage mentioned above are activated. The total

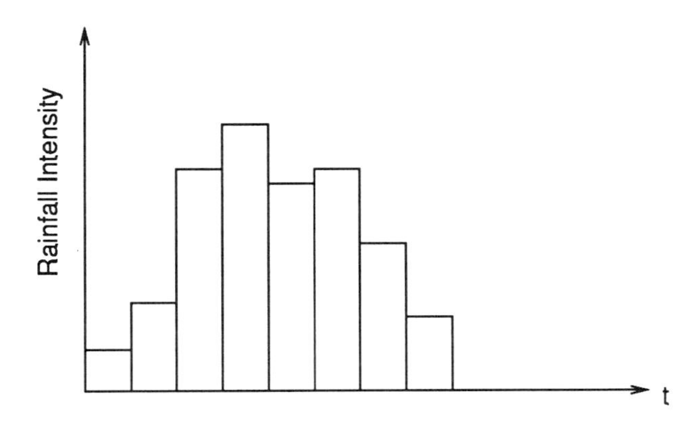

Figure 3.1: Storm Hyetograph

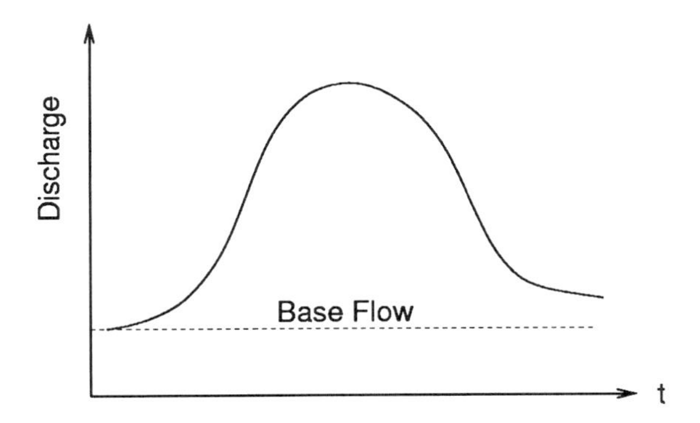

Figure 3.2: Runoff Hydrograph

amount of water taken from the rainfall to supply these elements is called *the initial abstraction*. Other *losses* are the infiltration of rainwater into subsurface storage.

What is left after these drains are filled is *rainfall excess*, and it is this amount which shows up at a gage station on the river draining the watershed.

A graph is constructed which shows the rainfall intensity as a function of time during the storm event. This graph is called the *storm hyetograph*, a representative version of which appears in Figure 3.1.

At a place on the river called a *gage station* the depth of the water is measured, and from a calibration equation for that place on that river, the volumetric rate of flow—the discharge—is obtained, also as a funtion of time. Of course, the normal depth of the river corresponds to its normal or base flow, which is a steady amount not related to the storm water runoff. The graph which shows the discharge of the river versus time is called the *runoff hydrograph*. See Figure 3.2.

Now it is possible to return to the concept of the black box model of this hydrologic system. The system input is the storm hyetograph. The output is the runoff hydrograph. And the functional which relates output to input is the

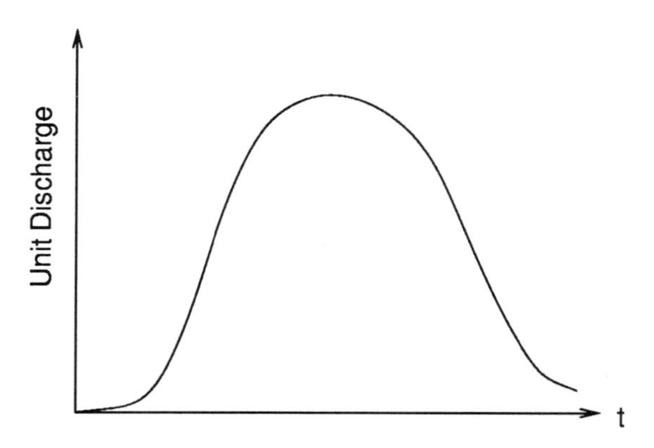

Figure 3.3: Unit Hydrograph

so-called ***unit hydrograph***, shown in Figure 3.3. It is an operator which converts rainfall into river runoff, and is naturally a function which depends explicitly on the peculiarities of that individual watershed. In effect it contains the fingerprint of the storm and that of the topology.

In more general systems theory the unit hydrograph would be called a ***transfer function***, which connotes the effect that the black box which is the system has had in converting input to output. It might also be called a ***transformation.***

In practice the data needed to convert the river depth to discharge, the relationships between initial abstration and losses to the geography and topology of the watershed, and the determination of the unit hydrograph are entrusted to databases and computational algorithms stored in computers. A typical process is one in which actual storm data are recorded by field instrumentation, after which the initial abstraction and losses are deducted, leaving the rainfall excess as a function of time. The runoff hydrograph data are likewise taken from field measurements.

From these measurements the unit hydrograph (transfer function) is computed and stored. This allows the watershed to be simulated by a black box system model. The engineer can then create a synthetic ***design storm***, allow the transfer function to operate upon it and produce a ***design hydrograph*** of the runoff. In other words, the computerized mathematical model of the watershed can be "run" over and over again as often as the engineer desires in order to optimize the design of catchment basins, impoundments, treatment facilities, and control structures.

3.3 LINKED COMPONENT SYSTEMS

Perhaps the most readily-identifiable "system" to students who have experienced a strong exposure to mechanics is the linked component configuration known as the "machine" or the "structure." In general the elements of these devices are solid links and members of specific geometry and physical properties. Well-developed techniques for carrying out primary force analyses using the equilibrium equations of statics and comprehensive motion diagrams for all links by the methods of kine-

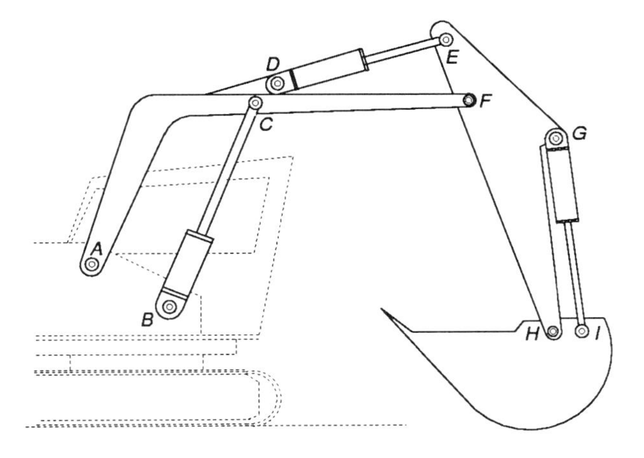

Figure 3.4: Backhoe

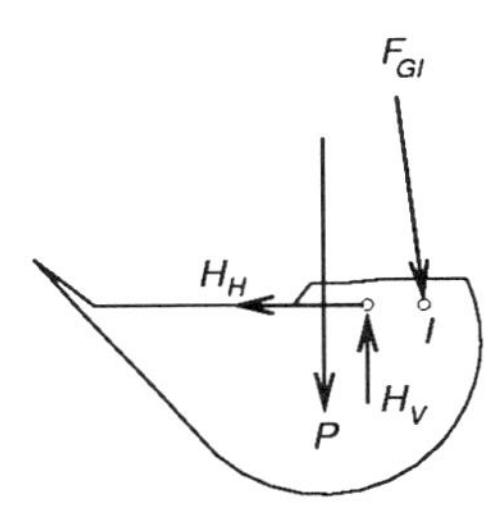

Figure 3.5: Bucket Forces

matics of machines have been studied and practiced to a state of high competence by most engineering students by the third year in the curriculum.

To demonstrate anew an appropriate method of modeling and analysis of one such system, consider the backhoe illustrated in Figure 3.4.

Let the bucket of the machine hold a load of 10,000 N in the position shown. The bucket is pinned to the arm *EFH* at point H and to the hydraulic actuator (cylinder) GI at point I. After the classical manner, remove these pins and replace them with the equivalent forces as shown in Figure 3.5. At H there are two components, H_H and H_V. The actuator constitutes a two-force member, so the pin forces at either end of it are equal, opposite, collinear, with the line of action lying along GI. The bucket is in equilibrium under this force system, and there are three equations available for a solution; namely, horizontal and vertical force sums, and a moment equation.

Take the moment sum about point H, since this eliminates the pin force components acting there, and there will be only one unknown in the equation—F_{GI}. Solve for it and obtain the result

$$F_{GI} = 40,000\,\text{N} \tag{3.2}$$

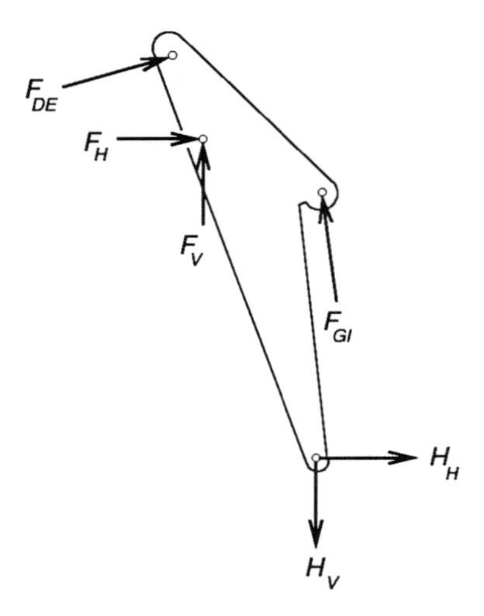

Figure 3.6: Arm *EFGH* Forces

Then enter this number into the horizontal and vertical force balance equations to get

$$H_H = 12,000\,\mathrm{N} \tag{3.3}$$

and

$$H_V = 50,000\mathrm{N} \tag{3.4}$$

Now consider the arm *EFGH* as an isolated member in equilibrium, as in Figure 3.6. Pins at E, F, G, and H have been removed and replaced by their force equivalents. Note that the forces at G and H are the opposites of those which act on the bucket in the preceeding step.

Again there are three equilibrium equations, two force balances and one moment sum. Take the moment about point F so as to negate the moments of the pin force components at F and leave only one unknown—the force in the actuator *DE*—in the equation. Consequently it is obtained directly as

$$F_{DE} = 16,000\mathrm{N} \tag{3.5}$$

Inject this value into the horizontal and vertical force balances and find

$$F_H = 16,000\,\mathrm{N} \tag{3.6}$$

and

$$F_V = 13,000\mathrm{N} \tag{3.7}$$

Move next to arm *ACF*, Figure 3.7, and replace the pins at those three points with equivalent forces. Write the three equilibrium equations for horizontal and vertical force sums and a moment sum. Let the moment be with respect to point A, as this will eliminate A_H and A_V and leave only one unknown, F_{BC}, in the

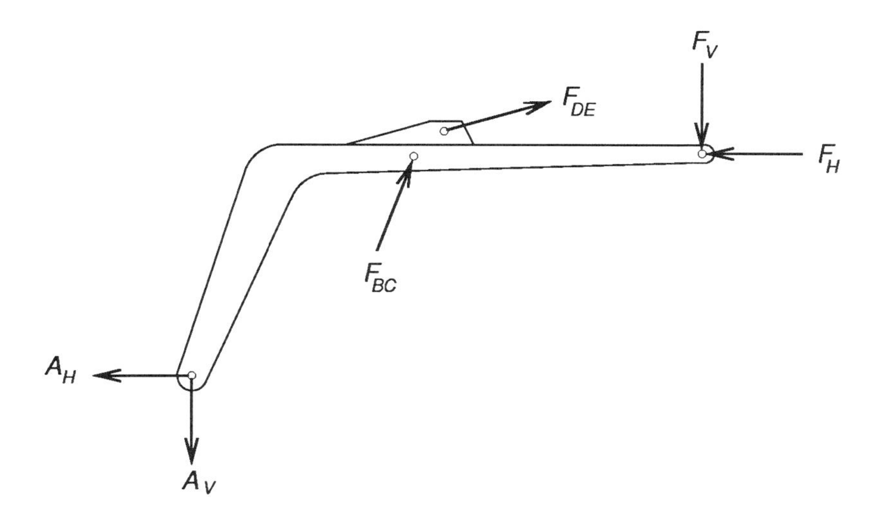

Figure 3.7: Arm ACF Forces

equation. The line of action of F_{BC} is known to be that through BC because the actuator is a two-force member. The moment equation gives

$$F_{BC} = 75,000\,\mathrm{N} \tag{3.8}$$

and using this number in the other two equations produces the result

$$A_H = 24,000\,\mathrm{N} \tag{3.9}$$

and

$$A_V = 60,000\,\mathrm{N} \tag{3.10}$$

The system of links has thus been "worked through" back to the frame of the crawler with all forces determined. The Chief Engineer had assigned the task of making this analysis to a member of the design staff, and would also have asked to have the maximum value of the pin forces at A and B for any position of the arms and bucket so as to be able to know the proper sizing of all elements for strength.

After the same fashion, equations of motion for each element of the system would be written as soon as their shapes, sizes and physical properties are selected. These equations would now include the dynamical pin forces at all joints, but the solution procedure would be analogous to that above.

3.3.1 MATRIX STRUCTURES MODELS

A prime example of the linked components system is the structure. Each element or link of a structure is securely attached to other elements to ensure that the structure is stable and deforms very little when subjected to a set of prescribed loads.

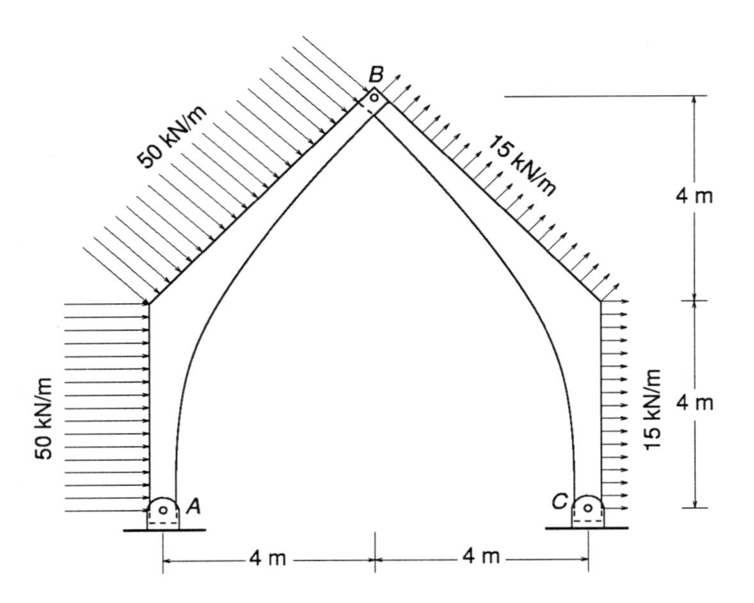

Figure 3.8: Central Gable Arch

Because of its prevalence in civil engineering practice, the structure has received abundant analytical attention, and there are many methods available for treating it. The current state of the art utilizes matrix-based computer programs, a number of which are classed as ***expert systems*** or ***artificial intelligence***. These approaches are highly specialized, and should properly be pursued as separate topics in their own right. A brief introduction from the systems perspective, however, may be appropriate.

Consider the central gable arch illustrated in Figure 3.8. This is a structure of choice for buildings such as auditoriums, churches, gymnasia, and vacation homes with cathedral ceilings or "great rooms." In the case shown, the loads are furnished by a wind blowing upon the walls and roof of the building, so that there is a positive pressure on the windward aspect and a vacuum or suction pressure on the lee side. These forces are described as ***distributed loads***, and in a preliminary analysis, at least, would be replaced by their concentrated force equivalents as in Figure 3.9. The arch is pinned to the foundation at A and C, and the two halves of the arch are also joined by a pin at B, the crest. When the free-body diagrams of the arch components are constructed as shown in Figure 3.9, it is seen that there are six unknown force components at the three pins. Since there are three equations of equilibrium for each half, the analysis of the pin forces is statically determinate. The equations are

$$\sum M_A = 0 = 8\,B_H + 4\,B_V - 2 \times 200 - 6 \times 282.84\tfrac{\sqrt{2}}{2} - 2 \times 282.84\tfrac{\sqrt{2}}{2} \tag{3.11}$$

$$\sum F_x = 0 = 200 + 282.84\tfrac{\sqrt{2}}{2} - A_H - B_H \tag{3.12}$$

$$\sum F_y = 0 = A_V + B_V - 282.84\tfrac{\sqrt{2}}{2} \tag{3.13}$$

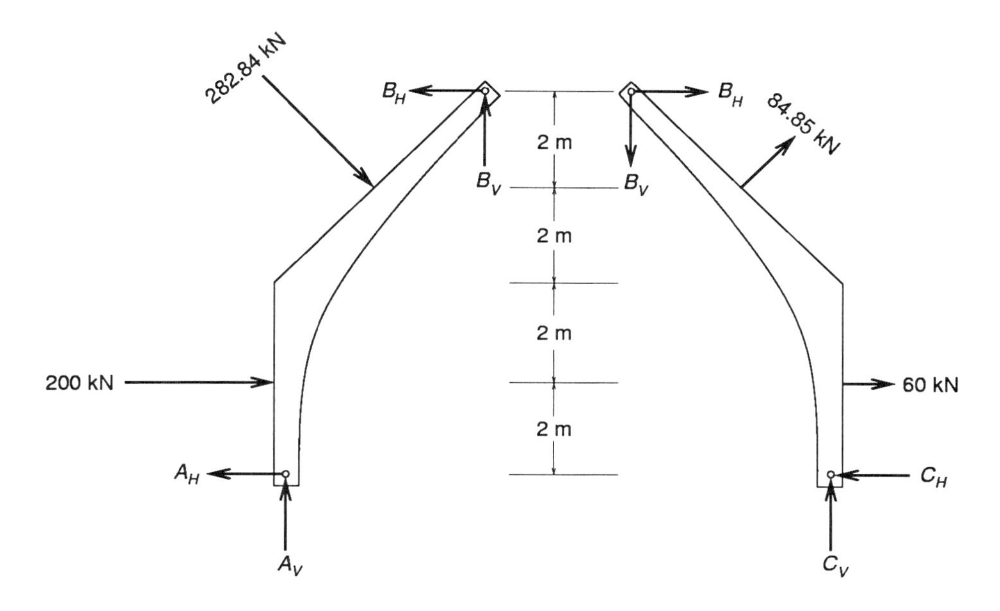

Figure 3.9: Gable Arch Freebody Diagrams

$$\sum M_B = 0 = 4\,C_V - 8\,C_H + 6 \times 60 - 2 \times 84.85\tfrac{\sqrt{2}}{2} - 2 \times 84.85\tfrac{\sqrt{2}}{2} \tag{3.14}$$
$$\sum F_x = 0 = B_H + 60 + 84.85\tfrac{\sqrt{2}}{2} - C_H \tag{3.15}$$
$$\sum F_y = 0 = C_V - B_V + 84.85\tfrac{\sqrt{2}}{2} \tag{3.16}$$

or

$$8\,B_H + 4\,B_V = 2 \times 200 + 6 \times 282.84\tfrac{\sqrt{2}}{2} + 2 \times 282.84\tfrac{\sqrt{2}}{2} = 2000 \tag{3.17}$$
$$A_H + B_H = 200 + 282.84\tfrac{\sqrt{2}}{2} = 400 \tag{3.18}$$
$$A_V + B_V = 282.84\tfrac{\sqrt{2}}{2} = 200 \tag{3.19}$$

$$8\,C_H - 4\,C_V = 6 \times 60 + 2 \times 84.85\tfrac{\sqrt{2}}{2} + 2 \times 84.85\tfrac{\sqrt{2}}{2} = 600 \tag{3.20}$$
$$C_H - B_H = 60 + 84.85\tfrac{\sqrt{2}}{2} = 120 \tag{3.21}$$
$$-C_V + B_V = 84.85\tfrac{\sqrt{2}}{2} = 60 \tag{3.22}$$

The solution of this set by successive eliminations gives

$$A_H = 312.5\,\text{kN} \quad A_V = -125\,\text{kN} \quad B_H = 87.5\,\text{kN} \quad B_V = 325\,\text{kN}$$
$$C_H = 207.5\,\text{kN} \quad C_V = 265\,\text{kN} \tag{3.23}$$

To arrive at a matrix model for this system, first write the equations of equilibrium as

$$0\,A_H + 0\,A_V + 8\,B_H + 4\,B_V + 0\,C_H + 0\,C_V = 2000 \tag{3.24}$$
$$1\,A_H + 0\,A_V + 1\,B_H + 0\,B_v + 0\,C_H + 0\,C_V = 400 \tag{3.25}$$
$$0\,A_H + 1\,A_V + 0\,B_H + 1\,B_V + 0\,C_H + 0\,C_V = 200 \tag{3.26}$$
$$0\,A_H + 0\,A_V + 0\,B_H + 0\,B_V + 8\,C_H - 4\,C_V = 600 \tag{3.27}$$
$$0\,A_H + 0\,A_V - 1\,B_H + 0\,B_V + 1\,C_H + 0\,C_V = 120 \tag{3.28}$$
$$0\,A_H + 0\,A_V + 0\,B_H + 1\,B_V + 0\,C_H - 1\,C_V = 60 \tag{3.29}$$

or

$$\begin{bmatrix} 0 & 0 & 8 & 4 & 0 & 0 \\ 1 & 0 & 1 & 0 & 0 & 0 \\ 0 & 1 & 0 & 1 & 0 & 0 \\ 0 & 0 & 0 & 0 & 8 & -4 \\ 0 & 0 & -1 & 0 & 1 & 0 \\ 0 & 0 & 0 & 1 & 0 & -1 \end{bmatrix} \begin{bmatrix} A_H \\ A_V \\ B_H \\ B_V \\ C_H \\ C_V \end{bmatrix} = \begin{bmatrix} 2000 \\ 400 \\ 200 \\ 600 \\ 120 \\ 60 \end{bmatrix} \tag{3.30}$$

Then $\mathbf{AF} = \mathbf{P}$ is the matrix equation, with

$$\mathbf{A} = \begin{bmatrix} 0 & 0 & 8 & 4 & 0 & 0 \\ 1 & 0 & 1 & 0 & 0 & 0 \\ 0 & 1 & 0 & 1 & 0 & 0 \\ 0 & 0 & 0 & 0 & 8 & -4 \\ 0 & 0 & -1 & 0 & 1 & 0 \\ 0 & 0 & 0 & 1 & 0 & -1 \end{bmatrix} \quad \mathbf{F} = \begin{bmatrix} A_H \\ A_V \\ B_H \\ B_V \\ C_H \\ C_V \end{bmatrix} \quad \mathbf{P} = \begin{bmatrix} 2000 \\ 400 \\ 200 \\ 600 \\ 120 \\ 60 \end{bmatrix} \tag{3.31}$$

This can easily be verified by performing the operation of $\mathbf{A}$ times $\mathbf{F}$ set equal to $\mathbf{P}$, and comparing with the set above.

The matrix equation can be solved by operating from the left on both sides by the matrix operator $\mathbf{A}^{-1}$, where $\mathbf{A}^{-1}\mathbf{A} = \mathbf{I}$, and the inverse matrix to $\mathbf{A}$ is defined by this relationship. $\mathbf{I}$ is the identity (or unit) matrix. Thus

$$\mathbf{A}^{-1}\mathbf{AF} = \mathbf{IF} = \mathbf{F} = \mathbf{A}^{-1}\mathbf{P} \tag{3.32}$$

$$\mathbf{F} = \mathbf{A}^{-1}\mathbf{P} \tag{3.33}$$

Clearly, it is only necessary to have the inverse matrix $\mathbf{A}^{-1}$ to solve for the forces at the pins.

In Appendix E it is shown that

$$\mathbf{A}^{-1} = \begin{bmatrix} -\frac{1}{16} & 1 & 0 & -\frac{1}{16} & \frac{1}{2} & \frac{1}{4} \\ -\frac{1}{8} & 0 & 1 & \frac{1}{8} & -1 & -\frac{1}{2} \\ \frac{1}{16} & 0 & 0 & \frac{1}{16} & -\frac{1}{2} & -\frac{1}{4} \\ \frac{1}{8} & 0 & 0 & -\frac{1}{8} & 1 & \frac{1}{2} \\ \frac{1}{16} & 0 & 0 & \frac{1}{16} & \frac{1}{2} & -\frac{1}{4} \\ \frac{1}{8} & 0 & 0 & -\frac{1}{8} & 1 & -\frac{1}{2} \end{bmatrix} \tag{3.34}$$

hence

$$\begin{bmatrix} A_H \\ A_V \\ B_H \\ B_V \\ C_H \\ C_V \end{bmatrix} = \begin{bmatrix} -\frac{1}{16} & 1 & 0 & -\frac{1}{16} & \frac{1}{2} & \frac{1}{4} \\ -\frac{1}{8} & 0 & 1 & \frac{1}{8} & -1 & -\frac{1}{2} \\ \frac{1}{16} & 0 & 0 & \frac{1}{16} & -\frac{1}{2} & -\frac{1}{4} \\ \frac{1}{8} & 0 & 0 & -\frac{1}{8} & 1 & \frac{1}{2} \\ \frac{1}{16} & 0 & 0 & \frac{1}{16} & \frac{1}{2} & -\frac{1}{4} \\ \frac{1}{8} & 0 & 0 & -\frac{1}{8} & 1 & -\frac{1}{2} \end{bmatrix} \begin{bmatrix} 2000 \\ 400 \\ 200 \\ 600 \\ 120 \\ 60 \end{bmatrix}$$

$$= \begin{bmatrix} -\frac{1}{16}(2000) + 1(400) + 0(200) - \frac{1}{16}(600) + \frac{1}{2}(120) + \frac{1}{4}(60) \\ -\frac{1}{8}(2000) + 0(400) + 1(200) + \frac{1}{8}(600) - 1(120) - \frac{1}{2}(60) \\ \frac{1}{16}(2000) + 0(400) + 0(200) - \frac{1}{16}(600) - \frac{1}{2}(120) - \frac{1}{4}(60) \\ \frac{1}{8}(2000) + 0(400) + 0(200) - \frac{1}{8}(600) + 1(120) + \frac{1}{2}(60) \\ \frac{1}{16}(2000) + 0(400) + 0(200) + \frac{1}{16}(600) + \frac{1}{2}(120) - \frac{1}{4}(60) \\ \frac{1}{8}(2000) + 0(400) + 0(200) - \frac{1}{8}(600) + 1(120) - \frac{1}{2}(60) \end{bmatrix}$$

$$= \begin{bmatrix} 312.5 \\ -125.0 \\ 87.5 \\ 325.0 \\ 207.5 \\ 265.0 \end{bmatrix} \tag{3.35}$$

Therefore

$$A_H = 312.5\,\text{kN} \tag{3.36}$$
$$A_V = -125\,\text{kN} \tag{3.37}$$
$$B_H = 87.5\,\text{kN} \tag{3.38}$$
$$B_V = 325\,\text{kN} \tag{3.39}$$
$$C_H = 207.5\,\text{kN} \tag{3.40}$$
$$C_V = 265\,\text{kN} \tag{3.41}$$

which confirms the solution previously obtained.

The matrix method has the powerful advantage that the matrices $\mathbf{A}$ and $\mathbf{A}^{-1}$ are functions only of the geometry of the structure, and not of the particular loadings. Consequently, once they are determined, they can be stored in memory and retrieved for each new case of loads at will. This permits the analyst-designer to investigate the structure for optimum conditions.

It is obvious that a disadvantage of the matrix model for structural analysis is the increasing complexity of the inverse matrix determination as the number of elements increases. However, computers are ideally suited to the task of producing $\mathbf{A}^{-1}$ when $\mathbf{A}$ is available.

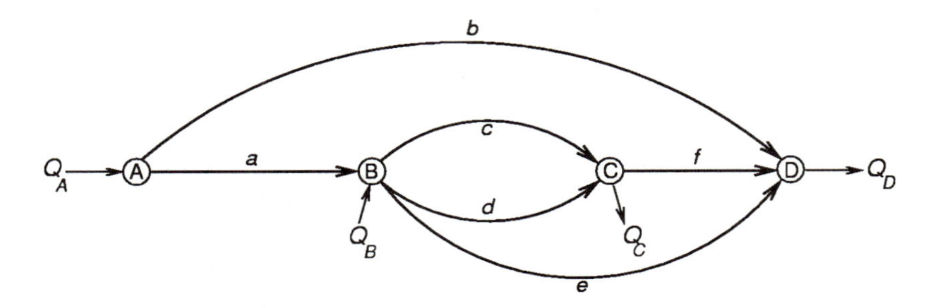

Figure 3.10: Pipeflow Network

3.3.2 PIPEFLOW NETWORKS

Another example of a system composed of linked components is that of a pipe network, although this system is also amenable to being modeled as a linear graph.

Figure 3.10 illustrates a small water distribution system which might be a portion of a public water supply. Water enters the system through nodes A and B and leaves at nodes C and D. The flowrates are Q_A, Q_B, Q_C, and Q_D, respectively. The pipes are named a, b, c, d, e, and f, with lengths $\ell_a = 200$ m, $\ell_b = 1200$ m, $\ell_c = 400$ m, $\ell_d = 500$ m, $\ell_e = 1000$ m, $\ell_f = 300$ m. Pressure gages at the nodes read $p_A = 608$ kPa $= 62$ m, $p_B = 560$ kPa $= 57$ m, $p_C = 510$ kPa $= 52$ m, and $p_D = 450$ kPa $= 46$ m, where the second figure in each case is the pressure converted to that at the base of a column of water that tall.

Some redundancy is built into the system, as is evident. This is good practice against the day (middle of the night, freezing weather, actually) when one of the pipes may fail. That failed pipe can be isolated while repairs are made, and the customers can still be served with water by the remaining routes in the interim.

The pipeflows are governed by the fundamentals of hydraulics. The Darcy-Wiesbach law describes the pressure drop along a pipe from intake to exit as

$$h = f \frac{\ell}{d} \frac{v^2}{2g} \tag{3.42}$$

in which h is the head loss (pressure drop) in meters of fluid flowing, ℓ is the length of the pipe, d is its diameter, v is the mean velocity in m/s, g is local gravity (m/s^2), and f is a friction factor that is dimensionless. Although f is actually a complex function of the pipe diameter, the fluid flow rate, and fluid viscosity, under many practical circumstances (turbulent flow in commercially available pipe of the size in question) it can be taken to be sensibly constant at approximately 0.02.

The continuity equation expresses the relationship between the mean velocity and flow rate or discharge as

$$Q = A v = \frac{\pi}{4} d^2 v \ \mathrm{m^3/s} \tag{3.43}$$

When this is solved for v, the result can be entered into the Darcy-Wiesbach law to convert it into

$$Q = \frac{\pi}{4}\left(\frac{2\,g}{f}\right)^{1/2}\left(\frac{h}{\ell}\right)^{1/2} d^{2.5} = 24.59\left(\frac{h}{\ell}\right)^{1/2} d^{2.5} \tag{3.44}$$

This last formula can be applied to each of the six pipes as

$$Q_a = 24.59\left(\frac{h_a}{200}\right)^{1/2}(0.3)^{2.5} = 0.085728251\, h_a^{1/2} \tag{3.45}$$

$$Q_b = 24.59\left(\frac{h_b}{1200}\right)^{1/2}(0.3)^{2.5} = 0.034998412\, h_b^{1/2} \tag{3.46}$$

$$Q_c = 24.59\left(\frac{h_c}{400}\right)^{1/2}(0.3)^{2.5} = 0.060619027\, h_c^{1/2} \tag{3.47}$$

$$Q_d = 24.59\left(\frac{h_d}{500}\right)^{1/2}(0.3)^{2.5} = 0.054219306\, h_d^{1/2} \tag{3.48}$$

$$Q_e = 24.59\left(\frac{h_e}{1000}\right)^{1/2}(0.3)^{2.5} = 0.038338839\, h_e^{1/2} \tag{3.49}$$

$$Q_f = 24.59\left(\frac{h_f}{300}\right)^{1/2}(0.3)^{2.5} = 0.069996824\, h_f^{1/2} \tag{3.50}$$

At each node of the network there is only one pressure; i.e., p_A, p_B, p_C, or p_D. Therefore

$$h_a = p_A - p_B = 62 - 57 = 5\,\text{m} \tag{3.51}$$
$$h_b = p_A - p_D = 62 - 46 = 16\,\text{m} \tag{3.52}$$
$$h_c = p_B - p_C = 57 - 52 = 5\,\text{m} \tag{3.53}$$
$$h_d = p_B - p_C = 57 - 52 = 5\,\text{m} \tag{3.54}$$
$$h_e = p_B - p_D = 57 - 46 = 11\,\text{m} \tag{3.55}$$
$$h_f = p_C - p_D = 52 - 46 = 6\,\text{m} \tag{3.56}$$

There is also a continuity requirement at each node; i.e.,

$$Q_A = Q_a + Q_b\,\text{m}^3/\text{s} \tag{3.57}$$
$$Q_B = Q_c + Q_d + Q_e - Q_a\,\text{m}^3/\text{s} \tag{3.58}$$
$$Q_C = Q_c + Q_d - Q_f\,\text{m}^3/\text{s} \tag{3.59}$$
$$Q_D = Q_b + Q_e + Q_f\,\text{m}^3/\text{s} \tag{3.60}$$

This system of equations can be solved by successive substitutions until only a set of simultaneous equations linear in $h_a^{1/2}$, $h_b^{1/2}$, $h_c^{1/2}$, $h_d^{1/2}$, $h_e^{1/2}$, and $h_f^{1/2}$ remain, and the set can then be solved for these pressure drops. Afterwards, the Qs are quickly found from the Darcy-Wiesbach equalities. However, it is also possible to use a matrix model of the system and solve it using matrix methods.

The Darcy-Wiesbach equations for the pipes can be shown as $\mathbf{Q} = \mathbf{kh}$ in which

$$\mathbf{Q} = \begin{pmatrix} Q_a \\ Q_b \\ Q_c \\ Q_d \\ Q_e \\ Q_f \end{pmatrix} \qquad \mathbf{h} = \begin{pmatrix} h_a^{1/2} \\ h_b^{1/2} \\ h_c^{1/2} \\ h_d^{1/2} \\ h_e^{1/2} \\ h_f^{1/2} \end{pmatrix} \tag{3.61}$$

and $\mathbf{k}$ is the matrix

$$\mathbf{k} = \begin{pmatrix} 0.08573 & 0 & 0 & 0 & 0 & 0 \\ 0 & 0.03500 & 0 & 0 & 0 & 0 \\ 0 & 0 & 0.06062 & 0 & 0 & 0 \\ 0 & 0 & 0 & 0.05422 & 0 & 0 \\ 0 & 0 & 0 & 0 & 0.03834 & 0 \\ 0 & 0 & 0 & 0 & 0 & 0.07000 \end{pmatrix} \tag{3.62}$$

The pressure relationships become $\mathbf{H} = \mathbf{Ap}$, where

$$\mathbf{H} = \begin{pmatrix} h_a \\ h_b \\ h_c \\ h_d \\ h_e \\ h_f \end{pmatrix} \quad \mathbf{p} = \begin{pmatrix} p_A \\ p_B \\ p_C \\ p_D \end{pmatrix} \quad \mathbf{A} = \begin{pmatrix} 1 & -1 & 0 & 0 \\ 1 & 0 & 0 & -1 \\ 0 & 1 & -1 & 0 \\ 0 & 1 & -1 & 0 \\ 0 & 1 & 0 & -1 \\ 0 & 0 & 1 & -1 \end{pmatrix} \tag{3.63}$$

The flow equations are $\mathbf{q} = \mathbf{B}\ \mathbf{Q}$, with

$$\mathbf{q} = \begin{pmatrix} Q_A \\ Q_B \\ Q_C \\ Q_D \end{pmatrix} \quad \mathbf{Q} = \begin{pmatrix} Q_a \\ Q_b \\ Q_c \\ Q_d \\ Q_e \\ Q_f \end{pmatrix} \quad \mathbf{B} = \begin{pmatrix} 1 & 1 & 0 & 0 & 0 & 0 \\ -1 & 0 & 1 & 1 & 1 & 0 \\ 0 & 0 & 1 & 1 & 0 & -1 \\ 0 & 1 & 0 & 0 & 1 & 1 \end{pmatrix} \tag{3.64}$$

The set of matrix equations

$$\mathbf{Q} = \mathbf{kh} \qquad \mathbf{H} = \mathbf{Ap} \qquad \mathbf{q} = \mathbf{BQ} \tag{3.65}$$

are the matrix model of the pipe network. Matrix $\mathbf{k}$ is a *constitutive formula* which describes the flow in each pipe in terms of the pipe parameters and fluid properties. Matrix $\mathbf{A}$ is an *incidence matrix* that defines the connectivity of the network, or its geometry, and matrix $\mathbf{B}$ expresses the *massflow balance* or *continuity* requirements for each of the nodes of the net. The validity of these matrix expressions can be demonstrated by performing the explicit operations indicated, and noting the outcomes as compared with the scalar equations of the original model.

To solve the matrix equations, note first that the pressure drops in each pipe are available, therefore the components of matrix **h** can be constructed as

$$h_a^{1/2} = \sqrt{5} = 2.23606 \tag{3.66}$$

$$h_b^{1/2} = \sqrt{16} = 4.00000 \tag{3.67}$$

$$h_c^{1/2} = \sqrt{5} = 2.23606 \tag{3.68}$$

$$h_d^{1/2} = \sqrt{5} = 2.23606 \tag{3.69}$$

$$h_e^{1/2} = \sqrt{11} = 3.31662 \tag{3.70}$$

$$h_f^{1/2} = \sqrt{6} = 2.44949 \tag{3.71}$$

and these allow the determination of the Qs as

$$Q_a = 0.08573\, h_a^{1/2} = 0.08573 \times 2.23606 = 0.19169\,\mathrm{m^3/s} \tag{3.72}$$

$$Q_b = 0.03500\, h_b^{1/2} = 0.03500 \times 4.00000 = 0.13999\,\mathrm{m^3/s} \tag{3.73}$$

$$Q_c = 0.06062\, h_c^{1/2} = 0.06062 \times 2.23606 = 0.13555\,\mathrm{m^3/s} \tag{3.74}$$

$$Q_d = 0.05422\, h_d^{1/2} = 0.05422 \times 2.23606 = 0.12124\,\mathrm{m^3/s} \tag{3.75}$$

$$Q_e = 0.03834\, h_d^{1/2} = 0.03834 \times 3.31662 = 0.12716\,\mathrm{m^3/s} \tag{3.76}$$

$$Q_f = 0.06700\, h_e^{1/2} = 0.06700 \times 2.44949 = 0.17146\,\mathrm{m^3/s} \tag{3.77}$$

Now the continuity requirement, $\mathbf{q} = \mathbf{B}\,\mathbf{Q}$, Equations 3.65, gives

$$\begin{aligned}
\mathbf{q} &= \begin{bmatrix} 1 & 1 & 0 & 0 & 0 & 0 \\ -1 & 0 & 1 & 1 & 1 & 0 \\ 0 & 0 & 1 & 1 & 0 & -1 \\ 0 & 1 & 0 & 0 & 1 & 1 \end{bmatrix} \begin{bmatrix} 0.19169 \\ 0.13999 \\ 0.13555 \\ 0.12124 \\ 0.12716 \\ 0.17146 \end{bmatrix} \\
&= \begin{bmatrix} 0.19169 + 0.13999 \\ 0.13555 + 0.12124 + 0.12716 - 0.19169 \\ 0.13555 + 0.12124 - 0.17146 \\ 0.13999 + 0.12716 + 0.17146 \end{bmatrix} \\
&= \begin{bmatrix} 0.33168 \\ 0.19226 \\ 0.08533 \\ 0.43861 \end{bmatrix} = \begin{bmatrix} Q_A \\ Q_B \\ Q_C \\ Q_D \end{bmatrix}
\end{aligned} \tag{3.78}$$

Accordingly,

$$Q_A = 0.33168\,\mathrm{m^3/s} \tag{3.79}$$

$$Q_B = 0.19226\,\mathrm{m^3/s} \tag{3.80}$$

$$Q_C = 0.08533\,\mathrm{m^3/s} \tag{3.81}$$

$$Q_D = 0.43861\,\mathrm{m^3/s} \tag{3.82}$$

3.4 LINEAR GRAPHS AND NETWORKS

Look at the fragment of a map showing a major portion of the Piedmont Crescent region of North Carolina as seen in Figure 3.11. At the lower left corner of the map is Charlotte, and on the right-hand edge about two-thirds of the way up is Raleigh.

The area framed by the map's borders is highly decorated with cities, towns, hamlets and ribbons of various capacity roadways leading from one place to another. This map displays Interstate highways, US-numbered routes and a large number of North Carolina state-numbered roads. If one is travelling between the extremes of this map's borders, it is very desirable to have this much degree of detail, and even more could be added if the county-numbered roads were also included.

However, for other purposes, such as analysis and design of transportation systems within the area, a somewhat different picture might be more helpful.

Suppose, for example, that one wishes to find the "best" route from Raleigh to Charlotte. The meaning of "best" is open to interpretation. It might mean "shortest distance," or least time of travel. In bad weather some routes may be preferable to others, and in some seasons one route may display scenic features which recommend it above others.

To address these questions in a more rational manner, an initial refinement of the routes map might be undertaken. Let all but the three principal avenues between the two cities be removed so that the optimization can focus upon the parameters associated with the three. See Figure 3.12. The high-level interstate route departs Raleigh as I-40, which continues to Durham, Burlington, and Greensboro. Then it becomes I-85 through Lexington, Concord, and Kannapolis into Charlotte.

The southern route leads from Raleigh on US 1 to Sanford, US 15 then goes to Carthage, and NC routes 24 and 27 lead to Biscoe, Troy, Albemarle, and Charlotte. The scenic route departs Raleigh as US 64 to Pittsboro, Siler City, and Asheboro, after which NC 49 goes to Harrisburg and into Charlotte.

A second round of refinement would be to suppress the literal dependence of the graph upon the geography of the region so as to present a schematic model seen in Figure 3.13. The cities are now depicted as nodes and the roads as straight-line segments. A legend which gives the relevant information associated with each segment can be attached to the segments.

Further amendments which could be made could include arrowheads on the line-segments to indicate the directions of traffic in each path, after which the nodes, links, and arrows would constitute what is called a *directed graph*. It would also be called a network.

A large concern of those engineers who plan and design road networks is the capacity of the net to carry a given number of vehicles per hour. In the case at hand, a question might be: "How many vehicles per hour can the primary highway network allow from Raleigh to Charlotte?" Assume that a speed limit has been established for each segment of road in the net, and that an inter-vehicle interval which corresponds to the safe stopping distance at that speed is maintained. The upper limit of number of vehicles per hour on a segment is thereby established, and is called the capacity of that segment. Of course, on a multilane highway, the

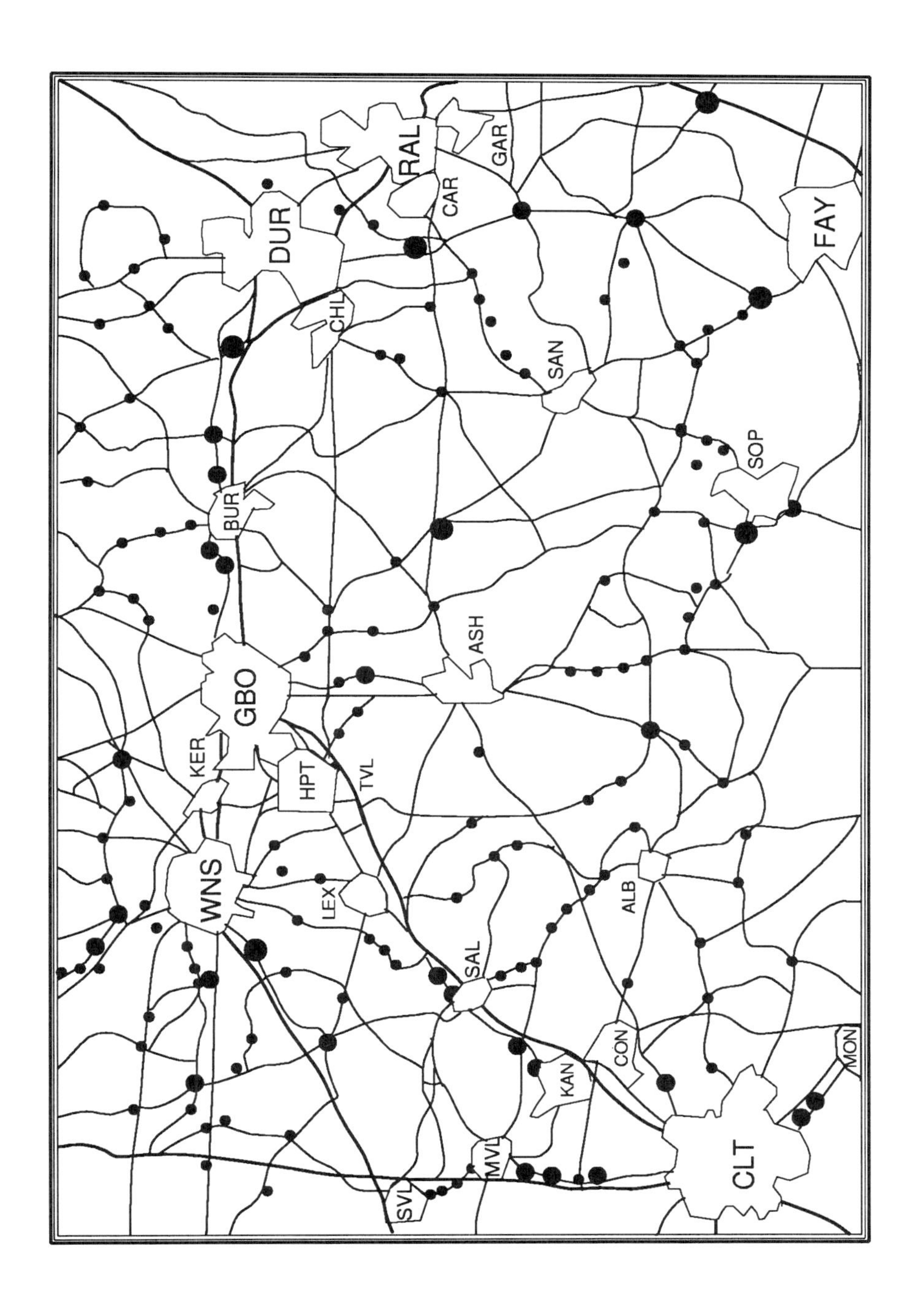

Figure 3.11: The Piedmont Crescent

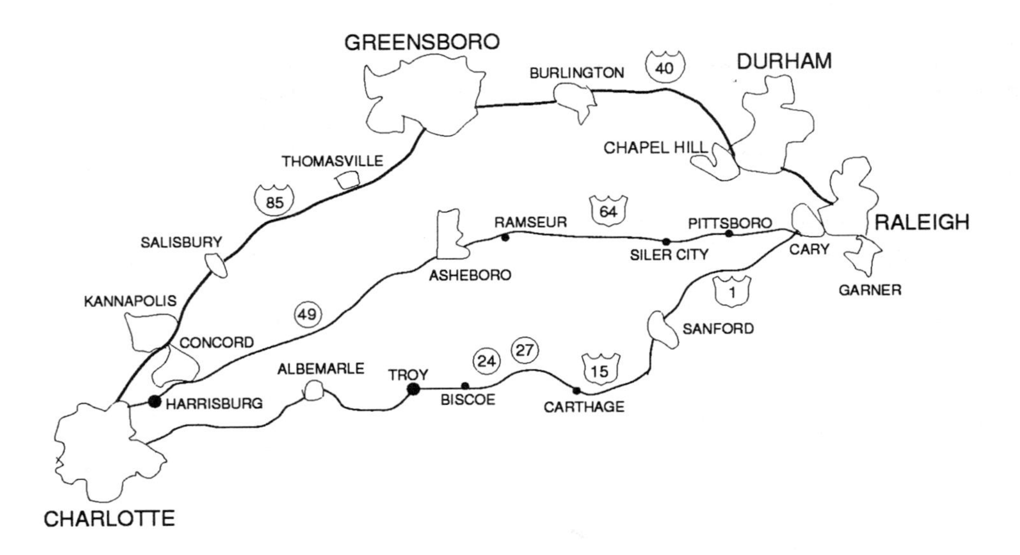

Figure 3.12: Schematic Routes Map

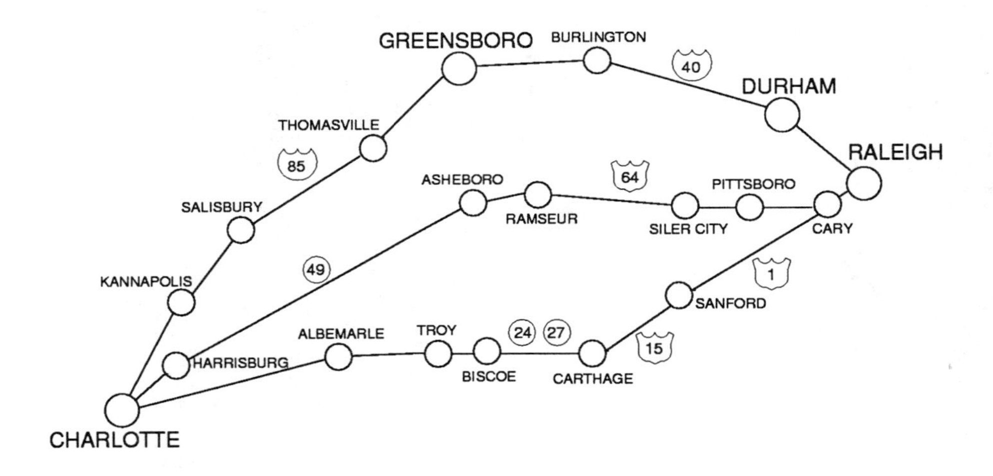

Figure 3.13: Crescent Network

lane capacity can be multiplied by the number of lanes to give the total traffic on the road in question.

Nowhere is the difference between the attribute of a system and its components made more evident than in the case of a network. Specifically, the capacity of the net is very much less than the sum of the capacities of the individual roads which comprise it. The sum of the traffics in each road which enters a node of the net must be balanced by the sum of the traffics in roads leaving the node, because there is no provision for accumulating traffic at a node. Therefore an equation can be written for each node expressing that the net balance of the actual traffic coming into the node against that leaving is zero. The *source* of traffic is the place where all traffic enters—Raleigh, in the network above. All network traffic exits at the ***sink***, or Charlotte.

The equations written at each node form the components of a system of linear algebraic equations with the actual traffic counts as variables. A similar equation written for either the source or the sink will give the net capacity if the system of equations is solved for the actual traffic counts in each of the roads. ***It will be seen that the form of the links and nodes—the network configuration—serves to "throttle" or regulate the total network flow.***

3.5 DECISION PROCESS APPROACH

The decision process approach to systems modeling is a method appropriate to the circumstances which pertain if little or nothing is known about the relationship between cause and effect operating between elements of the system, and is especially effective if the system elements include decision points.

Consider the following situation. A contractor is doing a job on the Outer Banks in late summer—i.e., well into the hurricane season. The contract calls for the job to be completed by the end of August, and at the middle of that month the contractor has begun to worry, because he is very well aware that historically the Banks have experienced bad storms during the time frame which he is now entering.

There are some options open to him. In the first place, he might hire additional crews and enter upon a crash program to finish the contract earlier than the expected completion date. This course of action would cost an extra amount, say $30,000, but would run no risk of a September storm damage.

A second option is to continue work as usual—to run the risk of there being no storm to delay his work. If he is lucky, there will be no additional cost incurred by this course of action. On the other hand, weather data indicate that the probability of a nor'easter during this period is 50%, and that of a full-blown hurricane is 10%. This makes the probability of acceptable working weather 40%.

If there is a storm, the contractor's damages will be, by his estimate, $50,000, and if there's a hurricane, he will lose all of his equipment and supplies and be out a sum of $100,000.

A diagram of the contractor's decision process model is shown below. A node consisting of a box represents his first decision—to inaugurate a crash program or not to do so. If he does crash it, the triangular terminal node is to be associated with an expenditure of $30,000. The alternate branch of the decision tree leads

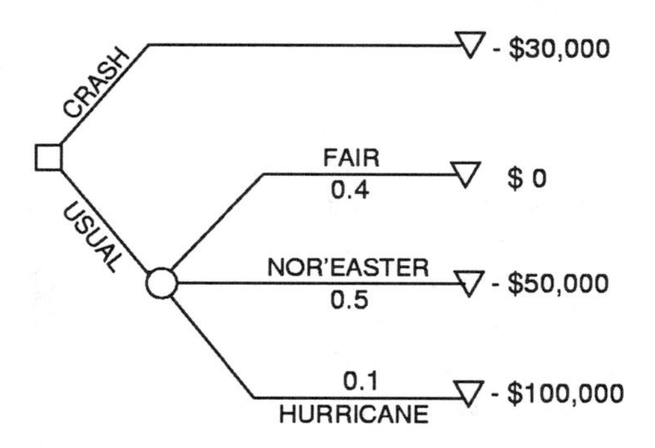

Figure 3.14: Contractor's Decision Tree

to a circular node, called a chance node, at which place in time the laws of chance (an absence of physical laws) take over.

Three "chances" are the branches of the tree following the chance node. He may experience good weather (0.4 probability, no loss), stormy conditions (0.5 probability, cash outgo of $\$50,000$) or a hurricane (0.1 probability and a loss of $\$100,000$) at the three triangular terminal nodes as shown in Figure 3.14.

The decision process method assumes that if these decisions are made many times by this contractor, on the average the combined outcome of his decision not to run a crash program will reflect the chance probabilities of the weather; namely,

$$\text{Cost} = 0.4 \times \$0.0 + 0.5 \times \$50,000 + 0.1 \times \$100,000 = \$35,000 \tag{3.83}$$

At this point the contractor can make an elightened decision that the prudent course of action will save him—on the average—$\$5,000$. He will begin the crash program without delay.

It is important to realize that this method can say nothing about what will happen, therefore it cannot predict the result for a particular case. The contractor's nearest competitor may very well be a braver person who will run the gauntlet, experience good weather for the whole month, and be up $\$30,000$ on the outcome.

3.6 STATE SPACE THEORY

A good bit was said earlier (in Chapter 1 on the Nature and Scope of Systems) about the fact that the elements may be very different in nature. The system as a fruit salad was an analogy put forward at that place. This comparison—though not to be taken literally, of course—suggests that there must be a common attribute of even so widely different elements.

One way in which to represent a collection of such otherwise incommensurate items is that of an affine vector and a vector space which is spanned or generated by a set of independent basis vectors. A position vector in such a space might be said to represent the state of the system which it contains. In other words, "take three units of tangerines on the tangerine axis, 2 apples on the apple axis, 1 pear on the pear axis, etc." The vector space so defined would contain within it the *state space* of the fruit salad.

In the general civil engineering systems problem, there aren't any of the delicious fruits mentioned above, but there are, for some, a set of numbers which represent the collection of values for the ordered array of a vector in a space wherein the state of the system is defined.

An important example of such a system is furnished by the problem of reservoir management. Let there be a dam on a river, created for a number of reasons, including flood control, water for irrigation, water supply for adjacent towns and for recreational uses such as swimming, boating, fishing, camping, picknicking, hiking, and nature enjoyment. The management problem associated with this impoundment is that of allocating water to these various purposes in some optimum way—to maximize the benefit to all the users.

It is a problem of competing interests, for if there is a drought condition during which the inflow to the reservoir from the river is slowed, there may be an increased demand for irrigation water as well, so there is less to furnish the water supplies of municipalities, and the recreational uses may also suffer if the water level drops too much.

The "well-being" of the system can be related to the state of the system, and one obvious measure of the state is the level of water in the reservoir. Therefore, choose the numbers 0, 0.1, 0.2, ..., 0.9, 1.0 to represent the range of empty, one-tenth full, two-tenths full, ..., nine-tenths full, to full that are the components of the state vector which describes the state of this reservoir system.

Historical records may be consulted for the average river flows in each month of the year, and these discharges can also be related to the volume of the impoundment as 0.1, 0.2, ..., 0.9, 1.0. The initial state of the reservoir (say at the beginning of the year) is a vector having the same range of numbers. So if the initial state is given, and the input is also furnished, the output state can be produced by the simple operation

$$\mathbf{X} = \mathbf{f}[\mathbf{X}_0; \mathbf{U}] \tag{3.84}$$

$$\mathbf{Y} = \mathbf{g}[\mathbf{X}_0; \mathbf{U}] \tag{3.85}$$

in which $\mathbf{X}_0$ is the initial state vector, $\mathbf{X}$ is the state vector, $\mathbf{U}$ is the input vector and $\mathbf{Y}$ is the output vector.

$\mathbf{X}_0$ is the vector whose components describe the fullness of the reservoir at the beginning of a year, $\mathbf{U}$, the driving function, is the vector whose components represent the inflows of the river, and $\mathbf{Y}$ is a vector whose components are the flows available to water supplies, irrigation and downstream river flow. The state vector of the reservoir is $\mathbf{X}$, whose components tell how full it is.

In this elementary model, one can say that, given its initial state vector, the input or driver vector and the state vector, a new state vector and an output vector can be calculated. The calculations are, in this instance, simple scalar arithmetics. In more general examples the operations will be carried out using matrix vector multiplications and general linear operators. A state-transition matrix will generate the new state from the initial state, given the input vector.

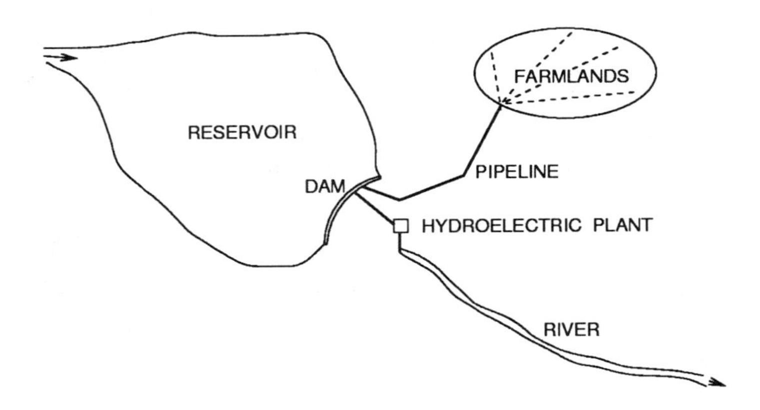

Figure 3.15: Stream Basin Features

In still more general applications, the entire family of vectors—$\mathbf{X}_0$, $\mathbf{U}$, $\mathbf{X}$, and $\mathbf{Y}$—are all time-varying vectors and $\mathbf{f}$ and $\mathbf{g}$ are hereditary functionals; i.e., integrals over the past history up to the present.

3.6.1 STATE SPACE THEORY EXAMPLE

Imagine that you are the manager who controls the flow from a large reservoir in a stream basin as shown in Figure 3.15. The project furnishes water for hydroelectric power generation and also for irrigation of farmlands.

Such a river system is typical of those found in countries such as Greece, Turkey, Spain, and other Mediterranean and Middle Eastern lands where water is not plentiful and fossil fuels are also scarce. These systems are also common in the western regions of the United States.

Describe the many factors you would need to consider when deciding how much water to release to irrigation and to power production.

Create a potential systems states model which would allow the reservoir manager to make such control decisions regarding water usage for this system. Given the initial condition and the Autumn, Winter, Spring, and Summer flows to the reservoir as shown in the table below, use your systems states model to create a graph which displays the annual system behavior based upon your decisions concerning water release. See Figure 3.16.

C_0 is the capacity of the reservoir. The initial reservoir condition is $\frac{1}{4}\,C_0$.

TABLE OF WATER FLOWS

Autumn:	Inflow to Reservoir $= 1/4\,C_0$	Water Released $= 1/4\,C_0$
Winter:	Inflow to Reservoir $= 1/2\,C_0$	Water Released $= 1/4\,C_0$
Spring:	Inflow to Reservoir $= 1/2\,C_0$	Water Released $= 1/4\,C_0$
Summer:	Inflow to Reservoir $= 1/4\,C_0$	Water Released $= 1/2\,C_0$

Let $\mathbf{X}_0$ be the initial state vector at the beginning of a season. Name the inflow vector during the season $\mathbf{U}$, and $\mathbf{X}$ is the system state vector for the period. The

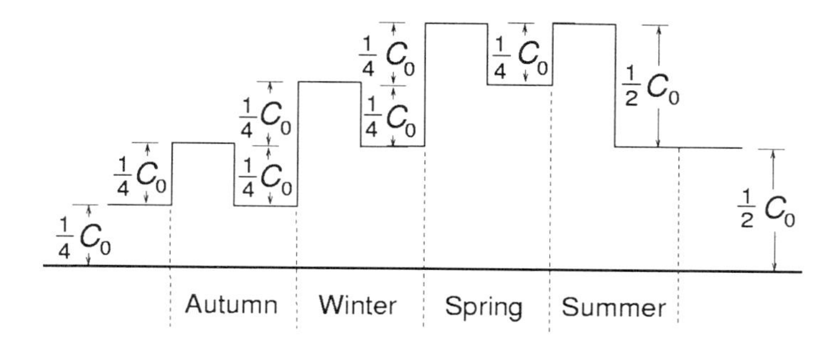

Figure 3.16: Reservoir Water Flows

outflow is called **V**, and the output vector is **Y** for the season in question.

AUTUMN

$$(\mathbf{X}_0)_A = \frac{1}{4}C_0\begin{bmatrix}0\\0\\0\\1\end{bmatrix} \quad \mathbf{U}_A = \frac{1}{4}C_0\begin{bmatrix}0\\0\\1\\0\end{bmatrix} \quad \mathbf{X}_A = (\mathbf{X}_0)_A + \mathbf{U}_A = \frac{1}{4}C_0\begin{bmatrix}0\\0\\1\\1\end{bmatrix}$$

$$\mathbf{V}_A = \frac{1}{4}C_0\begin{bmatrix}0\\0\\1\\0\end{bmatrix} \quad \mathbf{Y}_A = \mathbf{X}_A - \mathbf{V}_A = \frac{1}{4}C_0\begin{bmatrix}0\\0\\1\\1\end{bmatrix} - \frac{1}{4}C_0\begin{bmatrix}0\\0\\1\\0\end{bmatrix} = \frac{1}{4}C_0\begin{bmatrix}0\\0\\0\\1\end{bmatrix}$$

WINTER

$$(\mathbf{X}_0)_W = \frac{1}{4}C_0\begin{bmatrix}0\\0\\0\\1\end{bmatrix} \quad \mathbf{U}_W = \frac{1}{4}C_0\begin{bmatrix}0\\1\\1\\0\end{bmatrix} \quad \mathbf{X}_W = \frac{1}{4}C_0\begin{bmatrix}0\\1\\1\\1\end{bmatrix}$$

$$\mathbf{V}_W = \frac{1}{4}C_0\begin{bmatrix}0\\1\\0\\0\end{bmatrix} \quad \mathbf{Y}_W = \frac{1}{4}C_0\begin{bmatrix}0\\0\\1\\1\end{bmatrix}$$

SPRING

$$(\mathbf{X}_0)_{SP} = \frac{1}{4}C_0\begin{bmatrix}0\\0\\1\\1\end{bmatrix} \quad \mathbf{U}_{SP} = \frac{1}{4}C_0\begin{bmatrix}1\\1\\0\\0\end{bmatrix} \quad \mathbf{X}_{SP} = \frac{1}{4}C_0\begin{bmatrix}1\\1\\1\\1\end{bmatrix}$$

$$\mathbf{V}_{SP} = \frac{1}{4}C_0\begin{bmatrix}1\\0\\0\\0\end{bmatrix} \quad \mathbf{Y}_{SP} = \frac{1}{4}C_0\begin{bmatrix}0\\1\\1\\1\end{bmatrix}$$

SUMMER

$$\mathbf{X}_0)_{SU} = \frac{1}{4}C_0 \begin{bmatrix} 0 \\ 1 \\ 1 \\ 1 \end{bmatrix} \quad \mathbf{U}_{SU} = \frac{1}{4}C_0 \begin{bmatrix} 1 \\ 0 \\ 0 \\ 0 \end{bmatrix} \quad \mathbf{X}_{SU} = \frac{1}{4}C_0 \begin{bmatrix} 1 \\ 1 \\ 1 \\ 1 \end{bmatrix}$$

$$\mathbf{V}_{SU} = \frac{1}{4}C_0 \begin{bmatrix} 1 \\ 1 \\ 0 \\ 0 \end{bmatrix} \quad \mathbf{Y}_{SU} = \frac{1}{4}C_0 \begin{bmatrix} 0 \\ 0 \\ 1 \\ 1 \end{bmatrix}$$

The system manager will need to balance the several requirements upon the system, among these include:

- The output must not fall below a prescribed amount at any time so as to maintain the concentration of pollutants below the statutory level for wildlife habitat and human recreational activities such as swimming, fishing, and boating in the river downstream from the dam.
- The output should return to its mean average annual level, at year's end; otherwise, the reservoir will not be in equilibrium.
- If the hydroelectric plant is the principal source of power in the region, then a minimum flowrate will need to be maintained to supply the electric needs of the subscribing customers without interruptions.
- A minimum flowrate will need to be sustained for the agricultural interests which depend upon the water for irrigation.
- A maximum discharge from the reservoir must not be exceeded so that flooding of the river below the dam will not occur.

3.7 DIFFERENTIAL EQUATIONS FOR SYSTEMS

Suppose that a system has been delineated by drawing its boundary and identifying some elements within the region of the boundary. For the sake of simplicity, let there be two elements, and assume that the relationship between them is such as to warrant the description "protagonists." That is, they are combatants or contestants or colleagues or conspirators or some other form of one-on-one involvement with each other. A system of this sort is Volterra's famous problem of prey-predator relationships.

Consider the situation in which civil engineering practice has evolved to the point where there are two principal types of construction (steel, reinforced concrete) or two kinds of materials (bituminous, concrete) in vogue. In general, any technology has such options, although there are usually more than just two. Let the first protagonist be named N_1 and the other be called N_2. The relationships between the two protagonists are such that the rate of growth of each is proportional to the amount of that protagonist present, but is inhibited by the amount of the other. Then the equations of the system can be written as

$$\frac{dN_1}{dt} = a_{11}N_1 - a_{12}N_2 \tag{3.86}$$

$$\frac{dN_2}{dt} = a_{21}N_1 - a_{22}N_2 \tag{3.87}$$

and, again for simplicity's sake, take $a_{11} = a_{22}$.

The system of differential equations above is linear, of the first order. Solve the first for N_2 in terms of N_1 and substitute in the second, and similarly, solve the second for N_1 in terms of N_2 and substitute in the first. This procedure separates the variables to give two second order equations, which, with

$$\Delta = a_{12}\,a_{21} - a_{11}^2 \geq 0 \tag{3.88}$$

has the solution

$$N_1 = N_{10}\,\cos(\Delta t + p) \tag{3.89}$$

$$N_2 = N_{20}\,\cos(\Delta t + q) \tag{3.90}$$

N_{10}, N_{20} are the initial values of the protagonists, and p and q are phase shifts.

By eliminating dt from both of the system's differential equations, the phase space equation can be obtained as

$$a_{21}\,N_1^2 - 2\,a_{11}\,N_1\,N_2 + a_{12}\,N_2^2 = K \tag{3.91}$$

in which K is the constant of integration. To show this, note that from Equations 3.86 and 3.87

$$\frac{dN_1}{a_{11}N_1 - a_{12}N_2} = dt = \frac{dN_2}{a_{21}N_1 - a_{22}N_2} \tag{3.92}$$

and by differentiating Equation 3.91

$$2\,a_{21}\,N_1\,dN_1 - 2\,a_{11}\,N_1\,dN_2 - 2\,a_{11}\,N_2\,dN_1 + 2\,a_{12}\,N_2\,dN_2 = 0$$

This latter equation can be collected into

$$\frac{dN_1}{a_{11}\,N_1 - a_{12}\,N_2} = \frac{dN_2}{a_{21}\,N_1 - a_{11}\,N_2} \tag{3.93}$$

which is obviously identical with Equation 3.92 for $a_{11} = a_{22}$.

Graphs of the protagonists versus time are displayed in Figure 3.17, and the elliptic phase space trajectory is shown in Figure 3.18.

These results show that the system behaves in a stable manner, with the protagonists waxing and waning about a common focus. Neither is annihilated, and although at a given time one seems to be in the ascendent, the performance of the other is observed to be delayed by the amount of the phase shift, and it will in due course become the dominant player. As time passes, the roles reverse again and again.

Other relationships or modes of behavior between the protagonists can be postulated, and the system performance can be examined, simply by changing

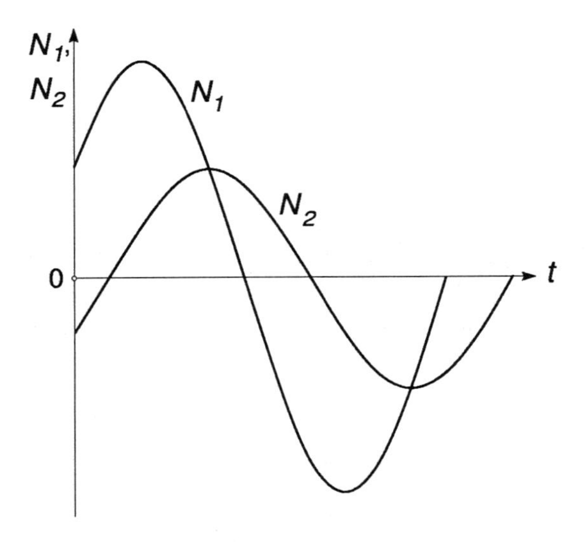

Figure 3.17: Protagonist Waveforms

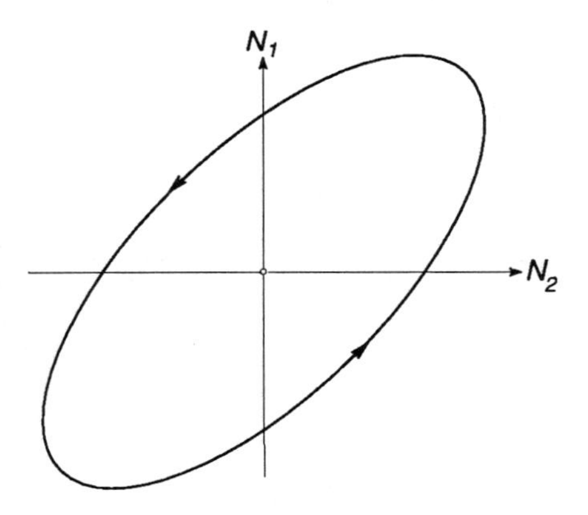

Figure 3.18: Phase Trajectory

the coefficients in the defining differential equations. A very common interaction is that of competition in a "zero-sum-game"; i.e., the two are vying for the same resource, so that whatever one gets, the other is denied. This condition is typical of the more general phenomenon of growth, which appears to be a system that is uniquely dependent upon systems models.

3.8 PROJECT #1

A. Organize a group of six engineers to accomplish this project.

B. Select one person of the group to be the Chief Engineer, and let another be designated Recorder (or Scribe).

C. At an early meeting of the group select one of the following as the subject of the team's work:

   a) The University of North Carolina System
   b) The Triangle J Council of Governments
   c) I-40 from Raleigh to Wilmington
   d) I-40 from Raleigh to Hillsborough
   e) City Government Buildings in Raleigh
   f) State Government Buildings in Raleigh
   g) Raleigh-Durham International Airport
   h) The Building Code of North Carolina
   i) Treyburn
   j) Lochmere
   k) Research Triangle Park
   l) Jordan Lake Watershed
   m) Falls Lake Watershed

D. Discuss your selection with your instructor to obtain concurrence in your choice and assistance with sources of information and other elements of organization of your work.

E. Obtain as much information about the selected topic as can reasonably be had within a few days time.

F. Prepare a report of up to 5 pages length (minus attachments) which describes the selected topic. The report should focus upon the systems nature of the topic, including especially the elements, their interrelationships and functionings. Point out the constraints which operate, and, if possible, suggest an optimizing criterion for the system. Try to set out the system's basis descriptors (decision variables).

G. Include with the report a brief summary of the contributions of each member of the group to the project. This can best be extracted from the log of the group's activities. Let all members of the group sign this contributions page.

CHAPTER 4

OPTIMIZATION

4.1 INTRODUCTION

Optimization is a word which identifies the central function served in the engineering design process. To optimize a design means to make it best by some criterion or standard. It implies that there is a myriad of solutions to a design situation, but that one or more are clearly better than all or most of the others. The identification of this best, or at least some of the better, designs is the first task in a serious creative procedure.

Optimization is thus a word "writ large," as John Hancock said, so that it cannot fail to be seen, across the face of the engineering business.

It may also be an elusive thing, something to be pursued at all costs when a smidgen of pragmatism might well be used to cut off the search at a more economically justified level. It can be, as with Jason and the Argonauts, a quest for the Golden Fleece or the Holy Grail—leading designers onward, ever onward, through perils and fogs and by sirens and hidden rocks, like a will-o'-the-wisp, to an imaginary but imagined goal.

Optimum results are frequently identified with the concept of minimum cost. Certainly this is a most important and a very frequent stipulation associated with the optimum. In the case of public works bid at auction in a competitive situation, it is the general concept that the "lowest" bidder wins the contract. In other words, given the same specifications, which presume a minimum satisfactory level of service and durability, that one which is least expensive is ranked highest. It becomes by definition the optimum.

There are, however, other considerations which enter the picture in many instances. To mention an obvious one, sometimes there are multiple criteria for an optimum. Another case occurs when a benefit is the quantity to be maximized, not a sum of money. Of course, it may be argued that benefits can be rendered in terms of money, and to a certain extent they can be, but there are instances in which the equivalence is most elusive. Subjective and personalized value structures intrude into the valuation schemes.

In present practice, one of the most frequently encountered criterions of optimization is the benefit-to-cost ratio, a common test especially when rating projects in the public sector where the whole public is assumed to gain from the proposed work.

However accomplished, by whatever yardstick, optimization seeks to formulate the design process in such a way that these issues can be rationally explored.

Since each endeavor may have its special features, perhaps a set of case studies drawn from civil engineering practice will be enlightening.

4.2 EARTHMOVING

An activity characteristic of many civil engineering projects is earthmoving. Soil is taken from one place, where it is either not needed or is unwanted, and carried to another place which is too low and must be filled in. The topological survey associated with the construction of a new superhighway, for example, will reveal a number of places where cuts must be made and another number where fills will be needed. Additionally, the "bread and butter" will sometimes not run out even, and a dump or third location will be indicated in which to park the excess soil.

Suppose in a particular case there are three cuts to be made at locations A, B, and C. The amounts of earth to be removed are: at A, 4670 m^3; at B, 7130 m^3; and at C, 8980 m^3. There are four places where fills are needed: at D, 1940 m^3; at E, 5930 m^3; at F, 8080 m^3; and at G, 3950 m^3. Because the total fill requirement is $19,900$ m^3 and the cuts produce a net of $20,780$ m^3, it is obvious that a dump will be required, say at a place named H.

Now create a transportation schedule or matrix array which shows the geographic distances from each cut to every fill as shown below.

DISTANCES (km)						
		FILLS				
		D	E	F	G	H
CUTS	A	0.5	0.2	0.4	0.2	0.8
	B	0.4	0.7	0.6	0.2	1.2
	C	0.2	0.1	0.6	0.7	0.9

Let the quantity of earth carried from the source i (A, B, C) to some destination j (D, E, F, G, H) be X_{ij}. There are then 15 of these Xs:

$$\begin{array}{ccccc} X_{AD} & X_{AE} & X_{AF} & X_{AG} & X_{AH} \\ X_{BD} & X_{BE} & X_{BF} & X_{BG} & X_{BH} \\ X_{CD} & X_{CE} & X_{CF} & X_{CG} & X_{CH} \end{array}$$

Evidently, these are the elements of the system, for if they were known, by whatever device, their numbers would constitute a solution to the earthmoving problem.

There are, however, certain constraints to which the X_{ij} are subjected; namely, all of the fill requirements must be satisfied, and all of the material produced at the cuts must be carried away to either the fills or the dump. These requirements produce the following set of constraint equations.

$$X_{AD} + X_{AE} + X_{AF} + X_{AG} + X_{AH} = 4670 \tag{4.1}$$
$$X_{BD} + X_{BE} + X_{BF} + X_{BG} + X_{BH} = 7130 \tag{4.2}$$
$$X_{CD} + X_{CE} + X_{CF} + X_{CG} + X_{CH} = 8980 \tag{4.3}$$
$$X_{AD} + X_{BD} + X_{CD} = 1940 \tag{4.4}$$
$$X_{AE} + X_{BE} + X_{CE} = 5930 \tag{4.5}$$
$$X_{AF} + X_{BF} + X_{CF} = 8080 \tag{4.6}$$
$$X_{AG} + X_{BG} + X_{CG} = 3950 \tag{4.7}$$

The criterion function, which is a measure of the best solution of the problem, is the total cost of the operation. To establish it, note that the cost of transporting the earth is proportional to the distance it must be moved. The contractor will have records which show the distance charges which have been incurred in similar operations in the past to which he can turn for a figure. Let it be called $\$M$/km. Then

$$\begin{aligned} C = {} & 0.5\,M\,X_{AD} + 0.2\,M\,X_{AE} + 0.4\,M\,X_{AF} + 0.2\,M\,X_{AG} + 0.8\,M\,X_{AH} \\ & + 0.4M\,X_{BD} + 0.7\,M\,X_{BE} + 0.6\,M\,X_{BF} + 0.2\,M\,X_{BG} + 1.2\,M\,X_{BH} \\ & + 0.2\,M\,X_{CD} + 0.1\,M\,X_{CE} + 0.6\,M\,X_{CF} + 0.7\,M\,X_{CG} + 0.9M\,X_{CH} \end{aligned} \tag{4.8}$$

In a shorthand notation this becomes

$$C = M \sum_i \sum_j d_{ij} X_{ij} \tag{4.9}$$

For this case study, the optimization can be formally stated as follows:

> Minimize the cost C while at the same time satisfying the constraint Equations 4.1 through 4.7. Further, all X_{ij} must be equal to or greater than 0, because negative values of the Xs would have no physical significance.

Another name for the cost function C is the ***objective function***; i.e., the function which gives an objective answer to the question, "What is the least cost of doing this operation?"

The Xs are called ***decision variables***, because when a decision is made concerning their values, this determines the cost of the operation.

One solution to the system of equations shown above is given by the matrix below:

QUANTITIES (m^3)							
		FILLS					
		D	*E*	*F*	*G*	*H*	
CUTS	*A*	940	950	950	950	880	4670
	B	1000	2000	3100	1000	0	7130
	C	0	2980	4000	2000	0	8980
		1940	5930	8080	3950	880	

The total cost associated with this set of numbers is easily computed to be $\$9,910\,M$. From the totals for each cut and each fill, it is seen that the solution does in fact satisfy all of the constraint requirements.

The best solution, however, is that contained in the table below:

QUANTITIES (m^3)							
		FILLS					
		D	*E*	*F*	*G*	*H*	
CUTS	*A*	0	0	4670	0	0	4670
	B	0	0	3180	3950	0	7130
	C	1940	5930	230	0	880	8980
		1940	5930	8080	3950	880	

Once again, the totals for rows and columns show that all of the constraints are fulfilled. In this case, however, the total cost function is $\$6477\,M$, which is 35% less than the first instance. If there were two contractors bidding on this job, and their bids were proportional to the indicated costs, that one with the optimal solution would certainly be the winning bidder, and should bring the job home within cost. It is easy to see who would get the business and stay in business.

4.3 CONCRETE PRODUCTS PLANT

Nearly every town and city of any size will have one or more industries which manufacture concrete products such as culverts, manhole junction boxes, building block, and prestressed structural members. A typical example is *N. C. CONCRETE PRODUCTS COMPANY*, founded, owned and operated for many years by Mr. Henry M. Shaw, Civil Engineering graduate of North Carolina State University. After Mr. Shaw died his family established the *SHAW LECTURE*

SERIES as a memorial to him. Each year the *SHAW LECTURE* brings outstanding international authorities on civil engineering practice to Raleigh to share their expertise with the faculty, staff, and students at NCSU.

Suppose that such a company is manufacturing concrete items in three shapes, S_1, S_2, S_3. The materials used in each are cement, fine aggregate, and coarse aggregate. (And water too, of course.) The amounts of cement and fine and coarse aggregate vary according to the shape. The table below gives the particulars. (All weights are in *Newtons.*)

Materials Schedule

	E L E M E N T		
	S_1	S_2	S_3
CEMENT	60	60	45
COARSE AGGREGATE	325	260	270
FINE AGGREGATE	145	180	170

Some features of the plant operation during a given time period are as follows.

Production takes place on three assembly lines. One of these is a rig originally purchased at the time the plant was established many years ago. When the business grew, a newer line was obtained, but the older one was kept as a topping unit and for backup in cases of breakdown. A third assembly was acquired and has been specialized to the production of S_3 shapes only, to obtain the benefits of this selectivity. The original rig is not used to make S_3 items. All three of the machines must be stopped and thoroughly cleaned after making a certain number of items—100 for the old rig, 150 for the newer one, and 200 for the specialized line. This maintenance is needed to remove cement from forms and loose aggregates and mix from about the forms and work areas, and to re-oil forms, lubricate bearings in extruders, etc.

The actual costs to manufacture each unit include those of the materials and those charged to the operation of the assembly lines. Materials supplies cost 2.5 cents/N for cement, 1 cent/N for large aggregate, and 1.5 cents/N for fine aggregate in bulk lots. Therefore, unit costs for materials are:

$$S_1 : 60 \times 2.5 + 325 \times 1.0 + 145 \times 1.5 = 693\,\text{cents}\,@ \tag{4.10}$$

$$S_2 : 80 \times 2.5 + 260 \times 1.0 + 180 \times 1.5 = 730\,\text{cents}\,@ \tag{4.11}$$

$$S_3 : 45 \times 2.5 + 270 \times 1.0 + 170 \times 1.5 = 638\,\text{cents}\,@ \tag{4.12}$$

In addition to materials costs, there are operational costs associated with each assembly line. These costs are for labor, utilities (water, electric power, telephone), and overhead, (including rent or debt service, cleaning, security, office staff expense, and similar items). These expenses can be lumped together for each member on each production line, and are shown in the table below.

Assembly Line Costs

	MEMBER	LINE COST PER UNIT
LINE 1	S_1	550 cents
	S_2	550 cents
LINE 2	S_1	900 cents
	S_2	800 cents
	S_3	600 cents
LINE 3	S_3	1100 cents

At this point it can be seen that an analysis of the plant's operation leading to an optimum condition will require that an identifying nomenclature be established to show which units are made on which assembly lines. So let the number of S_i items produced on line j be X_{ij}. There are then the following Xs:

$$X_{11} \quad X_{21} \quad X_{12} \quad X_{22} \quad X_{32} \quad X_{33}$$

Obviously, these are the decision variables—the elements of this system.

The total unit costs can now be assembled as:

$$X_{11} : 693 + 550 = 1243 \text{ cents @} \tag{4.13}$$

$$X_{12} : 693 + 900 = 1593 \text{ cents @} \tag{4.14}$$

$$X_{21} : 730 + 550 = 1280 \text{ cents @} \tag{4.15}$$

$$X_{22} : 730 + 800 = 1530 \text{ cents @} \tag{4.16}$$

$$X_{32} : 638 + 600 = 1238 \text{ cents @} \tag{4.17}$$

$$X_{33} : 638 + 1100 = 1738 \text{ cents @} \tag{4.18}$$

There are some constraints involved in the plant's operations. The available land and storage facilities limit the amounts of cement and aggregates which can be accomodated to cement, 12,000 N; coarse aggregate, 55,000 N; and fine aggregate, 35,000 N. To satisfy the demands of customers, the plant needs to supply at least 22 S_1, 24 S_2, and 18 S_3 units. The selling prices are 2200 cents for S_1 shape, 1700 cents for S_2, and 1800 for S_3.

Unit profits can now be determined as:

$$X_{11} : 2200 - 1243 = 957 \text{ cents @} \tag{4.19}$$

$$X_{12} : 2200 - 1593 = 657 \text{ cents @} \tag{4.20}$$

$$X_{21} : 1700 - 1280 = 420 \text{ cents @} \tag{4.21}$$

$$X_{22} : 1700 - 1530 = 170 \text{ cents @} \tag{4.22}$$

$$X_{32} : 1800 - 1238 = 562 \text{ cents @} \tag{4.23}$$

$$X_{33} : 1800 - 1738 = 62 \text{ cents @} \tag{4.24}$$

An equation for the total profit becomes the objective function for this problem and is as follows:

$$P = 957\, X_{11} + 657\, X_{12} + 420\, X_{21} + 170\, X_{22} + 562\, X_{32} + 62\, X_{33} \tag{4.25}$$

The constraints are expressed in the following equations.

Demand Constraints

$$X_{11} + X_{12} \geq 22 \tag{4.26}$$

$$X_{21} + X_{22} \geq 24 \tag{4.27}$$

$$X_{32} + X_{33} \geq 18 \tag{4.28}$$

Production Line Constraints

$$X_{11} + X_{21} \leq 100 \tag{4.29}$$

$$X_{12} + X_{22} + X_{32} \leq 150 \tag{4.30}$$

$$X_{33} \leq 200 \tag{4.31}$$

Stock Constraints

$$60\,X_{11} + 60\,X_{12} + 80\,X_{21} + 80\,X_{22} + 45\,X_{32} + 45\,X_{33} \leq 12,000 \tag{4.32}$$

$$325\,X_{11} + 325\,X_{12} + 260\,X_{21} + 260\,X_{22} + 270\,X_{32} + 270\,X_{33} \leq 55,000 \tag{4.33}$$

$$145\,X_{11} + 145\,X_{12} + 180\,X_{21} + 180\,X_{22} + 170\,X_{32} + 170\,X_{33} \leq 35,000 \tag{4.34}$$

The formal statement of the optimization becomes:

MAXIMIZE $P(X_{ij})$

SUBJECT TO CONSTRAINT EQUATIONS 4.26–4.34

ALL DECISION VARIABLES $X_{ij} \geq 0$

Whereas in the earthmoving problem the constraints were equalities, in this case they are inequalities of both less-than and greater-than types. The successful solution of the above set of equations will yield values for the decision variables such that the profit from the plant is maximized. The plant operator can accordingly schedule the number of units to be made on each assembly line in the confidence that no better result could be had within the limitations imposed.

One obvious solution is:

$$X_{11} = 11 \quad X_{12} = 11 \quad X_{21} = 12 \quad X_{22} = 12 \quad X_{32} = 9 \quad X_{33} = 9 \tag{4.35}$$

This satisfies all nine constraints, and the profit function gives

$$P = 30,450\,\text{cents} = \$304.50 \tag{4.36}$$

Another solution ($X_{11} = 100$, $X_{22} = 54$, $X_{32} = 24$) is one for which the profit is computed to be P = 118,368 cents = \$1183.68, nearly four times larger than the former! Can you think of a better one? (The optimum P is \$1428.86.)

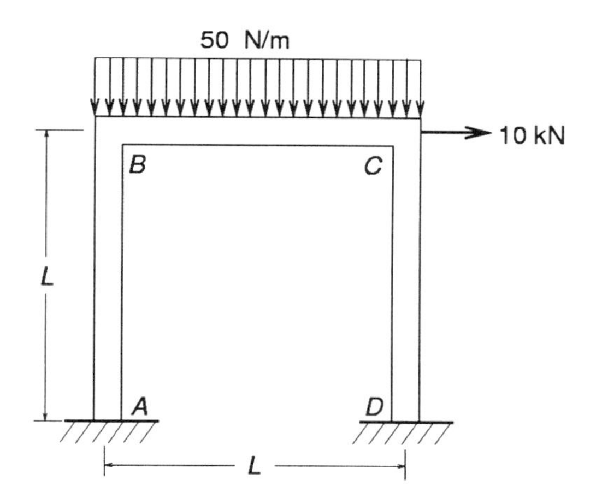

Figure 4.1: Portal Frame

4.4 PORTAL FRAME STRUCTURE

Figure 4.1 shows a steel structure called a *portal frame* because it is similar to the frame surrounding a door or other entryway. Members AB and CD are columns, and BC is a beam. All are supposed to be standard rolled steel sections, and all may be of different cross-sectional area. The usual assumption is made concerning the corners, to wit, the angle is preserved (at 90 degrees) during elastic rotations. The beams are built in at the foundations at A and D.

One method of analysis of the behavior of this frame when subjected to loads is that known as "plastic hinge analysis." In this theory, a steel member is stressed to the point of plastic yielding at some location, which thereafter cannot support any additional moment and so yields, allowing rotation to take place at that location. Since the remainder of the beam or column is still elastic, with the concomitant small deformations, the yielded place behaves much like a door-hinge would. That is, large angular rotation is permitted at that point, and not elsewhere. The resistance to bending is a bending moment developed at the hinge.

Two loads are placed on the frame. One is a distributed load of 50 N/m across the beam, on top. The other is a lateral force of 10 kN applied to point C at the top of the column CD. A factor of safety (factor of ignorance!) of 2.0 is specified for this design, and an easy way to supply this is merely to multiply the loads by this factor at the outset. Hence, the values are taken as 100 N/m and 20 kN, respectively. The lengths L are 3 m each.

Six possible collapse mechanisms can readily be identified. Two are beam failures, two are side-sways, and two are combined modes. Since it is not known in advance whether the beam or the columns are weaker, there are two cases to be considered for each type of collapse. The six cases are depicted schematically in Figure 4.2.

Associated with each collapse mode is a relationship between the external work done by the applied loads and the energy dissipated in plastic deformations. For example, in case (ii), the beam across the top yields at its center where there is a plastic hinge created, and the columns yield at their tops, where there are also plastic hinges. All of the external work is done by the beam load, and it is equal to the moments at the beam center times the (small) angles of rotation at the center; i.e.,

$$W = 2 \times \tfrac{1}{8}(100)(3)2 \text{ kN} \times \theta = 225\,\theta \text{ kN} \cdot \text{m} \tag{4.37}$$

The energy equivalent to this work is balanced against the work of the fully-plastic moments in the two hinges at the tops of the columns and the work of the fully-plastic hinge moments at the center of the beam. The former is

$$U = 2 \times M_2 \times \theta = 2\,M_2\,\theta \text{kN} \cdot \text{m} \tag{4.38}$$

and the latter is

$$U = M_1 \times 2\theta = 2\,M_1\,\theta \text{ kN} \cdot \text{m} \tag{4.39}$$

So if the structure is not to collapse, the energy stored in the plastic hinges must be equal to or greater than the work done by the external load during the deformation. That is,

$$2\,M_1\,\theta + 2\,M_2\,\theta \geq 225\,\theta \tag{4.40}$$

or

$$2\,M_1 + 2\,M_2 \geq 225 \tag{4.41}$$

Similar relationships hold for the other five collapse modes, and are shown on the diagram.

Now, the cost of steel is proportional to its weight (or to its volume, assuming a constant density). The lengths of the members of the structure are specified in the pre-design conditions, as this is a part of the functional requirements of the portal, so only the cross-sectional areas are left open as variables. Therefore the volume is

$$V = 6\,A_1 + 2 \times 3\,A_2 \text{ m}^3 \tag{4.42}$$

A simple analysis of a steel member with a W-section shows that the plastic moment is given approximately by

$$M_P \cong 0.5\,s_o\,h\,(A - 0.5\,A_w) \tag{4.43}$$

in which s_o is the yield stress, h is the depth of the web, A_w is the web area and A is the cross-sectional area. This suggests that there is an approximately linear relationship between the fully-plastic bending moment and cross-sectional area, as shown in Figure 4.3.

That is,

$$A = a + b\,M \tag{4.44}$$

so

$$V = 6(a + b\,M_1) + 2 \times 3\,(a + b\,M_2) = 12\,a + b(6\,M_1 + 6\,M_2) \tag{4.45}$$

In other words, the cost function is proportional to

$$C = M_1 + M_2 \tag{4.46}$$

Now the optimization of the structure for minimum cost can be formulated as follows:

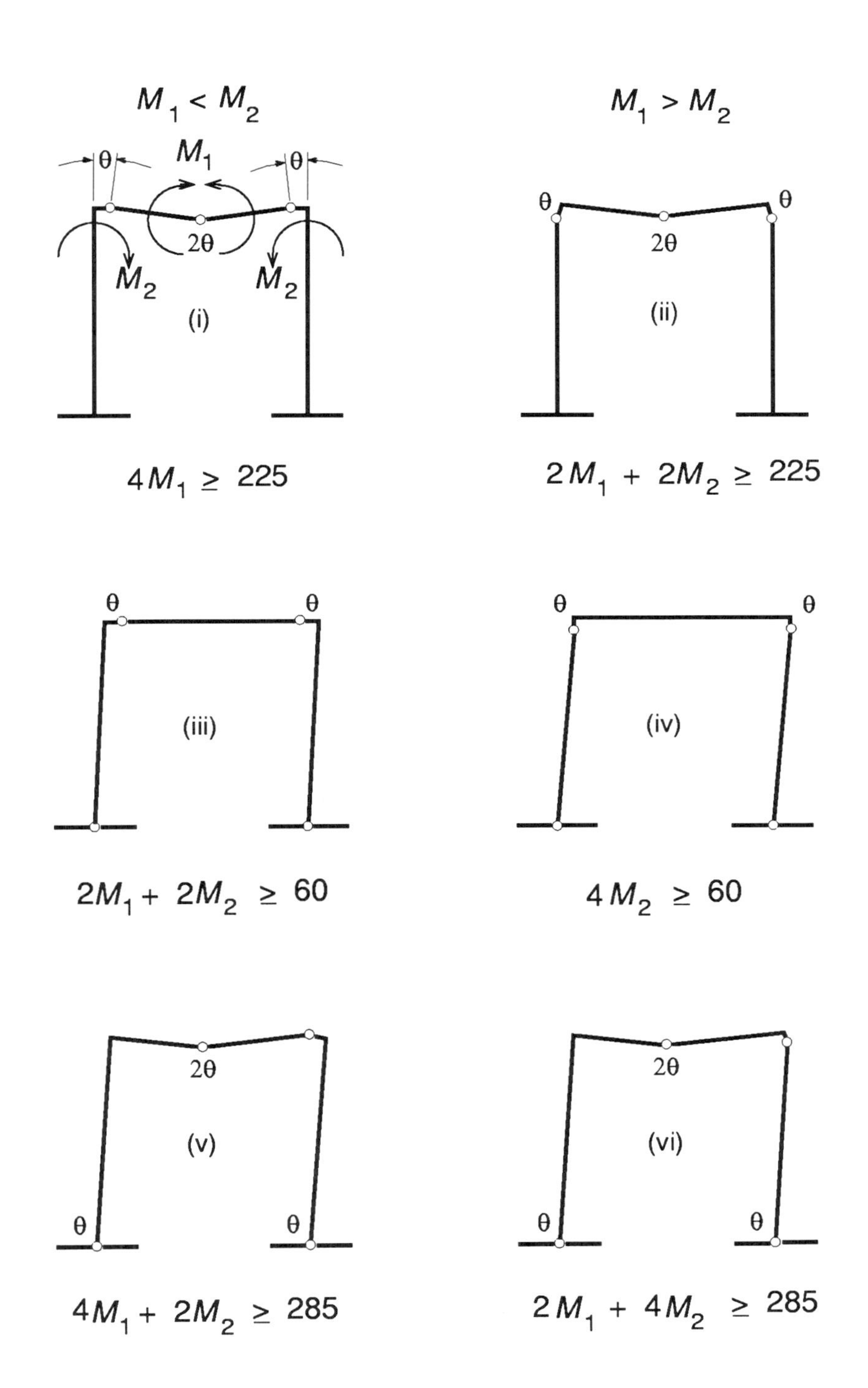

Figure 4.2: Portal Frame Collapse Mechanisms

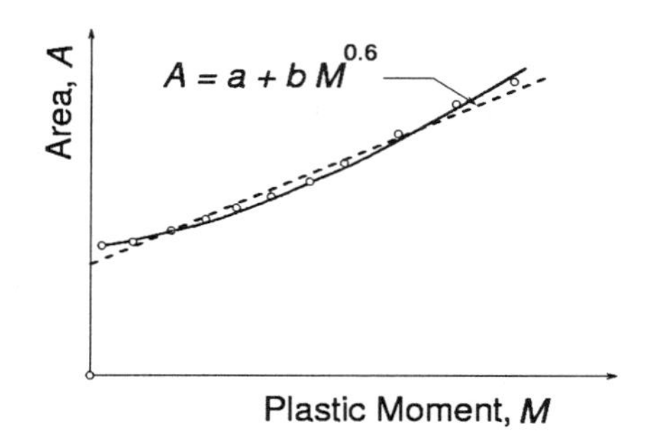

Figure 4.3: Fully-Plastic Moments

$$\text{Minimize} \quad C = M_1 + M_2 \tag{4.47}$$
$$\text{subject to} \quad 4\,M_1 \geq 225 \tag{4.48}$$
$$2\,M_1 + 2\,M_2 \geq 225 \tag{4.49}$$
$$2\,M_1 + 2\,M_2 \geq 60 \tag{4.50}$$
$$4\,M_2 \geq 60 \tag{4.51}$$
$$4\,M_1 + 2\,M_2 \geq 285 \tag{4.52}$$
$$2\,M_1 + 4\,M_2 \geq 285 \tag{4.53}$$
$$M_1, M_2 \geq 0 \tag{4.54}$$

Here is a system characterized by an objective function C and six constraint relationships that are all greater-than type inequalities. The usual additional requirement is that both of the resisting moments shall be larger than zero, otherwise there is no load on the system, therefore no problem. Notice that the system elements here are bending moments. These moments have become the decision variables for the problem.

The optimum moment values are $M_1 = 56.25\,\text{kN} \cdot \text{m}$ and $M_2 = 56.25\,\text{kN} \cdot \text{m}$.

4.5 EWASA

The Eno River is a small stream in north central North Carolina, almost entirely in Orange and Durham counties. Although it is short and relatively narrow and shallow, it is the principal stream in some very prime real estate and is one source of the much greater Neuse River. As it happens, the watershed drained by the Eno contains some of the most historic country in the region as well as some of the most attractive lands for development.

The dual emphases on conservation and growth have combined to produce enormous pressures upon this modest river and its basin. The Eno Water and Sewer Authority (EWASA) was created by the North Carolina Legislature to manage the multitude of demands upon the Eno by the many municipalities, agencies, and development interests within its purview.

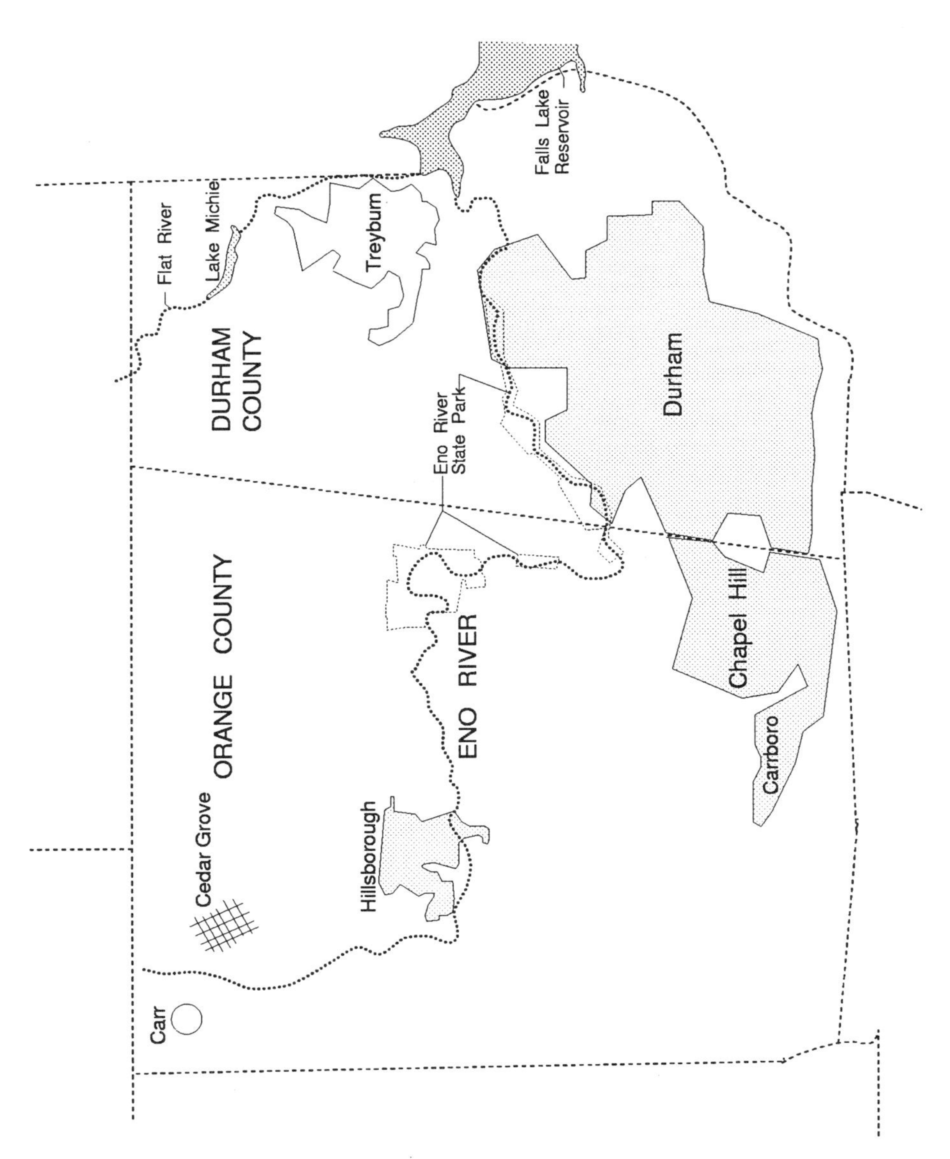

Figure 4.4: The Eno River Basin

Figure 4.4 shows some of the features along the Eno. The upper reaches lie within Orange County, an area established and populated from earlier colonial times. The county seat is Hillsborough, a historic town with courthouse and numerous main street buildings on the National Register of Historic Places. Naturally, there are very strict zoning and preservation district regulations, and, moreover, this rubric extends to most of the County. The lower portion of the river passes through Durham County, largely of 20th Century vintage. About half of the County is taken up by the City of Durham and its environs, and there is a large industrial component dating to the early post-Civil War days. Union soldiers found cigarette-making prevalent in the Durham Station (Bennett House) community, and this became the nucleus of a tobacco-manufacturing industry and its affiliated services.

Chapel Hill in Orange County is the site of the University of North Carolina, and Duke University and North Carolina Central University are located in Durham. These, together with North Carolina State University in Raleigh, only 20 miles southeast, anchor the Research Triangle, with its contemporary scientific, technological, business, educational, and governmental features.

The Eno empties into Falls Lake Reservoir at the head of the Neuse River, as does the Flat River, through Lake Michie. These lakes supply water to the Cities of Raleigh and Durham and a number of other towns in the region. Carr, Cedar Grove, and Hillsborough all depend upon the Eno itself for water supplies, and discharge treated wastes into the river. Carrboro, Chapel Hill, and Durham likewise take some water from the Eno, and several waste treatment facilities discharge into it as well.

Along much of its common boundary with the City of Durham the little stream is the basis for the Eno River State Park, a very popular recreational strip which, in turn, has probably motivated other nearby establishments such as the Museum of Life and Science and Treyburn, a planned urban development (PUD) of residences, businesses, and upscale industries.

The EWASA professional staff are directed to produce a model of the Eno basin which can serve as the means of optimizing the river's utilization of its multiple roles of water supply, waste removal, and recreational and conservational resource. The model generated by the staff is displayed in Figure 4.5.

Just upstream from Cedar Grove the river has a flowrate or discharge of $Q_1\,\mathrm{m}^3/\mathrm{s}$, and there is a pollutant load of $L_1\,\mathrm{N}/\mathrm{m}^3$, mainly from fertilizer and animal wastes runoff in the largely agricultural northern portion of Orange County. Cedar Grove draws off $Q_2\,\mathrm{m}^3/\mathrm{s}$ of water, which passes through its treatment plant and is then distributed to users through its public water supply. Approximately 41% of this water is lost to leakage in the system and other causes, but 59% is passed through the town's waste treatment system and returns to the river a short distance downstream from the intake, bearing a pollutant load $L_2\,\mathrm{N}/\mathrm{m}^3$. It is assumed that this pollutant is completely mixed with the river's flow at the effluent outfall.

Along the reach between Cedar Grove and Hillsborough about 22% of the pollutant load in the river is removed by the actions of natural causes such as sunlight. The river's flow rate also increases to $Q_3\,\mathrm{m}^3/\mathrm{s}$ because of the groundwater inflow by the time it reaches Hillsborough.

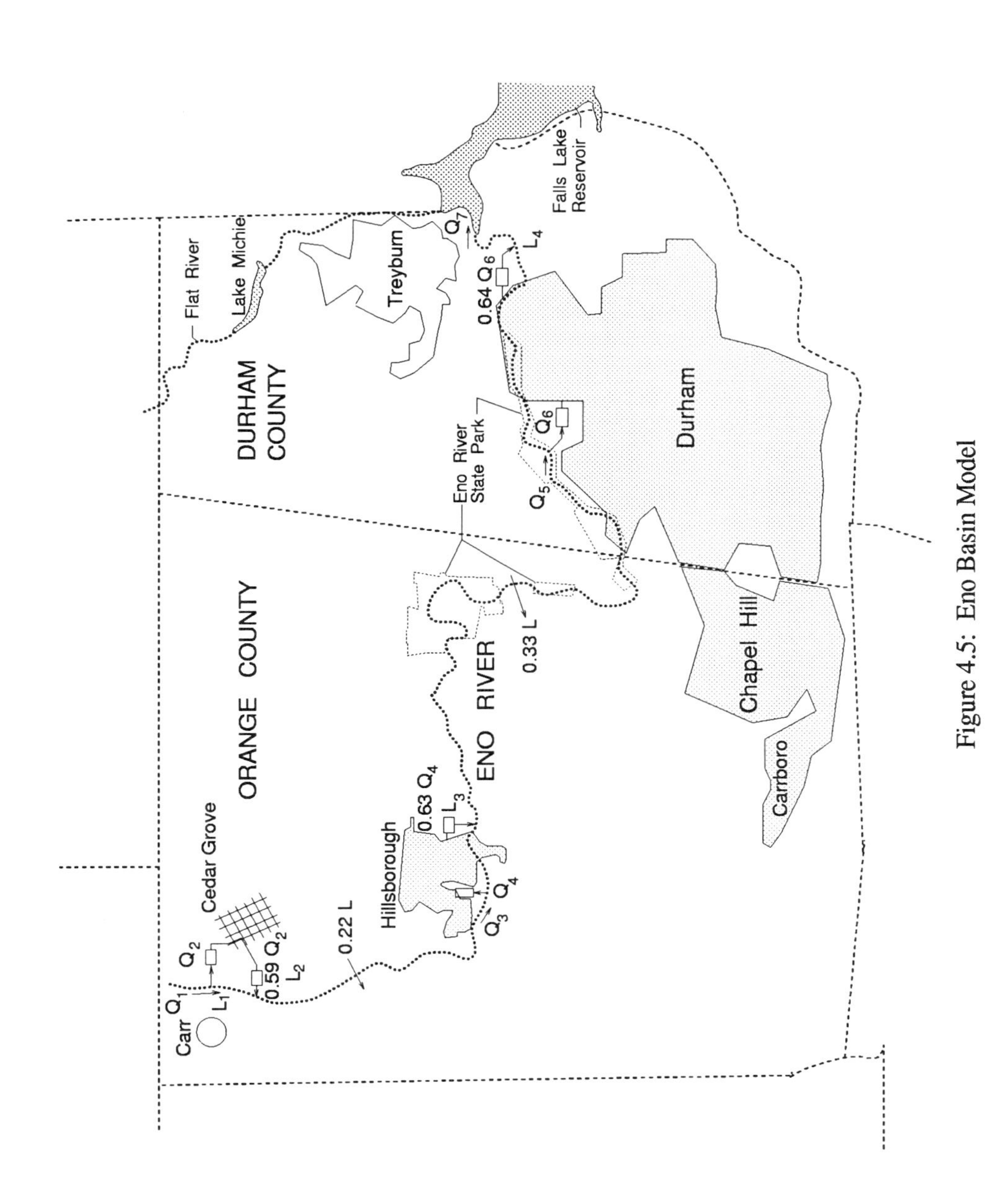

Figure 4.5: Eno Basin Model

Hillsborough abstracts Q_4 m$^3/s$ for its water supply use, of which 37% is lost to leakage, etc., and the remainder is returned to the Eno bearing a pollutant load L_3 N/m^3 just downstream from the town after passing through the treatment plant. Again assuming that the pollutant is well-mixed in the river's flow, natural causes are the agent for removing 33% of the pollutant by the time the flow reaches Durham's intake. Also, the river carries a discharge of Q_5 m^3/s after augmentation by groundwater flow before Durham's intake.

The water removed by the City of Durham is Q_6 m^3/s, which is treated and circulated through the water system. About 36% is lost, but the remainder passes through the waste treatment plant and is returned to the Eno with a pollutant load L_4 N/m^3. The outfall is near the entrance to Falls Lake Reservoir, at which place the river's flowrate is Q_7 m^3/s.

There are federal and state laws governing the allowable amounts of the pollutant L in rivers and streams. There is an overall requirement that L must be less than L_L for the sake of fish and other wildlife and so that fishing, boating, and recreational activities such as swimming will not be hazardous. In other words, throughout the EWASA jurisdiction this upper limit must be met to satisfy water quality standards.

Water for public consumption is held to a higher standard: drinking water codes require that there shall be no more than L_D N/m^3 of this pollutant.

Water and waste treatment plants at Cedar Grove, Hillsborough, and Durham are for the purpose of removing some of the pollutant loads in order to maintain the levels within the statutory requirements.

Let T_1 denote the amount of pollutant removed in the water plant at Cedar Grove, and let T_2 signify the amount removed in its waste treatment plant. Similar labels can be put on the amounts removed at Hillsborough, T_3 and T_4, and in Durham, T_5 and T_6. These treatments cost money. Some of the plants are older and less efficient than others, and, of course, the flow rates in each municipality are different. Designate the costs as C_1 \$/N and C_2 \$/N for Cedar Grove, and C_3 \$/N and C_4 \$/N at Hillsborough, and C_5 \$/N and C_6 \$/N at Durham.

Now the model can be given an algebraic form. First there are the constraints on water quality in the river:

Just above Cedar Grove $$L = \frac{L_1 - T_o}{Q_1} \leq L_L \tag{4.55}$$

Just below Cedar Grove $$L = \frac{L_1 + L_2}{Q_1 - 0.41\,Q_2} \leq L_L \tag{4.56}$$

Just before Hillsborough $$L = \frac{0.78(L_1 + L_2)}{Q_3} \leq L_L \tag{4.57}$$

Just after Hillsborough $$L = \frac{0.78(L_1 + L_2) + L_3}{Q_3 - 0.37\,Q_4} \leq L_L \tag{4.58}$$

Just before Durham $$L = \frac{0.67[0.78(L_1 + L_2) + L_3]}{Q_5} \leq L_L \tag{4.59}$$

Just after Durham $$L = \frac{0.64\{0.67[0.78(L_1 + L_2) + L_3]\} + L_4}{Q_7} \tag{4.60}$$

Second, there are the constraints on the drinking water:

At Cedar Grove
$$L_1 - \frac{T_1}{Q_2} \leq L_D \tag{4.61}$$

At Hillsborough
$$L_1 + L - 2 - \frac{T_3}{Q_4} \leq L_D \tag{4.62}$$

At Durham
$$0.67[0.78(L_1 + L_2) + L_3] - \frac{T_5}{Q_6} \leq L_D \tag{4.63}$$

A third set of relationships relates to the removal of pollutant from the wastes:

At Cedar Grove
$$L_1 - \frac{T_1}{Q_2} - \frac{T_2}{0.59\,Q_2} = L_2 \tag{4.64}$$

At Hillsborough
$$0.78(L_1 + L_2) - \frac{T_3}{Q_4} - \frac{T_4}{0.37\,Q_4} = L_3 \tag{4.65}$$

At Durham
$$0.67[0.78(L_1 + L_2) + L_3] - \frac{T_5}{Q_6} - \frac{T_6}{0.36\,Q_6} = L_4 \tag{4.66}$$

A fourth set describes the removal of pollutant by natural causes in the river reaches between cities:

Between Cedar Grove and Hillsborough

$$0.78(L_1 + L_2) = 0.78\left[\frac{L_1 + L_2 + L_1(Q_1 - Q_2)}{Q}\right] \tag{4.67}$$

Between Hillsborough and Durham

$$0.67[0.78(L_1 + L_2) + L_3] = 0.67\left[0.78(L_1 + L_2) + L_3 + \frac{(L_1 + L_2)(Q_3 - Q_4)}{Q_3}\right] \tag{4.68}$$

At this juncture it can be seen that the decision or control variables are the Ts—the amount of pollutant removed by treatment at each of the plants. That is, the T_1, T_2, T_3, T_4, T_5, and T_6 are quantities, which, when known, determine the state of the system. Accordingly, a cost function can be presented as

$$Z = C_o\,T_o + C_1\,T_1 + C_2\,T_2 + C_3\,T_3 + C_4\,T_4 + C_5\,T_5 + C_6\,T_6 \tag{4.69}$$

and the optimization becomes a task of minimizing Z while still satisfying the constraint relationships Equations 4.55 through 4.68.

4.6 OPTIMIZATION SUMMARY

In each of the case studies above, the systems have been characterized by a collection of decision variables which play the role of system elements. Each system has contained several constraint relationships, either as equalities or as greater-than or less-than type inequalities. The constraints have also been expressed in terms of the decision variables.

Finally, it may be observed that in each of the examples the format of the optimization problem has been of the same genre. This format will furnish the basis of a standard frame of reference for solving such problems, and will be developed further in a later chapter.

4.7 PROJECT #2

1. The partners of an engineering firm are considering whether or not to expand. If they add two engineers to the staff, they can undertake $100,000 more business each year, but the new salaries and overhead will cost $92,000. If they add three engineers, the new business will be $120,000, and the additional salaries and overhead will be $122,000. The present gross annual income is $205,000 and the salaries and overhead are $90,000. What should they do?

2. A developer has acquired a tract of land 1000 m × 400 m bordering a lake which he plans to transform into a Planned Urban Development (PUD). The 1000 m side lies along the lakefront. The lot sizes will be 100 m × 200 m, and those on the lake will sell for $100,000, those not on the lake will bring $80,000. It will cost the developer $20,000 for the lakeside lots and $5,000 for the others to put in the infrastructures. How many of the lakeside lots should he make to maximize his return? How many of the others will there be?

3. Your engineering firm owns a site upon which it plans to erect a new building for its headquarters. The lot is 20 m × 30 m, and local building codes prohibit structures more than 4 floors tall in this zone. The "ground floor" and "penthouse" level (top floor) could be rented to other organizations for premium income, but would also cost more to build and equip, as shown in the schedule below. The firm will need two floors for itself. How many floors should the building be?

Rental Schedule

	Cost	Return
Ground Floor	$250,000	$350,000
2nd Floor	$175,000	$160,000
3rd Floor	$220,000	$215,000
Top Floor	$330,000	$340,000

4. KONKRETE COMPANY, Inc., operates three sand pits to furnish material for its two concrete batch plants. Plant *A* requires 24 units of sand, and Plant *B* needs 28 units. Unfortunately, the sand pits are nearing depletion, and the estimates are that Pit *C* has only 8 units left, Pit *D* has only 23, and Pit *E* only 21 units.

   The table below gives the hauling costs from the pits to the plants.

   Write the equations which, when solved, will allow KONKRETE to determine how much to haul from each pit to each plant in order to minimize the total haulage cost.

HAULING COSTS ($/unit)			
		PLANT	
		A	*B*
PIT	*C*	$17	$11
	D	$8	$10
	E	$6	$19

5. ROX COMPANY, Inc., sells stone from its quarry to ready-mixed concrete companies. Based upon previous experience, the table below gives the sales, costs, and available time estimates for the next 4 months' operations.

SALES, COSTS, AND AVAILABLE TIMES ROX COMPANY				
	MONTH			
	1	2	3	4
Stone required (1000 m^3)	1200	2500	2400	2700
Normal cost ($/1000 m^3)	100	100	110	110
Overtime cost ($/1000 m^3)	110	116	120	124
Normal available time (hours)	2400	2400	2400	2400
Overtime (hours)	990	990	990	990

There is no stone in stock at the start of the first month, and there is to be none remaining at the end of the fourth month. Each 1000 m^3 of stone produced requires 1.5 hours, and the cost to store each 1000 m^3 from one month to the next is $5.

Write the systems equations which, when solved, will enable ROX to construct a production schedule that will meet the sales demands, stay within the available times, and minimize the cost.

6. A building materials processing plant requires 300,000 m^3 of filtered water per week. The water must be chlorinated and softened before it can be used. Two different brands of additive contain certain amounts of a chlorinating chemical, *Clor*, and a softening agent, *Soff*.

U. S. CHEMCO supplies a product which contains 8 kg of Clor and 3 kg of Soff per package, while GEN KHEM, Ltd., furnishes packages with 4 kg of

Clor and 9 kg of Soff. CHEMCO's package costs \$8, while that of KHEM is \$10.

The recommended levels of softness and chlorination are maintained by adding 150 kg of Clor and 100 kg of Soff per week.

Write the equations which, when solved, will allow the plant manager to specify how many packages of additive from U. S. CHEMCO and how many from GEN KHEM are to be used in order to minimize the cost.

7. Marshville and Claytown are two municipalities which discharge treated wastes into Napse River. Marshville produces 24 units of pollutants daily and Claytown generates 18 units. Some of these pollutants are removed by the waste treatment facilities at each town. Each unit of pollutant costs \$1000 to remove at Marshville and \$800 at Claytown.

   Just before the sewer outfall at Marshville the river has a flowrate of 120 million cubic meters per day, and there are negligible pollutants there. Marshville discharges 5 million m^3/day into the river, and Claytown's flow is 2 million m^3/day.

   Twenty percent of the pollutants discharged into the river at Marshville will be removed by natural processes (i.e., oxidation by air, breakdown by sunlight, etc.) before they reach Claytown.

   Water quality standards require that the number of units of pollutants per million m^3 of flow should not exceed 2 for the river.

   Write the system equations the solution of which will enable the Napse River Water Quality Control Authority to require of Marshville and Claytown as the number of units of pollutants each must remove at their treatment plants for the lowest total cost.

8. Swiftcreek region has three towns named Nutbush, Six Oaks, and Pretty Point. Solid waste from these towns is transported to either of two landfills at Owls Roost and Locust Grove. The daily amounts of waste produced at the towns are: Nutbush, 20,000 kg; Six Oaks, 15,000 kg; Pretty Point, 24,000 kg. The landfills have the following daily capacities: Owls Roost, 37,000 kg; Locust Grove, 42,000 kg. The distances between towns and landfills is given by the table below. Assume that the cost for hauling waste is uniform for all routes. Create a model for minimizing the total regional cost of solid waste management for Swiftcreek.

	NUTBUSH	SIX OAKS	PRETTY POINT
OWLS ROOST	2 km	4 km	7 km
LOCUST GROVE	9 km	3 km	5 km

CHAPTER 5

SYSTEMS EVALUATION

5.1 INTRODUCTION

A minimum necessity associated with the process of optimization is a fair and equitable standard by means of which systems may be compared with each other in order to say which is "best." Candidate systems thus compete on the optimum (level) playing fields and vie for this preeminent title and the garlands of victory that accompany it.

As was previously discussed in Chapter 4, there may be any number of ingredients in the formula for "best," and some may be quite subjective. Individual tastes may enter the judges' lists.

Since almost every real engineering project requires that resources be allocated to its accomplishment, one obvious measure of the goodness of a proposal is how much it will cost. The measure of cost, naturally, is money, so the expenditure of money becomes a criterion. For those systems in which most of the elements are easily rendered in terms of money, this will be the criterion of choice.

The other major test of a system against other candidates for the best pennant is that called the *Benefit-to-Cost Ratio*. Actually, present practice generally prefers the *Incremental Benefit-to-Cost Ratio*, because only the good designs make it to the formal evaluation stage, so the more reliable figure of merit is how they compare in a differential sense.

In this Chapter there will be developed specific criteria which can be applied to perform systems evaluations. These are rooted in basic economic theory, and utilize concepts from the field of engineering economics.

5.2 PLANNING HORIZON

Some, in fact most, civil engineering works last not only decades but for periods such as "twenty years," "thirty years," or "fifty years." A few favored ones—the survivors—last a century of more, and some endure for millennia, in various states of functioning. The ruins of the classical Greek and Roman civilizations come to mind, although it must certainly be admitted that the original design purposes have been replaced by their current utility as archaeological monuments and tourist attractions.

Centennial objects such as The Statue of Liberty and the Brooklyn Bridge have been maintained or refurbished so that they still function in the same mode as that for which they were designed. Of course, the refurbishing may have cost many times that of the item in its mint condition, and rightfully ought to merit recognition as a new design or project in its own right.

But the customary ***planning horizon*** is for a delimited time span—one which can be seen, in the same way that objects can be observed "all the way to" the edge of the world at the horizon. A planned service life of "twenty years" is a very common time frame for many things. Actually, it is an expression which might better be translated as "for a long time." The roof on one's house is expected to need replacement in about twenty years, or the furnace, or, indeed, any household appliance. The moral is, as one wag was heard to say, "When the refrigerator goes, sell the house, quick! You know it's all coming apart after that."

Major civil engineering works have realistic lifetimes—in service at the intended function—in the range upward from 30 years to 50 years, 100 years, or even longer without major refurbishments.

A moment's reflection will serve to convince one that a system created decades or centuries ago cannot reasonably be compared as to cost with an alternative scheme to be obtained at present costs. More appropriately, it must be realized at the time of comparison of two or more systems, that the medium of comparison—money—has different values at different times. So works being evaluated over large time spans must take this phenomenon into account.

Moreover, two systems under evaluation will most likely have different life spans, so, in fairness, it will likely be necessary to find a common annual measure of their relative merits, and this will require taking into account the way in which money varies in value from year-to-year.

5.3 THE TIME VALUE OF MONEY

To investigate the nature of the time value of money calls into play the concept of *equivalence*. It is to ask (and answer) the question, "What amount of money *then* is equivalent to a certain amount of money *now*?"

Everyone is familiar with the idea of interest in which a sum of money is loaned to another to use on the premise that the sum itself plus a fee for its use will be returned on a date certain in the future. The fee is called the *interest* or the *rate of interest*, and is normally quoted as a percentage of the amount loaned.

The concept described above is *simple interest*. When the fee is to be computed at some regular interval called the *period of compounding*, the fee is called *compound interest*. In other words, interest in a succeeding period is paid on the principal sum plus the accumulated interest at the beginning of the period.

A similar notion is that of the *discount rate*, which is a percentage return on a sum of money that functions in the same manner as compound interest. It is the time value of money used in equivalence computations. By definition, it

> *"reflects the opportunity cost, equal to the return which would have been realized had the money been invested in other available options."*

There are several important methods of discounting money, and there are corresponding *discount factors* associated with each of these methods. The methods have particular applicability to certain forms of money exchange. In each case, however, the discount factor serves as a *shifter*, to move money from one location in time to another time. The move can be forward or backward, as the case requires.

The discount factor becomes, therefore, the device to use to ensure that monies from different epochs are compared in an equitable manner.

5.4 DISCOUNT FACTORS

There are six discount factors in common use. These are:

1. Single Payment Compound Amount Factor.

   This operator is a device to compute the future sum at the end of n years which would accrue at a discount rate i. Let the present amount be named P. At the end of one year, it would be worth $P(1+i)$, at two years its value would be $[P(1+i)](1+i) = P(1+i)^2$, and at n years the sum would be $P(1+i)^n$. Hence, the future amount, F, is given by

   $$F = P\,(CAF, i, n) \tag{5.1}$$

   where the notation (CAF, i, n) stands for the factor $(1+i)^n$.

2. Single Payment Present Worth Factor.

   The inverse operator to that above is the factor by which a future amount must be multiplied to yield a present sum. Obviously, it is the inverse of $(1+i)^n$, so that if F is the amount of the future sum, its present worth P is given by

   $$P = F\,(PWF, i, n) \tag{5.2}$$

   in which $(PWF, i, n) = 1/(1+i)^n$ is the present worth factor. In other words, it is a shifter which brings back to the present a sum of money named in the future year.

3. Uniform Series Present Worth Factor.

   Using the Single Payment Present Worth Factor described above permits the computation of the present worth of a uniform series of future sums of money. Let the amount of the future money be named A. Then the present worth of that amount A at the end of one year is $A/(1+i)$. The present worth of a second sum A at the end of two years is $A/(1+i)^2$, and for sum A at the end of n years it is $A/(1+i)^n$. The total present worth of the series of sums A is

   $$P = \frac{A}{(1+i)} + \frac{A}{(1+i)^2} + \cdots + \frac{A}{(1+i)^n} = A\,\frac{(1+i)^n - 1}{i(1+i)^n} \tag{5.3}$$

   Then $P = A\,(PWUS, i, n)$ in which $(PWUS, i, n)$ is the shifter that moves a series of sums A located at the end of each year for n years back to the present and accumulates them into one present amount P.

4. Capital Recovery Factor.

   This factor is the inverse of the Uniform Series Present Worth Factor. It allows the determination of the annual sums paid at the end of each year for n years, with a discount rate i prevailing over the time span, so that the total sum will be equivalent to a present amount P. The operator is used especially to compute the annual payment needed to retire a debt incurred at the present time.

$$A = P\,(CRF, i, n) \tag{5.4}$$

   where $(CRF, i, n) = [i(1+i)^n]/[(1+i)^n - 1]$ is the spreader that distributes the present sum P over n years ahead, each year with its amount A. The discount rate i is taken to be constant over the period.

5. Uniform Series Compound Amount Factor.

   This operator annualizes a total money amount F which occurs at the end of the period. In other words, it tells how much money should be invested each year (at the same discount rate) so as to accumulate F dollars at a date certain in the future. The computation is as follows:

$$F = A + A(1+i) + A(1+i)^2 + \cdots + A(1+i)^n = A\,\frac{(1+i)^n - 1}{i} \tag{5.5}$$

   Then $F = A\,(USCAF, i, n)$, where $(USCAF, i, n) = [(1+i)^n - 1]/i$. The $(USCAF)$ shifts a uniform series of annual sums forward to a time in the future, where it accumulates the then total F of the shifted sums.

6. Sinking Fund Factor.

   The functional inverse of the Uniform Series Compound Amount Factor is called the Sinking Fund Factor. It is used by financial planners to determine how much money should be set aside (and invested, of course) each year so that a future sum will be available. The name comes from the language "sinking money into a fund for the future" to replace capital items or other major expenditures of that ilk. Thus,

$$A = F\,(SFF, i, n) \tag{5.6}$$

   where $(SFF, i, n) = i/[(1+i)^n - 1]$ is the annual amount sunk.

5.5 CASH FLOW DIAGRAMS

Associated with the shifter and spreader operators described above are graphical devices called ***cash flow diagrams*** which are intended to give a visual display of the particular case which is being carried out by the operation. Since the operators are paired so that their pair partner is the inverse operator, a single diagram will suffice to demonstrate the effect for a given pair. In other words, three basic cash flow diagrams are defined.

Figure 5.1 shows the diagram for the Single Payment Compound Amount Factor and for the Single Payment Present Worth Factor. A time line coincident

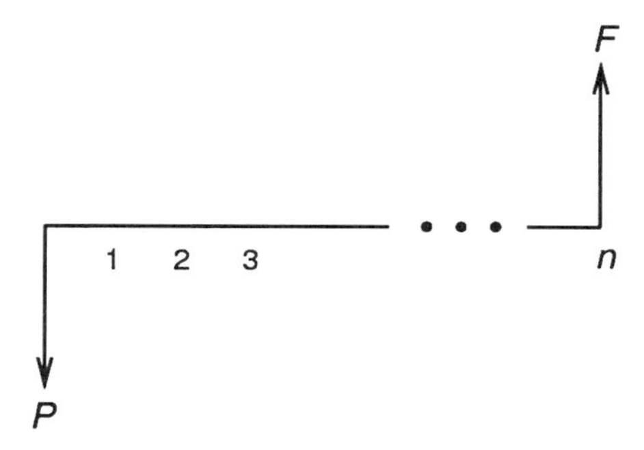

Figure 5.1: Single Payment Compound Amount and Present Worth Factors

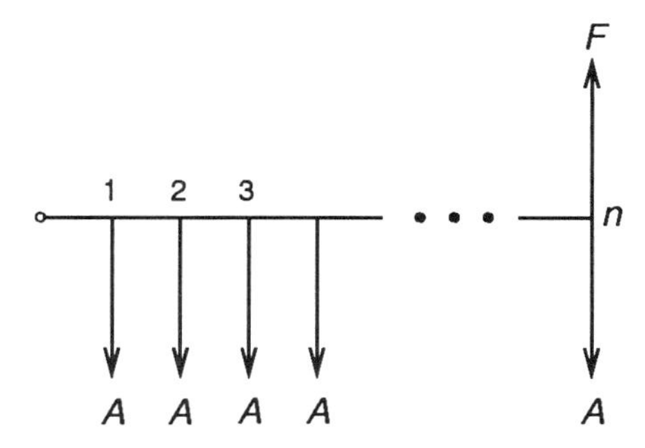

Figure 5.2: Uniform Series Compound Amount and Sinking Fund Factor Factor

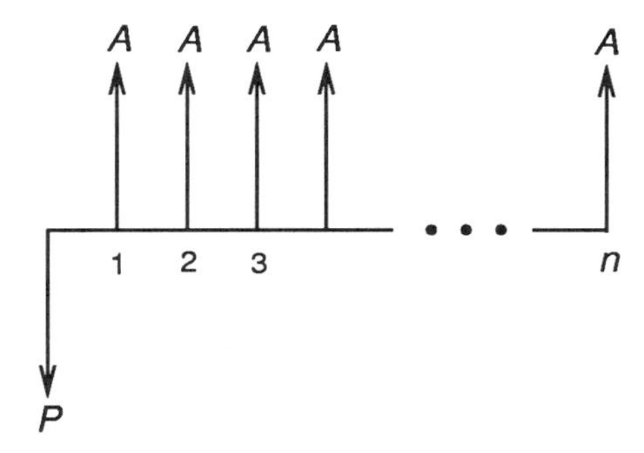

Figure 5.3: Uniform Series Present Worth Factor and Capital Recovery Factor

with the horizontal is graduated in years from the present time to the end of the period in question as represented by the year n. At time zero, the left origin of the time line, is the present value P. Notice that by custom it is plotted downward, to represent an outgo or negative quantity. Presumably the vertical scale is proportional to the amount of P, although frequently either no scale is given or it is ignored.

At the right-hand end of the time line is F, the future amount. It is by custom plotted upward, so as to suggest a positive quantity or income. Again, the scale is presumably proportional to the amount of F, but no axis with numbers is to be seen.

The diagram represents the transformation from P to F or its inverse F to P, depending upon which operator is selected. It is correct to think of these operators as ***shifters*** which serve to translate the quantities F and P from one end of the time line to the other.

Figure 5.2 shows the cash flow diagram for the pair Uniform Series Compound Amount Factor and Sinking Fund Factor. In this case, regular annual outgo amounts A at the end of each year beginning with year 1 are realized in a return of the amount F at the end of n years. The operators can be considered as spreaders which shift the sum F backwards in time uniformly over the years from 1 to n. It is said that F is ***annualized*** by this device.

The cash flow diagram corresponding to the Uniform Series Present Worth Factor and the Capital Recovery Factor pair is seen in Figure 5.3. In this instance, an outgo amount P is recovered by annual repayments at the end of years 1 through n. The operators serve to annualize P, or to spread P uniformly across the time span.

5.6 INFLATION

Throughout history there have been periods of time during which a high rate of inflation has been experienced, although there have also been periods during which the inflationary rate has been virtually zero. Inflation greatly affects the costs of goods and services. Those who lived through the early years of the 1930s will remember the dramatic examples furnished by the runaway German inflation of those times. Some newspaper and magazine photographs showed Germans rolling wheelbarrows full of inflated ***Deutsche Marks*** to market to buy one loaf of bread, and stamp collectors of that period recall seeing German postage stamps surcharged with black overprint to establish their new values which were often several million times greater than the originals.

A more recent episode is the double-digit inflation of the early 1980s in the United States. The rates reached nearly the 20% level, and are within the near-range memory of most persons today.

In the United States during the Great Depression years up until the outbreak of World War II the rate of inflation was essentially zero, and interest rates were very low—as little as fractions of 1% for a number of years in a row.

The nature of inflation is revealed by the fact that if the future rate of inflation is i, then an item which costs one dollar today will cost $(1 + i)$ dollars at one year from now, or $(1 + i)(1 + i) = (1 + i)^2$ at two years hence, and $(1 + i)^n$ at the end of n years. In other words, it functions exactly as does the discount rate. Therefore, inflation can be accounted for by the same mathematical devices elucidated previously.

5.7 EXAMPLE 1

The municipality of Grenville has a one-time revenue windfall of $1,082,673. Two proposals are being considered for the utilization of this money. The first alternative is to purchase 30-year Federal Reserve Bonds which will earn 7.85% compounded annually. The second proposition is to build a water treatment plant which will produce more revenue directly. It is estimated that the net income from the plant will be similar to the following schedule. First year, $12,000; second year, $18,000; third year, $26,000; fourth year, $38,000; fifth year, $54,000; sixth year, $72,000; seventh year, $98,000; eighth year $116,000; ninth year $130,000; tenth through thirtieth years, $146,000.

The Planning Department is asked to make a recommendation on which alternative to pursue. As a basis for comparison, it is decided that if the present value of all the expected benefits from the treatment plant exceeds $1,082,673, then the plant should be the option of choice. A discount rate of 7.85% will be assumed. An *Internal Rate of Return* indexinternal rate of return for the project will be computed for the sake of comparison. Presumably, the IRR will be something more than the 7.85%, otherwise the bonds will be more attractive.

A tabular computation for the first 9 years is illustrated below.

GRENVILLE REVENUE

Year	Income	($PWSP$,7.85,n)	Present Worth $i = 7.85\%$	($PWSP$,8,n)	Present Worth $i = 8.00\ \%$
1	$12,000	0.927221	$11,126.56	0.92593	$11,111.11
2	18,000	0.85973	15,475.06	0.85734	15,432.10
3	26,000	0.79715	20,725.88	0.79383	20,639.64
4	38,000	0.73913	28,086.85	0.73503	27,931.13
5	54,000	0.68533	37,007.78	0.68058	36,751.49
6	72,000	0.63545	45,752.16	0.63017	45,372.21
7	98,000	0.58919	57,741.10	0.58349	57,182.06
8	116,000	0.54631	63,371.91	0.54027	62,671.19
9	130,000	0.50655	65,850.94	0.50025	65,032.37
			345,138.24		342,123.30

Since the revenues for the years 10 through 30 are assumed to be equal at \$146,000, the present value is computed using the $PWUS$ factor. The two numbers $(PWUS, 7.85, 30)$ and $(PWUS, 7.85, 9)$ are needed. Thus

$$(PWUS, 7.85, 30) = \frac{(1+0.0785)^{30}-1}{0.0785(1+0.0785)^{30}} = \frac{9.65171-1}{0.0785(9.65171)} = \frac{8.65171}{1.31985} = 11.4190$$

$$(PWUS, 7.85, 9) = \frac{(1+0.0785)^{9}-1}{0.0785(1+0.0785)^{9}} = \frac{1.97416-1}{0.0785(1.97416)} = \frac{0.97416}{0.15497} = 6.28604$$

Then

$$\begin{aligned}
\text{Total Present Value} &= \$342,123.30 + \$146,000[(PWUS, 7.85, 30) - (PWUS, 7.85, 9)] \\
&= \$345,138.24 + \$146,000(11.4190 - 6.2860) \\
&= \$345,138.24 + \$146,000(5.1330) \\
&= \$345,138.24 + \$749,411.85 \\
&= \$1,094,550.09.
\end{aligned}$$

Since this is larger than \$1,082,673, the water treatment plant option should be chosen.

To compute the IRR, two other numbers are needed.

$$PWUS, 8, 30 = \frac{(1+0.08)^{30}-1}{0.08(1+0.08)^{30}} = \frac{10.06266-1}{0.08(10.06266)} = \frac{9.06266}{0.80501} = 11.25778$$

$$PWUS, 8, 9 = \frac{(1+0.08)^{9}-1}{0.08(1+0.08)^{9}} = \frac{1.99900-1}{0.08(1.99900)} = \frac{0.99900}{0.15992} = 6.24689$$

Then

$$\begin{aligned}
\text{Total Present Value} &= \$342,123.30 + \$146,000[(PWUS, 8, 30) - (PWUS, 8, 9)] \\
&= \$342,123.30 + \$146,000(11.25778 - 6.24689) \\
&= \$342,123.30 + \$146,000(5.01089) \\
&= \$342,123.30 + \$731,590.73 \\
&= \$1,073,714.03.
\end{aligned}$$

Therefore

$$\begin{aligned}
IRR &= 0.0785 + (0.0015)\frac{1,094,550.09 - 1,082,673.00}{1,094,550.09 - 1,073,714.03} \\
&= 0.0785 + (0.0015)\frac{11,857.09}{20,836.06} \\
&= 0.0785 + (0.0015)(0.56907) \\
&= 0.0785 + 0.000854 \\
&= 0.07935 \text{ or } 7.935\%
\end{aligned}$$

This last computation has interpolated between 7.85% and 8% and between $1,094,550.09 and $1,073,714.03. 7.93% is thus seen to correspond to the amount $1,082,673.00.

The Internal Rate of Return is a measure of the discount rate paid by the municipality to itself on the money loaned by it to itself as an investment in the treatment plant.

5.8 EXAMPLE 2

A construction company which works in low-lying areas is contemplating the purchase of a dragline equipment to use in its ditching operations. The bucket will cost $82,000 at present. At the time of purchase the estimated useful life is 12 years, and the expected annual maintenance is $8,000; the expected annual revenue is $22,000, and the salvage estimate is $9,000. Assuming a uniform annual inflation rate of 6% for the period, develop the cash flow diagram.

The inflation rate of 6% is treated as if it were a discount rate of the same amount, so that the future maintenance costs and the future incomes will be larger than the stated amounts based on present dollars. The problem, then, is to move the present estimates forward to the several years within the life of the dragline and display the new numbers on the cash flow diagram with the present values as year 0. The operator to invoke is the *PWSP* factor in the form $F = P\,(PWSP, 6, i)$, where i takes on the numbers 1 to 12 in turn.

This may be an appropriate place to mention that maintenance costs are traditionally shown at the end of the year, rather than at other times. This practice evidently originated with the customs prevalent during an earlier era—a time when no such things as service contracts existed, so that if something broke, a serviceman was called, he came, fixed it, and the owner-operator was later billed for the repair services at the end of the relevant accounting period. In other words, the actual cost was associated with the later time (at the end of the period or year).

Nowadays, a service contract would be negotiated at the time of the equipment purchase, would be paid in advance of the service period with the service on-call as needed as well as at regular maintenance intervals. In fact, the invoice for the service contract would be rendered forthwith upon purchase of the equipment and would be paid in (essentially) present dollars. The owner-operator is no longer allowed the fine luxury of being able to pay off this item in inflated future dollars as in a more genteel bygone era, when the inflation rate was probably negligible, anyway!

This situation illustrates the proof test of money for time valuation. If one takes out one's checkbook and writes a check for some goods or services, signs and dates it, delivers it to the party being paid for the goods or services, and receives the cancelled check in the monthly statement, then the amount of money stated upon the face of the check is, by definition, present value. Actually, the check will have cleared the bank in a matter of a very few days, by law, and some transactions are actually carried out at electronic speeds, so the transfer meets a literal test of simultaneity. Any other functional equivalent to the above process likewise furnishes prima facie evidence of present value.

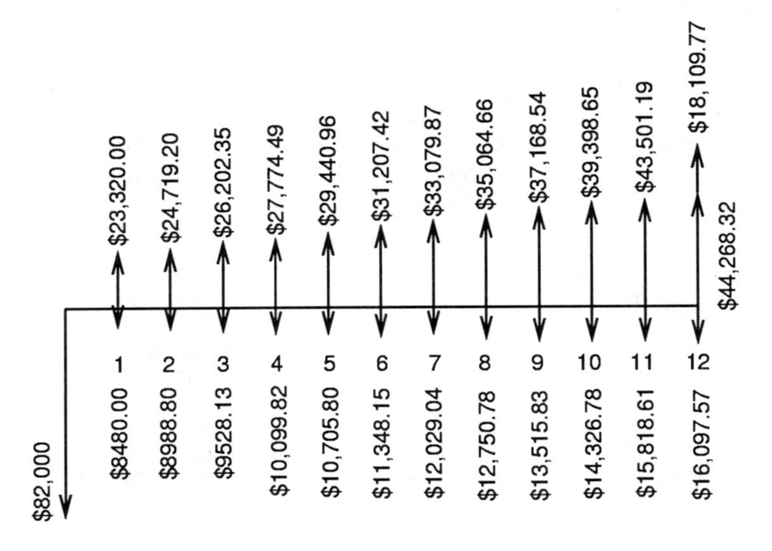

Figure 5.4: Dragline Cash Flow Diagram

The computation of the data needed to produce the cash flow diagram for the dragline purchase is illustrated in the table below. The diagram itself is displayed in Figure 5.4.

D R A G L I N E C A S H F L O W

Year	$(PWSP,6,i)$	Maintenance ($/year)	Revenue ($/year)	Salvage ($)
0	1.0000	8000.00	22,000.00	
1	1.0600	8480.00	23,320.00	
2	1.1236	8988.80	24,719.20	
3	1.1910	9528.13	26,202.35	
4	1.2625	10,099.82	27,774.49	
5	1.3382	10,705.80	29,440.96	
6	1.4185	11,348.15	31,207.42	
7	1.5036	12,029.04	33,079.87	
8	1.5938	12,750.78	35,064.66	
9	1.6895	13,515.83	37,168.54	
10	1.7908	14,326.78	39,398.65	
11	1.9773	15,818.61	43,501.19	
12	2.0122	16,097.57	44,268.32	18,109.77

5.9 THE PRINCIPLE OF REPLACEMENT

One of the most fundamental aspects of the concept of equivalence is that involved in comparing two or more systems which have different service lives. Equivalence in such an instance requires that there shall be some parity over time as the systems are weighed against each other.

When the lifetimes of two systems are not equal, a device that is often employed is the principle of replacement by means of which the shortest-lived system is replaced by another one just like itself, or several ones, until the lives of the shorter and the longer systems are thereby made equal. In case the life of the longer-lived system is not an exact multiple of that of the shorter-lived one, a replacement in kind of each system may be assumed until there is a commensurate time for both. Both groups of systems will thereby end at a common date, so that any or all operations of shifting, spreading and comparison can be done equitably, even if that common date is fictitious.

5.10 EXAMPLE 3

Your engineering firm is considering replacing its antique surveying equipment with modern laser-based instruments. Two particular brands are under investigation. Koppel & Escher's system will cost $100,000, can be expected to serve for 25 years, and will allow for annual earnings of $40,000, but will require maintenance of $15,000 per year. The present estimate of its salvage value is $20,000.

Goettner's apparatus will cost $60,000, will probably be obsolete in ten years, may bring in earnings of $20,000 per year, and is estimated to require $5,000 per year in maintenance. Its salvage value at ten years is negligible.

Assume that either system will be replaced by its equivalent at the end of its service life. Take an appropriate discount rate to be 8%. Draw the cash-flow diagrams for both options. Based upon present worth considerations, which is the better proposal?

The cash-flow diagrams are illustrated in Figures 5.5 and 5.6, respectively. Note that the commensurate period of service life is 50 years, which is two K & E lives and five of Goettner's.

PWF shifters are employed to move the initial cost amounts for both systems back to the present, and $PWUS$ factors bring the distributed annual earnings and annual maintenance amounts back, also to the present. PWF shifters also relocate the $20,000 salvage amounts of the K & E systems to the present; there are no such salvage items for the other system.

The present worth computation for each system is rendered as follows.
For the Koppel & Escher:

$$\begin{aligned}
PW &= -\$100,000 + (\$40,000 - \$15,000)(PWUS, 8, 50) \\
&\quad +(-\$100,000 + \$20,000)(PWF, 8, 25) + \$20,000(PWF, 8, 50) \\
&= -\$100,000 + \$25,000(12.233) - \$80,000(0.146) + \$20,000(0.0213) \\
&= -\$100,000 + \$305,837 - \$11,681 + \$426 = \$194,582
\end{aligned}$$

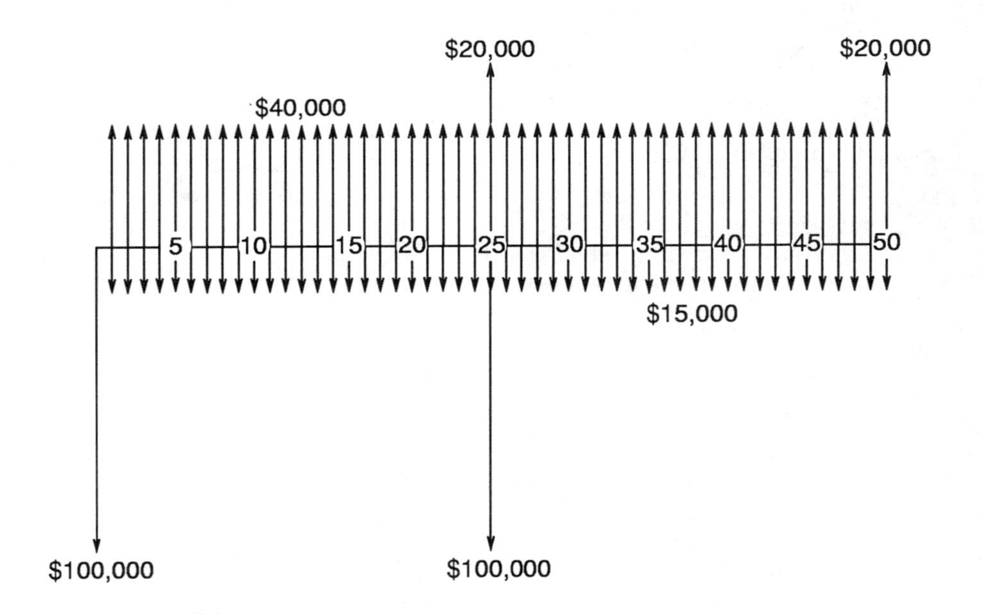

Figure 5.5: Koppel & Escher Cash-Flow Diagram

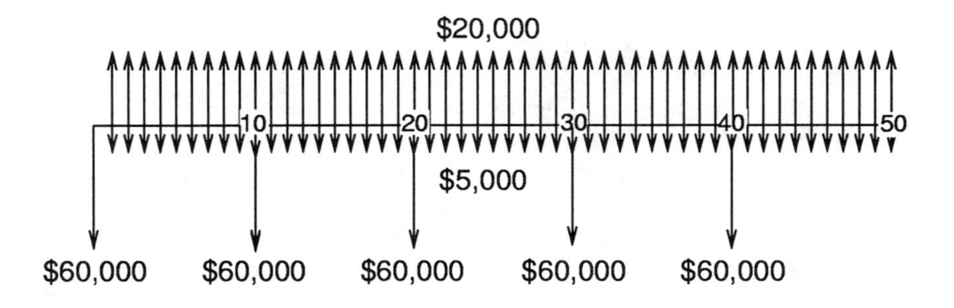

Figure 5.6: Goettner Cash-Flow Diagram

For the Goettner:

$$\begin{aligned} PW &= -\$60,000 - \$60,000(PWF, 8, 10) - \$60,000(PWF, 8, 20) \\ &\quad -\$6,000(PWF, 8, 40) + (\$20,000 - \$5,000)(PWUS, 8, 50) \\ &= -\$60,000 - \$60,000(0.463) - \$60,000(0.215) - \$60,000(0.0994) \\ &\quad -\$60,000(0.046) + \$15,000(12.233) \\ &= -\$60,000 - \$27,791 - \$12,872 - \$5,963 - \$2,762 + \$183,502 \\ &= \$74,113 \end{aligned}$$

It is readily seen that the Koppel & Escher system is a better choice than the Goettner, although it is much more expensive initially and in terms of annual maintenance.

In the calculations above, the factors were computed as follows:

$$(PWF, 8, 25) = \frac{1}{(1.08)^{25}} = 0.146 \tag{5.7}$$

$$(PWF, 8, 50) = \frac{1}{(1.08)^{50}} = 0.0213 \tag{5.8}$$

$$(PWF, 8, 10) = \frac{1}{(1.08)^{10}} = 0.463 \tag{5.9}$$

$$(PWF, 8, 20) = \frac{1}{(1.08)^{20}} = 0.215 \tag{5.10}$$

$$(PWF, 8, 30) = \frac{1}{(1.08)^{30}} = 0.0994 \tag{5.11}$$

$$(PWF, 8, 40) = \frac{1}{(1.08)^{40}} = 0.0460 \tag{5.12}$$

$$(PWUS, 8, 50) = \frac{(1.08)^{50} - 1}{0.08(1.08)^{50}} = 12.233 \tag{5.13}$$

5.11 BENEFIT/COST RATIOS

The increasing amount and total cost of public works—many of which are carried out by Federal, State, and Local agencies and combinations of these—has engendered corresponding measures of accountability on the determination of worthiness of the several projects in question. This is only natural, since public works are intended to be demographically and politically and even geographically equitable. Since there are always more needs than can be met with the available dollars in governmental budgets, it comes down to establishing a rational basis for initiating one project ahead of another.

In fact, the first order of business is to demonstrate that the project is worthy on its own terms; that is, that more benefit will accrue than will be the equivalent cost of the work in the first place. Otherwise, the project is not profitable in the simple sense. Unfortunately, there have been such projects in the history of governmentally-sponsored undertakings. Some have been referred to as "boondoggles," and other similar but less complimentary appellations.

Congressional requirements, and similar policies of state and local legislative bodies, have mandated that proposed systems be justified by documented demonstrations that the Benefit-to-Cost Ratio exceeds unity. Since a great many do project a benefit greater than their indicated cost, the next step is logically that the ones with larger ratios be placed in higher priorities that those of lesser merit. It is not enough that a proposal should be profitable, it must also be more profitable than others which are also viable candidates for funding.

Accordingly, the concept of Incremental Benefit-to-Cost Ratio has evolved. In times of stringent budgets, such criteria are growing in favor with governmental bodies who appropriate sums of money, with agencies which propose and administer the works, and with the public which uses (benefits from) and pays for the programs. The requirements of polity are more evenly met when such measures are the basis for action.

5.12 EXAMPLE 4

A transportation planning engineer has been notified that a budget of about $500,000 will be made available for an accident reduction program to improve the safety of the highways. There is a list of accident-prone locations which are candidates for amelioration, each of which would require its own level of investment and would return its particular benefit in the form of reduced user costs charged against accidents. There would also be annual maintenance costs incurred at each site.

The engineer is asked to develop a strategy for allocating the funds to maximize the total benefit. She can assume that each project will have a service lifetime of 25 years, and a suitable discount rate of 8.125% is indicated.

The first order of business is to rank the projects and the second is to evaluate them in an increasing order of cost. The working equation she will use is

$$(B/C)_{\text{alternative i to existing condition}} =$$

$$\frac{(\text{user costs})_{\text{existing}} - (\text{user costs})_{\text{i-th}}}{(\text{annualized investment cost})_{\text{i-th}} + (\text{maintenance})_{\text{i-th}} - (\text{maintenance})_{\text{existing}}}$$

$$= \frac{\text{incremental annual user costs}}{\text{annualized investment} + \text{incremental annualized maintenance}}$$

It is important to understand that each ingredient in this formula is of a commensurate nature. That is, all elements are computed on an annualized basis, so that if the project which is being compared to the existing condition has a different life-span than the existing condition, the inequity of testing two different items from different time frames is removed.

The investment in each project is a sum of money which is expended at the project's beginning, although this may sometimes extend over a significant length of time. It is an annual outlay when compared to the much longer time spans over which the project life stretches. Therefore this capital expenditure is spread over each year for purposes of comparison.

ACCIDENT REDUCTION PROGRAM

Project Number	Investment ($)	Existing Annual Maintenance	Existing Annual User Costs	New Annual Maintenance Costs	New Annual User Costs
1	60,000	600	40,000	500	10,000
2	75,000	800	70,000	700	35,000
3	80,000	900	60,000	800	20,000
4	95,000	950	70,000	1000	25,000
5	110,000	950	100,000	800	50,000
6	120,000	1000	118,000	1200	60,000

The spreader is the Capital Recovery Factor (CRF) which is computed as follows:

$$\begin{aligned} CRF &= \frac{i(1+i)^n}{(1+i)^n - 1} \\ &= \frac{(0.08125)(1+0.08125)^{25}}{(1+0.08125)^{25} - 1} \\ &= \frac{(0.08125)(1.08125)^{25}}{(1.08125)^{25} - 1} \\ &= \frac{(0.08125)(7.049413)}{7.049413 - 1} \\ &= 0.094681053 \end{aligned}$$

The factor is to be applied to each project's investment cost to produce its annualized investment cost. The table below shows the computation of the Benefit-to-Cost Ratios for six of the projects.

Each of these proposed projects has a Benefit/Cost Ratio that is greater than 1.0, so each of them would be an appropriate activity to be funded with the assurance that it would contribute to highway safety. However, some are obviously more "profitable" than others, so the next step is to determine the Incremental Benefit-to-Cost Ratio. The working formula for this computation is the following:

$$(B/C)_{i-j} = \frac{\text{incremental benefits}}{\text{incremental costs}}$$

$$= \frac{(\text{difference in user costs})_{i-j}}{(\text{difference in annualized investment})_{i-j} + (\text{difference in maintenance})_{i-j}}$$

BENEFIT/COST RATIOS

Project Pair	Annualized Investment	Incremental Annualized Maintenance	Incremental Annualized User Costs	Benefit/Cost Ratio
1-exist	5,680.86	− 100	30,000	5.376
2-exist	7,101.08	− 100	35,000	4.999
3-exist	7,574.48	− 100	40,000	5.352
4-exist	8,994.70	+ 50	45,000	4.975
5-exist	10,414.92	− 150	50,000	4.871
6-exist	11,361.73	+ 200	58,000	5.017

INCREMENTAL BENEFIT/COST RATIOS

$$(B/C)_{2-1} = \frac{35,000 - 30,000}{(7,101.08 - 5,680.86) + (-100 + 100)} = 3.521$$

$$(B/C)_{3-2} = \frac{40,000 - 35,000}{(7,574.48 - 7,101.08) + (-100 + 100)} = 10.562$$

$$(B/C)_{4-3} = \frac{45,000 - 40,000}{(8,994.70 - 7,574.48) + (50 + 100)} = 3.184$$

$$(B/C)_{5-4} = \frac{50,000 - 45,000}{(10,414.92 - 8,994.70) + (-150 - 50)} = 4.098$$

$$(B/C)_{6-5} = \frac{58,000 - 50,000}{(11,361.73 - 10,414.92) + (-200 + 150)} = 8.921$$

PROJECT PRIORITIES

#1	Project 6	$120,000
#2	Project 5	$110,000
#3	Project 4	$95,000
#4	Project 3	$80,000
#5	Project 2	$75,000
#6	Project 1	$60,000
		$540,000

Then the projects can be ranked in order of preference, together with their costs.

The transportation planner can now adjust the number of projects and the total cost to whatever budget is allocated. In fact, she will no doubt beg, plead, cajole—even fight—to obtain the whole program. But whatever decision is made, she will be in a position to tailor the operation to suit the available resources, and can do so in good conscience that the best possible good has been done under the circumstances. That is usually a very good feeling.

It is important to emphasize that the Benefit-to-Cost Ratio method of evaluating systems—particularly the Incremental Benefit-to-Cost Ratio procedure—is susceptible of manipulation. That is, a differing order of the outcomes is possible depending upon the discount rate chosen. Therefore, it is mandatory that some care be taken not to choose a rate that is either too high or too low, but is a realistic number supported by the best analysis available at the time of its application.

5.13 COMPUTING THE DISCOUNT RATE

The examples illustrated above have assumed that a discount rate is known or easily obtainable. Sometimes, however, the situation will be reversed. If the present amount P and the annual payment A are the known quantities, and it is desired to know the discount rate, then an inverse procedure is indicated. The spreader operator is the $PWUS$ factor, so that

$$P = A\,(PWUS) \tag{5.3}$$

or

$$PWUS = P/A = \frac{(1+i)^n - 1}{i\,(1+i)^n} \tag{5.14}$$

Let this be solved as follows:

$$\frac{P}{A}\,i\,(1+i)^n = (1+i)^n - 1 \tag{5.15}$$

$$f(i) = \frac{P}{A}\,i\,(1+i)^n - (1+i)^n + 1 \tag{5.16}$$

Then the desired discount rate i is the root of the equation $f(i) = 0$.

The most expeditious route to the answer is probably through the application of a Newton-Raphson procedure. Take the derivative

$$f'(i) = \frac{P}{A}\,n\,i\,(1+i)^{n-1} + \frac{P}{A}\,(1+i)^n - n\,(1+i)^{n-1} \tag{5.17}$$

and choose an initial or starting value of i called i_o. The next (better) approximation is

$$i_1 = i_o - \frac{f(i_o)}{f'(i_o)} \tag{5.18}$$

and the process can be iterated until

$$i_m = i_{m-1} - \frac{f(i_{m-1})}{f'(i_{m-1})} \tag{5.19}$$

converges to whatever pre-set accuracy is desired.

As an example, suppose that an investment of present value \$100000 yields an annual return of \$10000 for a period of 12 years. It is desired to know the rate of return. Say that an initial estimate is 6.0%. The computation proceeds as follows.

$$f(i_o) = \frac{100000}{10000}(0.06)(1+0.06)^{12} - (1+0.06)^{12} + 1 = 0.195121411$$

$$f'(i_o) = 10(12)(0.06)(1.06)^{11} + 10(1.06)^{12} - 12(1.06)^{11} = 11.010131638$$

$$i_1 = i_o - \frac{f(i_o)}{f'(i_o)} = 0.06 - \frac{0.195121411}{11.010131638} = 0.042278012$$

$$\begin{array}{lll} f(i_1) = 0.051266601 & f'(i_1) = 5.513258584 & i_2 = 0.032979228 \\ f(i_2) = 0.010744470 & f'(i_2) = 3.268365026 & i_3 = 0.029691814 \\ f(i_3) = 0.001166695 & f'(i_3) = 2.566125072 & i_4 = 0.029237161 \\ f(i_4) = 0.000021306 & f'(i_4) = 2.472538082 & i_5 = 0.029228543 \\ f(i_5) = 0.000000007 & f'(i_5) = 2.470772402 & i_6 = 0.029228540 \\ f(i_6) = 0.000000000 & f'(i_6) = 2.470771771 & i_7 = 0.029228540 \end{array}$$

The iteration is stopped and the value of i is 0.029228540 or 2.9228540%.

Of course, it is a simple matter to program a machine computation for the above, instead of the hand-calculation which has been shown. A hand-held programmable calculator can easily do it, and a suggested algorithm is shown in Appendix A.

In fact, the entire business of time-value-of-money calculations is best done on a machine. The formulas are certainly simple enough to program in any and all languages, and a repertory of them, probably menu driven, should be done by all serious students of this subject. A graphics compendium of the cash flow diagrams should accompany it. The customary lengthy set of tables for various discount rates and several time intervals is totally unnecessary and highly wasteful. This should be avoided.

5.14 PROJECT #3

1. Maintenance costs for a new bridge with an expected life of 50 years are estimated to be \$1,000 each year for the first five years, followed by a \$10,000 expenditure in the 15th year and another \$10,000 in year 30. If the discount rate is 10% per year, what is the equivalent uniform annual cost over the entire 50-year period? Draw the cash-flow diagram.

2. An engineering firm spends \$100,000 for a minicomputer which it leases to other organizations for \$10,000 per year for 12 years. Compute their rate of return on this investment.

3. A drainage contractor is considering buying new equipment which will allow him to expand his operations. Two models of dragline are under consideration, with the data pertaining to each given in the table below.

ECONOMIC PARAMETERS FOR TWO DRAGLINES

Model	Cost ($)	Annual Benefits	Scrap Value	Life (years)
Midorii	100,000	20,000	10,000	10
Viggin	100,000	15,000	10,000	20

The contractor asks his daughter, a Civil Engineer, to do an economic analysis and advise him which to buy. He asks her to use four criteria for the comparison; namely, Net Present Value, Equivalent Annual Worth, Incremental Benefit/Cost Ratio, and Internal Rate of Return. She is to explain any differences.

The annual benefits for each machine are estimates of the annual revenue less the annual operating costs. The life of each model is the time after which it is either obsolete or too expensive to keep running. The scrap value is the estimated resale value at the end of its life. An appropriate discount rate for the analysis is 9%. Assume that a Midorii will be replaced by another Midorii at the end of its life.

A cash-flow diagram is to be drawn to support the conclusions of the analysis.

4. Civil Engineering Systems Society of America is considering moving its national headquarters to Research Triangle Park, North Carolina. It has identified two proposed locations, one of which is an existing building about 12 years old and the other is a site upon which a new building would be erected for the Society to its specifications.

The existing building would cost $980,000 and would require $116,000 worth of repairs to make it suitable for the Society's occupancy. Two upper floors might be rented out to other tenants at an estimated annual income of $62,000. Utilities, taxes, janitor services and other expenses would cost an estimated $39,000 per year. A major renovation can be foreseen in ten years' time which will cost (at present estimate) $200,000. After a lifetime of 20 years the Society expects it would sell the building and relocate. The sale value (present estimate) is $920,000.

The new building would cost $540,000, would have no rental or other income space, and would cost an estimated $26,000 annual upkeep. The lifetime is thought to be 30 years. A present estimate of its sale value at 30 years is $700,000.

Based upon a present worth criterion and a discount rate of 6%, which option would be better for the Society? Draw the cash flow diagrams for each.

5. Nulty Tyier Stone Company operates quarries which supply gravel to contractors. The Company has invested $1,200,000 in new quarry sites which

they estimate will produce (at present value) $2,000,000 by the end of ten years' time. What is the *IRR* (Internal Rate of Return) to the Company on this investment?

6. Your office has been informed by the Fire Department that its alarm system is antiquated and in need of replacement. Two possible makes of security systems are potential candidates for the new installation. You are asked to determine, on the basis of present worth considerations, which is the better one. Take the discount rate to be 6%.

   Sentry's initial cost will be $89,000, and its expected life is 25 years. The annual maintenance cost is anticipated to be $1,200. A salvage value estimate is $2,000.

   The capital outlay for *Sentinel* will be $60,000, its annual maintenance is estimated at $1,500, and its service life is thought to be ten years. No salvage value can be attached to this system.

   Be sure to draw cash-flow diagrams for both systems.

7. The City of Gotham is examining its bus service. The Mayor has appointed a task group to review operations, and has directed the City Manager to furnish the task group with initial cost, operating expenses, the annual benefits and a salvage estimate. These data are incorporated in the table below.

   BATMAN© has been asked to do the economic analysis (ROBIN© has not yet taken CE Systems.)

INITIAL COST OF BUSES X1	ANNUAL OPERATING COSTS X2	ANNUAL BENEFITS X3	SALVAGE VALUE OF BUSES X4
$6,000,000	$400,000	$1,500,000	$700,000

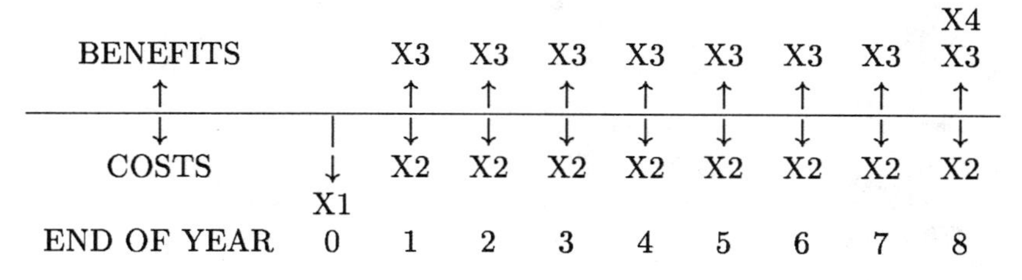

$$Present\ Worth = -X1 + (X3 - X2)\left[\frac{(1+i)^n - 1}{i(1+i)^n}\right] + X4\left[\frac{1}{(1+i)^n}\right]$$

Using a discount rate of 7.00%, the present worth is $975,835.

BATMAN needs to find the equivalent annual worth for the 8-year period.

Using the data from above, (1) help BATMAN find the Internal Rate of Return (*IRR*) for the bus service described above and (2) plot the present worth versus discount rate. Use a computer routine such as a spreadsheet (*e.g.*, XESS on *eos*) to confirm your answer to (1).

8. BATMAN© must decide how to dispose of the nuclear waste from The Bat Cave. He has obtained bids from 2 organizations which handle this type of job. Based upon a present worth analysis, which company is the better choice for a project that will last 200 years?

COMPANY	INITIAL COST ($)	ANNUAL EXPENSES ($)	ANNUAL BENEFITS ($)	LIFE OF THE SYSTEM (YEARS)
A	60,000,000	4,000,000	9,000,000	50
B	100,000,000	3,000,000	11,000,000	100

Use a discount rate of 4% for your analysis. In your report include (1) a cash flow diagram for each company and (2) present worth values for each.

Do the Benefit/Cost Ratio for Companies *A* and *B*. Include in the report (1) Benefit/Cost Ratio for both companies against the do-nothing option; (2) the Incremental B/C ratio for the companies against each other.

9. Stoneway Structural Engineering anticipates that they will need to update their computer software in 9 years at a then cost of $155,000. The firm desires to make annual deposits into a sinking fund account at an annual interest rate of 8% so that at the end of nine years $155,000 will be available to purchase the new software.

   - How much should the annual payments be?
   - Draw a cash-flow diagram that represents this situation.

10. You are Ms. or Mr. Civil, the owner and CEO of *I-BEAM, Inc.* The business is doing well in these hard times. You anticipate that you may retire in 12 years, and turn the business over to your children. Accordingly, you would wish to reap the largest benefit during this next 12 years as a nest-egg for your post-retirement years.

    One idea is to expand the business, which would cost $4,000,000 to enlarge the facilities and update the machines. At present the annual operating cost is $990,000 and the annual benefit is $1,225,000. The best estimates for the proposed expanded business are: annual operating cost, $1,550,000; annual benefit, $2,200,000.

    Your decision is made more difficult when the County Supervisors learn from one of their staff (who is your close friend and confidant) of your thoughts about expanding. They see in the proposed expansion more much-needed jobs for the community, and are offering you a $500,000 per year tax credit for the first six years after the business expands.

What should be your decision to make the most money for the next 12 years? Base your decision on the following criteria: Net Present Value, Equivalent Annual Worth, Incremental Benefit/Cost Ratio, Internal Rate of Return.

Show a cash-flow diagram for both options. Use a discount rate of 7%.

11. Trexxler Engineers, PA, are considering replacing their asphalt core testing apparatus with more modern equipment. Two particular brands are under investigation. An ultrasonic system will cost $100,000, can be expected to serve for 25 years, and will allow annual earnings of $40,000, but will require maintenance of $15,000 per year. The present estimate of its salvage value is $20,000. The radioactive tracer device will cost $60,000, will probably be obsolete in 10 years, may bring in earnings of $20,000, and is estimated to require $5,000 per year in maintenance. Its salvage value at ten years is negligible.

    Assume that either system will be replaced by its equivalent at the end of its service life. Take an appropriate discount rate to be 8%. Draw the cash flow diagrams for both options. Based upon present worth considerations, which is the better proposal?

12. A new freeway is proposed to run between two cities. The present road is 50 km long and requires an annual maintenance cost of $3,000/km. It currently needs a major overhaul costing $5,000,000.

    The new freeway will cost $60,000 to build. It will be 40 km long and will have an annual maintenance cost of $6,000/km.

    Benefits of the proposed highway include savings in travel time and operating costs and reduction of accident costs. These can be summarized as follows: $2,330,000 (commercial users); $270,000 in accident reduction.

    DOT planners suggest a life of 30 years and a constant 8% discount rate as approporiate to this proposed project.

    Use Net Present Value and Incremental Benefit-to-Cost Ratio methods to determine the economic feasibility of the new road (compared to the existing one).

13. A city engineer is faced with a decision on one of the municipality's trunk water mains. The existing pipe is old and leaking badly. The losses in fact exceed 40% of the water entering the distribution system. The engineer can either replace the pipe altogether with new pipe, or insert a liner which will stop the leaks but also decrease the diameter and hence the capacity. The new pipe will have an estimated life of 50 years, will cost $10,000 per year and will achieve an annual benefit of $12,000 per year. The lined pipe can be expected to last 20 years, to cost $7,500 per year and to return a benefit equivalent to $8,500 per year. For the sake of comparison, use an annual discount rate of 6% (which can be assumed to remain constant throughout the period of evaluation.) Draw the cash flow diagrams and use a Benefit/Cost ratio method to decide which option is best.

14. A water company will pump from its well-field to the community 3 km away where the water will be sold. To lay the pipeline will cost \$45 D per running meter, in which D is the diameter, and the annual pumping cost will be \$200 h per year, where h is the head loss or drop in pressure given by the Darcy-Wiesbach law

$$h = 0.02 \frac{L}{D} \frac{V^2}{2g}$$

with L, the pipe length in meters, V, the mean velocity of the flow in meters per second, and g, the local gravity—9.806 m/s^2.

Assume that the pipe will last 10 years, and that the discount rate is a constant 10% over this time period. Pipe can be obtained in increments of 0.2 m diameter, beginning with 0.2 m.

If the design flow is to be 1.2m^3 per second, what size pipe should the company use to minimize its annual costs?

15. The Counties of SILAS, DEMONE, and MURTHA have created an agency called OCHOBE WATERSHED AUTHORITY which they hope will better regulate the OCHOBE River Basin. The OWA has proposed to build a multi-purpose reservoir to control flooding, generate some electric power, and augment the existing recreational resources of the region. The initial cost is estimated to be 25 million dollars, and annual maintenance will be \$300,000. The projected annual benefits are \$1,000,000 in flood control, \$500,000 in electricity, and \$1,500,000 for recreation. Assume that the life of the project is to be 40 years, and that a constant discount rate of 7.5% will prevail over that time.

   - Sketch the cash-flow diagram
   - Compute the Benefit/Cost Ratio

16. A new town is planned to service a proposed weapons test establishment in a remote area. Two alternative sources of water supply for the town are being considered.

   - Groundwater pumped from a nearby aquifer
   - Construction of a dam in a range of hills some distance from the town and a pipeline which connects to the town

Relevant information for the two alternatives is given below.

	Groundwater Scheme	Dam and Pipeline
Initial Cost (\$)	150,000	1,000,000
Annual Operating Cost (\$)	100,000	40,000
Annual Benefits (\$)	200,000	240,000
Life (years)	10	30

It is anticipated that the project will last for 20 years, after which the town, test facility, and water supply system will become surplus.

- Use a discount rate of 7% and compare the two alternatives in terms of net present value.
- Calculate the Internal Rate of Return of the dam and pipeline project.

17. As manager of a building supplies yard, you are considering improving the operation of your facility by purchasing a new heavy-duty forklift. You are evaluating two different machines, each of which has the same rated capacity.

    Machine *A* costs $30,000, has a life of 20 years, annual maintenance costs of $1,500, and a salvage value estimated at $5,000. Machine *B* costs $37,000, has an expected life of 15 years, annual maintenance costs of $1,200, and its estimated salvage value is $5,700.

    Use an annual discount rate of 6%, and assume that initial costs, annual maintenance, and discount rates stay sensibly constant throughout the period of evaluation. Draw a cash flow diagram for each machine, and use a Benefit/Cost Ratio method to determine the most economical choice.

18. Sandhills City is considering a multipurpose impoundment that will control flooding, provide water supply, and furnish some recreational needs. The project will cost $23,000,000 to build and $550,000 per year to maintain. Annual benefits are thought to be $,1050,000 for flood control, $650,000 in water revenues, and $1,250,000 from recreation sources. The expected life is 50 years, and the assumed discount rate is 8%.

    - Sketch the cash-flow diagram for this project.
    - Compute the Benefit/Cost Ratio.

19. BWANG Aircraft Company has just received an order for 36 747-350 series jets from JAYPUN AIRLINES. It has 7 years in which to fill this order.

    The order presents a large problem for the CEO of BWANG in terms of the manufacturing facilities. He accepts that either an older existing hanger must be used, or a new one must be built.

    If the existing hanger is remodeled, it will require one year to accomplish, whereas constructing the new facility will take three years. The remodeled hanger can produce 6 planes per year, while the new hanger can yield 9 planes per year. Staff studies supply the data below to assist in making the decision. An appropriate discount rate for the comparison is 8.8%.

    What should the CEO recommend to the Directors of BWANG, Inc.?

	Remodeled Hanger (millions of $)	New Facility (millions of $)
Initial Cost	230	500
Annual Operating Cost	75	20
Annual Benefits	240	360

    The recommendation is to be based upon the following:

    - Net Present Value

- Equivalent Annual Worth
- Incremental Benefit/Cost Ratio
- Internal Rate of Return

and cash-flow diagrams should be included.

20. You are planning to purchase a personal computer which will cost $1,575 with the peripherals you desire. You expect to pay for the machine in three equal installments: one at the time of purchase, and annually in each of the succeeding years. At the end of three years you intend to trade in the computer in order to upgrade to a newer model, and the present estimate of its trade-in value is $200. Your annual maintenance costs are expected to be $42. Draw the cash-flow diagram if the discount rate is expected to be 7% over the period in question.

21. An engineer who was preparing to attend Vth International Congress on Acoustics in Liège, Belgium, in August, 1965, decided to purchase some Travelers Cheques to take along on the trip.

    While at the conference, he spent most of this money for meals, local travel, presents for his wife and children, and souvenirs for the office staff.

    When he returned home, he had one cheque in the amount of $50 left over. He decided to keep this cheque as a reserve in the event of an unexpected out-of-town trip or week-end emergency. (No doubt he was influenced by the advertizing campaign of the Travelers Cheques Company, which had been promoting this very idea.)

    In fact, the engineer did use the cheque as extra security on a number of occasions, but never actually cashed it. As time passed, he gradually forgot about it, until one day he was cleaning out his desk and found it in the back of a drawer under an envelope where he had hidden it.

    Using the annual discount rates available in the literature, compute the amount of money the engineer lost on his $50 cheque (CX113-933-410) from August 1965 until August 1989 at which time he cashed it (for $50 in 1989 money, of course).

CHAPTER 6

LINEAR GRAPHS

6.1 INTRODUCTION

A device which is of high utility in transportation engineering is the ***linear graph***. These words are used here in their technical sense, not as a reference to drawings of the usual kind. The theory of linear graphs is a branch of mathematics which is closely related to logic, and has a wide-ranging applicability far beyond the elementary form employed in analyzing traffic networks. Nevertheless, it is of very great value in studying traffic flows in highway systems and similar structures in Civil Engineering practice.

For the most part, the form of this theory which is most often used in traffic studies is that called ***directed graphs***. An introductory example was employed in Chapter 3, where a descriptive model was mentioned in connection with routes from Raleigh to Charlotte.

In the present exposition, imagine that one might wish to go from downtown Raleigh (say Union Square, where the Capitol is, and from whence the original layout of the city took its origin) to the center of Cary, where the dogwood emblem is located at the intersection of Chatham and Academy Streets. This is a typical urban-suburban traffic situation, common to many areas of the country. For the sake of simplicity, assume that the trip is associated with the evening rush hour traffic created by persons leaving their employment in, say, State Government buildings, and going to their homes in the "bedroom" suburb which is now the third largest city of the Triangle.

For reference refer to the excerpt of a road map which shows the highways and streets of the area. See Figure 6.1. Following the same scheme used in Chapter 3, a first step is to focus upon the major routes leading from the source (downtown Raleigh) to the sink (downtown Cary). One road, *Hillsborough Street*, leads from the Capitol to the State Fairgrounds and thence as *NC 54* (*Chapel Hill Road*) to *Academy Street* just North of Cary Center. Alternatively, the motorist could depart downtown Raleigh through Boylan Heights and take *Western Boulevard* to its intersection with *Old US 1* beyond *Jones-Franklin Road* on into Cary on *Old US 1*. A third possibility for the commuter would be *Hillsborough Street* to the Fairgrounds area and thence over to *Western Boulevard* and *Old US 1*.

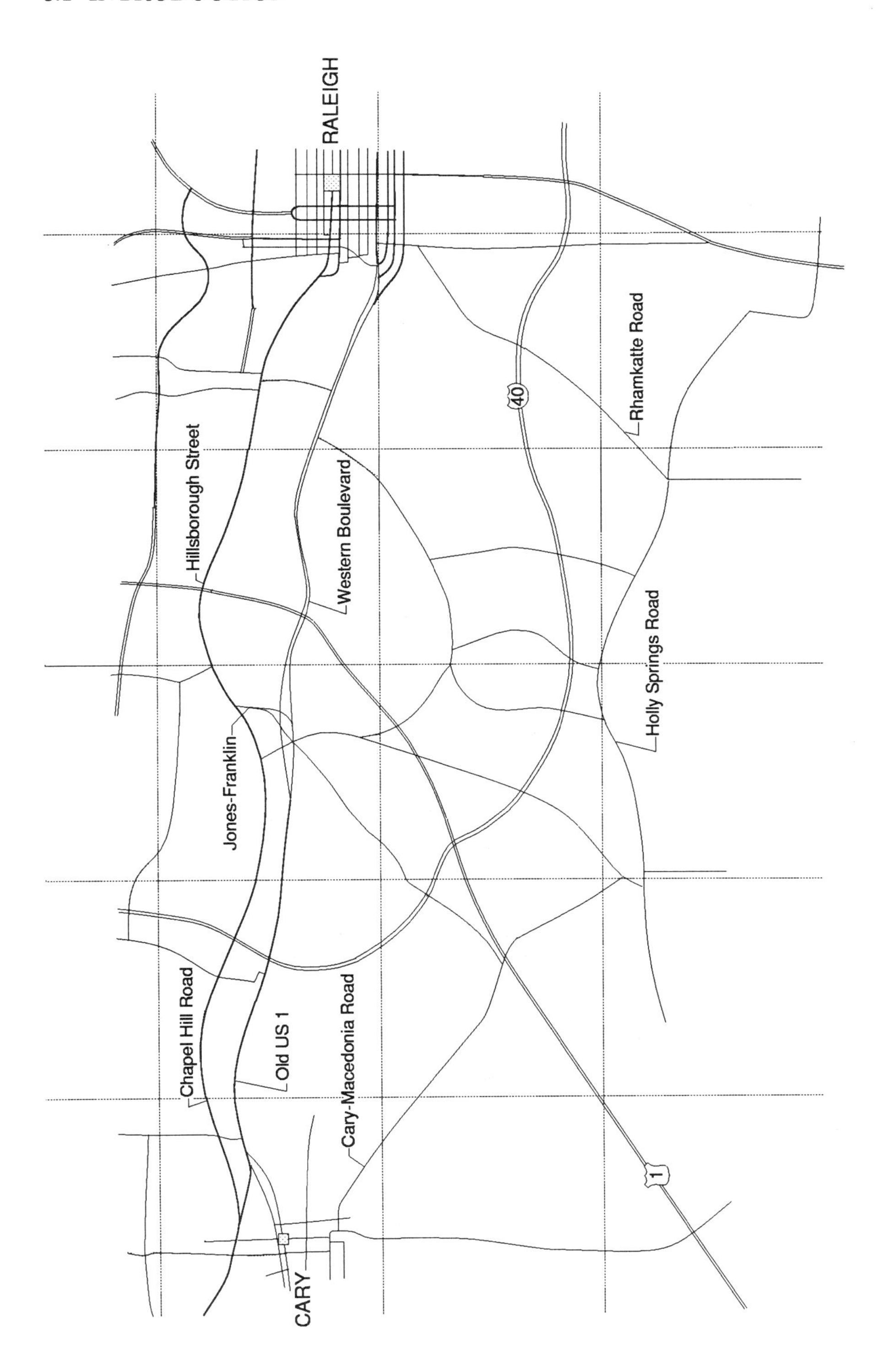

Figure 6.1: Cary-Raleigh Area Map

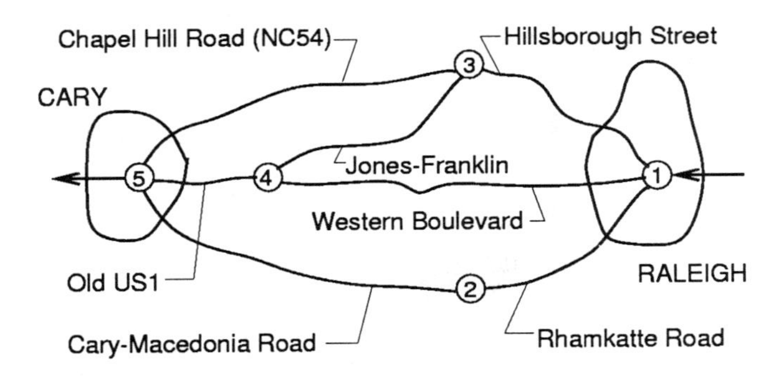

Figure 6.2: Simplified Cary-Raleigh Area Map

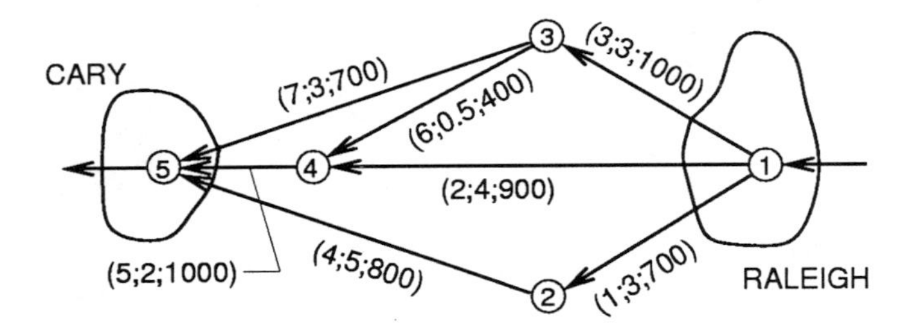

Figure 6.3: Schematic Cary-Raleigh Area Map

The scenic route winds through *Salisbury Street* to *South Saunders Street* to *Rhamkatte Road* to *Tryon Road* (formerly *Holly Springs Road*) to *Cary-Macedonia Road* to *Walnut Street* onto *Academy* and into the town center.

Even this much simplification leaves some pretty tortured routes to contend with, as shown in Figure 6.2. So the next step would be to purify the map and convert it into a schematic diagram with only straight line segments as shown in Figure 6.3. Let the critical junctions be designated with numbered nodes. Node 1 is at the Capitol, Node 3 is at the State Fairgrounds, Node 4 denotes the Jones-Franklin confluence, and Node 2 is Macedonia. Cary Center is Node 5, of course. All of the traffic is assumed to enter the net at Node 1, and all of it will exit at Node 5. There is no provision for exiting traffic or traffic "storage" at any point within the network. Arrowheads delineate the assumed direction of traffic flow in each of the roads of the net.

A legend can be attached to each route to display the route number, the road's length, the capacity in vehicles per hour, and, possibly, number of lanes or other information deemed important to the study.

Probably the single most important thing the traffic engineer would wish to know is the total network capacity in vehicles per hour, since this would illuminate the problems associated with moving people from their employments to their domiciles in good order. Since it has been assumed that all traffic is moving from Node 1 toward Node 5 at the evening rush hour, counterflow traffic will

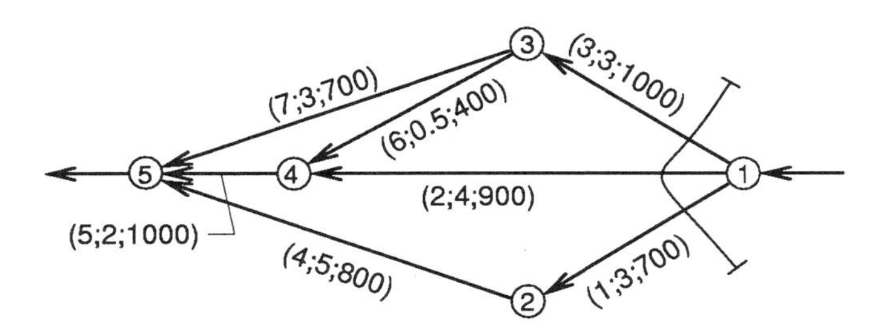

Figure 6.4: First Cut Set

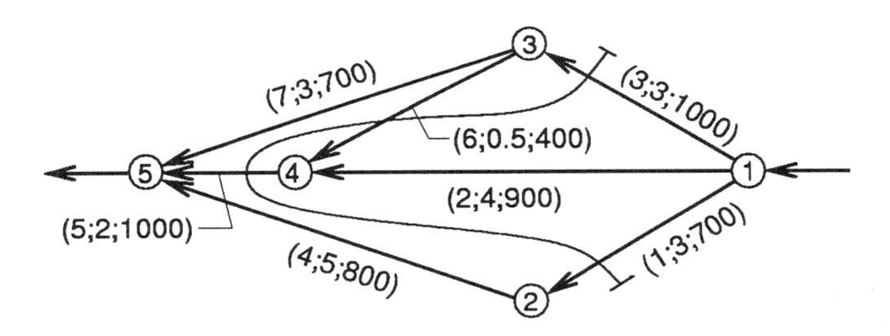

Figure 6.5: Second Cut Set

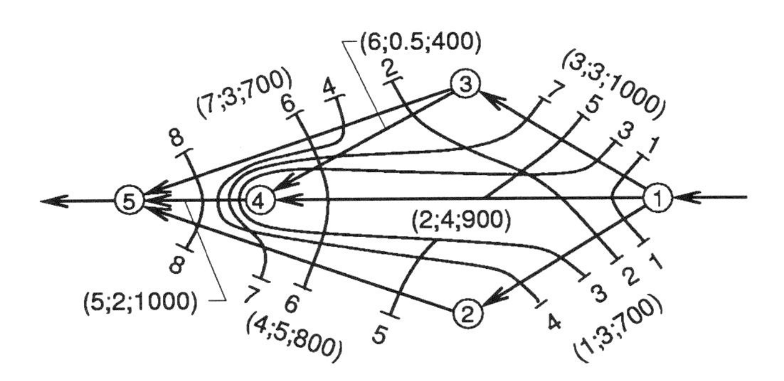

Figure 6.6: Assembled Cut Sets

be assumed to be prohibited during this period. It would be possible to enforce this by making the streets one-way for the time being, and reversing the one-way directions next morning to accomodate the inrush of people going to work.

However, it will be noticed immediately that Road 6 is unknown as to direction. Actually, it is preferable to say that it could be either direction, and a key finding of the analysis will be to determine which should be that direction in order to maximize the net capacity. Two solutions are obviously called for, with the direction of traffic in that road reversed between them.

6.2 CUT SETS

One obvious way to ascertain how much traffic leaves Node 1 is to put a traffic counter across the three routes which depart from there and actually measure the number of vehicles per hour. Almost everyone is familiar with the black cables across the road surface with a counter box on the side of the road, perhaps secured to a post or other firm object. These devices are pressure activated, so that when the cars's wheels cross them a pressure pulse is delivered to the counter, which registers on a number wheel.

Figure 6.4 illustrates the placement of such a counter across the routes departing downtown Raleigh. Such a counter could be placed across any of the network locations which separate the *Source* (Node 1) from the *Sink* (Node 5). Such a counter is displayed in Figure 6.5, which cuts across Roads 1, 5, 6, and 3. In fact, every possible set of roads should be provided with such a counter mechanism in order to determine that one which limits the total capacity.

All of the possible sets are seen in Figure 6.6, and it may be noted that there are eight of them.

The operative word which describes the function of these eight named paths is "cut." Instead of actually placing a piece of hardware on the roads, a mathematical cut is mentally and graphically put there to serve the same function—that of counting the flow of vehicles in that section. "Section" is the other name of importance. Specifically, the topology of the network is severed into two parts, one of which contains the source, and the other of which has the sink. No cuts are made around such Nodes as 2 and 3, for example, since these would not measure network capacities.

The cutting operation described above bears the label "proper." A proper cut set is one which fulfills the requirements laid out there, to wit: one side of the cut holds the source, the other the sink, and no cuts which leave out either source or sink, or which contain both source and sink on the same side of the cut are proper.

6.3 NETWORK CAPACITY

When all of the proper cut sets have been identified for a network, they can be investigated individually to determine the capacity associated with each. When this is accomplished, the capacity of the entire net is revealed. A theorem states this carrying power as follows:

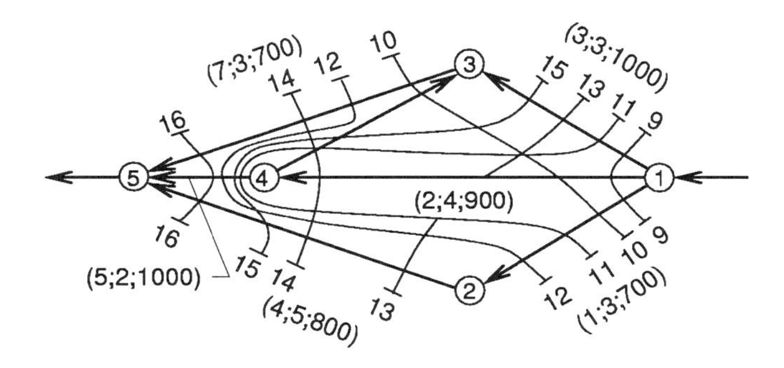

Figure 6.7: Road Network with Road 6 Traffic Reversed

For a source-sink connected network the maximum flow capacity from source to sink is equal to the minimum of the proper cut-set capacities for all of the proper cut-sets separating the source from the sink.

In the illustrated network, each cut set has its capacity labeled at its terminus. The eight values are seen to be 2600, 2700, 2700, 2800, 2700, 2400, 2800, 2500. Obviously the minimum and therefore determining one is 2400, the network capacity.

Figure 6.7 depicts the same network with Road 6 reversed in direction, and the cut-sets all redone. The numbers are now 2600, 3100, 2700, 3200, 2300, 2400, 2400, 2500, so the network capacity for this case is 2300 vehicles per hour (vph). The traffic control officer would make Road 6 one-way leading from Node 3 to Node 4 in order to maximize the netflow.

A note of special importance pertains to the determination of the capacity of individual cuts in respect to the direction of traffic in the roads which are cut. The whole purpose of the analysis is to determine the maximum capacity of the network, therefore any counterflow traffic to that which would proceed from source to sink is counterproductive. Specifically, if a road that is cut by a cut set has a direction such that the traffic in it would go from sink side to source side, then that road cannot be allowed to have any traffic at all, since it would buck the desired flow and diminish the net capacity. Accordingly, its vehicle-per-hour count is defined to be zero; i.e., the road is blocked by a barricade.

In Figure 6.8, for example, cut sets numbered 3, 6, and 7 cut Road Number 6. The traffic direction in this road is such that for Cuts 3 and 7 it would flow from sink-side to source-side, and is therefore zero by definition. On the other hand, for Cut 6 it would flow from source to sink, and the value 400 vph is included in the capacity 2800 vph of that cut. Figure 6.9 shows the detail.

Similarly, when the direction of traffic in Road 6 is reversed, as shown in Figure 6.10, Cuts 14 and 10 sever this road in such a manner that the flow would be from sink to source, which is not allowed, so the vehicle count is zero, not included in the capacity of those cut sets. The other two cuts which include this road, 11 and 15, would have the flow from source to sink, and so are allowed and included in the cut set capacities. See Figure 6.11.

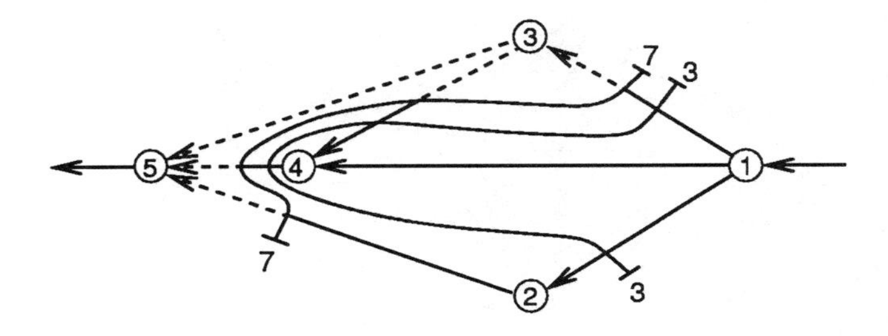

Figure 6.8: Cuts Sets Sign Convention

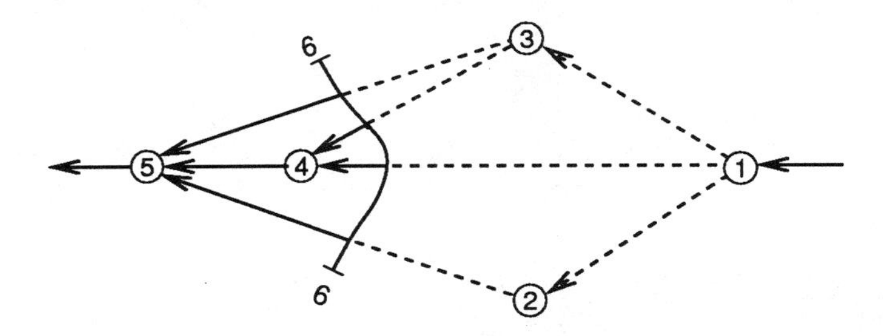

Figure 6.9: Detail of Cut Sets Sign Convention

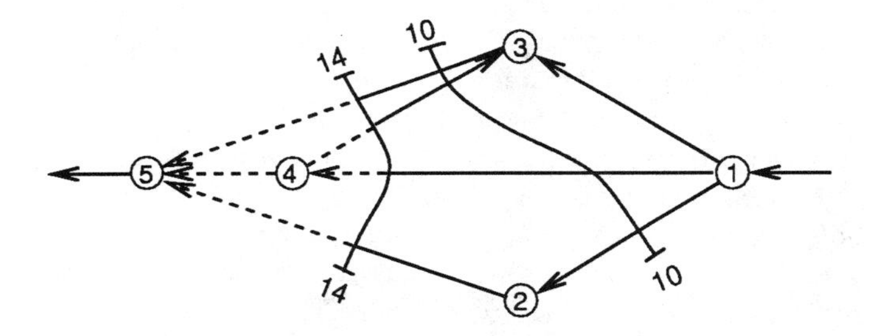

Figure 6.10: Road 6 Traffic Reversed, Cuts 10 and 14

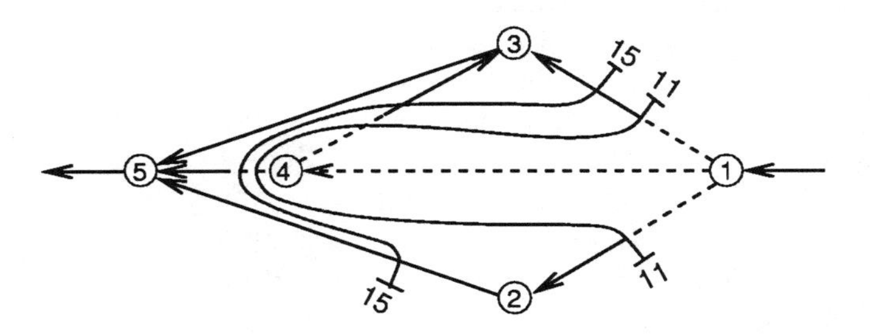

Figure 6.11: Road 6 Traffic Reversed, Cuts 11 and 15

One might think of the criterion above as similar to a sign convention, and it should be applied to each road in a cut set as the analysis proceeds. One might imagine actually taking scissors and making a physical cut on the paper graph of the net, and noting, by making arrows at the cut where it intersects each roadway, the traffic direction in each road. If the flow enters from the source side of the cut paper, it is allowed, whereas otherwise it is not.

6.4 TABULAR FORM OF CUT SET ANALYSIS

A convenient method for doing a systematic analysis of the network in order to determine its capacity is the tabular form shown below. In it are listed the highway number, its capacity, the cut set number and some marks (the number one) which define whether or not a particular highway is cut by a particular cut set. If the number is +1, the capacity of that road is included. If, on the other hand, the number is −1, it indicates that the traffic flow in that road when cut by that set is from sink-side to source-side, so the traffic count in it is disallowed and the capacity is not included in the set. The absence of a number or a zero indicates that the road is not cut by the set.

HIGHWAY		CUT-SET NUMBER							
Number	Capacity	1	2	3	4	5	6	7	8
1	700	1	1	1	1				
2	900	1	1			1	1		
3	1000	1		1		1		1	
4	800					1	1	1	1
5	1000			1				1	1
6	400		1	− 1			1	− 1	
7	700		1		1		1	1	
Cut-Set	Capacity	2600	2700	2700	2400	2700	2800	2800	2500

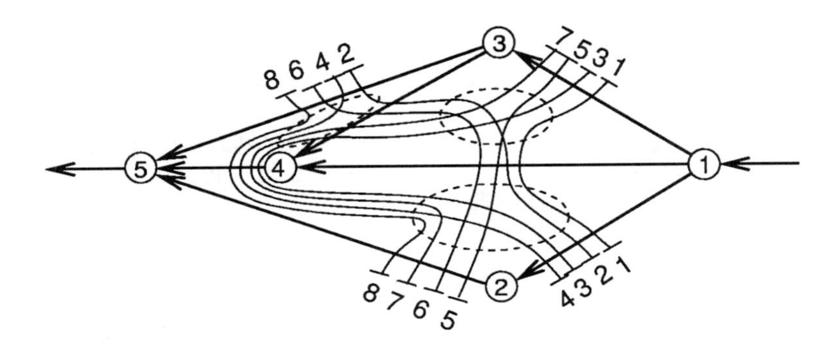

Figure 6.12: Metamorphosis of Cut Sets

6.5 DUAL NETS

The process described above worked well for the over-simplified problem of Raleigh-Cary evening rush hour traffic, but one reason why it did so was that the diagram had been severely sculptured. As soon as there are more nodes, and more roads, the picture becomes very cluttered. In a real-world situation, the network will contain not just five nodes and six routes, but something more like a thousand times that number. On the face of it, it would be impossible to draw the cuts, and likewise impossible to put the cut capacities on the figure. Finding that one which has the minimum capacity by inspection from the plot would be beyond the possible.

A construction which alleviates this problem for larger nets (and smaller ones, too) is called the *dual net*. This device takes the same net, shown in Figure 6.12, and tidies it up a bit. The cuts are bundled up together, in much the same manner as would be done with a cable composed of many wires, and bound as with cable ties, strategically located in each of the "open" spaces of the network structure. The bundles are, so-to-speak, tied down in the triangular internal regions of the network topography. One such tie-down would be assigned to each such triangle.

Now each group of cut sets in a cable is replaced by a chain. A dummy source and a dummy sink are provided, and as shown in Figure 6.13, the source-sink axis is transverse to that of the original net. That is, the flow in the chains of the contrived network is from bottom to top. The tie-down points are also provided with fictitious nodes and the whole chain system is given a new nomenclature. The nodes are lettered alphabetically and the chains are designated with capital Roman numerals.

Chain I replaces Cuts 1, 2, 3, and 4; Chain II is for Cuts 1, 2, 5, and 6; Chain III contains Cuts 1, 3, 5, and 7; Chain IV stands for Cuts 5, 6, 7, and 8; Chain V takes in Cuts 3, 4, 7, and 8; Chain VI bundles Cuts 2, 3, 6, and 7; and Chain VII has Cuts 2, 4, 6, and 8. Each chain cuts one road of the network only, and its "length" is made equal to the capacity of the cut road in vph.

The total chain length by any path leading from the dummy source to the dummy sink is the sum of the lengths of the individual chains comprising the source-sink path, and is the capacity of the network by that path.

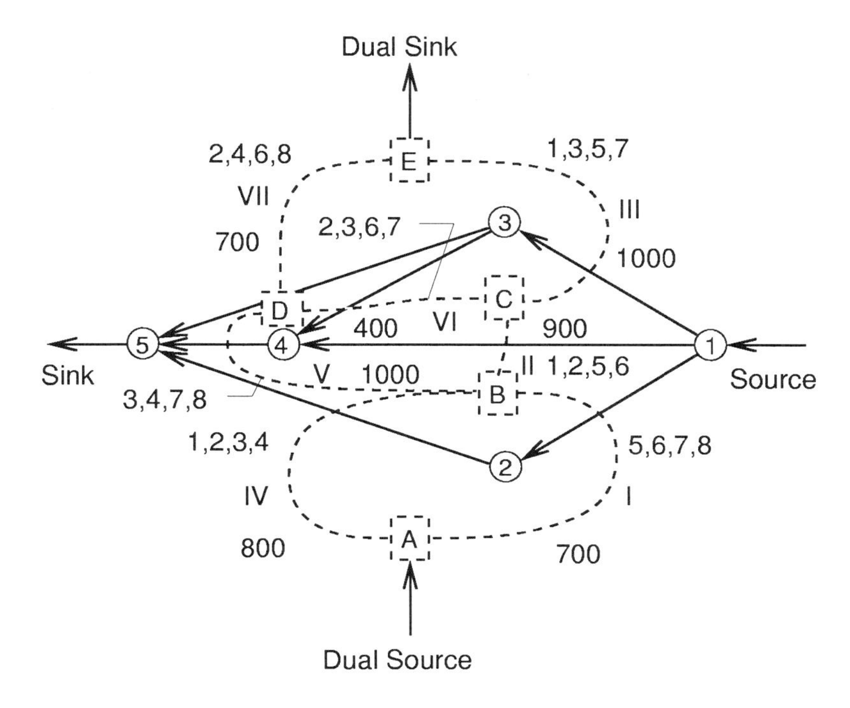

Figure 6.13: Dual Net

The topology of the original network and its cut sets has suffered a complete metamorphosis. In recognition of this fact, the new construction is given the new name *dual net*, or the net dual to the original. The source becomes the *dual source*, the sink is the *dual sink*.

The merit of the dual net construction is that it can be drawn and seen and analyzed, and, of course, is useful as a check in computations for at least those nets for which the cut set method can be carried out.

A further advantage of the dual net method by comparison with the cut sets analysis is that there is only one chain which cuts each roadway, therefore the map is not cluttered with multiple cuts. This advantage becomes increasingly important as the number of roads in the net increases. In fact, for even a small network the number of cut sets can become so large as to cloud the understanding and comprehension of the analyst.

It is necessary to recall the convention that was established for the *primal net*, as the original version is called; namely, that when a chain cuts a roadway and the flow in the road is such that it would be from sinkside to sourceside, that traffic count is disallowed in the chain capacity determination. When the traffic direction in the road that is cut by the chain is such that it proceeds from source to sink, the count is included.

The tabulation below summarizes the chain analysis for the dual net displayed in Figure 6.13.

CHAIN ANALYSIS

Chain	Length	Corresponding Primal Cut-Set
I-II-III	700+900+1000 = 2600	1
I-II-VI-VII	700+900+400+700 = 2700	2
I-V-VI-III	700+1000+1000 = 2700	3
I-VI-VII	700+1000+700 = 2400	4
IV-II-III	800+900+1000 = 2700	5
IV-II-VI-VII	800+900+400+700 = 2800	6
IV-V-VI-III	800+1000+1000 = 2800	7
IV-V-VII	800+1000+700 = 2500	8

Maximum Capacity = Minimum Chain Length = 2400

The schedule below shows a comparison of the Primal and Dual Analyses.

Primal Cut-Set Analysis Cut 3		Dual Chain Analysis Chain I-V-VI-III	
Road	Capacity	Branch	Length
1	700	I	700
5	1000	V	1000
6	0	VI	0
3	1000	III	1000
Total	2700	Total	2700

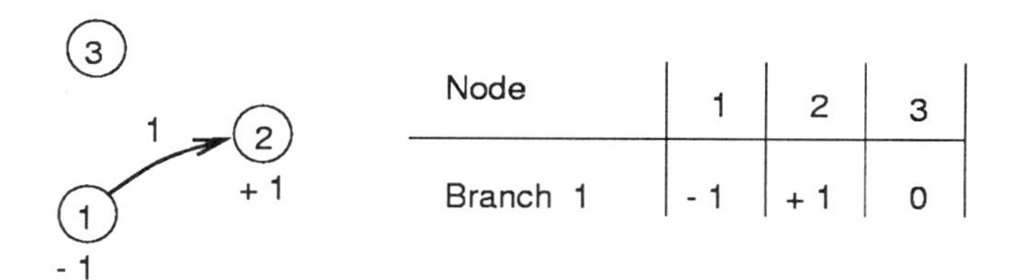

Figure 6.14: Generic Nodes

6.6 THE LINK-NODE INCIDENCE MATRIX

The elementary networks treated above serve to display the fundamental techniques by means of which the carrying capacities of road systems can be ascertained. It is immediately clear, however, that real road nets are highly complicated, with many links and many nodes. The problem of analyzing such complex systems is a classical example of a venue in which computers can be of enormous assistance. The dual net is, of course, only a first and very limited approach to an efficient method for treating roadway nets by applying computers to the task.

Return for a moment to the tabular version of the cut-sets shown in Section 6.4. Note that the symbols $+1$, -1, and 0 are employed to indicate structure and connectivity for the network in question. This suggests that a scheme based upon such a representation might be employed in the larger-scale systems to prepare them for computer analysis. Accordingly, imagine that there are three generic nodes as shown in Figure 6.14, and a branch or link which connects two of the nodes. The link is a directed line segment, since there is an arrowhead which shows that traffic proceeds from Node 1 to Node 2 along the link. Node 3 is not connected to either of the other two nodes.

Now establish an array which lists the node numbers on a row heading and the Link or Branch 1 as the first column element of the array. The entries in this matrix are to be either $+1$, if the branch enters the node, -1 if it leaves a node, or 0 if the branch is not connected to the node. From the picture it is clear that there must be a -1 in the cell pertaining to Node 1 and Branch 1, a $+1$ under Node 2 and across from Branch 1, and 0 where Branch 1 falls under Node 3. Obviously a larger such matrix can be constructed for a network of any larger size.

Take the simple network illustrated in Figure 6.15. There are four nodes, n_1, n_2, n_3, n_4, and five branches, b_1, b_2, b_3, b_4, b_5. Lay out the matrix array with the node numbers on the first row and the branch numbers in the first column. Make entries of $+1$, -1, and 0 at each cell within the array as shown. For example, branch b_2 leaves the source node, n_1, and enters node n_3, so there will be $+1$ in the b_2–n_3 cell, -1 under b_2–n_1, and 0 in b_2–n_2 and b_2–n_4, the sink node.

A moment's thought will suffice to allow the conclusion that what enters the network at its source, node n_1, must come out through the sink node, n_4, so that if one knows the traffic through either of these nodes, it is known for the other also. The meaning of this is that no new information can be expected to be learned by including both of them in the study of the network. It is said that one of them is redundant. Accordingly, either Column 1 or Column 4 of the matrix is deleted.

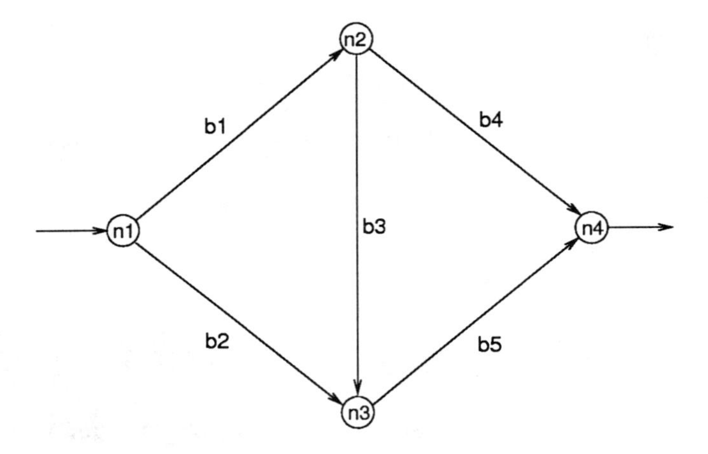

Figure 6.15: Simple Network

		NODE			
		n_1	n_2	n_3	n_4
	b_1	−1	+1	0	0
	b_2	−1	0	+1	0
BRANCH	b_3	0	−1	+1	0
	b_4	0	−1	0	+1
	b_5	0	0	−1	+1

Figure 6.16: Link-Node Matrix

In Figure 6.16 it is Column 1 that is taken out, leaving the formal matrix

$$\mathbf{A} = \begin{bmatrix} +1 & 0 & 0 \\ 0 & +1 & 0 \\ -1 & +1 & 0 \\ -1 & 0 & +1 \\ 0 & -1 & +1 \end{bmatrix} \tag{6.1}$$

which is called the ***link-node incidence matrix***, and which carries within it sufficient information to reconstruct the topology of the network when **A** and its elements a_{ij} are furnished.

The formation of this matrix is a crucial first step leading to the introduction of computer methods to network analysis, for it is well-known that machines can easily accomodate such digitized information as is contained in the cells of the matrix. Moreover, computers can easily be programmed to carry out routines associated with matrix operations. So it is seen that the link-node incidence matrix bears the entire structure of the network in a form readily utilized by a computer.

For instance, the link-node incidence matrix corresponding to the tabular description in Section 6.4, Tabular Form of Cut Set Analysis, is, as taken directly from the table itself,

$$\mathbf{A} = \begin{bmatrix} 1 & 1 & 1 & 1 & 0 & 0 & 0 & 0 \\ 1 & 1 & 0 & 0 & 1 & 1 & 0 & 0 \\ 1 & 0 & 1 & 0 & 1 & 0 & 1 & 0 \\ 0 & 0 & 0 & 0 & 1 & 1 & 1 & 1 \\ 0 & 0 & 0 & 0 & 1 & 1 & 1 & 1 \\ 0 & 1 & -1 & 0 & 0 & 1 & -1 & 0 \\ 0 & 1 & 0 & 1 & 0 & 1 & 1 & 0 \end{bmatrix} \tag{6.2}$$

In accordance with the continuity principle expressed above, either the first column or the last column might be deleted without loss of generality, because the traffic flow entering the source node equals that exiting through the sink at Node 5. In the matrix of Equation 6.2 neither the source node nor the sink node has been omitted.

6.7 AN EXAMPLE

Figure 6.17 shows a traffic net with 6 nodes and 12 branches. Notice how rapidly the number of branches increases with the increase of just one node. The source is Node 1, and the sink is Node 6.

The direction of traffic flow in the branches connecting Nodes 2 and 3 and Nodes 3 and 4 is unknown, or rather may be assumed to be in either sense. The link-node incidence matrix method allows for the possibility of either direction and therefore of both in the same analysis.

The means for accomplishing this is simply to include two branches between the two nodes in question, one branch going from Node 2 to Node 3, and another going from Node 3 to Node 2. Similar paths would be constructed between Nodes 3 and 4, as shown in Figure 6.18.

With these alterations made in advance, the matrix becomes

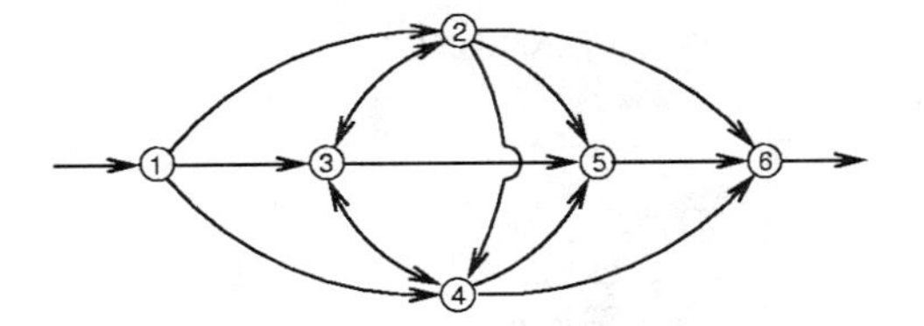

Figure 6.17: Example Traffic Net

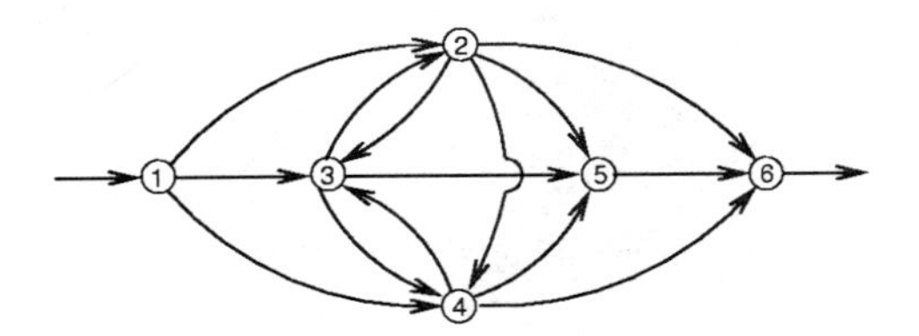

Figure 6.18: Augmented Traffic Net

LINK-NODE INCIDENCE MATRIX

LINK	NODE 1	2	3	4	5	6
1-2	− 1	+ 1				
1-3	− 1		+ 1			
1-4	− 1			+ 1		
2-3		− 1	+ 1			
3-2		+ 1	− 1			
2-4		− 1		+ 1		
2-5		− 1			+ 1	
2-6		− 1				+ 1
3-4			− 1	+ 1		
4-3			+ 1	− 1		
3-5			− 1		+ 1	
4-5				− 1	+ 1	
4-6				− 1		+ 1
5-6					− 1	+ 1

In this example the last column will be deleted as redundant so that the formal mathematical matrix becomes

$$\mathbf{A} = \begin{bmatrix} -1 & +1 & 0 & 0 & 0 \\ -1 & 0 & +1 & 0 & 0 \\ -1 & 0 & 0 & +1 & 0 \\ 0 & -1 & +1 & 0 & 0 \\ 0 & +1 & -1 & 0 & 0 \\ 0 & -1 & 0 & +1 & 0 \\ 0 & -1 & 0 & 0 & +1 \\ 0 & -1 & 0 & 0 & 0 \\ 0 & 0 & -1 & +1 & 0 \\ 0 & 0 & +1 & -1 & 0 \\ 0 & 0 & -1 & 0 & +1 \\ 0 & 0 & 0 & -1 & +1 \\ 0 & 0 & 0 & -1 & 0 \\ 0 & 0 & 0 & 0 & -1 \end{bmatrix} \tag{6.3}$$

The state vector of the system is composed of 14 elements—the traffic flows in each of the (now) fourteen branches of the net. These are X_{12}, X_{13}, X_{14}, X_{23}, X_{32}, X_{24}, X_{25}, X_{26}, X_{34}, X_{43}, X_{35}, X_{45}, X_{46}, X_{56}, hence the state vector is represented by the matrix

$$\mathbf{X} = \begin{bmatrix} X_{12} \\ X_{13} \\ X_{14} \\ X_{23} \\ X_{32} \\ X_{24} \\ X_{25} \\ X_{36} \\ X_{34} \\ X_{43} \\ X_{35} \\ X_{45} \\ X_{46} \\ X_{56} \end{bmatrix} \tag{6.4}$$

and the input vector is a single column matrix

$$\mathbf{I} = \begin{bmatrix} C \\ 0 \\ 0 \\ 0 \\ 0 \end{bmatrix} \tag{6.5}$$

wherein C is the network capacity—the traffic flowing into Node 1. The zeros in this column matrix represent traffic which flows into each of the other nodes, except the sink node, which has already been taken out by virtue of its redundancy with the input node. Note the implication that networks with multi-node inputs can be treated by furnishing numbers other than zeros here.

Now the state space theory model of the traffic network can be presented as

$$\mathbf{I} = \mathbf{A}^T\mathbf{X} \tag{6.6}$$

in which $\mathbf{A}^T$ is the transpose of the link-node incidence matrix. That is,

$$\mathbf{A}^T = \begin{bmatrix} -1 & -1 & -1 & 0 & 0 & 0 & 0 & 0 & 0 & 0 & 0 & 0 & 0 & 0 \\ +1 & 0 & 0 & -1 & +1 & -1 & -1 & -1 & 0 & 0 & 0 & 0 & 0 & 0 \\ 0 & +1 & 0 & +1 & -1 & 0 & 0 & 0 & -1 & +1 & -1 & 0 & 0 & 0 \\ 0 & 0 & +1 & 0 & 0 & +1 & 0 & 0 & +1 & -1 & 0 & -1 & -1 & 0 \\ 0 & 0 & 0 & 0 & 0 & 0 & +1 & 0 & 0 & 0 & +1 & +1 & 0 & -1 \end{bmatrix} \tag{6.7}$$

is the matrix generated from $\mathbf{A}$ by transposing its rows and columns; i.e., row number one of $\mathbf{A}$ becomes column number one of $\mathbf{A}^T$, etc.

The equation $\mathbf{I} = \mathbf{A}^T\mathbf{X}$ is an operator equation in which matrix $\mathbf{A}^T$ operates upon, in the sense of multiplying, the matrix $\mathbf{X}$ to produce the new matrix $\mathbf{I}$, which in this case is a vector whose components are the external inputs to the nodes of the network. That is to say, $\mathbf{A}^T$ transforms $\mathbf{X}$ into $\mathbf{I}$. Matrix $\mathbf{X}$ is the vector whose components are the traffics in the several branches of the net. In the model above, it is $\mathbf{A}^T$, the system topology, which is given in advance, and the problem is to find $\mathbf{I}$ and $\mathbf{X}$. Not your usual little arithmetic exercise!

The key to the solution is the phrase enunciated above; namely, "operates upon, in the sense of multiplying." Clearly, this means matrix multiplication in the usual way. To follow the customary rules, this then requires that $\mathbf{A}^T$ must be commensurate with $\mathbf{X}$. There must be the same number of elements in the first row of $\mathbf{A}^T$ as there are elements in the column matrix that is $\mathbf{X}$. Moreover, the number of column elements in $\mathbf{A}^T$ must equal the number of column elements in $\mathbf{I}$. The matrices pass these tests on all counts, and the operations can proceed.

The literal expression of the matrix multiplication is shown below. Row 1 of $\mathbf{A}^T$ multiplies Column 1 (the only column) of $\mathbf{X}$, element-by-element, and the products are added to produce Column 1 element of $\mathbf{I}$.

$$\begin{bmatrix} -1 & -1 & -1 & 0 & 0 & 0 & 0 & 0 & 0 & 0 & 0 & 0 & 0 & 0 \\ 1 & 0 & 0 & -1 & 1 & -1 & -1 & -1 & 0 & 0 & 0 & 0 & 0 & 0 \\ 0 & 1 & 0 & 1 & -1 & 0 & 0 & 0 & -1 & 1 & -1 & 0 & 0 & 0 \\ 0 & 0 & 1 & 0 & 0 & 1 & 0 & 0 & 1 & -1 & 0 & -1 & -1 & 0 \\ 0 & 0 & 0 & 0 & 0 & 0 & 1 & 0 & 0 & 0 & 1 & 1 & 0 & -1 \end{bmatrix} \begin{bmatrix} X_{12} \\ X_{13} \\ X_{14} \\ X_{23} \\ X_{32} \\ X_{24} \\ X_{25} \\ X_{26} \\ X_{34} \\ X_{43} \\ X_{35} \\ X_{45} \\ X_{46} \\ X_{56} \end{bmatrix} = \begin{bmatrix} C \\ 0 \\ 0 \\ 0 \\ 0 \end{bmatrix} \tag{6.8}$$

The result of this series of operations is the set of linear equations

$$C = -X_{12} - X_{13} - X_{14} \tag{6.9}$$

$$0 = X_{12} - X_{23} + X_{32} - X_{24} - X_{25} - X_{26} \quad (6.10)$$
$$0 = X_{13} + X_{23} - X_{32} - X_{34} + X_{43} - X_{35} \quad (6.11)$$
$$0 = X_{14} + X_{24} + X_{34} - X_{43} - X_{45} - X_{46} \quad (6.12)$$
$$0 = X_{25} + X_{35} + X_{45} - X_{56} \quad (6.13)$$

An inspection of this set of equations reveals that it is of the same form as those found in Chapter 4; to wit, there is a criterion function, Equation 6.9, and constraint Equations 6.10, 6.11, 6.12, and 6.13. In this case the constraints are "equalities," rather than less-than or greater-than inequalities as in many of the previous examples.

Physically, the set of Equations 6.9 through 6.13 expresses the condition at each node that all traffic entering the node is balanced by that which is leaving it. Equation 6.9 applies to the source node, Equation 6.10 is for Node 2, Equation 6.11 characterizes Node 3, Equation 6.12 relates to Node 4, and Equation 6.13 describes Node 5. Node 6 is redundant (the same as Node 1), and therefore not present. One might in fact go directly to each node and write these balance equations at once on purely physical grounds without recourse to the model equation $\mathbf{I} = \mathbf{A}^T\mathbf{X}$, but it is comforting to know that a general model theory comprehends all such networks.

An explicit solution of the network equations must await a general methodology which will be described in complete detail in Chapter 10. In the interim, the methods of cut-sets and dual networks will suffice.

The practical outcome of the exercise is the determination of the network capacity, C, and the individual branch traffics, X_{ij}, for all branches.

Note especially that the direction of flow is also determined for those links which were ambiguous at the start. This is a bonus feature of the analysis.

6.8 MULTIPLE SOURCES AND SINKS

Throughout the previous portions of this chapter it is assumed that the networks which are being analyzed are such that there is only one source and only one sink. All traffic entering the net comes through one particular node and leaves by a single node. Such an assumption is convenient for the purposes of learning the procedures of network analysis by any of the methods which were exposed in previous sections, but the reality is that most roadway nets have many points of entry and many places where the traffics exit the network.

In fact, the likelihood is that a network which is to be analyzed is a portion abstracted from some more extensive one—one which has multiple entry nodes and many exit points. Actually, it would be impossible to construct a network which had only a single source and a single sink in isolation from the surrounding world.

Figure 6.19 shows a simple net with two sources and two sinks. This net can be converted into one with a sole source and sole sink by creating the *supersource* and *supersink* shown in Figure 6.20, where it is necessary to allow the links connecting the supersource and the actual entry nodes to have as large capacities as may be needed to carry the traffics indicated. A similar assumption would need to hold for the links joining the actual exit points to the supersink.

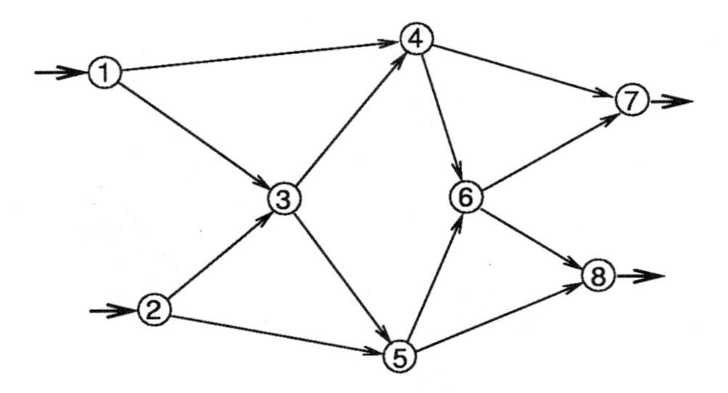

Figure 6.19: Multi-Source, Multi-Sink Traffic Net

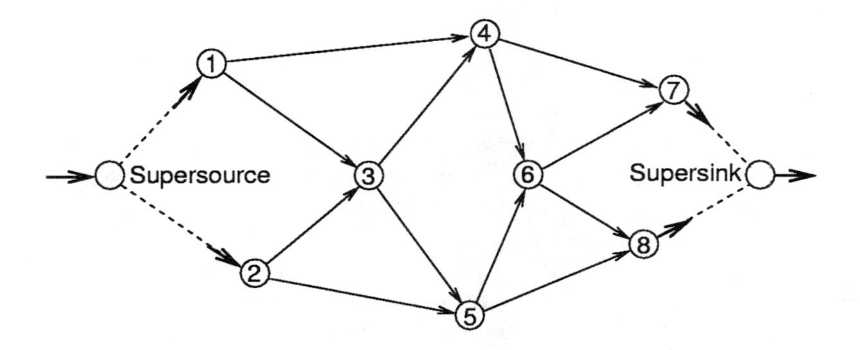

Figure 6.20: Supersource, Supersink Traffic Net

With these adaptations, the analysis of the supersource-to-supersink network is carried out as previously described for a single-source, single-sink net.

6.9 PROJECT #4

1. The linear graph model of a highway network is illustrated below. The legend $(a; b; c)$, in which a is the road number, b is the road length in *kilometers* and c is the road capacity in *vehicles per hour*, is applied to each of the roads in the network.

   - Use the cut-set method to determine the capacity of the net. Check your work by using the dual net and chain analysis.
   - Suppose the traffic in Road 5 is reversed. What will be the capacity of the net in this case?
   - What is the net capacity if the traffic in Road 4 is reversed?

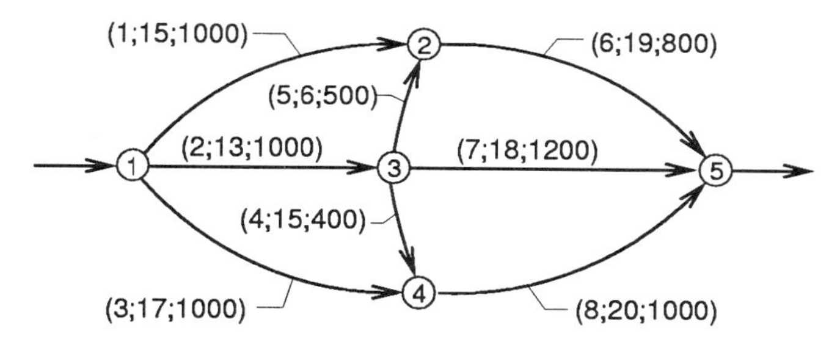

2. A road network is illustrated below, together with the capacities of each road in the net.

   - Use a cut-set method to determine the capacity of the net.
   - Confirm your solution by using a dual net and chain analysis.
   - Discuss the question of whether or not the capacity of the net would be increased by changing the direction of traffic in road number 6.
   - What effect on the net capacity would occur if traffic in road number 4 were reversed?

Road Number	1	2	3	4	5	6	7	8	9
Capacity (vph)	1100	300	900	600	700	500	1200	800	900

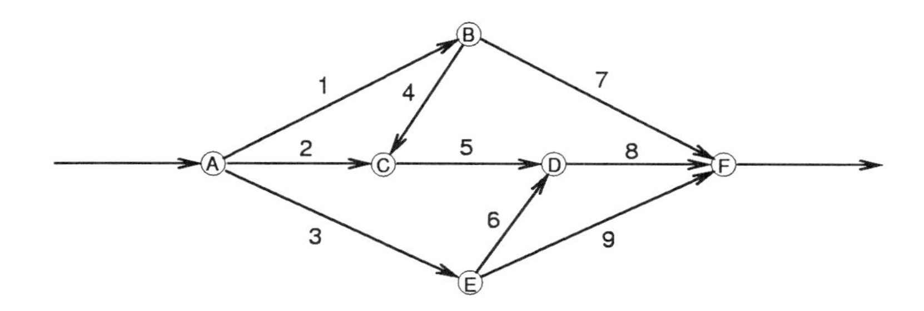

3. Use the method of cut-sets to determine the capacity of the traffic net illustrated. Verify your answer using the dual net and chain analysis.

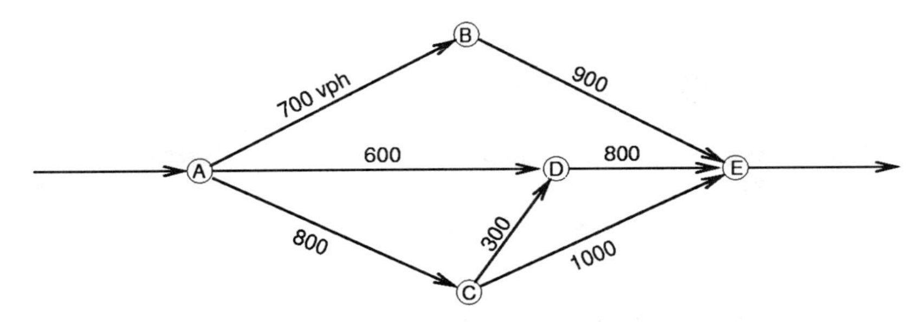

4. Use the cut-set method to find the capacity of the road network shown. Confirm your answers by using the dual network and a chain analysis.

Road Number	1	2	3	4	5	6	7
Capacity (vph)	3000	2300	2000	1700	2500	1700	2000

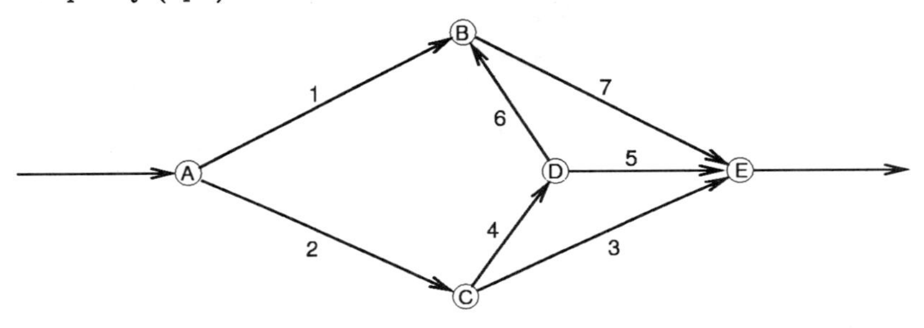

5. Write the matrix model equation for the road net illustrated. This model will be similar to Equation 6.8. Then, obtain the set of linear equations comparable to those of Equations 6.9, 6.10, 6.11, 6.12, and 6.13, the solution of which establishes the traffics in each road of the net.

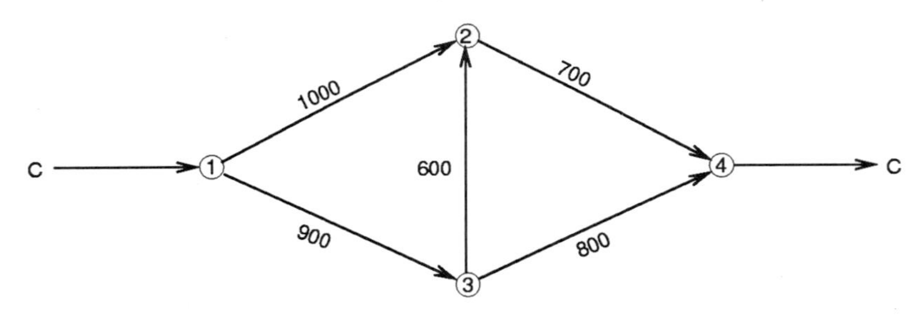

6. The linear graph model of a highway network is illustrated in the figure. The legend (a;b;c), in which *a* is the road number, *b* is the road length in *kilometers*, and *c* is the road capacity in *vehicles per hour*, is applied to each of the roads in the network.

   - Use the cut-set method to determine the capacity of the network. Check your work by employing the dual net and a chain analysis.

- Suppose that the traffic in Road 6 is reversed. What will be the capacity of the net in this case?

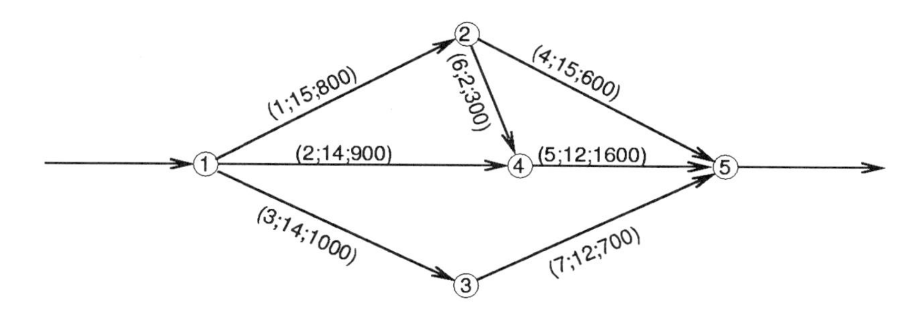

7. Use the cut-set method to find the capacity of the road network shown. Confirm your answers by using the dual network and a chain analysis.

Road Number	1	2	3	4	5	6	7
Capacity (vph)	3000	2300	2000	1700	2500	1700	2000

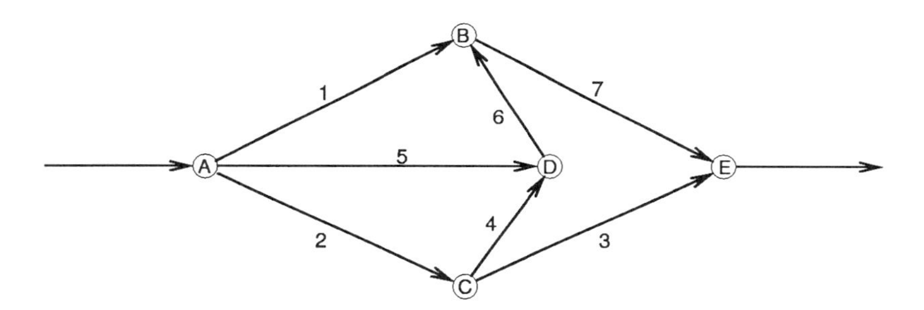

8. Write the matrix model equation for the road net illustrated in 1, above. This model will be similar to Equation 6.8. Then, obtain the set of linear equations comparable to those of Equations 6.9, 6.10, 6.11, 6.12, and 6.13, the solution of which establishes the traffics in each road of the net.

9. Write the matrix model equation for the road net illustrated in 2, above. This model will be similar to Equation 6.8. Then, obtain the set of linear equations comparable to those of Equations 6.9, 6.10, 6.11, 6.12, and 6.13, the solution of which establishes the traffics in each road of the net.

10. Write the matrix model equation for the road net illustrated in 3, above. This model will be similar to Equation 6.8. Then, obtain the set of linear equations comparable to those of Equations 6.9, 6.10, 6.11, 6.12, and 6.13, the solution of which establishes the traffics in each road of the net.

11. Write the matrix model equation for the road net illustrated in 4, above. This model will be similar to Equation 6.8. Then, obtain the set of linear equations comparable to those of Equations 6.9, 6.10, 6.11, 6.12, and 6.13, the solution of which establishes the traffics in each road of the net.

12. Write the matrix model equation for the road net illustrated in 6, above. This model will be similar to Equation 6.8. Then, obtain the set of linear equations comparable to those of Equations 6.9, 6.10, 6.11, 6.12, and 6.13, the solution of which establishes the traffics in each road of the net.

CHAPTER 7

CPM AND PERT

7.1 INTRODUCTION

CPM and PERT are acronyms for *Critical Path Method* and *Program Evaluation and Review Technique*. These are devices which have been created to facilitate the management of large projects. Essentially, they arise out of the logistics problems associated with complex systems and the need to know the status of and allow control of such mammoth works. Their origins can be traced to times of national emergency in World War I and World War II when it was necessary to create, train, supply, transport, deliver to far-flung theaters of operation, re-supply, maintain, and command large field armies, great task-force fleets, and monster air armadas. Both are network models.

Following World War II, the successful aspects of prior methodologies were incorporated into refined schemes applicable to industrial enterprises as well as weapons systems. The exegencies of the Cold War, especially the missile gap (real or imagined), the space race, new global military crises in Korea, Cuba, Vietnam, the Middle East, and elsewhere, gave further impetus to their introduction into the fabric of contemporary society.

Precisely because huge projects require enormous resources drawn from a limited supply of the same, and just because real-time operations within such projects are governed by the clock, there needs to be a management scheme which will insure that the outcome is satisfactory according to criteria established in advance. Failing this, the ongoing management methods must provide for and allow mid-course corrections to assure an optimum performance.

PERT and CPM purport to fulfill these roles. Both are drawn from the repertory of that larger, more encompassing realm called Operations Research, a term used to refer to the discipline which specializes in the study of logistic systems.

7.2 CPM

The Critical Path Method emerged out of a program begun at E. I. DuPont de Nemours Company in 1956. Mr. M. R. Walker, a DuPont engineer, and Mr. J.

E. Kelley, Jr., a computer specialist from Remington-Rand, collaborated in the development of the method, which they reported in two papers in the late 1950s.[1]

Civil Engineers have adopted the Critical Path Method with something of a vengeance. It is ideally suited to many of the activities associated with planning, design, construction, and operation in fields where Civil Engineers specialize. Some even consider it to be the Civil Engineer's own engineering science. Some have gone too far in claiming more for it than it can reasonably deliver. As with all technologies, there are desirable attributes and weaknesses.

There are the following specific purposes associated with CPM:

1. To encompass the definition, planning, and scheduling of projects in such a way as to proscribe the overall effort that must be committed to them.

2. To facilitate the detailed planning of particular activities and sequences through which the projects will be accomplished.

3. To elucidate the resources required and to specify the time frames through which special or scarce items will be needed or employed.

4. To frame policies under which resource allocations can be committed to tasks crucial to the project.

These purposes naturally support the evolution of the method into a sequence of time-related logics. In other words, a schematic algorithm whose elements are tasks (or activities) and times (beginning, ending).

Times, both beginning and ending, are represented by nodes (small circles), and tasks by directed line segments that proceed from one node to another one. In this sense, the CPM diagram somewhat resembles a traffic network. When the model is constructed in this fashion, it is called "activity on the path" method.

Alternatively, the amount of an activity can be accumulated and associated with the node itself, so that the path or line segment connecting the nodes then becomes purely a logic scheme to tell what precedes what (or what follows what). This latter form, which is called "activity on the node," is generally preferred by those who work with computers, because it lends itself so much more readily to machine utilization.

Tradespeople and craftsworkers, on the other hand, often prefer the "activity-on-the-path" method, since it is somewhat more inclined to display tasks graphically.

There are a great many excellent books available which describe both activity-on-the-path and activity-on-the-node versions of CPM in exhaustive detail, and it is not the purpose of the present treatment to duplicate those other works. Our purpose is, rather, to display the essence of the method for the sake of an introduction to it. Therefore, only the activity-on-the-node method will be pursued further here.

---

[1]M. R. Walker and J. E. Kelley, Jr., "Critical Path Planning and Scheduling," *Proceedings of the Eastern Joint Computer Conference*, Boston, Massachusetts, 1959, pp. 160-173, and M. R. Walker and J. S. Sayer, "Project Planning and Scheduling," Report No. 6059, E. I. Dupont de Nemours and Company, Wilmington, Delaware, March 1959.

7.3 ACTIVITY ON THE NODE

One who stands aside from the progress of a construction project and observes the scene of activity taking place—perhaps from the vantage point of a slight rise or hillock or better yet through a knothole viewport in the traditional sidewalk superintendent's stance—will note that various works take place in their special seasons, each appropriate to the nature of the task. Sometimes the progress seems slow, while at other times there appears to be a veritable swarm of people and machines, so many in fact that one has to wonder how can interference possibly be avoided.

It is eminently clear that a mastermind—someone totally and thoroughly familiar with the several tasks which must be done to bring the project from nothing at the start into something at the end—has been at work, long before anyone else set foot on the premises. The project planner must indeed be an experienced veteran of many similar previous projects, or at least someone who can visualize in advance a probable sequence of operations and the relative amounts of time needed for their completions.

At the heart of the planner's function is the necessity of establishing those things which must come first, those things which must wait upon other things for completion before they themselves can be begun, and a third category, those things which may proceed in a parallel manner with other things without interference.

The master-planner will enumerate the work tasks and reduce the list to hard-copy form; i.e., put it down on paper. Of course, this will be done at some rather high level in the hierarchy of systems within the total project. At subsequent levels of the hierarchy, other planners will do a similar thing, always within the framework of the appropriate level in question, and always in increasingly finer detail. *The choice of the appropriate hierarchical level for the initial structure of the project plan is perhaps the overall planner's finest and most demanding function.* The right sculpturing of this plan at this phase of the project may have much to do with its successful completion in an effective and efficient manner.

In general, the plan will include two other columns in the hard-copy listing; namely, the estimated duration of each of the work tasks, and the cost associated with each. The usual custom at this stage is to assume that the resources will somehow be furnished to carry out the tasks, and when the schematic of the total project plan is in hand, separate added schedules of these resources will be furnished, as based upon the master diagram.

7.4 BOOK TOWER PROJECT

Consider the example of a nine-story book-tower addition to the main library on a university campus. The existing library has an original wing (two stories above ground with an imposing entrance, all built to monumental proportions, as is customary with named structures at a university), a separate wing (also two levels above grade, with a third terrace level below grade on one end but at grade on the other because of steeply-sloping terrain, formerly a student union, in a very contemporary design, used as an undergraduate reading library), and an existing

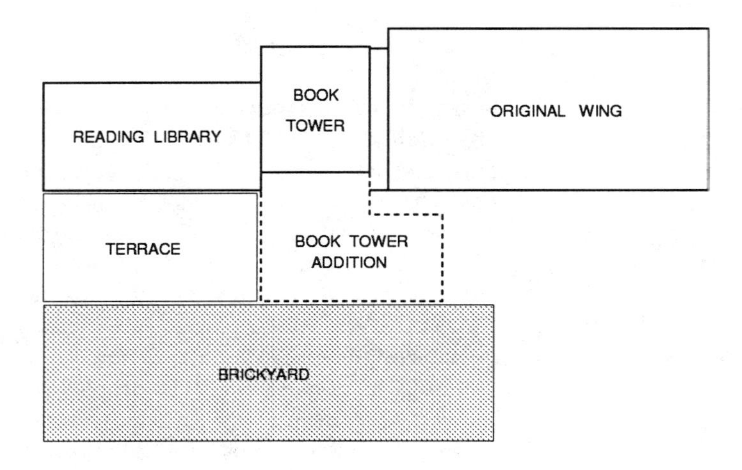

Figure 7.1: Plan View of Book Tower

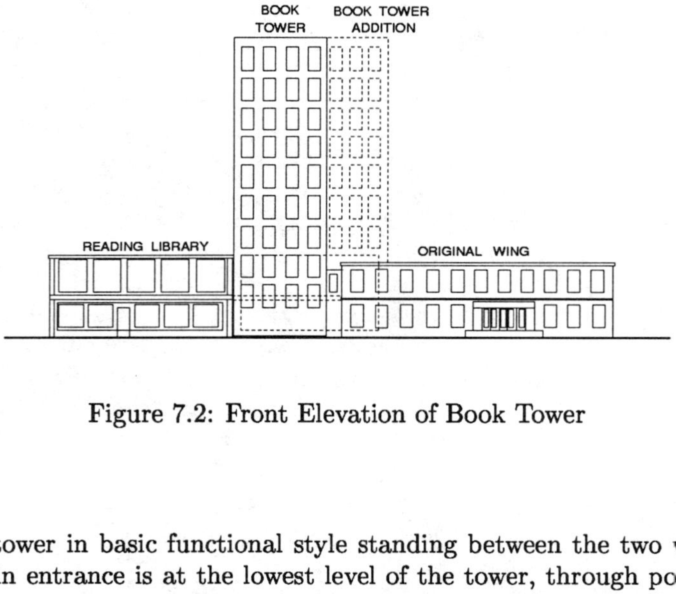

Figure 7.2: Front Elevation of Book Tower

nine-story tower in basic functional style standing between the two wings. The present main entrance is at the lowest level of the tower, through portals facing the site where the new tower will be added. See Figures 7.1 and 7.2.

A complication of the project is that there is a subterranean branch flowing directly under the site of the new tower, and there will need to be pilings driven to support the foundation. The tower also faces a plaza paved in patterned brick, and appropriately called "the brickyard," a place where celebrations following athletic victories, polemic harangues, and major congregations occur. Permission has been granted to use the brickyard as the construction site venue.

The planner for this project might list the following major activities associated with its completion.

	Work Task	Duration (months)	Cost ($\$\times 10^3$)
1.	Create construction traffic entryway, fence-in the brickyard, remove and stack brick, locate office trailer	1.0	8.0
2.	Excavate for drainage and main utility tunnels	2.0	18.0
3.	Locate pilings, bring in pile driver, drive pilings, take out pile driver	1.5	95.0
4.	Excavate, form, install rebars, pour and cure foundation	1.0	80.0
5.	Form, pour, cure 1st level columns, lay reinforcement, place concrete, cure 1st level floor slab	1.0	40.0
6.	Install crane and elevator	0.5	50.0
7.	Form, pour, cure 2nd level columns, lay reinforcement, place concrete, cure 2nd level floor slab	1.0	40.0
8.	Make 3rd level columns, floor slab, 1st level walls	1.0	40.0
9.	Make 4th level columns, floor slab, 2nd level walls, 1st level utilities	1.5	50.0
10.	Make 5th level columns, floor slab, 3rd level walls, 2nd level utilities, 1st level windows and doors	2.0	56.0
11.	Make 6th level columns, floor slab, 4th level walls, 3rd level utilities, 2nd level windows and doors, 1st level interior finish	2.5	68.0
12.	Make 7th level columns, floor slab, 5th level walls, 4th level utilities, 3rd level windows and doors, 2nd level interior finish	2.5	70.0
13.	Make 8th level columns, floor slab, 6th level walls, 5th level utilities, 4th level windows and doors, 3rd level interior finish	2.5	70.0
14.	Make 9th level columns, floor slab, 7th level walls, 6th level utilities, 5th level windows and doors, 4-th level interior finish	2.5	70.0
15.	Make 8th level walls, 7th level utilities, 6th level windows and doors, 5th level interior finish	1.5	58.0
16.	Make 9th level walls, 8th level utilities, 7th level windows and doors, 6th level interior finish	1.5	55.0
17.	Do 9th level utilities, 8th level windows and doors, 7th level finish, construct roofing	2.0	69.0
18.	Install heating, ventilating, air conditioning (HVAC), remove crane	1.0	180.0
19.	Do 8th level windows and doors, 7th level finish	0.5	40.0
20.	Do 9th level windows and doors, 8th and 9th level finishing	0.5	44.0
21.	Remove elevator, clean exterior walls, grade site, cleanup, replace brickyard, landscape	2.0	100.0
		31.5	1301.0

In the example above, the work tasks are listed in the order in which they would need to be done, and it is easy to understand that for a linear vertical

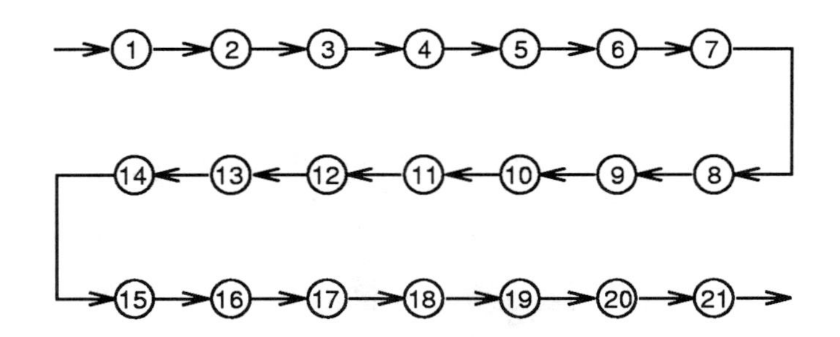

Figure 7.3: Book Tower Construction Project Logic

construction project such as this, the tasks are also in a rigid order of process. The seventh floor, for example, cannot simply be made and left hanging there in air while the fourth floor, for instance, is being made. That is, these tasks must follow each other in a *seriatim* manner. Accordingly, the schematic diagram representing the logic of the construction is simply a series of nodes joined by links in a pearl-necklace model string. See Figure 7.3 for details.

On the face of it, the earliest completion time for the project is 31.5 months. The total project cost is $1,301,000.

7.5 THE CRITICAL PATH

Suppose the project is one which allows several activities to proceed at the same time, but has several other tasks which must await the completion of others. The project planner will again list these tasks, together with their cost estimates and times of accomplishment, but will also name those other tasks which must preceed them, and those which will follow each. The schematic diagram is once more laid out, showing these orders of precedence and succession. Say that the diagram is given by Figure 7.4, below.

A Start node is provided, which corresponds to the time that a "stopwatch" for the project is clicked on. No length of time or passage of time is associated with this node. Node 1 is affiliated with Activity 1, and Node 2 follows on upon the completion of the activity at Node 1. Activities 3 and 4 can not begin until 2 is completed, but then both may begin and proceed in parallel. Activity 5 must await completion of both Activities 3 and 4, however long they may take, and irrespective of whether they require equal or differing lengths of time.

Seven follows 3 and 6 waits upon 5 to finish. Now activity 8 must hold up until both 6 and 7 are done. Tasks 11, 12, 14, and 15 must all wait until Number 8 is finished. Work 9 cannot begin until 11 and 15 have ended, and likewise 10 holds up until 11 and 12 are consummated. 13 follows 10 and 16 comes after 15 and 13. When Activity 16 has been finished, the project is also done.

Clearly, in this latest example, one cannot simply add up the activity times in order to determine the earliest date of completion of the project. As a matter of fact, each work task will have its own earliest time of initiation (time elapsed since the stopwatch started running), and it will necessarily be the longest of the

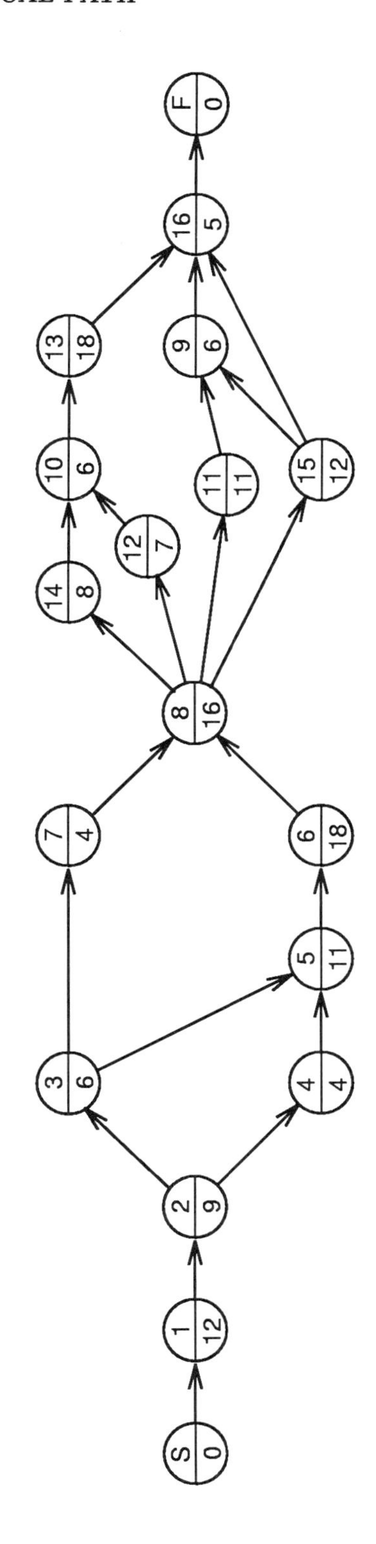

Figure 7.4: Construction Project Schematic Diagram

sums of the activities on all preceeding paths to it. Its earliest time of completion can then be had by adding the task duration to the earliest start time.

In a similar fashion, each work task must be completed by a certain latest time in order not to hold up those tasks which succeed it. That much is clear from the diagram.

The total project duration is evidently the longest time from Start Node to Finish Node by any path through the diagram. This longest path is obviously more important than any other path through the activities, simply because it does indicate the earliest project completion. It is accordingly named the *Critical Path*, and is usually emboldened, printed in red, or otherwise emphasized graphically; it is also named by the sequence of nodes which define it.

The only way in which the total project time to completion can be shortened is to reduce the length of the Critical Path. Correspondingly, any delay of a task on the Critical Path will attenuate the total project. These facts serve to increase the importance of this trajectory.

Moreover, the knowledge that there is a critical trace through the network of tasks raises the importance of knowing the latest start times of each task, lest one of these drag out long enough to delay the start of a Critical Path work. To establish the latest start times, begin at the Finish Node and work backward through the diagram, subtracting the duration of a task from the succeeding task's latest start time. Do this for all possible paths that lead backward from the Finish Node to the Start Node.

Of course, all of those nodes on the Critical Path will have the same earliest start time and latest start time. That is, there is no slack along this path, whereas with parallel paths there may be an opportunity for some tasks to wait for a season, so long as it is not later than the latest start time.

Actually, there are four time marks of importance for each task in the project. These are its *Earliest Start Time* (EST), *Latest Start Time* (LST), *Earliest Finish Time* (EFT), and *Latest Finish Time* (LFT). The Latest Start Time is obtained from the Latest Finish Time for a given task by subtracting its duration from the LFT, and the Earliest Finish Time comes from adding the duration to the Earliest Start Time for the task. All of these times would like to be known by the project manager, because they provide the working margins through which progress can be maintained.

It is the usual practice to create a legend which can be applied to each node in the network so as to show the numbers which are relevant to that node. In addition to EST, EFT, LFT, and LST, the task duration, the preceding task on the earliest start path, and the total slack are wanted for each node. This working room for tasks not on the critical path is called the *float*; i.e., that task is left "floating" between its earliest start time and its latest finish time. It might be adjusted to begin and end anywhere within this region. The diagram has thickened like a sauce in no time at all.

Some of the float at each node may be free to adjust without delaying the total project completion. This amount is designated *free float*. There is also a portion of the slack which may interfere with other work tasks, and is thus called the *interfering float*. The elementary formulas below define the total float, free float and interfering float for an activity in the diagram in terms of its own earliest start

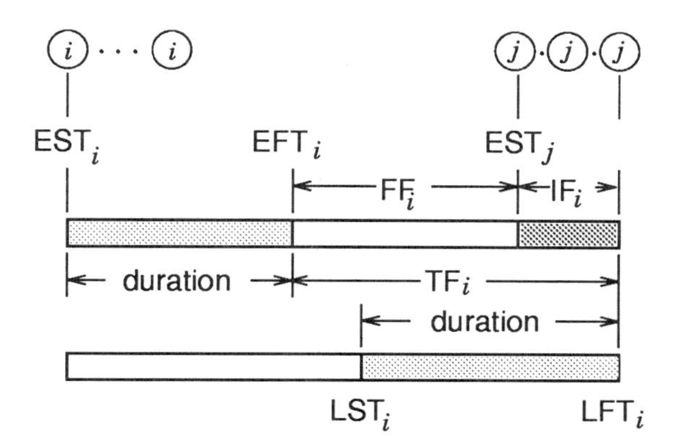

Figure 7.5: Float Diagram

time and latest start time, and the earliest start time and latest start time for the successor activity. Call these activity i and activity j. As the little diagram, Figure 7.5, shows,

TOTAL FLOAT	$TF_i = LFT_i - (EST_i + DUR_i)$
INTERFERING FLOAT	$IF_i = EST_j - LFT_i$
FREE FLOAT	$FF_i = TF_i - IF_i$

so these numbers can be included in the legend box at each node. When there are multiple successor activities, each must be searched to determine that one which has the *earliest* start time.

Figure 7.6 illustrates the current state of the diagram.

Figures 7.7 and 7.8 give the project schedule.

7.6 NETWORK COMPUTER PROGRAMS

It seems self-evident that computers are ideally suited to the task of determining the various network parameters enumerated in the section above. Indeed, various forms of computer programs have been invoked for this purpose for about four decades, now. These have been developed and refined to a very high state of utility, and almost every design office and construction site office trailer will be equipped with workstations and PCs and their relevant softwares for accomplishing the tasks of network analysis.

There are many commercially available programs for CPM work. Two of the more prominent ones are *PRIMAVERA*© and *QUIKNET*©. Such programs do a lot more than just compute the times in question for given networks, of course. They provide the user with the ability to investigate effects of altered durations of selected activities, so that the managerial function can be greatly enhanced. A project supervisor can estimate the effect on the overall network critical path of a change in the duration of one or more activities by the application of additional

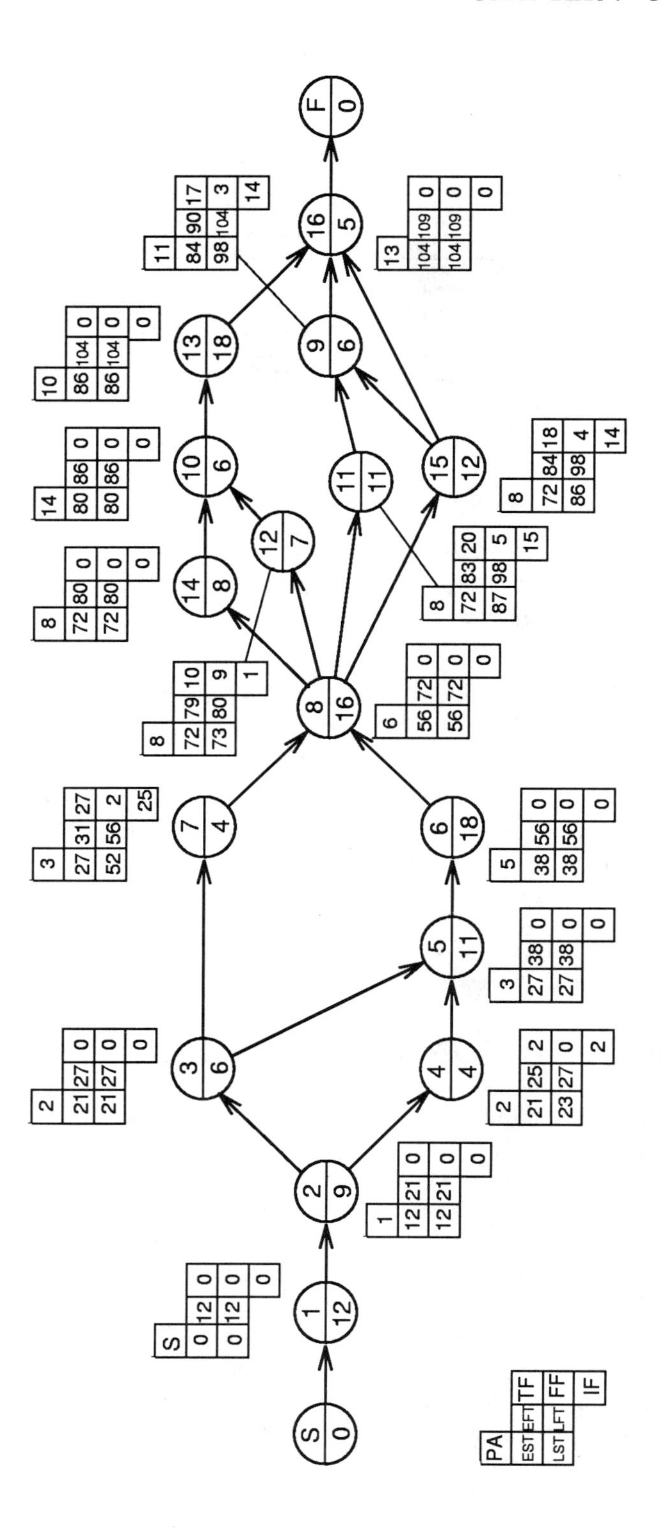

Figure 7.6: CPM Diagram

PROJECT SUMMARY

Task No.	Duration Days	Cost $	Start Time Days
1	12	10,000	0
2	9	6,000	12
3	6	12,000	21
4	4	8,000	21
5	11	10,000	27
6	18	30,000	38
7	4	50,000	27
8	16	60,000	56
9	6	60,000	84
10	6	15,000	80
11	11	14,000	72
12	7	2,000	72
13	18	48,000	86
14	8	30,000	72
15	12	78,000	72
16	5	6,000	104
		$439,000	104

Figure 7.7: Schedule

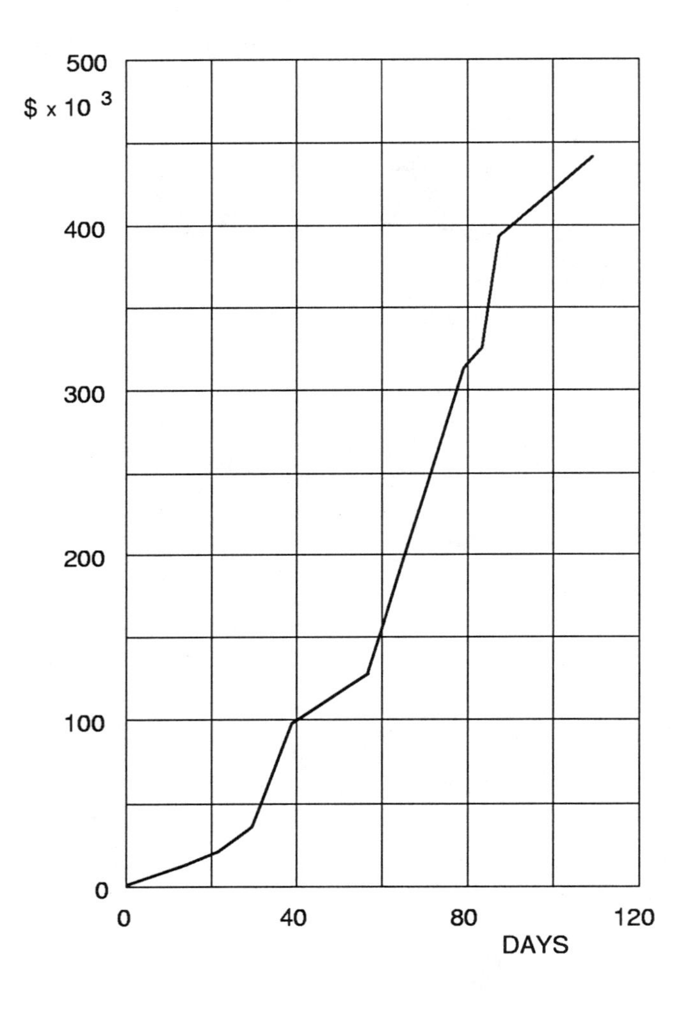

Figure 7.8: Cash-Flow Schedule

resources. Or the results on the project's finish time caused by stoppages and delays from weather, cash-flows, labor problems, and the like can be investigated.

In fact, the project manager will no doubt maintain a daily (or even hourly) update of the network's parameters as the actual work progresses.

The available programs present the net status in a readily understandable mode, accompanied by colorful visual effects and appropriate printouts. There are many "bells and whistles," and "dog and pony shows" for visiting personages are enabled.

For initial student learning of CPM methods, a much more modest computer capability is indicated. A simple program, *cpm.go*, has been prepared and placed in the locker *mcdonald_src* on *eos* net for this purpose. It is keyboard interactive and "user friendly." When called, the program asks how many activities (nodes) there are, the duration of each activity in turn, and the predecessors of the node. These data inputs are then echoed in tabular form on the monitor, so that they can be confirmed. As the run continues, the complete network analysis is furnished, again in tabular form on the monitor, and the user is asked if a hard copy is desired, and whether or not another run may be wanted.

The printout is made to the file *cpm.dat*, which must be printed in order to see the results on paper. The file *cpm.dat* must be erased before a new call of the program will be successful.

The printout for the network of Figure 7.6 is displayed in Figure 7.9.

7.7 THE GANTT CHART

The technological ancestor of CPM and PERT is a device called the "Gantt Chart," which was invented by Henry Gantt about the turn of the century. It was extensively used to schedule production processes in industrial manufacturing in the early decades of this century, and even down to the present time.

The Gantt chart for the example in the preceding section is shown in Figure 7.10. Actually, it is a bar chart organized in such a way that a project manager or other person can look at it and gain an overall view of the project's scope, duration, and progress. It is intended to furnish an explicit guide to the review of the work as it progresses, and during a time when corrections can be made which will ensure its on-schedule completion. The scope and schedule of the resources (personnel, materials, capital equipment, and money) are visually highlighted.

The Gantt chart is ideally suited to contemporary desk-top computer displays. It is often included in software packages which are designed to run CPM analyses, so that the added visual emphasis makes a stronger impact on the user.

However, the Gantt chart does not display the interdependencies between the activities represented by the individual bars, nor does it allow for changes of plan or slippages in schedule because it is based upon a calendar scale. Further, it does not permit the inclusion of uncertainty or variability in duration of the several activities which compose the project. These weaknesses came to the forefront during the 1950s, which was a time of great industrial expansion and initiation of defense systems on a global scale.

Accordingly, the pressures associated with management of large-scale projects in industry and defense agencies produced CPM and PERT.

NODE	PREDECESSORS				SUCCESSORS				DUR	EST	LFT	LST	EFT	TF	IF	FF
1	0	0	0	0	2	0	0	0	12	0	12	0	12	0	0	0
2	1	0	0	0	3	4	0	0	9	12	21	12	21	0	0	0
3	2	0	0	0	5	7	0	0	6	21	27	21	27	0	0	0
4	2	0	0	0	5	0	0	0	4	21	27	23	25	2	2	0
5	4	3	0	0	6	0	0	0	11	27	38	27	38	0	0	0
6	5	0	0	0	8	0	0	0	18	38	56	38	56	0	0	0
7	3	0	0	0	8	0	0	0	4	27	56	52	31	25	25	0
8	6	7	0	0	11	12	14	15	16	56	72	56	72	0	0	0
9	11	15	0	0	16	0	0	0	6	84	104	98	90	14	14	0
10	12	14	0	0	13	0	0	0	6	80	86	80	86	0	0	0
11	8	0	0	0	9	0	0	0	11	72	98	87	83	15	15	0
12	8	0	0	0	10	0	0	0	7	72	80	73	79	1	1	0
13	10	0	0	0	16	0	0	0	18	86	104	86	104	0	0	0
14	8	0	0	0	10	0	0	0	8	72	80	72	80	0	0	0
15	8	0	0	0	9	16	0	0	12	72	98	86	84	14	14	0
16	9	13	15	0	0	0	0	0	5	104	109	104	109	0	0	0

PROJECT DURATION = 109

Figure 7.9: Computer Printout

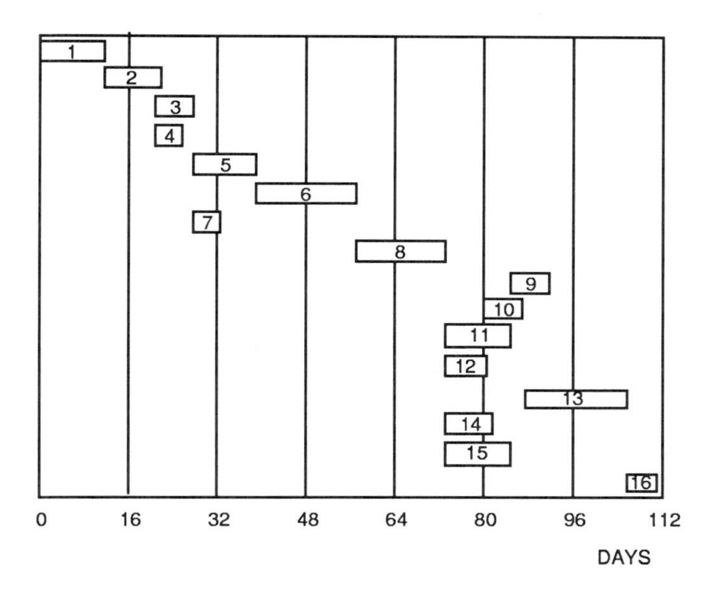

Figure 7.10: Gantt Chart

7.8 PERT

In 1957 the Special Projects Office, Bureau of Ordnance, U.S. Navy, faced the daunting challenge of developing the POLARIS missile system or Fleet Ballistic Missile program. This initiative called for the creation of a huge, complex weapon system employing many elements at or beyond the then state-of-the-art of technology, and distributed over literally hundreds of scientific-technological organizations engaged in research, design, and manufacturing, as well as testing, certifying, and deploying the system. Worst of all, the basic theme of the project was "confluent development," which means that various elements would be produced in parallel streams, then joined together at some later stages of the project.

Obviously, such a program requires a scheduling and management method which accentuates uncertainty and variability of activity times or durations. In addition, the enormous draft of resources looming over such a project requires that every effort be made to optimize the entire program. Consequently, the Special Projects Office set up a group composed of its members and Booz, Allen, and Hamilton, which is a management consulting firm, and charged the team to produce the method. The result was PERT, described in two papers.[2] These reports were widely distributed throughout the military-industrial establishment, and made a definitive impact reflected in the burgeoning number of applications reported by the early 1960s.

PERT is in many ways synonymous with CPM. Both are network methods,

[2]"Program Evaluation Research Task Summary Report, Phase I, and Phase II," Special Projects Office, Bureau of Ordnance, Department of the Navy, Washington, D.C., July, 1958, pp. 646-669.

both incorporate descriptive times for project activities, and both seek to optimize the project length and resource utilization. CPM is the progeny of well-established processes in the manufacturing and construction industries where activities are accurately defined as to both time and resources. PERT, on the other hand, was born of national emergency, featured in military-industrial-research-technology venues, with uncertain—oftentimes unknown—times and resources. It therefore emphasizes uncertainty and variability in scheduling.

7.9 EXPECTED TIMES

The central problem in PERT is the determination of the *expected time* for an activity, and, of course, the compounding of this variability with those of the other activities in the network. What it comes down to is assessing the *most probable time*, given a number of estimates made in advance by, perhaps, a number of different persons of varying experience. In the prototypical case, there will be *no* experience of record, so that judgments of activity times must be extrapolated from similar or related ones.

A fundamental postulate of the PERT algorithm is that, in the face of uncertainty, the duration of an activity can be given an *optimistic*, or a *pessimistic*, or a *most frequent* completion time estimate. The optimistic estimate assumes the shortest time for completion of the activity, without any delays, under ideal conditions. The pessimistic estimate furnishes the longest time required to finish, and assumes that everything which can go wrong, will do so. It is customary to exclude hurricanes, strikes, insurrections, and other catastrophies of this ilk from consideration in arriving at the pessimistic estimate.

The most frequent estimate lies somewhere within the range defined by the optimistic and pessimistic ones. The assumption here is that some things go wrong, as they almost always seem to do, but that some things go well, as they sometimes do, and overall, things are *normal*. Underlying the business of estimating the expected time is the concept of probability. If there are repeated acts of estimating this time, say by a group of individuals, then there will be a distribution of these times which reflects the spectrum of the estimators' bases for the estimates. Such a distribution is shown in Figure 7.11. This *histogram* is typical of statistical or probabilistic frequency distribution samples.

The histogram is the discrete embodiment of the continuous frequency distribution of the universe behind the estimating phenomenon. That is, if a very large number (theoretically an infinite number) of people were asked to make such an estimate for a given activity, there would be a graph similar to that shown in Figure 7.12.

If such a distribution were available, statistical methods could be applied to the determination of the most probable completion time estimate. As it is, the sample statistics are not available in quantity, so that a frequency distribution must be *assumed*, rather than computed. The originators of PERT assumed that the *beta distribution* of statistics is the model to use, so they made that selection.[3]

[3]Appendix B, "PERT Summary Report, Phase I," *Special Projects Office, Bureau of Naval Weapons, Department of the Navy*, Washington, D.C., U.S. Government Printing Office, 1962-640679.

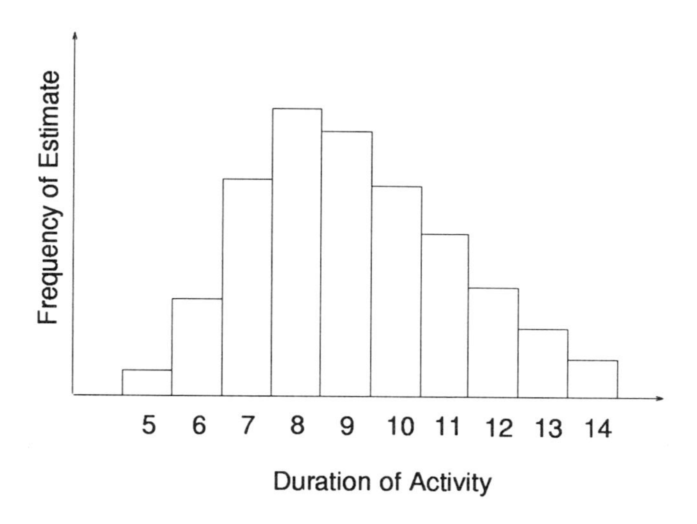

Figure 7.11: Estimates of Activity Duration

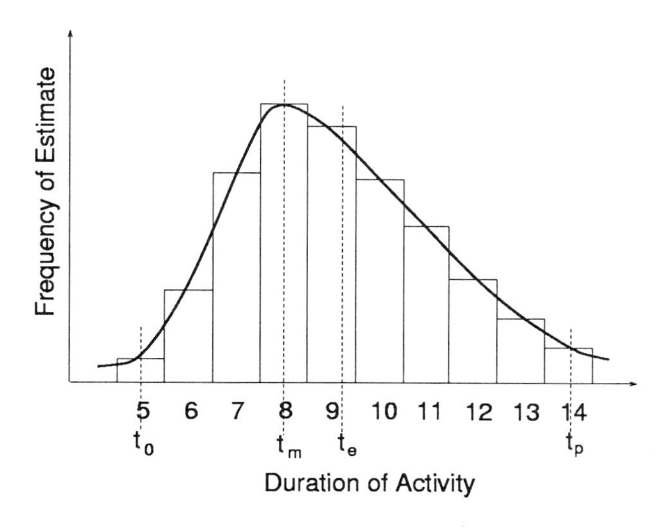

Figure 7.12: Estimators' Frequency Distribution

The derivation of the expected completion estimate is outlined in Appendix D. A more complete discussion of probability and frequency distributions is found in Chapters 8 and 13, as well as Appendix C. The simple equation

$$t_e = \frac{t_o + 4\,t_m + t_p}{6} \tag{7.1}$$

has come to be known as *the PERT formula* for determining the activity duration.

7.10 RELIABILITY OF EXPECTED TIMES

A second parameter available in statistics is a measure of the *variance*, *standard deviation*, or "spread" of the expected time, as determined above. Again, this parameter can be computed for a given frequency distribution, if its equation is known. In PERT the calculation of the standard deviation is approximated by

$$\sigma = \frac{t_p - t_o}{6} \tag{7.2}$$

and the variance is then its square

$$S = \left(\frac{t_p - t_o}{6}\right)^2 \tag{7.3}$$

Each activity in a network will have its own expected time of completion, as calculated above. If the activities are independent of each other, as they are assumed to be, then the *expected length* of a sequence of such activities is the sum of their individual expected times. The project then has an expected duration T_e, which is this sum.

Likewise, each activity will have its individual variance for its expected time. The variance of the project time, T_e, equals the sum of the individual variances of the sequence of activities in the project. This, or the standard deviation, is a measure of the uncertainty which clouds the expected time computation.

The expected length and variance of a sequence of activities as the sum of individual expected lengths and sum of individual variances is a consequence of the *Central Limit Theorem* of statistics.[4]

It should be pointed out at once that the project expected time T_e may sometimes fail to be the most accurate estimate. This may occur, for example, when a particular sequence of activities, with particular activity times and variances, may yield a computed variance greater than that of the designated critical path. If there is bad luck on this "near-critical" path, and good fortune with most of the activities on the critical path, then the actual length of time on the near-critical path will exceed that on the critical path. Therefore, it should be assumed that for these circumstances, alternate critical paths are present. This situation is similar to that in which additional resources are applied to certain activities in a network to shorten the critical path, with the result that a near critical path then becomes the critical path.

---

[4]J. J. Mader, C. R. Phillips, and E. W. Davis, "Project Management with CPM, PERT, and Precedence Diagramming," Van Nostrand Reinhold Company, New York, 1983, pp. 276, 277.

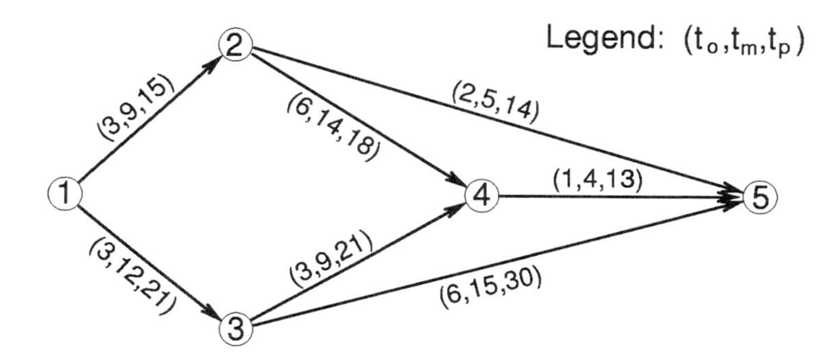

Figure 7.13: Example of PERT

Experience also shows that the simple PERT procedures give estimates which are often overly optimistic.

To illustrate these concepts, refer to the simple network shown in Figure 7.13.

The expected times, standard deviations, and variances of the activities are computed from PERT formulas 7.1, 7.2, and 7.3, and entered in the following table.

Activity	Expected Time t_e	Standard Deviation σ	Variance S
1→2	9	2	4
1→3	12	3	9
2→4	10	2	4
3→4	10	3	9
4→5	5	2	4
2→5	6	2	4
3→5	16	4	16

The critical path is defined by the longest sequence of expected times, which is activities 1→3 followed by 3→5. Its expected length is $T_e = 12 + 16 = 28$. The standard deviation and variance on the critical path are $\sigma = 5$ and $S = 9+16 = 25$.

Path 1→2, 2→4, 4→5 has $T_e = 9+10+5 = 24$, $S = 4+4+4 = 12$, $\sigma = \sqrt{S} = \sqrt{12}$, and expected slack of 4. Path 1→3, 3→4, 4→5 has $T_e = 12 + 10 + 5 = 27$, $S = 9 + 9 + 4 = 22$, $\sigma = \sqrt{22}$, and expected slack of 1. Path 1→2, 2→5 has $T_e = 9 + 6 = 15$, $S = 4 + 4 = 8$, $\sigma = \sqrt{8}$, and expected slack of 13.

7.11 NETWORK OPTIMIZATION

Activity networks can be presented in a format quite similar to those exposed in the previous chapters for purposes of optimization. Normally, this will mean that the length of the critical path is to be minimized. The process is very much like that associated with the link-node incidence matrix in Chapter 6.[5]

[5] J. E. Kelley, Jr., "Critical-Path Planning and Scheduling: Mathematical Basis," *Operations Research*, Volume 9, Number 3 (May-June 1961), pp. 296-320.

For example, let x_i denote the Early Start Time of activity i. If there are k nodes in the network, then the optimization seeks to minimize the quantity $x_k - x_1$, assuming that there is a single source and single node network.

For each activity in the network, there is a constraint; namely, the time from node i to node j must be equal to or greater than the duration of the activity, d_{ij}. Then the following statement incorporates the optimization:

$$\text{MINIMIZE} \quad x_k - x_1 \tag{7.4}$$

$$\text{SUBJECT TO} \quad x_j - x_i \geq d_{ij} \tag{7.5}$$

$$i, j = 1, \ldots, k$$

The solution of the optimizing equations above will customarily produce the result $x_1 = 0$, since the start node is usually assumed to be of zero duration. If, as is sometimes the case, it is desired to introduce planning or layout time before physical activities begin, then this may be included by specifying a definite value other than zero for x_1.

Two other forms of the optimization can be employed to yield important results. In the first case, suppose that, as before, x_i represents the Early Start Time of activity i, and it is desired to force *each* activity to its EST, then the minimization statement is

$$\text{MINIMIZE} \quad x_k - x_1 + \sum_{i=2}^{k-1} x_i \tag{7.6}$$

$$\text{SUBJECT TO} \quad x_j - x_i \geq d_{ij} \tag{7.7}$$

$$i, j = 1, \ldots, k$$

which can be written

$$\text{MINIMIZE} \quad \sum_{i=1}^{k} x_i \tag{7.8}$$

$$\text{SUBJECT TO} \quad x_j - x_i \geq d_{ij} \tag{7.9}$$

$$i, j = 1, \ldots, k$$

In the second case, should it be desired to force each activity to its LST, the optimization statement is

$$\text{MINIMIZE} \quad m\, x_k - \sum_{i=1}^{k-1} x_i \tag{7.10}$$

$$\text{SUBJECT TO} \quad x_j - x_i \geq d_{ij} \tag{7.11}$$

$$i, j = 1, \ldots, k$$

Herein, the multiplier m serves to translate all x_i to their upper bounds while still holding x_k to its smallest value as found in the first application of the optimization, above. Using m allows the product $m\, x_k$ to always be larger than $\sum_{i=1}^{k-1} x_i$.

When the results of the three applications of the optimization shown above are available, the slacks associated with each activity in the network can be determined. The latter two optimization forms constitute "forward pass" and "backward pass" operations as set out in the prior section on CPM.

All of these calculations can be effected in a computer program in a most expeditious manner, of course.

7.12 PROJECT #5

1. The table below supplies some estimates of the logic elements proposed for the creation of a regional airport.

   Draw the network (activity-on-the-nodes version) that describes the project. For each activity determine the EST, EFT, LST, LFT, and FLOAT.

   What is the Critical Path, and what is the time to finish the project?

Activity Number	Activity Name	Duration (Months)	Predecessor Node Numbers
1	Draw plans and specifications	3	
2	Hold public hearings on project	2	1
3	Incorporate modifications	1	2
4	Clear runway areas	3	3
5	Clear terminal area	2	3
6	Construct drainage systems	2	3
7	Grade and pave runways	8	4
8	Construct terminal building	12	5
9	Construct control tower	2	4
10	Build parking area	1	7
11	Install lighting and fencing	1	6
12	Construct service roads	2	9

   Confirm your solution by using the program *CPM.go* on *eos* net. You may wish to recall that the program writes to an output file *CPM.dat*, and this file cannot be overwritten, so you will need to remove it each time (after printing, of course!) before a new run is made.

2. S. H. E. Diggs, P. E., Contractor, is preparing for a construction project with the following activities:

Activity Number	Duration (Days)	Predecessor Activities
1	6	0
2	3	1
3	4	1
4	5	2,3
5	8	3,4
6	11	4,5
7	14	6,9
8	10	6
9	12	6
10	7	8
11	8	9
12	9	7,10,11

M. Diggs wishes to employ the CPM method to determine the minimum length of time for completion of the project. She also wishes to find the earliest start times, the earliest finish times, the latest start times, the latest finish times, the total float, the interfering float, and the free float for each of the project activities.

Use the activity at the node logic to depict the network of these activities. Construct a label for each node (with a legend in a convenient place) that gives EST, EFT, LST, LFT, TF, IF, and FF for the node. Identify the critical path on the diagram of the network.

Confirm your solution by using the software *CPM.go* in the *eos* locker *mc-donald_src*. Include a print-out with your report. (Note that the program writes to a file *CPM.dat*, and that the existing *CPM.dat* file, if any, needs to be erased before each new run.

3. Listed below are the activities involved in a certain project, together with their durations and predecessor activities. Use this information to construct a CPM diagram based upon activities-on-the-node. Compute total, free, and interfering floats for each activity, and the project duration.

   Please include a legend with your diagram. Please also include a bar chart.

   The program *CPM.go* in the locker *mcdonald_src* on *eos* net may be used to assist in confirming your computations. Please include a print-out with your report.

Activity Number	Duration (Days)	Predecessor Activity
1	0	1
2	10	1
3	8	1
4	6	2,3
5	16	3
6	24	4
7	4	5,6
8	4	7
9	10	7
10	4	7
11	14	8
12	16	9,10,11
13	8	4
14	24	13
15	4	12,14
16	0	15

4. Given durations for the activities in the project below and the immediately-preceeding activities, construct an activity-on-the-node diagram showing: the critical path, activity durations, ESTs, LSTs, EFTs, LFTs, Total Floats, and Free Floats. Be sure to include a legend with your diagram.

Activity Name	Duration (Days)	Predecessor Activities
A	1	
B	5	D
C	3	D
D	6	A
E	2	B
F	1	E, H
G	7	A
H	2	C, I
I	3	G

5. The Black Ankle Reservoir will supply water, provide flood control and recreational activities, and generate electricity. Listed below are the activities and their estimated durations involved in the project.

   Construct the activity-on-the-node network diagram and determine the minimum length of time to finish the work. Show the critical path and compute the slack times for the activities.

Activity Number	Predecessor Activities	Activity Name	Duration (Months)
1	7	Clear dam site	2
2	7	Clear other reservoir area	34
3	1	Construct earthen dam	40
4	1	Construct concrete spillway	20
5	3	Build powerhouse structure	18
6	5	Install generators	10
7	S	Construct access roads	3
8	1	Construct other roads	8
9	8,2	Prepare recreation facilities	34
10	7	Administration unit, utilities	5
11	10	Water treatment	10
12	11	Install plant equipment	12

6. Shown below is an early start network for a construction project. Complete the diagram using the notation $T_i^E = b, c$ in which T_i^E is the earliest start time at node i, b is the earliest start time at the node, and c is the preceding node on the earliest start path. X is the node number, Y is the activity duration at the node.

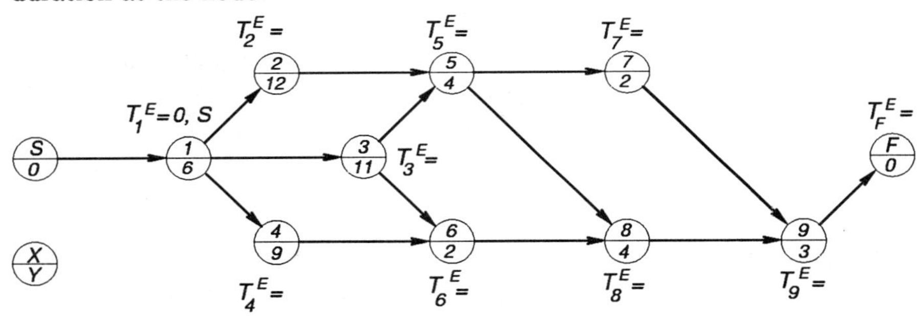

CHAPTER 8

DECISION ANALYSIS MODELS

8.1 INTRODUCTION

Civil engineering projects, unlike those of other branches of engineering, are generally carried out in a theater or venue which is susceptible to all of the vagaries of the weather. There are also indeterminate elements associated with materials and labor supplies, with the maintenance of machinery and with shifting fiscal policies, although these latter are not, of course, unique to civil engineering work.

The point is, that there are many aspects of such projects which, by their very nature, cannot be planned for in an expectable sense, as with the laws of mechanics and other engineering sciences. Some allowance must be made for those things which can never be known in advance, and for those contingencies which may arise during the course of the work. Although some processes are known to be stochastic (fluctuating about a steady or level value) and can be treated with a full degree of statistical confidence, there are also those which can only be assigned certain probabilities—which is to say, the laws of chance pertain.

The methods of Decision Analysis are particularly designed to be applicable to especially such situations. They allow a manager to work at managing a project in the face of both uncertainty, ignorance, and unpredictability. They can be adapted as well to the personal style of the manager.

8.2 ELEMENTS OF PROBABILITY

Probability is a measure of the likelihood that an event will occur, when given a universe of opportunities for the event to be possible. For example, if weather records are kept continuously for, say, 85 years, and during 34 of those years there are no major storms in a given locality, then it would be said that "the probability of no storms is 0.40000," a number obtained as the ratio of 34 to 85. Similarly, if in 42 of the years there is a mild storm, the probability is 0.49412 of a mild storm. Further, suppose that in nine of the 85 years there is a hurricane. The probability of a hurricane is therefore 0.10588.

It will be understood immediately that, as 34 years with no storms, 42 years with mild storms and 9 years with a hurricane add up to the total of 85 years

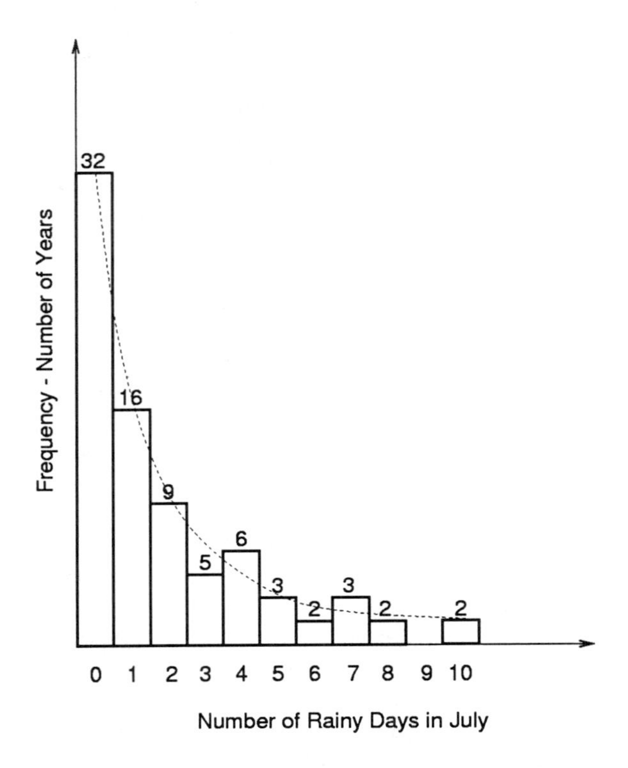

Figure 8.1: Frequency of Rainy Days During July

altogether, so also must the probabilities 0.40000, 0.49412, and 0.10588 add up to 1.00000, which is therefore a measure of the total. In other words, the probability 1.0 is associated with the certainty that, in a given year, there will either be no storm, a mild storm, or a full-blown hurricane.

Figure 8.1 shows a bar graph that depicts the frequency of rainy days during the month of July at a particular locale. An eighty-year history of this phenomenon is available, from which it is noted that during 32 of the 80 years no rain occurred in July. In 16 of the years, there was one rainy day, 9 years had 2 rainy days, 5 had 3 days of rain, etc. One might reasonably construct a curve through the tips of the bars of the bar graph and say that this represents the law of distribution of rainy days in July, for that area.

In general, probability is a continuous function of the parameter called the *argument*. For the example above, the parameter is the number of rainy days in July. The frequency is then said to be a *frequency distribution*. The one shown is often called a "J-distribution," because it looks like the capital letter "J" lying on its side.

Figure 8.2 illustrates a generic frequency distribution for the probability of an argument x. The distribution is given by the continuous function $f(x)$, and as the graph indicates, the probability of finding x in the range from x_1 to x_2 is

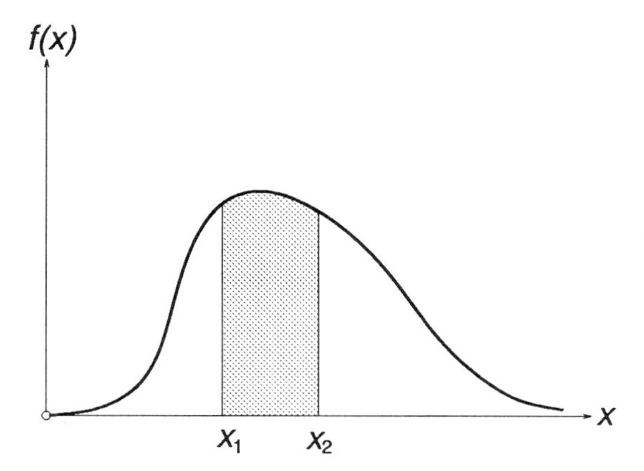

Figure 8.2: Frequency Distribution

$$P(x_1 \leq x \leq x_2) = \int_{x_1}^{x_2} f(x)\, dx \tag{8.1}$$

Obviously, probability is represented by an area under a frequency curve. This is true even if the probability is over a differential or discrete argument, as the rainy-days in July example demonstrates. Moreover, it is clear that the total area under the curve represents the totality of all possibilities; i.e., certainty or a probability of 1.0. Therefore,

$$1.0 = \int_{0}^{\infty} f(x)\, dx \tag{8.2}$$

represents this condition for the continuous distribution and

$$1.0 = \sum_{i=1}^{n} P(x_i) \tag{8.3}$$

holds for the discrete case in which the integral sum is replaced by the discrete summation appropriate to a histogram.

8.3 DECISION TREES

In the Decision Analysis process the decisioneer acts in a frame of reference which is independent of place and time, but is governed by events. One event is followed in the event stream by another event, and at each event there is the possibility of making a decision among certain alternatives available, or of leaving the decision to chance, associated with definite probabilities, or of quitting the game at that point. The structure of the logic thus enumerated resembles that of a tree with many branches, which is why it received that name.

The decision tree begins at its roots with an initial event called a decision node, because the decision maker is presented at that point with a decision between some alternatives. At each such opportunity for decision-making thereafter, the node will be classified as a decision node, and by custom the node is identified as a decision node by using a small square figure to represent it.

Following each decision node the branches lead to other nodes, some of which are called chance nodes, because at those places chance enters the picture to create branches which depend upon the probabilities associated with each possible outcome. Chance nodes and decision nodes may follow each other in whatever order the logic may require for that stage. Chance nodes are customarily represented by small circles.

At the end of the process for each branch of the tree there is an event called a terminal node. It is here that a decision is accepted as final: there is an outcome. Terminal nodes are taken to be small triangles, to distinguish them from the other event nodes.

8.4 ON THE BEACH

A dredging contractor has a job making some boat slips on the sound-side of one of the Outer Banks for a housing development. The slips will open to a common canal which in turn leads to the Inland Waterway at the Sound. This area is notorious for its susceptibility to hurricanes and lesser disturbances such as nor'easters and other gales. The season of maximum frequency for such weathers is August and September each year. The contractor is working in mid-to-late summer, and is thoroughly familiar with weather patterns in this region.

An additional factor facing the contractor is the usual and customary penalty clause in the contract which provides for the payment of a sum by the contractor to the client as a penalty for the delay or default in not completing the work on an agreed-upon schedule. In the present case the penalty is stated as $1,000 per week for the first five weeks, and $2,000 per week for the next five after that.

As the contractor reviews the status of the project on August 1, there are several weeks of work ahead at the present pace. The contractor's options appear to be the following:

1. Lay on extra dredges and crews so as to finish the job before the prime season for storms begins.
2. Accept that there may come bad weather during the weeks ahead, but plan to repair any damages, make up as much lost time afterward as possible, and pay the penalty for the remainder.

Accordingly, the contractor is presented with an event and a decision at this juncture. The decision is either 1 or 2, above. The decision may be guided by a knowledge of the extra cost to initiate a crash program and finish before the storm season, by the probable nature and corresponding cost of repair in the event that a storm does occur and by the known penalty clause provisions of the contract. The weather history can be consulted for the long-term probabilities of storms in this season, also.

Further, the experienced contractor will have a good idea of the cost of cleanups following rough weathers, and of the time which can be made up, as well. A storm

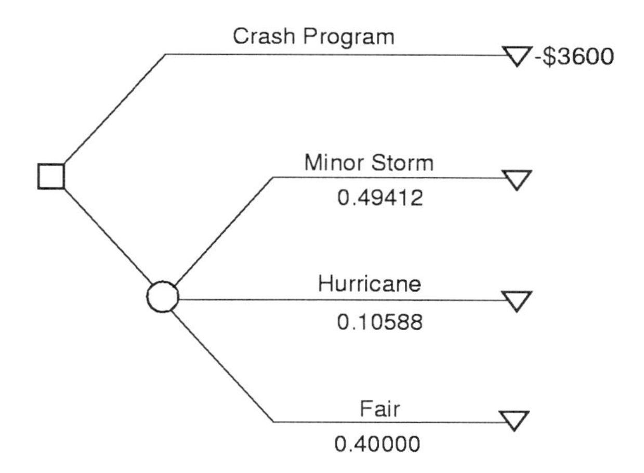

Figure 8.3: Contractor's Decision Tree

will make an estimated delay of 4 weeks, and the hurricane will delay the project 10 weeks.

Let us say that the contractor's estimate of the cost of the crash program is \$3,600. This is represented by a negative number, $-\$3,600$—a cost or loss or outgo—at the terminal node on this branch of the decision tree in Figure 8.3.

The next event on the alternate branch of the tree, as represented by Option 2, above, is a chance node. The contractor, having made a decision to run the risk of adverse weather, stares chance in the face with the inexorable probabilities of weather history. Take these to be 0.40000 for no storms, 0.49412 for a minor storm and 0.10588 for the hurricane, as before.

Should there be a storm, either a hurricane or a lesser one, the contractor then comes to another decision point; namely, to enter upon a crash program to reclaim some of the lost time, or to simply accept the delay penalty necessitated by the storm's damage.

The costs involved in carrying on work at the normal pace and paying the delay penalties are easy to compute. After the gale-force storm, four weeks delay at \$1,000/week yields \$4,000 total. For the hurricane, ten weeks delay at \$2,000/week amounts to \$20,000 altogether.

The contractor, from past experience, judges the probabilities of making up for lost time after a minor storm as 0.58321 for one week, 0.31047 for two weeks and 0.10632 for three weeks. The consequences of a hurricane will be much greater, of course. The estimate is 0.68858 probability to make up 2 weeks, 0.20089 for 3 weeks and 0.11053 for 4 weeks saved. Each week made up following one of these storms will entail an additional expense of \$500.

The contractor could now construct a schedule of the weekly costs which will be incurred after an unfavorable weather event due to delay and extra effort.

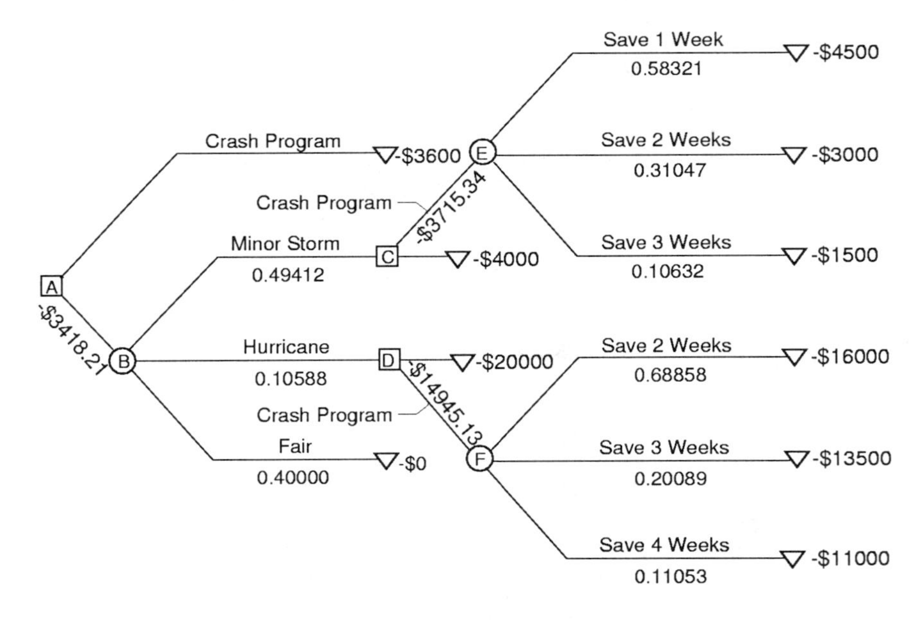

Figure 8.4: Contractor's Decision Tree

1st Week	2nd Week	3rd Week	4th Week	5th Week	6th Week	7th Week	8th Week	9th Week	10th Week
1000	1000	1000	1000	2000	2000	2000	2000	2000	2000
500	500	500	500	500	500	500	500	500	500

This schedule could then be applied to the chance occurrence of storms as follows.

Action Taken	Probability	Added Cost
	Nor'easter Event	
Save 1 Week	0.58321	$4,500
Save 2 Weeks	0.31047	$3,000
Save 3 Weeks	0.10632	$1,500
	Hurricane Event	
Save 2 Weeks	0.68858	$16,000
Save 3 Weeks	0.20089	$13,500
Save 4 Weeks	0.11053	$11,000

At this juncture the contractor is prepared to finish the details of the decision tree as shown in Figure 8.4. Each of the terminal nodes has a dollar value associated with it, according to the price which would be paid should that be the chance outcome. Note that all amounts are negative, which is to say they are costs or outgoing monies.

How does the contractor decide what to do, now that the consequences of the possible decisions are evident? There is a mechanism called *Expected Monetary*

Value (EMV) which can be invoked to reflect backward to the point of decision at Node A the relative costs associated with the two decision branches there. This procedure assumes that at chance nodes the cost of each branch will be proportional to the probability of that branch occurring. The node total cost will then be the sum of these branch amounts.

Therefore, the EMV can be calculated for Nodes E and F of the contractor's decision tree in the following way.

$$\begin{aligned}(\text{EMV})_E &= 0.58321(-\$4,500) + 0.31047(-\$3,000) + 0.10632(-\$1,500) \\ &= -\$3,715.34 \qquad (8.4)\\ (\text{EMV})_F &= 0.68858(-\$16,000) + 0.20089(-\$13,500) + 0.11053(-\$11,000) \\ &= -\$14,945.13 \qquad (8.5)\end{aligned}$$

Now the contractor can see that at Decision Node C, after a gale, it would be less costly to undertake a crash effort which could cost \$3,715.34 than to pay the delay penalty of \$4,000. Naturally, the path of least cost would be chosen, and the other branch of the tree would be pruned away. In the same manner, at Decision Node D, after a hurricane, a comparison shows that it would be less expensive to inaugurate the crash effort at a price of \$14,945.13 rather than pay the higher penalty of \$20,000 for the delay. The normal pace branch is pruned in this instance.

This brings the process back to Chance Node B. Using only the figures retained after the indicated prunings at the previous decision points, above, the EMV here computes as

$$\begin{aligned}(\text{EMV})_B &= 0.40000(\$0.0) + 0.49412(-\$3,715.34) + 0.10588(-\$14,945.13) \\ &= -\$3,418.21\,. \qquad (8.6)\end{aligned}$$

Finally, the contractor can compare the price tag for running the chance gauntlet compared to crashing his program before the storms come. The reflected value, $-\$3,418.21$, of the program done at normal pace stands against the crash program estimated cost which is $-\$3,600$. The decision is easy.

8.5 THE EMV CRITERION

The decision may be easy—as easy as comparing the two numbers and selecting that one which is less negative. Living with the decision may be more difficult. The nature of the system being modeled is that it obeys the laws of chance and no other laws. The simple procedure enunciated above can in no way make definite or certain the outcome of the chance events embedded in the decision tree. That is to say, it cannot supply a deterministic law to replace the law of chance.

For this reason, it is terribly important to understand what is happening in this process. It is seen that the one who is making the decisions is relying on the probabilities. Over the long haul—over the full range of the history of the chance universe in question—decisions made in accordance with these probabilities will produce a favorable result. *In a given particular trial (a certain year) the outcome may be entirely different.*

It must be emphasized in the strongest terms that the risk of an unfavorable outcome of the decision analysis process is not removed by applying the rationale of EMV. A contractor who does everything right in the example above could lose his shirt, simply because "nature," through the caprice of chance, dealt him a losing hand. A contractor who knows nothing of decision analysis might continue to work right through the hurricane season and suffer no delay, no penalty, from a September storm and pocket the profit. Neither virtue nor cunning nor greed or avarice have anything to do with the operation of the laws of chance. Only in Greek, Roman, Aztec, and Maya lore do the gods interfere to save or punish human enterprise by removing the work of chance.

If a contractor applies decision analysis methods to decide what to do during hurricane season over a period of a number of years, then over the long run it can be expected that the total outcome will be favorable, assuming the length of time (number of applications) is "sufficiently large."

Another limitation of the EMV procedure as described above is that it fails to take into account the personality of the decisioneer. Some persons will risk more than others in the same circumstances. Some can afford to risk amounts which are beyond the reach of others facing the same chances. Some are gamblers, and some will gamble with smaller amounts but become conservative as the stakes rise.

These personal proclivities are nevertheless very real, and should logically be accounted for to the extent that is possible in a decision process.

8.6 UTILITY

It should not be surprising that money will have different values to different people, anymore than that it should vary in value from time-to-time, as was discussed in Chapter 5. The expression "Easy come, easy go" refers to someone who has money easily, and who spends it recklessly. Persons who make money the old-fashioned way (They earn it!) are generally considered to be more conservative with its expenditure.

This and other aspects of the variable value of money are reflected in the concept drawn from economics which is called *Utility*. The idea was originally introduced by one of the patron saints of engineering science, Daniel Bernoulli (1700-1782), the same for whom Bernoulli's equation of fluid mechanics is named. He was a member of a very famous family of nine or more scholars of Basel(Bâle), Switzerland, all of whom contributed much to human knowledge. The synergetic activities of these and their colleagues at Basel and in other European academies allowed for a number of spillovers from one branch of learning into other fields.

Further impetus was given to the idea of Utility by another engineer, Vilfredo Pareto, a century or more later. His *Manuël d'économie politique*, Second Edition, Paris, 1927, treats the subject on page 556. Pareto drew upon the concept of an energy function, similar in nature to potential energy or virtual work, to describe the "potential-like" nature of Utility. He devoted most of his academic life to economics and sociology, and became a leader in the development of both of these social sciences at Lausanne, Switzerland.

Utility starts with the premise that money has different value to different

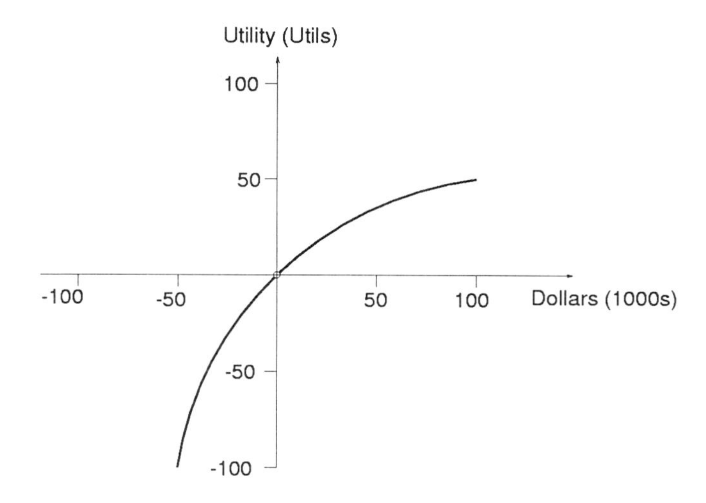

Figure 8.5: Utility Graph

people under different circumstances. A convenience market on the beach will sell a loaf of bread or a carton of milk for prices considerably higher than would a supermarket well-inland because people will pay these higher prices for the same items becuse of their utility. They are "at hand," convenient, so that one does not need to break one's vacation to travel far to get them. This is an example of the utility of location.

Or consider the study group which has worked late into the night, preparing for an examination. Someone says, "Let's go for pizza" at 3 a.m., and they go off to an all-night establishment which happens to charge more than the places which close at more reasonable hours. They will pay the higher amounts because of the greater utility associated with having it at 3 a.m. In other words, Utility includes a temporal element.

Utility for the decision maker is represented by a graph of that individual's historical monetary risk decisions converted to an arbitrary scale called ***utils***. The nonlinearity of this graph reflects the degree to which the value of money varies for this individual under different risk situations. The picture is given in Figure 8.5, where the representation shows clearly that as the sums grow larger (both positive and negative) the saturation effect sets in. Having a few more dollars when one already has many is less utile than being given the same dollars if one has none at all. Also, if one loses large sums, what difference does it make to lose a few more dollars? As was mentioned earlier, the Utility graph is calibrated in utils, which is an arbitrary scale devoid of any association with a particular currency. The actual graph data points are obtained by going to the historical record of decisions which have been made by that individual in times past. Suppose, for example, that it is the dredging contractor of the example above. And suppose that it should cover a range of monies comparable to that which are involved in dredging operations, covering possible losses in the range 0 to $-\$20,000$.

The curve should pass through the origin of the coordinate system, obviously.

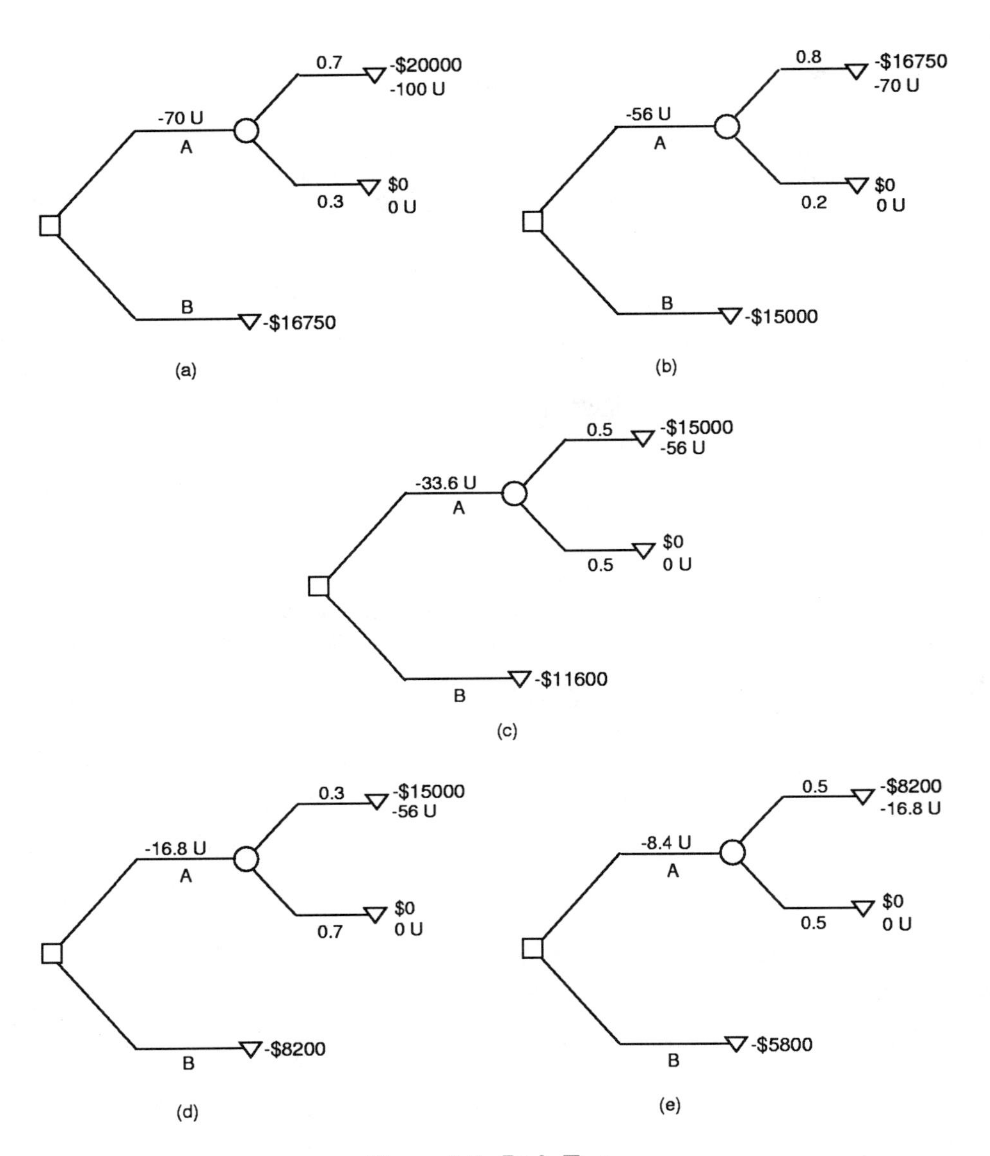

Figure 8.6: Risk Tests

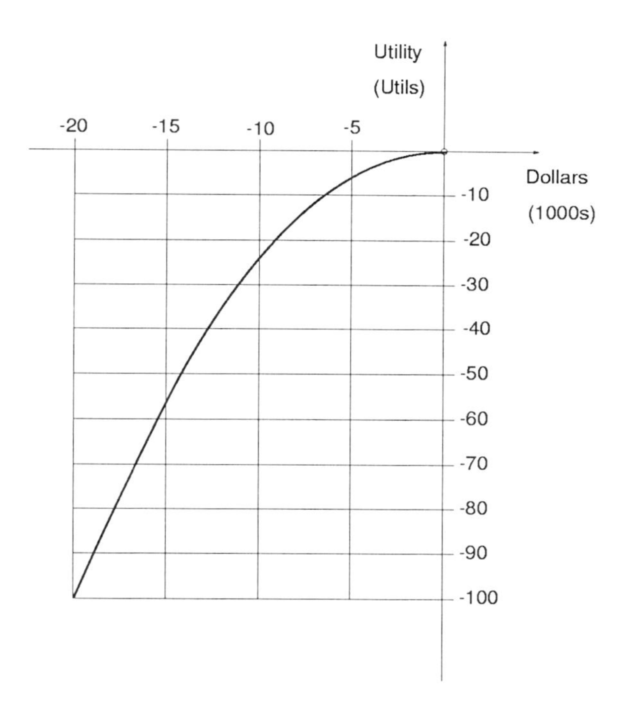

Figure 8.7: Contractor's Utility Function

That establishes one point. Let the sum $-\$20,000$ be given the arbitrary value -100 utils. This creates another data point. Perhaps a half-dozen more plotting points would be adequate to permit a smooth curve to be drawn. Hence, let there be generated the elementary decision trees that represent four or five of the contractor's actual decisions of record, with the data drawn from his files. These are shown in Figures 8.6 (a), (b), (c), (d), and (e).

For instance, on one occasion the contractor took a loss of $16,750 by following an Alternative B instead of the other one, Alternative A, which led to a chance node with two branches. One branch with a probability of 0.7 would have entailed a loss of $20,000, while the other, with lesser probability of 0.3, would have lost nothing. As the sketch of this test shows, see Figure 8.6 (a), the sum of $-\$16,750$ was of equal measure with -70 U, since $-\$20,000$ has been given the arbitrary value -100 U and $0 equals 0 U.

The other four cases are similar: only the numbers are changed. -56 U corresponds to $-\$15,000$; -33.6 U is equivalent to $-\$11,600$; -16.8 U equals $-\$8,200$; and -8.4 U becomes $-\$5,800$.

With these data points the graph can be constructed as in Figure 8.7. Using the methods displayed in Appendix B, a least-squares fit of the data gives the equation

$$U = -0.037441096 - 0.000017849\,D - 0.000000250\,D^2 \tag{8.7}$$

for converting dollars, D, to utils, U.

Return now to the decision tree of the dredging contract treated above. This

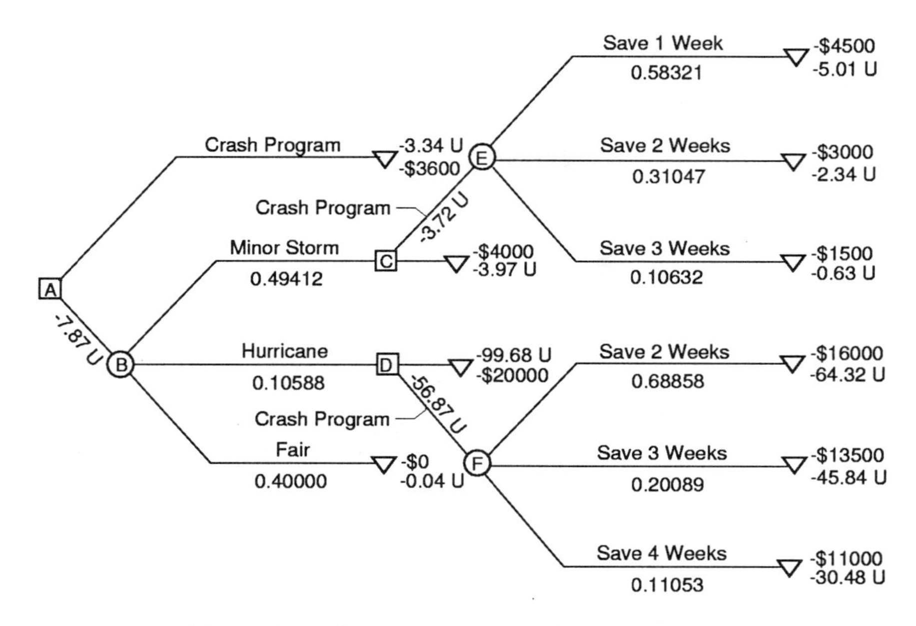

Figure 8.8: Contractor's Decision Tree in Utils

time let the decisions be made based upon utils. The first step will be to use Equation 8.7 or Figure 8.7 to convert all terminal node dollar amounts into utils, as in Figure 8.8. Working backwards through the chance branches and their respective probabilities, it is found that the EUV (Expected Utility Value) at Node E (the crash program following a gale) is −3.72 U. This compares with −4.11 U, the cost of a normal pace, following the gale. The normal program alternative will be rejected, and its branch cut off.

After a hurricane, the crash program cost, in Utils, is found to be −56.87 U, while the normal pace cost is −99.68 U. So the normal pace branch is lopped off, leaving only the crash program amount, −56.87 U.

At Chance Node B, there are these options: no delay, no loss, −0.04 U, probability 0.40000; a nor'easter, 5 weeks delay, −3.72 U cost, probability 0.49412; and a hurricane, 10 weeks delay, −56.87 U cost, probability 0.10588. These are combined to yield an EUV of −7.87 U at Node B.

The contractor, facing Decision Node A, now compares the crash program cost of −3.34 U to the normal pace cost of −7.87 U and selects the lowest cost, which is that of the crash program which will finish the job before the storm season can begin. In this case, EMV and EUV give different answers to the same decision tree.

Indeed, it is not at all necessary or to be expected that the two approaches will give the same indication. When they differ, it is because of the inherent nonlinearity of the utility curve for the individual who is making the decisions.

The utility curve may be considered as a fingerprint-of-sorts for the individual whose risk-taking established it. In many cases the individual will purchase insurance against the losses in order to diminish the impact against personal bank accounts. The curve could also be thought of as a constitutive law which defines the individual's risking behavior. And there are some executives of financial institutions who risked too much, although the monies in question were those of the institution (its depositors and its creditors) and, now, of the public which must bail them out through taxes and in other ways. *Sic transit gloria mundi.*

8.7 EXTENDED PROBABILITY

Formal studies of probability were begun by the famous French scientists Blaise Pascal and Pierre de Fermat, who were induced to aid some of their friends who were gamblers by determining the odds at various games of chance. A major work, *Ars Conjectandi* ("The Art of Conjecture"), by Jakob Bernoulli, appeared in 1713, and De Moivre's "The Doctrine of Chance" followed in 1718.

The formative treatise *Théorie analytique des probabilitiés*, by Pierre-Simon, marquis de Laplace, in 1812, established the modern basis of the subject.

In the classical theory of probability a system of events is assumed in which the

1. events are mutually exclusive,
2. occurrence of one event is inevitable, and
3. events are all equally probable.

Then the probability $P(E)$ of an event E is

$$P(E) = \frac{a}{b} \tag{8.8}$$

in which a is the number of "favorable" occurrences in a set of b possible events of the system. Thus

$$0 \leq P(E) \leq 1 \tag{8.9}$$

where $P(E) = 0$ corresponds to the impossibility of event E's occurrence, and $P(E) = 1$ denotes the certainty that E will take place.

The definition in Equation 8.8 is obviously based upon a discrete set of occurrences. As the system of events is allowed to increase in complexity, the occurrence of events is transformed into a random or statistical behavior such that probability is defined by

$$P(E) = \lim_{B \to \infty} \frac{A}{B} \tag{8.10}$$

where A is the number of occurrences of E in a random number of trials B.

Karl Pearson (see Section 13.6), one of the great contributors to modern statistics, actually conducted an experiment to demonstrate Equation 8.10. He tossed a coin 24,000 times and recorded 12,012 heads, so that

$$P = \frac{12012}{24000} = 0.50050 \tag{8.11}$$

which is certainly approaching the theoretical $P = 0.50000$ or fifty-fifty probability for a coin toss.

8.7.1 CONDITIONAL PROBABILITY

There are more complex levels of probability associated with chance events which can affect decision making. For instance, to employ the customary illustrations drawn from the gaming table, the probability of throwing a die and obtaining a two is the same as that of throwing a five, or any other number from one to six; namely, 1/6. This is because there are six sides to the die, and the proposition is that one and only one side will come up. But the real game is based upon throwing the die twice, once to establish a point, and again to make a second point, with the premium result associated with making the second point equal to the first point.

Thus, the mathematical question is:

"What is the probability of event A being followed by event B?"

Such a probability is called a *conditional probability*, because B is to occur upon condition that A has already occurred. Obviously, this is a probability of higher order than the simpler one associated with the occurrence of A alone. In other words, it is like throwing a die and having a three come up on the second try after throwing a three on the first roll. The conventional representation of this probability is

$$P(A|B) \tag{8.12}$$

in which the *random* event is A, and the *definite* event is B, even if it does occur in a gambling venue.

Note that $P(A|B)$ is very different than $P(B|A)$, which is the probability that the definite event A will follow after the random event B. The difference can be clarified by taking, for example, the failure of a reinforced concrete beam as A, and a defective rebar as B. $P(A|B)$ is the probability that the beam will fail when it is known that the rebar is bad, whereas $P(B|A)$ is the probability that the rebar is bad when it is known that the beam has failed.

Computation of a conditional probability can be illustrated by using a die, again. Consider the following contingencies of event A, which is rolling a four, followed by event B, which is rolling

(a) a number greater than four,

(b) an even number greater than four,

(c) an odd number greater than four,

(d) a prime number greater than three and smaller than five,

(e) a prime number greater than four,

(f) a number greater than four which solves the equation,

$$x^2 - 11\,x + 30 = 0$$

(g) a number greater than four, if the day is Friday.

These several probabilities are displayed in the table below.

Case	First Roll	Possible 2nd Roll	Favorable 2nd Roll	Conditional Probability
(a)	4	1, 2, 3, 4, 5, 6	5, 6	$P(A\|a) = \frac{2}{6} = \frac{1}{3}$
(b)	4	2, 4, 6	6	$P(A\|b) = \frac{1}{3}$
(c)	4	1, 3, 5	5	$P(A\|c) = \frac{1}{3}$
(d)	4	4	none	$P(A\|d) = \frac{0}{1} = 0$
(e)	4	1, 3, 5	5	$P(A\|e) = \frac{1}{3}$
(f)	4	5, 6	5, 6	$P(A\|f) = \frac{2}{2} = 1$
(g)	4	1, 2, 3, 4, 5, 6	5, 6	$P(A\|g) = \frac{2}{6} = \frac{1}{3}$

Case (g) demonstrates the very important notion of independence of two events, because it is clear that neither Friday nor any other day of the week has anything to do with tossing the die. The definition of independence is taken to be

$$P(A|B) = P(A) \tag{8.13}$$

8.7.2 JOINT PROBABILITY

There is also the situation in which a *random* event B will follow after a *random* event A. This *joint probability* is denoted

$$P(A, B) \tag{8.14}$$

where a comma replaces the vertical stroke as in conditional probability. If the two events are independent, the probability of their simultaneous occurrence equals the product of the probabilities of their separate occurrences, or

$$P(A, B) = P(A)P(B) \tag{8.15}$$

8.7.3 VENN DIAGRAM

A very convenient and helpful method of illustrating and demonstrating combinatorial aspects of probability theory is the *Venn Diagram*. This device employs the fundamental property of probability as an area to do its work.

Take an area S, shown in Figure 8.9, to represent the system of all possible events associated with a complex of circumstances. This area is customarily represented by a plane rectangular figure, for convenience sake. Let the area A represent a random event of the complex of circumstances. Then the ratio of areas represents the probability of the random event as

$$P = \frac{\text{area}\, A}{\text{area}\, S} \tag{8.16}$$

Some immediate extensions to other circumstances are evident, as illustrated in Figures 8.10 a, b, c, d, e, f, g, h, i, and j.

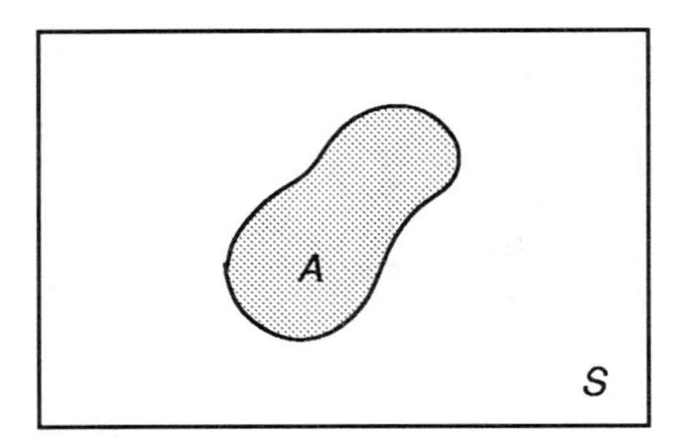

Figure 8.9: A Random Event Venn Diagram

The Venn diagram geometrical representation makes possible the easy demonstration of additional elements of probability theory. In Figure 8.11, for instance, the probability of event A is $P(A) = \text{area}\,A/\text{area}\,S$, that of event B is $P(B) = \text{area}\,B/\text{area}\,S$, and the shaded area is the joint occurrence of events A and B, $P(A, B)$.

Now think of the conditional probability $P(A|B)$. In this affair, B is a fixed outcome, so nothing about A or the rest of the complex system of occurrences can alter that fact. Hence, the field of possibilities shrinks from the rectangular area S down to the circle that represents B. The only effect remaining to A is the shaded area A, B. The conditional probability $P(A|B)$ is then formulated as

$$P(A|B) = \frac{\text{area}\,A,B}{\text{area}\,B} = \frac{\text{area}\,(A,B)/\text{area}\,S}{\text{area}\,B/\text{area}\,S} = \frac{P(A,B)}{P(B)} \tag{8.17}$$

This can be solved as the *theorem of joint probability*

$$P(A,B) = P(A|B)P(B) = P(B|A)P(A) \tag{8.18}$$

which, with Equation 8.15, yields the *multiplication theorem*

$$P(A,B) = P(A)P(B) \tag{8.19}$$

Events A and B in this equation must be *independent* of each other, of course.

When there are a number of independent events, the formula can be generalized by adding an event at a time to get

$$P(A_1, A_2, \ldots, A_n) = P(A_1)P(A_2)\ldots P(A_n) \tag{8.20}$$

8.7.4 TOTAL PROBABILITY

Consider the areas A_1, A_2, ..., A_{11} in Figure 8.12. Since none of these overlap each other, they represent mutually exclusive occurrences. Now look at area B, any place of which is also a place in either A_2, A_3, A_4, A_5, or A_6. Event B cannot occur without one of the events A_2, A_3, A_4, A_5, or A_6 likewise occurring.

The area (B, A_3), shown shaded in Figure 8.12, is the joint occurrence of B and A_3, and similar overlapping areas of B and A_2, A_4, A_5, and A_6 represent the

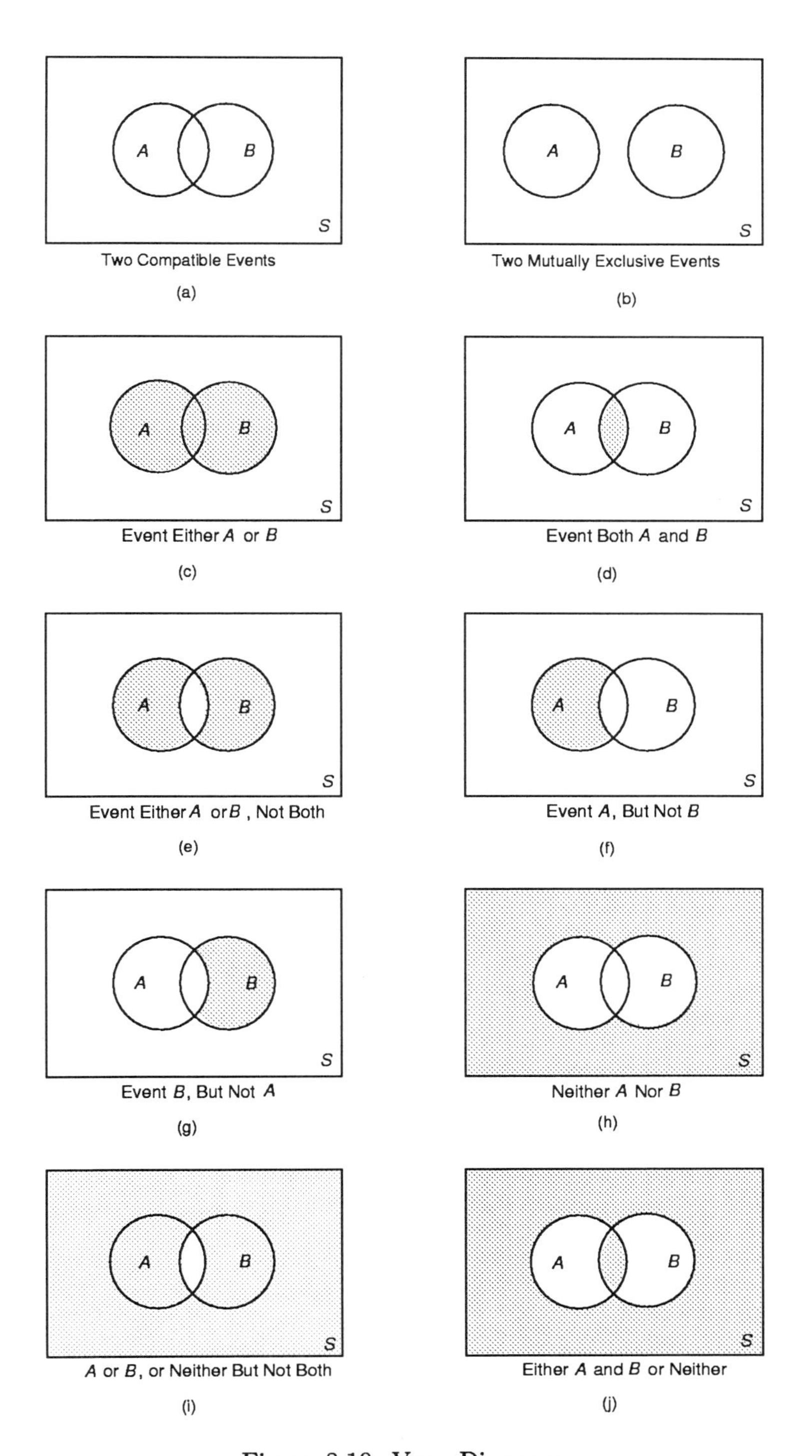

Figure 8.10: Venn Diagrams

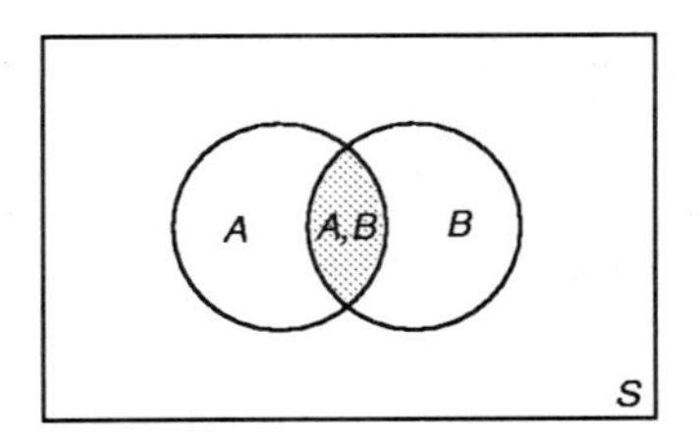

Figure 8.11: Venn Diagram Showing Joint Probability

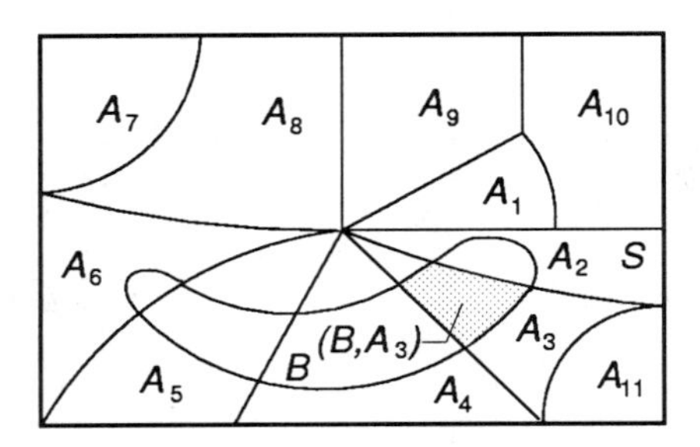

Figure 8.12: Joint Probability

joint occurrences of B with A_2, A_4, A_5, and A_6, respectively. Now it is evident that, since the total area of B is the sum of the overlapping areas of B with the As,

$$P(B) = \frac{\text{area}\,B}{\text{area}\,S} = P(B, A_2) + P(B, A_3) + P(B, A_4) + P(B, A_5) + P(B, A_6)$$

$$= \sum_{i=1}^{n} P(B, A_i) \tag{8.21}$$

But from Figure 8.11 and Equation 8.18,

$$P(B, A_i) = P(B|A_i)\,P(A_i) \tag{8.22}$$

therefore

$$P(B) = \sum_{i=1}^{n} P(B, A_i)\,P(A_i) \tag{8.23}$$

which is called the ***theorem of total probability***.

8.7.5 BAYES' THEOREM

The Venn Diagram allowed the theorem of joint probability to be presented, in Equation 8.18, as

$$P(A, B) = P(A|B)\,P(B) = P(B|A)\,P(A)$$

This can be solved for the conditional probability $P(A|B)$

$$P(A|B) = P(B|A)\,\frac{P(A)}{P(B)} \tag{8.24}$$

after which $P(B)$ from Equation 8.23 can be inserted to yield

$$P(A|B) = \frac{P(A)}{\sum_{i=1}^{n} P(B, A_i)\,P(A_i)}\,P(B|A) \tag{8.25}$$

This important formula is called *Bayes' Theorem.*

Bayes' Theorem is very useful in failure analysis. That is, $P(A|B)$, the probability that event B follows event A, is desired to be known when $P(B|A)$ is available. The question to be answered is, "Given that a random event, A, a cause, has happened, what is the probability that it is followed by definite event B, a failure?" The analyst, working after the fact, can examine a hypothesis that the failure was caused by a specific source.

8.8 ON THE BEACH AGAIN

The concepts of extended probability can be invoked in connection with the contractor's decisions of the previous example. Rather than relying on elementary estimates, the contractor could decide to consult a meterologist who is expert at predicting weather patterns. A call might be placed to the Weather Center at Channel 5 (WRAL) for Greg (Super) Fishel, whose predicting skills are widely respected since he accurately forecast the Blizzard of '89.

The meteorologist will have a personal record of forecasting weather events that can be summarized in Table 8.1.

Table 8.1
WEATHER PREDICTION PROBABILITIES

		Actual Event		
		Good Weather	Nor'easter	Hurricane
Predicted Event	Good Weather	0.80122	0.30004	0.11072
	Nor'easter	0.19860	0.60039	0.20516
	Hurricane	0.00018	0.09957	0.68412

It is readily seen that these listings are conditional probabilities. The expert has predicted good weather which has actually been followed by good weather 80.122% of the time, for example.

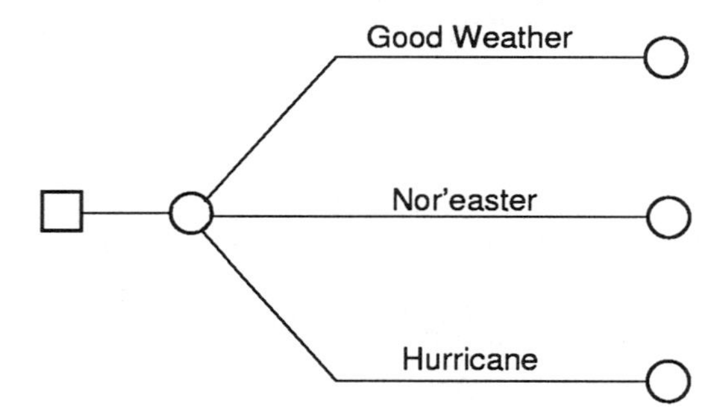

Figure 8.13: Expert's Forecast

Introduction of the weather expert alters the contractor's decision tree, because any decisions (other than the one to hire the expert) will be delayed until *after* the expert's predictions; otherwise it would make no sense to call upon the expert for advice. The expert will forecast one of three events: good weather, a nor'easter, or a hurricane. There is a chance node with distinct probabilities associated with each weather event, and the probabilities remain to be determined. Therefore, the decision tree appears as in Figure 8.13 after the forecast.

To make a beginning at determining these probabilities, note that a tree can be constructed for the expert as in Figure 8.14. There will be the three actual events—good weather, a nor'easter, or a hurricane—with their *actual historical* probabilities; 0.40000, 0.49412, and 0.10588, respectively, and the expert's prognostications associated with each together with their conditional probabilities. Next, the joint probabilities that the expert will predict an event, A, and that it will be followed by another event, B, are computed from each case from

$$P(B, A) = P(B|A)\,P(A)$$

and these are associated with the appropriate terminal event as in Figure 8.14. The total of the latter will sum to 1.000000000, naturally.

Then the total probability that the expert will forecast a particular event is

$$P(B) = P(B, A_1) + P(B, A_2) + P(B, A_3) \tag{8.26}$$

or

$$P(B_1) = 0.320488000 + 0.148255764 + 0.011723033 = 0.480466797 \tag{8.27}$$
$$P(B_2) = 0.079440000 + 0.296664706 + 0.217223400 = 0.397827046 \tag{8.28}$$
$$P(B_3) = 0.000072000 + 0.049199528 + 0.072434625 = 0.121706153 \tag{8.29}$$

$P(B_1)$, $P(B_2)$, and $P(B_3)$ are the probabilities for good weather, a nor'easter, or a hurricane, respectively.

At this stage the conditional probabilities can be computed for the expert to forecast an event which will be followed by an actual event. The formula to be employed is that of Bayes' Theorem

$$P(A, B) = P(B|A)\,\frac{P(A)}{P(B)}$$

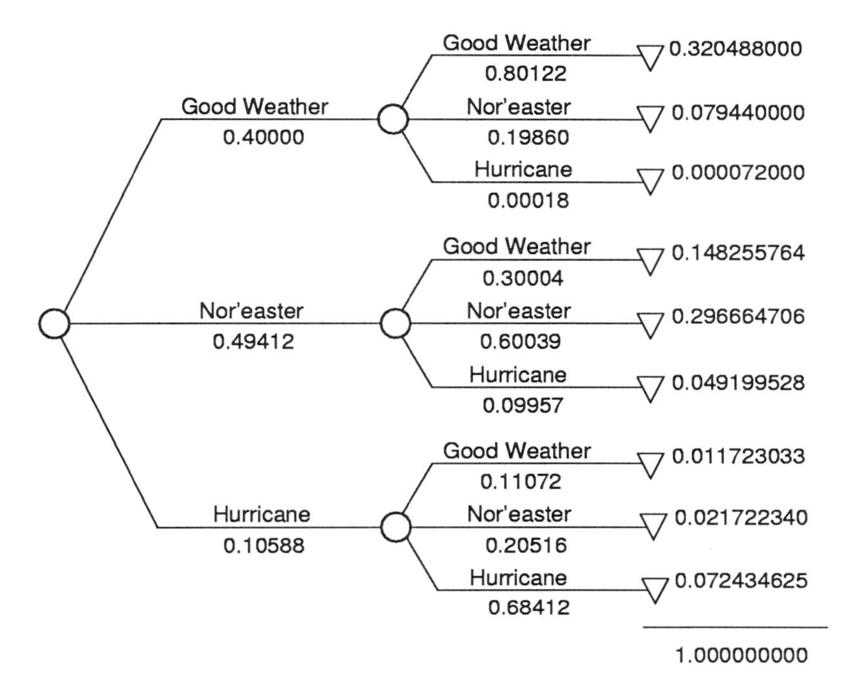

Figure 8.14: Conditional Weather Tree

Thus

$$P(A_1|B_1) = P(B_1|A_1)\frac{P(A_1)}{P(B_1)} = 0.80122\frac{0.40000}{0.48047} = 0.66703 \tag{8.30}$$

$$P(A_2|B_1) = P(B_1|A_2)\frac{P(A_2)}{P(B_1)} = 0.30004\frac{0.49412}{0.48047} = 0.30857 \tag{8.31}$$

$$P(A_3|B_1) = P(B_1|A_3)\frac{P(A_3)}{P(B_1)} = 0.11072\frac{0.10588}{0.48047} = 0.02440 \tag{8.32}$$

$$P(A_1|B_2) = P(B_2|A_1)\frac{P(A_1)}{P(B_2)} = 0.19860\frac{0.40000}{0.39783} = 0.19968 \tag{8.33}$$

$$P(A_2|B_2) = P(B_2|A_2)\frac{P(A_2)}{P(B_2)} = 0.60039\frac{0.49412}{0.39783} = 0.74571 \tag{8.34}$$

$$P(A_3|B_2) = P(B_2|A_3)\frac{P(A_3)}{P(B_2)} = 0.20516\frac{0.10588}{0.39783} = 0.05460 \tag{8.35}$$

$$P(A_1|B_3) = P(B_3|A_1)\frac{P(A_1)}{P(B_3)} = 0.00018\frac{0.40000}{0.12171} = 0.00059 \tag{8.36}$$

$$P(A_2|B_3) = P(B_3|A_2)\frac{P(A_2)}{P(B_3)} = 0.09957\frac{0.49412}{0.12171} = 0.40425 \tag{8.37}$$

$$P(A_3|B_3) = P(B_3|A_3)\frac{P(A_3)}{P(B_3)} = 0.68412\frac{0.10588}{0.12171} = 0.59516 \tag{8.38}$$

Now the complete decision tree can be constructed and the EMV analysis can be carried out as before. The result is shown in Figure 8.15. It shows that the decision to use this expert saves \$3,418.21 – \$2,591.04 = \$827.17 under the circumstances. This amount would need to include a payment to the meterologist, of course. A similar analysis could be done using EUV in addition to EMV.

8.9 PROJECT #6

1. A dredge is working to clear a channel through an inlet on the Outer Banks when the weather takes an unexpected turn for the worse. The crew elects to suspend operations and head for their usual berth at a mooring in the nearby sound. As the day wears on, the weather becomes a full-blown gale, and the dredge is torn from its moorings. The crew are unable to make way against the storm and tides, and the vessel—out of control—collides with the pilings of a long bridge which spans the inlet.

   When the storm abates and DOT personnel are able to make an inspection of the damage to the bridge, they are concerned that the damaged pilings could fail.

   The cost to replace the damaged section is \$26,000, but if failure should occur while the bridge is in service the cost of rebuilding the bridge would be \$850,000.

   One inspector suggests that a specialist in nondestructive testing by ulttrasonic methods can determine if there is any significant damage. The tests would cost \$2,000.

   If the pilings are replaced, the estimate is that there is a probability of 0.99999 that no failure will occur, but if they are not replaced, this probability drops to 0.90. If tests are made, they will indicate either a significant reduction in service strength or no appreciable loss. In the latter case, the probability of failure is the same as if the pilings had been replaced (i.e., 0.99999.) However, if the tests show a strength reduction, the probability of no failure drops to 0.99. If the probability is 0.864 that the tests will indicate no reduction of strength, use an EMV procedure to select an optimum course of action for the engineers to follow.

2. Your engineering firm is invited to bid on a contract to furnish design services for a nearby municipality. The prospective client (municipality) plans to a) proceed with work during the current year, if the budget allows, or b) begin in the next fiscal year if the current budget is too tight or c) delay the project a second year if the cost requires a capital (bond) campaign.

   You estimate the probability of these contingencies as 0.2, 0.3, 0.5, respectively.

   The firm will need extra staff if its bid is successful, the cost of which will be \$60,000 in case a), \$30,000 b), and \$20,000 c).

   Bidding will take place in two stages; selected bidders in the preliminary round will be invited to submit more detailed proposals in a final competition.

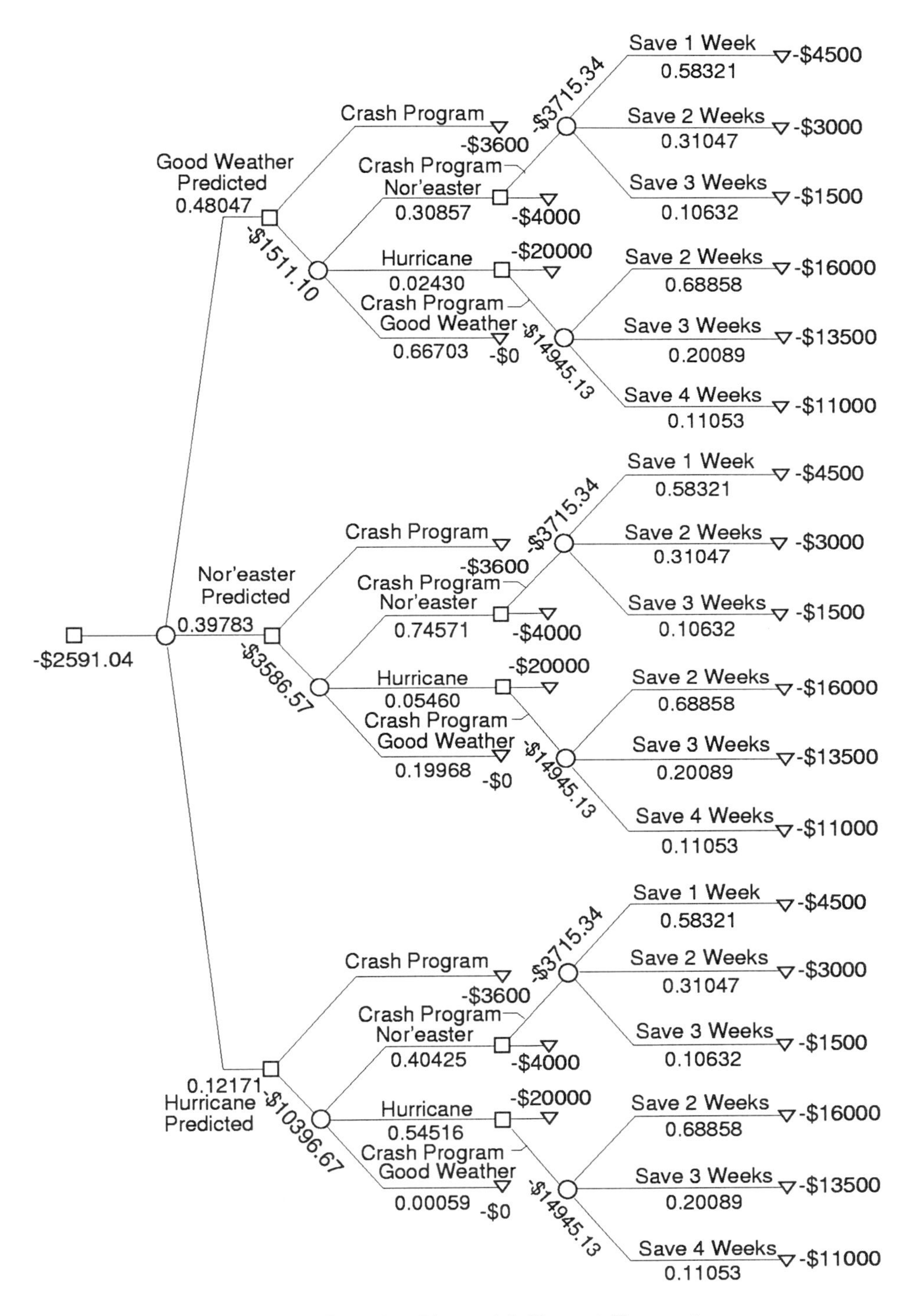

Figure 8.15: Decision Tree with Expert Forecast

Your estimate of your chances of surviving the initial cut is 50-50, and of the final bid is 1 in 5. It will cost $5,000 to prepare the first-round bid, and $15,000 for the second.

If you are successful, the income to the firm will be $120,000.

What should be your response to the bid invitation?

3. An old bridge is showing signs of increasing unworthiness. The Bridge Division has under consideration a decision concerning what should be done. The options which have been delineated are:

    (a) Condemn and replace the bridge at a cost of $6,000,000.

    (b) Repair the bridge at a cost of $600,000.

    (c) Do nothing.

    If the bridge is repaired, there is a 90% chance that it will prove to be satisfactory, but if not, the structure must be removed and replaced at a cost of $6,500,000.

    If nothing is done, there is an 82% probability that the bridge will remain serviceable, a 14% chance that it will become unserviceable, in which case it will be replaced ($6,000,000), and a 4% chance that it will collapse. In the latter instance, the total cost may well reach $20,000,000.

    What should be done?

4. A CE student expects to obtain his degree at the Spring Commencement and is exploring some options for his employment thereafter. He has a standing offer to work for his father in the family business at a salary of $40,000. He has interviewed potential employers from the North Shore Alaska oilfields, an Australian Outback Radar installation, and Waterways Experiment Station in Mississippi. He estimates the chances of offers from these as 50%, 20%, and 30%, respectively.

    If he accepts the Alaska job, he estimates that, following an initial training period, his salary potential will be $62,000 (0.1), $54,000 (0.6), and $45,000 (0.3). The comparable figures for Australia are $70,000 (0.2), $60,000 (0.4), and $50,000 (0.4), and for Mississippi they are $40,000 (0.1), $32,000 (0.4), and $27,000 (0.5).

    The student builds for himself an utility curve as shown below.

    Construct the decision tree which represents the student's options, and use it to assist him in determining what to inform his father about his intentions. Evaluate the student's choices using both EMV and EUV procedures.

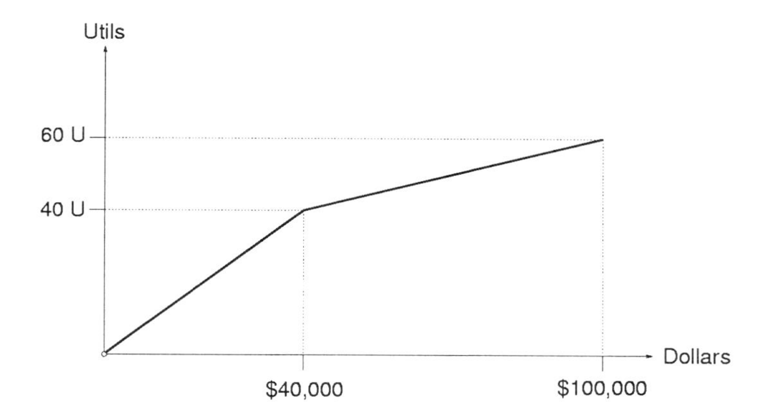

5. As Chief Engineer, one of your duties is to hire new structural analysts as the need arises. There is presently a vacancy, which has been duly advertized, and there are a number of applicants. The list has been narrowed down to two promising candidates, each of whom has stated that they will accept the job, if it is offered.

   The qualifications of both applicants are similar, so the interview becomes important. Your plan is to meet with applicant A first, followed by contender B in due course. The plan is to place each person in a category "outstanding," "very good," or "good" following the interview. In previous hirings, you have rated 30% of the applicants as outstanding, another 30% as very good, and the remaining 40% as good.

   Your normal procedure would be to interview both candidates before making a final decision, but after talking with A, you learn that she has other offers, and faces a deadline from one of them. You need to decide whether to offer her the job or wait and hire B.

   Decide upon a hiring strategy based upon your rating of candidate A. Would your strategy change if you restrict your hiring to only those rated better than "good?"

6. The Chief Engineer asks you to design a small, isolated shed to be used for storing equipment. One aspect of the design is the selection of the design wind which the structure must be expected to resist. The options are:

   (a) Design the shed to resist a five-year wind W_5 (a wind which, on the average, is exceeded only once in five years).

   (b) Design it to resist a 25-year wind, W_{25}.

   (c) Design for resistance to the 100-year wind, W_{100}.

   The estimated cost of the structure for the three cases cited above is $10,000, $14,000, and $20,000, respectively, but if there is failure, the cost (including equipment losses, etc.,) will be $40,000.

   You identify the following possibilities:

(a) The maximum wind velocity $W_{\max}$ does not exceed W_5.

(b) $W_{\max}$ exceeds W_5 but is less than W_{25}.

(c) $W_{\max}$ exceeds W_{25} but remains less than W_{100}.

(d) $W_{\max}$ exceeds W_{100}.

From meteorological data the probabilities of these four are 0.6, 0.2, 0.18, and 0.02, respectively.

Draw a decision tree which represents the design situation.

Use EMV analysis to determine the best option under the given circumstances.

7. An engineer is in charge of building a bridge utilizing a prestressed concrete girder. Because of cash-flow limitations and bad weather, the construction has been delayed, and she has noted that there is some rusting of the prestressing strand. This could cause a reduction in the fatigue resistance due to local pitting of the strand surface.

   The cost to replace the strand is $22,000, but if fatigue failure should occur while the bridge is in service, the cost of rebuilding the bridge would be $800,000.

   Fatigue tests could be carried out on samples of the strand at a cost of $2,000.

   If the strand is replaced, the engineer estimates that there is a probability of 0.99999 that no failure will occur, but if it is not replaced, this probability drops to 0.90. If tests are made, they will indicate either a significant reduction in fatigue life or no appreciable loss of fatigue life. In the latter case, the probability of failure is the same as if the strand had been replaced (i.e., 0.99999.) However, if the tests show a fatigue life reduction, the probability of no failure drops to 0.99. If the probability is 0.864 that the tests will indicate no reduction of fatigue life, use an EMV procedure to select an optimum course of action for the engineer to follow.

8. An engineer is about to begin a special rush project. To assure meeting the desired schedule, he has decided to back up the usual equipment with an item from another job, and is considering whether to overhaul this equipment before putting it into operation.

   The cost to overhaul is $300, but if he puts it on line and it breaks down, it will cost $1,500 to cover the cost of repair and lost time. He estimates a 66% chance that the equipment is reliable, but is assured that it will be reliable if it is overhauled. A bench test costs $100 but will only indicate good or bad with a 10 percent chance of an invalid test. There is a 70% chance that the bench test will indicate a reliable equipment.

   Model the decision problem faced by the engineer. If the decision is based upon an EMV policy, what should be the optimum strategy?

9. A professor of engineering is considering early retirement. He plans to become a builder of single-family residences or two-story townhomes in his second career.

Based upon his knowledge of building practices, he has constructed the schedule below governing the two types of structures.

Type I Single-Family Residence

Estimated Cost	Probability	Estimated Return	Probability
$160,000	0.6	$275,000	0.5
$120,000	0.3	$200,000	0.4
$100,000	0.1	$160,000	0.1

Type II Two-Story Townhouse

Estimated Cost	Probability	Estimated Return	Probability
$155,000	0.7	$250,000	0.5
$130,000	0.2	$220,000	0.3
$110,000	0.1	$200,000	0.2

The professor estimates that in order to maintain his present life-style, he will require an income of $80,000, but that any profit beyond this will be pleasant but not necessary. Accordingly, he has mapped the utility curve shown below to reflect his personal valuation of his money at risk.

Use decision analysis (based upon utils) to assist the professor in his choice to build either residences or townhomes to optimize his income if he decides to go into this business.

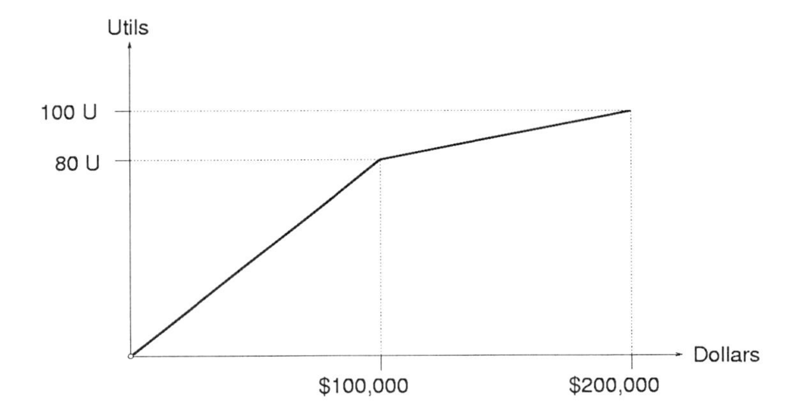

CHAPTER 9

CALCULUS OPTIMIZATION

9.1 INTRODUCTION

It is very likely that from the time of ancient civilization, even back to prehistoric times, mankind had some logical methods of optimization to use. However, at the time of the emergence of Western European civilization from the Dark Ages, the rational logical processes began to yield better results, and finally Sir Isaac Newton created and codified the procedures of the Calculus in his famous monumental work, the *Principia Mathematica Philosophia Naturales, etc.*, in 1687.

The modern versions of this subject are rigorously taught to Freshmen and Sophomores in every college of engineering, and it is assumed that thereafter each of these persons is proficient and skilled in its use. That presumption will be taken here.

A case study method will be employed to apply the calculus to examples of optimization drawn from civil engineering practice. The examples will start with some which are especially simple, and progress through increasing degrees of difficulty to build a base for extending these principles to large-scale problems.

9.2 BLACK ANKLE ONE

Chatham, Montgomery, Moore, and Randolph Counties and the Soil Conservation Service, the Corps of Engineers, the Fish and Wildlife Service, and the National Park Service have collaborated to create the BLACK ANKLE AUTHORITY, an agency chartered under the laws of North Carolina for the purpose of planning a reservoir on Deep River with a dam, probably near Highfalls, to control flooding, provide a water supply, generate some hydroelectric power, and furnish a recreational facility in the Black Ankle Region, famous for its bootleggers, rattlesnakes, and world-class pottery manufacturing.[1]

The Authority has commissioned studies which show the costs and benefits for various sizes of reservoirs. These studies have resulted in a report which can be

[1] A convenient access to the Black Ankle is ***Black Ankle Road***, at Exit #41, I73-I74, North of Ether, South of Seagrove, near the Montgomery-Randolph County line.

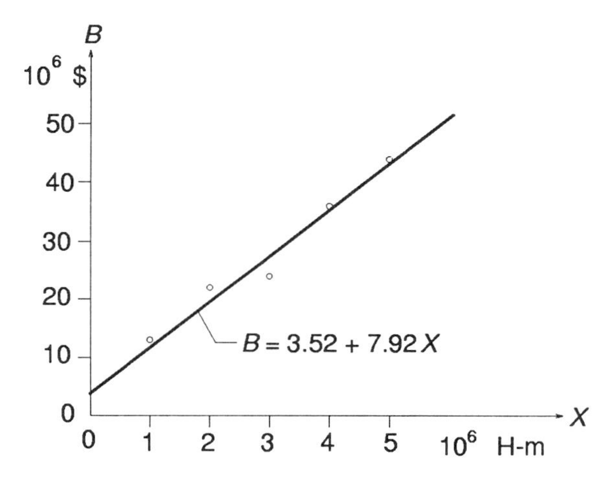

Figure 9.1: Benefits versus Reservoir Size

summarized in the two graphs shown. These are Figures 9.1 and 9.2, which plot the gross benefits (in millions of dollars) versus the size of reservoir in hectare-meters, and the cost, also in millions of dollars, against the reservoir size. In addition to the graphs themselves, a least-squares fitting of the data (see Appendix B) has produced the equations

$$B = 3.52 + 7.92\,X \qquad C = 2.16 + 2.89\,X + 1.46\,X^2 \tag{9.1}$$

for the benefit and cost in terms of the size, X.

The Authority has a simple optimization requirement: to maximize the net benefit from the enterprise.

There is also a constraint; the reservoir cannot exceed 4.72 million hectare-meters in size, because within the region there is no space larger than this without disrupting whole municipalities and other major facilities.

The constraint can be given mathematical form as $X \leq 4.72$ where X is the size in million hectare-meters. One hectare-meter is a volume equal to one hectare in area covered to a depth of one meter by the water impounded.

The net benefit is the gross benefit minus the cost. That is,

$$\begin{aligned} P = B - C &= (3.52 + 7.92\,X) - (2.16 + 2.89\,X + 1.46\,X^2) \\ &= 1.36 + 5.04\,X - 1.46\,X^2 \end{aligned} \tag{9.2}$$

so the formal statement of the optimization is as follows.

$$\text{Maximize} \quad P = 1.36 + 5.04\,X - 1.46\,X^2 \quad \text{subject to} \quad 0 \leq X \leq 4.72 \tag{9.3}$$

This is the case of a function P of a single variable X which is to be maximized. It is an elementary problem of the calculus, where the procedure is to take the derivative of the function with respect to the variable, set the result equal to zero, and solve for the value of X at that condition. Accordingly,

$$\frac{dP}{dX} = 5.04 - 2\,(1.46)\,X = 0, \qquad X = \frac{5.04}{2(1.46)} = 1.72 \times 10^6 \ \text{hectare-meters} \tag{9.4}$$

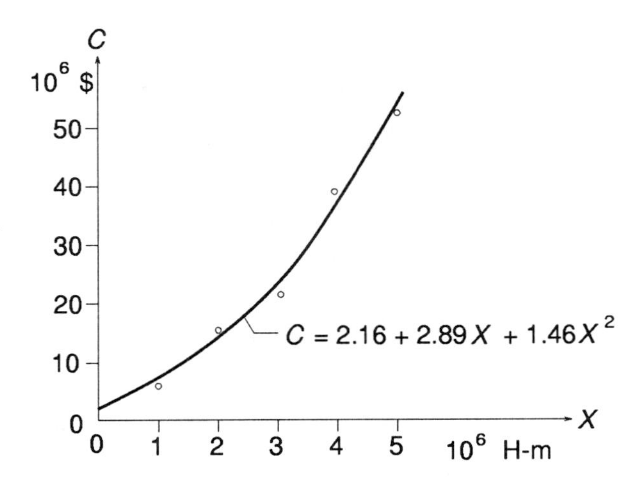

Figure 9.2: Cost versus Reservoir Size

The maximum net benefit itself can be determined to be

$$P_{\max} = 1.36 + 5.04(1.72) - 1.46(1.72)^2 = \$5.71 \times 10^6 \tag{9.5}$$

Of course, this result must be checked to determine that it is a maximum, rather than a minimum, as the procedure up to this point only establishes an extremum or stationary value.

Taking the second derivative of P with respect to X shows that

$$\frac{d^2P}{dX^2} = -2.92 \tag{9.6}$$

and since this is negative, the extremum is a maximum. A graph of the function versus X also reveals the character of the optimum as seen in Figure 9.3. In this case the maximum is both a local and a global maximum.

The Authority might have elected to utilize the Benefit-to-Cost Ratio as the yardstick for measuring the optimum. In this case, the formal statement is

$$\text{Maximize} \quad R = \frac{B}{C} = \frac{(3.52 + 7.92\,X)}{(2.16 + 2.88\,X + 1.46\,X^2)} \quad \text{subject to} \quad X \leq 4.72 \tag{9.7}$$

Then

$$\begin{aligned}\frac{dR}{dX} &= \frac{C(dB/dX) - B(dC/dX)}{C^2} \\ &= \frac{(2.16 + 2.88\,X + 1.46\,X^2)(7.92) - (3.52 + 7.92\,X)(2.88 + 2 \times 1.46 \times X)}{(2.16 + 2.88\,X + 1.46\,X^2)^2} \\ &= \frac{17.15 + 22.87\,X + 11.58\,X^2 - 10.16 - 10.29\,X - 22.87\,X - 23.16\,X^2}{(2.16 + 2.88\,X + 1.46\,X^2)^2} \\ &= \frac{6.99 - 10.29\,X - 11.58\,X^2}{(2.16 + 2.88\,X + 1.46\,X^2)^2} = 0\end{aligned} \tag{9.8}$$

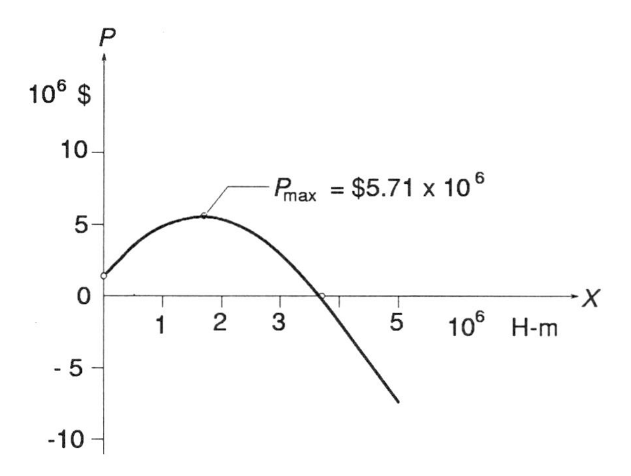

Figure 9.3: Global Maximum

Therefore, for real values of X,

$$6.99 - 10.29\,X - 11.58\,X^2 = 0 \tag{9.9}$$

and using the quadratic formula gives

$$\begin{aligned} X &= \tfrac{-(-10.29)}{2(-11.58)} \pm \tfrac{1}{2(-11.58)}\sqrt{(-10.29)^2 - 4(-11.58)(6.99)} \\ &= -1.339172878,\ 0.450834330 \end{aligned} \tag{9.10}$$

The negative value has no physical significance, so the other root is the one used. With it,

$$R_{\max} = \frac{3.52 + 7.92(0.450834330)}{2.16 + 2.88(0.450834330) + 1.46(0.450834330)^2} = 1.88542 \tag{9.11}$$

and

$$P_{\max} = 1.36 + 5.04(0.450834330) - 1.46(0.450834330)^2 = \$5.6619 \times 10^6 \tag{9.12}$$

9.3 BLACK ANKLE TWO

The previous case was an especially simple one in which there was only one independent variable; namely, the reservoir size. To be sure, there was a constraint on that variable, since room in the Black Ankle Region could not be found for a lake of more than 4.72 million hectare-meters. This element might very well govern the stated objectives of flood control, power generation, water supply, and water recreation.

However, another of the objectives is to develop further the recreational potential of the site by creating a park which will allow hiking, camping, picnicking, and nature and wildlife preservation and enjoyment. The area is known to be inhabited by a diverse variety of wildlife, including eagles.

The Authority will naturally try to focus upon the lake first, as it is central to the entire plan, but before there is a firm decision on that alone, the problem will be reformulated and restudied with the park land entered as a new variable. Let the new variable, hectares of parkland, be named Y. Y is given in thousands of hectares. The estimate is that a maximum of 3860 hectares can be found in the Authority's jurisdiction, so this becomes a constraint also.

Professional staff studies commissioned by the Authority reveal that there will be an interaction between the lake and the park, such that some people who come for water activities will also engage in land-based recreation, and some who would be primarily attracted to the park will also use the lake because it is readily available. The synergism between lake and park will reflect in both the cost and benefit functions. The studies suggest the following analytical couplings:

$$B = 3.52 + 7.92\,X + 0.88\,XY \qquad C = 2.16 + 2.88\,X + 1.46\,X^2 + 1.03\,Y^2 \tag{9.13}$$

and it is clear that the optimization must now treat a two-variable problem.

The formal statement of this example is

Maximize

$$P = B - C = (3.52 + 7.92\,X + 0.88\,XY) - (2.16 + 2.88\,X + 1.46\,X^2 + 1.03\,Y^2)$$

$$= 1.36 + 5.04\,X + 0.88\,XY - 1.46\,X^2 - 1.03\,Y^2 \tag{9.14}$$

$$\text{subject to} \quad 0 \le X \le 4.72 \quad \text{and} \quad 0 \le Y \le 3.86 \tag{9.15}$$

The calculus procedure for this problem is

$$\frac{\partial P}{\partial X} = \frac{\partial}{\partial X}(1.36 + 5.04\,X + 0.88\,XY - 1.46\,X^2 - 1.03\,Y^2)$$
$$= 5.04 + 0.88\,Y - 2(1.46)\,X = 0 \tag{9.16}$$

$$\frac{\partial P}{\partial Y} = \frac{\partial}{\partial Y}(1.36 + 5.04\,X + 0.88\,XY - 1.46\,X^2 - 1.03\,Y^2)$$
$$= 0.88\,X - 2(1.03)\,Y = 0 \tag{9.17}$$

and from the second equation $Y = \frac{0.88}{2(1.03)}\,X$, so upon inserting this into the first equation there is

$$5.04 - 2.92\,X + 0.38\,X = 0 \quad \text{or} \quad 2.55\,X = 5.04 \quad X = 1.98 \tag{9.18}$$

Then

$$Y = \frac{0.88}{2(1.03)}X = \frac{0.88}{2(1.03)}1.98 = 0.845$$

so that the optimum result is

$$X = 1.98 \times 10^6 \,\text{hectare-meters}, \quad Y = 845\,\text{hectares}$$

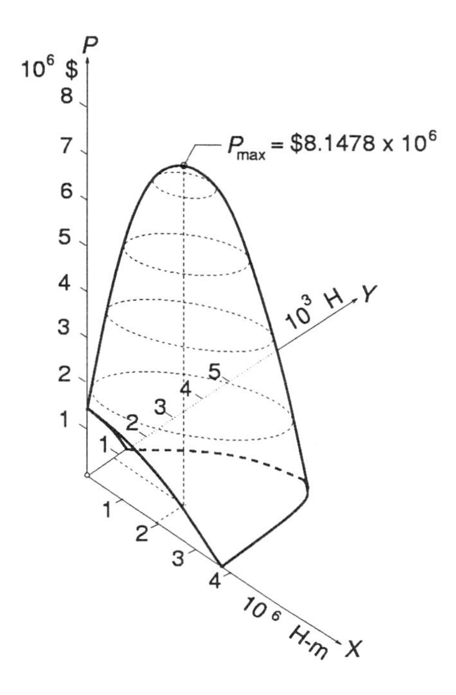

Figure 9.4: Global Maximum

Of course, it is not known yet whether this may be a maximum or a minimum, only that it is an extremum. To settle the question, the second derivatives must be investigated.

$$\frac{\partial^2 P}{\partial X^2} = \frac{\partial}{\partial X}(5.04 + 0.88\,Y - 2(1.46)\,X) = -\,2(1.46) = -2.92 \quad (9.19)$$

$$\frac{\partial^2 P}{\partial Y^2} = \frac{\partial}{\partial Y}(0.88\,X - 2(1.03)\,Y) = -2(1.03) = -2.06 \quad (9.20)$$

$$\frac{\partial^2 P}{\partial X \partial Y} = \frac{\partial}{\partial X}\left(\frac{\partial P}{\partial Y}\right) = \frac{\partial}{\partial X}(0.88\,X - 2(1.03)\,Y) = 0.88 \quad (9.21)$$

$$\left(\frac{\partial^2 P}{\partial X \partial Y}\right)^2 - \frac{\partial^2 P}{\partial X^2}\frac{\partial^2 P}{\partial Y^2} = 0.7744 - (-2.92)(-2.06) = -\,5.24 \quad (9.22)$$

All criteria for a maximum have been fulfilled, and so it can thus be reported that $X = 1.98$, $Y = 0.85$ is a maximum. The conclusion can be confirmed by constructing a graph of the behavior of the function P as shown in Figure 9.4. In fact, the picture shows that, within the region bounding the problem solution as defined by the constraints, there is both a local and a global maximum.

It is easy to see that the result also fulfills both of the constraints, and so is declared the answer. Thereafter the maximized net benefit can be computed as

$$\begin{aligned} P_{\max} &= 1.36 + 5.04(1.98) + 0.88(1.98)(0.85) - 1.46(1.98)^2 - 1.03(0.85)^2 \\ &= \$8.1478 \times 10^6 \end{aligned} \quad (9.23)$$

Obviously, this is an improvement over the previous case, and, obviously, the establishment of it as a maximum has been more complex. The simple criterion of checking one second derivative has been replaced by the computation of several partial derivatives, following which a criterion function was evaluated. This test, or litmus of the gender of the optimum, will evidently become increasingly complicated as the number of independent variables increases. This factor is of great importance, in consideration of the fact that real problems usually do have a large number of decision variables.

9.4 STATIONARY POINTS

To obtain the maximum or minimum of a function of a single variable it is only necessary to compute the first order derivative of the function, set this equal to zero (i.e., make the slope equal to zero so that the tangent to the curve at the point in question is a horizontal line), solve for the special value—the optimum—of the variable at that condition, and substitute this value into the functional equation. Immediately thereafter, test whether the value is a maximum or a minimum by computing the second dervivative of the function with respect to its argument to see if it is positive (minimum), negative (maximum) or zero (inflection point).

As soon as the number of decision variables is greater than one, it is convenient to employ the space vector $\mathbf{x} = \mathbf{x}(x_1, x_2, \ldots, x_n)$, and the criterion function $\mathbf{y} = \mathbf{y}(x_1, x_2, \ldots, x_n)$ is then described as a surface constructed in a space over the "plane" defined by the decision variables. "Plane" needs to be interpreted as "hyperplane" when the number of variables exceeds two. The question of maximum or minimum for the criterion function at the point of optimum becomes one of testing whether or not the curvature of the surface is positive or negative or some other condition.

Indeed, the question of the existence of an extremum takes on a larger meaning for multi-variable surfaces. The vanishing of the first derivatives is a necessary but not a sufficient condition for a point to be a local minimum. The tangent is also horizontal for an inflection point or other flat place. Any place where the gradient of the surface vanishes is called a ***stationary point***.

In optimization work it is not at all uncommon to encounter a set of adjacent local extrema. If these are maxima, there is a ***ridge***, or a ***valley*** if they are minima. At a ***saddle point*** the conditions for an extremum are violated in every feasible neighborhood of the point.

For the one-dimensional problem, Maclaurin gave the rules for distinguishing between stationary points, whether they are a maximum, a minimum or an inflection point. These conditions were extended to the multidimensional case by Lagrange. Basically, the procedure starts at some place named $\mathbf{x}_o$, a supposed stationary point of the space vector $\mathbf{x}$, and proceeds to evaluate $\mathbf{y}$, the criterion function, at a place nearby, given by $\delta\mathbf{x}$, to determine if $\mathbf{y}$ is greater than, less than or on a level with its value at the stationary point. Thus

$$\mathbf{y}(\mathbf{x}_o + \delta\mathbf{x}) = \mathbf{y}(\mathbf{x}_o) + \delta\mathbf{y}_o \tag{9.24}$$

or

$$\delta\mathbf{y}_o = \mathbf{y}(\mathbf{x}_o + \delta\mathbf{x}) - \mathbf{y}(\mathbf{x}_o) \tag{9.25}$$

so that

$$\delta \mathbf{y}_o = \mathbf{y}(\mathbf{x}_o + \delta\mathbf{x}) - \mathbf{y}(\mathbf{x}_o) = \sum_{i=1}^{n} \left(\frac{\partial \mathbf{y}}{\partial x_i}\right) \delta x_i + \frac{1}{2}\sum_{i=1}^{n}\sum_{j=1}^{n} \left(\frac{\partial^2 \mathbf{y}}{\partial x_i \partial x_j}\right) \delta x_i \delta x_j + O(\delta\mathbf{x}^3)$$

$$= \nabla \mathbf{y}_o(\delta\mathbf{x}) + \frac{1}{2}(\delta\mathbf{x})^T \mathbf{H}_o(\delta\mathbf{x}) + O(\delta\mathbf{x}^3) \tag{9.26}$$

$\mathbf{H}_o$ is the Hessian of the matrix of $\mathbf{y}(\mathbf{x}_o)$, evaluated at the place $\mathbf{x}_o$, and $\delta\mathbf{x}^T$ is the transpose of the differential vector $\delta\mathbf{x}$. (That is, the row of the matrix of $\delta\mathbf{x}$ becomes the column of $\delta\mathbf{x}^T$.) If $\delta\mathbf{x}^T\mathbf{H}_o\delta\mathbf{x}$ is positive for all possible $\delta\mathbf{x}$, then the differential quadratic form $\frac{1}{2}(\delta\mathbf{x})^T\mathbf{H}_o(\delta\mathbf{x})$ (and $\mathbf{H}_o$ also) is *positive definite*, a sufficient condition for $\mathbf{x}_o$ to be a local minimum. At local maxima $\mathbf{H}_o$ is *negative definite*. In valleys (ridges) the form is nonnegative (nonpositive), vanishing for $\delta\mathbf{x}$ along the valley (ridge) where $\mathbf{H}_o$ is semidefinite. At saddle points, the quadratic form is positive, negative or zero, depending on $\delta\mathbf{x}$. $\mathbf{H}_o$ is said to be *indefinite* under this condition.

From the matrix vector equation for $\delta\mathbf{y}_o$, above, the Hessian is seen to play the role of a matrix operator (multiplier) which, operating upon $\delta\mathbf{x}$ and subsequently being operated upon in turn by $\delta\mathbf{x}^T$, produces the differential quadratic form $\delta\mathbf{x}^T\mathbf{H}_o\delta\mathbf{x}$. Thus, if $\mathbf{y} = \mathbf{f}(x_1, x_2, \ldots, x_n)$ is the criterion function, a vector-valued functional of the decision variables $x_1, x_2, \ldots, x_n$, the Hessian is, explicitly,

$$\mathbf{H} = \begin{bmatrix} \dfrac{\partial^2 \mathbf{f}}{\partial x_1^2} & \dfrac{\partial^2 \mathbf{f}}{\partial x_1 \partial x_2} & \cdots & \dfrac{\partial^2 \mathbf{f}}{\partial x_1 \partial x_n} \\ \dfrac{\partial^2 \mathbf{f}}{\partial x_2 \partial x_1} & \dfrac{\partial^2 \mathbf{f}}{\partial x_2^2} & \cdots & \dfrac{\partial^2 \mathbf{f}}{\partial x_2 \partial x_n} \\ \cdot & \cdot & \cdots & \cdot \\ \cdot & \cdot & \cdots & \cdot \\ \cdot & \cdot & \cdots & \cdot \\ \dfrac{\partial^2 \mathbf{f}}{\partial x_n \partial x_1} & \dfrac{\partial^2 \mathbf{f}}{\partial x_n \partial x_2} & \cdots & \dfrac{\partial^2 \mathbf{f}}{\partial x_n^2} \end{bmatrix} \tag{9.27}$$

The test to determine the character of a stationary point then becomes a trial to ascertain whether or not the criterion function is convex, concave, or neither. The function is said to be convex if the determinant of the Hessian matrix is greater than zero, and if all of its diagonal elements are also greater than zero. There will be a minimum at the stationary point. On the other hand, if the determinant of the Hessian is less than zero, and all of the diagonal elements are likewise less than zero, the function is concave, and there will be a maximum at the stationary point.

For example, with two decision variables:

$$\det\mathbf{H} = \det \begin{bmatrix} \dfrac{\partial^2 \mathbf{f}}{\partial x_1^2} & \dfrac{\partial^2 \mathbf{f}}{\partial x_1 \partial x_2} \\ \dfrac{\partial^2 \mathbf{f}}{\partial x_2 \partial x_1} & \dfrac{\partial^2 \mathbf{f}}{\partial x_2^2} \end{bmatrix} = \frac{\partial^2 \mathbf{f}}{\partial x_1^2}\frac{\partial^2 \mathbf{f}}{\partial x_2^2} - \frac{\partial^2 \mathbf{f}}{\partial x_2 \partial x_1}\frac{\partial^2 \mathbf{f}}{\partial x_1 \partial x_2} \tag{9.28}$$

which is immediately seen to be the test applied in Black Ankle Two to determine if a maximum or minimum existed at the optimum point.

The notion of convexity will be encountered again in Chapter 11, where it becomes important to the discussion of pattern search procedures for nonlinear programming work. Meanwhile, the analysis of other forms of complexity in the optimization process can continue.

9.5 BLACK ANKLE THREE

Let the Black Ankle project be revisited with the additional feature that the Authority now estimates that the project must accomodate 20,000 people on major holidays such as Memorial Day, Independence Day (July 4th), and Labor Day. The staff studies suggest that each one million hectare-meters can accept 15,000 persons in water-related activities such as swimming, boating, and fishing; that each 1,000 hectares of park can handle 1,000 individuals in camping, picnics, and hiking. These figures suggest a new constraint for the project. It is

$$15,000\,X + 1,000\,Y = 20,000 \tag{9.29}$$

or $15\,X + Y = 20$. The previous constraints, still operative, are of the less-than type inequality, while this new one is an equality. The consequences of this type of constraint will be seen to be very significant.

The formal optimization statement is

$$\text{Optimize} \quad P = 5 + 5.14\,X + 0.88\,X\,Y - 1.46\,X^2 - 1.03\,Y^2 \tag{9.30}$$

$$\text{Subject to} \quad 15\,X + Y = 20 \quad 0 \le X \le 4.72 \quad 0 \le Y \le 3.86 \tag{9.31}$$

The new constraint forces an additional relationship between X and Y. The obvious thing to do is to eliminate Y in favor of X, which will leave the criterion function P in terms of X only, thereby transforming the problem to that of a single variable again. Thus $Y = 20 - 15\,X$ and

$$P = 1.36 + 5.04\,X + 0.88\,X\,(20 - 15\,X) - 1.46\,X^2 - 1.03\,(20 - 15\,X)^2$$

$$= -410.64 + 640.64\,X - 246.41\,X^2 \tag{9.32}$$

Now the optimum can be found by differentiation of the function P, with the result set equal to zero.

$$\frac{dP}{dX} = 640.64 - 2 \times 246.41\,X = 0 \qquad 492.82\,X = 640.64 \qquad X = 1.30 \tag{9.33}$$

and

$$Y = 20 - 15(1.30) = 0.50 \tag{9.34}$$

$$\frac{d^2P}{dX^2} = -\,492.88 \tag{9.35}$$

Since the second derivative is negative, the optimum is a maximum.

The solution to this problem is, then,

$$X = 1.30 \times 10^6 \text{ hectare-meters}, \; Y = 0.50 \times 10^3 \text{ hectares}$$

and the net benefit is

$$P = 5 + 5.04(1.30) + 0.88(1.30)(0.50) - 1.46(1.30)^2 - (0.50)^2 = \$9.16 \times 10^6 \tag{9.36}$$

The inequality constraints are also obviously satisfied.

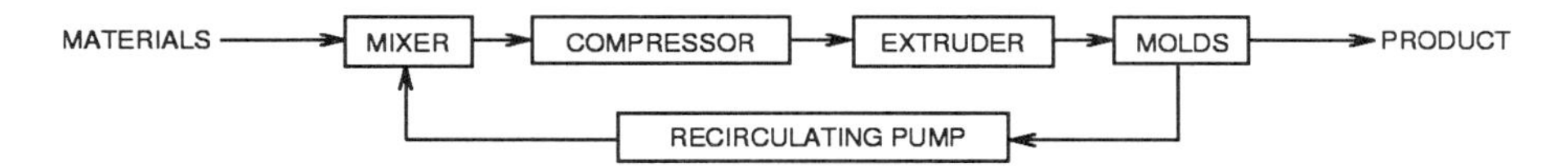

Figure 9.5: Concrete Products Plant Schematic Diagram

9.6 CONCRETE PRODUCTS PLANT REDUX

The operation of a concrete products plant was discussed at some length in Chapter 4. Another view of the optimization of such a plant can be based upon the stationary points concepts of the calculus. Let the production process of the plant be as shown schematically in Figure 9.5. The materials—coarse aggregate, fine aggregate, cement, and water—are introduced into a mixer, following which the mixture passes through a compressor and an extruder into the molds, after which the shapes set-up, are cured, and are moved out as finished product.

A significant fraction of the extruded concrete fails to enter the molds as a part of the product, and can be salvaged (if recovered in time) and recirculated to the mixer.

Two important parameters dominate the cost of running the production line: x_1 is the operating pressure of the compressor, and x_2 is the recycle fraction of concrete. Suppose that the available compressors have an upper pressure limit of X_1 MPa, and suppose further that, in accordance with an industry-wide rule-of-thumb, the concrete quality will be adversely affected if the recycle fraction exceeds X_2. These limits can be translated into constraint relationships

$$0 \le x_1 \le X_1 \,\text{MPa} \tag{9.37}$$

$$0 \le x_2 \le X_2 \tag{9.38}$$

(Negative pressures and recycle fractions would have no meaning in this model.)

Next, an objective function can be created to express the annual operating cost. The expense of running the compressor is a direct function of the pressure, as more power and maintenance are required as the pressure is increased. The records of the past operation of the plant show that the cost is $\$a$ per MPa on an annualized basis. Because the extruder experiences the effects of both first-run and recirculated product, the records indicate that the cost of this component is $\$b/x_1\,x_2$, annualized. Running the recirculating pump costs $\$c\,x_2$ per year.

Some expenses are, practically speaking, independent of the variable parameters, and are omitted from the objective function for purposes of optimizing analysis. That is, these costs must be paid irrespective of the values of x_1 and x_2.

The objective function is then written

$$C = a\,x_1 + \frac{b}{x_1\,x_2} + c\,x_2 \tag{9.39}$$

The optimization requires that

$$\frac{\partial C}{\partial x_1} = 0 = a - \frac{b}{x_1^2\,x_2} \tag{9.40}$$

$$\frac{\partial C}{\partial x_2} = 0 = -\frac{b}{x_1 \, x_2^2} + c \tag{9.41}$$

from which a simultaneous solution yields

$$\overset{\star}{x}_1 = \left(\frac{b\,c}{a^2}\right)^{1/3} \qquad \overset{\star}{x}_2 = \left(\frac{a\,b}{c^2}\right)^{1/3} \tag{9.42}$$

$\overset{\star}{x}_1$ and $\overset{\star}{x}_2$ are thus the optimum values for the operating conditions of the plant. The optimum operating cost is computed using these values as

$$\begin{aligned}
\overset{\star}{C} &= a\left(\frac{b\,c}{a^2}\right)^{1/3} + \frac{b}{\left(\dfrac{b\,c}{a^2}\right)^{1/3}\left(\dfrac{a\,b}{c^2}\right)^{1/3}} + c\left(\frac{a\,b}{c^2}\right)^{1/3} \\
&= \frac{a\,(b\,c)^{1/3}}{a^{2/3}} + b\left(\frac{a^2}{b\,c}\right)^{1/3}\left(\frac{c^2}{a\,b}\right)^{1/3} + \frac{c\,(a\,b)^{1/3}}{c^{2/3}} \\
&= (a\,b\,c)^{1/3} + (a\,b\,c)^{1/3} + (a\,b\,c)^{1/3} = 3\,(a\,b\,c)^{1/3}
\end{aligned} \tag{9.43}$$

The remaining task is to determine if the optimum cost is a minimum, or possibly a maximum. The test of the stationary point will involve the examination of the determinant of the Hessian matrix for convexity of the cost function.

In this example the decision variables are x_1 and x_2, and the objective function $\mathbf{y} = \mathbf{f}(x_1, x_2)$ is the cost

$$C = a\,x_1 + \frac{b}{x_1\,x_2} + c\,x_2$$

The determinant of the Hessian is therefore

$$det\,\mathbf{H} = \frac{\partial^2 C}{\partial x_1^2}\,\frac{\partial^2 C}{\partial x_2^2} - \frac{\partial^2 C}{\partial x_2\,\partial x_1}\,\frac{\partial^2 C}{\partial x_1\,\partial x_2} \tag{9.44}$$

so three derivatives must be obtained. These are

$$\frac{\partial^2 C}{\partial x_1^2} = \frac{\partial}{\partial x_1}\left(\frac{\partial C}{\partial x_1}\right) = \frac{\partial}{\partial x_1}\left(a - \frac{b}{x_1^2\,x_2}\right) = -\frac{b}{x_2}\left(-\frac{2\,x_1}{x_1^4}\right) = \frac{2\,b}{x_1^3\,x_2} \tag{9.45}$$

$$\frac{\partial^2 C}{\partial x_2^2} = \frac{\partial}{\partial x_2}\left(\frac{\partial C}{\partial x_2}\right) = \frac{\partial}{\partial x^2}\left(c - \frac{b}{x_1\,x_2^2}\right) = -\frac{b}{x_1}\left(-\frac{2\,x_2}{x_2^4}\right) = \frac{2\,b}{x_1\,x_2^3} \tag{9.46}$$

$$\frac{\partial^2 C}{\partial x_1\,\partial x_2} = \frac{\partial}{\partial x_2}\left(\frac{\partial C}{\partial x_1}\right) = \frac{\partial}{\partial x_2}\left(a - \frac{b}{x_1^2\,x_2}\right) = -\frac{b}{x_1^2}\left(-\frac{1}{x_2^2}\right) = \frac{b}{x_1^2\,x_2^2} \tag{9.47}$$

Now

$$det\,\mathbf{H} = \left(\frac{2\,b}{x_1^3\,x_2}\right)\left(\frac{2\,b}{x_1\,x_2^3}\right) - \left(\frac{b}{x_1^2\,x_2^2}\right)^2 = \frac{4\,b^2}{x_1^4\,x_2^4} - \frac{b^2}{x_1^4\,x_2^4} = \frac{3\,b^2}{x_1^4\,x_2^4} \tag{9.48}$$

This quantity is, perforce, positive for all values of b, x_1, and x_2, including particularly the optimum point $\overset{\star}{x}_1$, $\overset{\star}{x}_2$. Therefore, the optimum $\overset{\star}{C}$ is a minimum.

9.7 PROJECT #7

1. The operation of a building supplies yard is described by the equations

$$Z = (X_1 - 6.58)^2 + (X_2 - 5.48)^2$$

$$5\,X_1 + 4\,X_2 = 10$$

$$14\,X_1 + 4\,X_2 = 20$$

   in which Z is a function to be minimized subject to the two constraints. a) Determine optimum values of X_1 and X_2. b) Check this solution by a graphical procedure. c) Suppose the Z-function is changed to

$$Z = (X_1 - 6.00)^2 + (X_2 - 5.48)^2$$

   Will this change the optimum values of X_1 and X_2? Show the change (if any) on the graph.

2. C. E. Good, PE, has a professional consulting engineering practice. He estimates that, based upon his last few years of experience, his income (in dollars) will be

$$I = 36,000 + 2,000\,X - 20\,X^2$$

   in which X is the number of projects he undertakes each year. He also recognizes that he and his small staff cannot do justice to any client's project if there are too many at any time, so he sets an upper limit of 32 projects per year.

   How many projects should he do to maximize his income? What will be this maximum income?

3. The County Manager of Excelsior County has acquired a property suitable for maintenance of the County's vehicles from an auto dealership which is going out of business. A portion of the building, which has a total area of 6,000 square meters, could also be allocated to supplies storage. He estimates that the annual internal net benefit (in dollars) to the county of this new space will be

$$B = 10,000 + 60\,X - X^2 + X\,Y - 4\,Y^2$$

   in which X is the area to be allocated to vehicle maintenance and Y is the area used for supplies storage, all in units of square meters.

   What is the optimum size of each area, and what is the benefit to be expected?

4. A developer has acquired a 246-hectare tract of land upon which he plans to build a shopping mall. He hires your consulting engineering firm to assist in the design of the mall.

   Experience suggests that, on a daily basis, the benefits are

$$B = 2\,X\,Y$$

and the costs are

$$C = -0.5X^2 + 70Y$$

in which X is the number of specialty shops and Y is the number of food shops in the mall. Experience also shows that the average specialty shop will have 12 customers at any time and 8 will patronize the average food shop.

The developer wishes to accomodate an average of 1000 people at all times.

Determine how many shops of each kind should be in the mall, and estimate the optimum daily profit.

5. PCS CONSTRUCTIONS, Inc., owns two loaders—an ACME SUPERLOAD and a SPACELY SPROCKET D-LUX LOAD-R. They are subcontractors for a job on which they move topsoil at a construction site. They are paid \$50 for each load of soil hauled by the ACME and \$40 for each from the SPROCKET.

   The characteristics of the two machines are:

	ACME	SPROCKET
Capacity (m^3/load)	10	8
Hauling rate (loads/hour)	40	20
Operating cost (dollars/load)	$(10 + 0.9X)$	$(5 + 0.5Y)$

   If the job requires moving 2,000 cubic meters of topsoil in a 10-hour work day, determine the number of loads, X, with the ACME and the number of loads, Y, with the SPACELY SPROCKET to optimize the income. What is the income? Also find the time required.

6. An engineering firm is considering replacing its ozalid printer and first-generation CAD minicomputer with state-of-the-art devices. The net benefit of the two new pieces of equipment over their useful life is estimated to be

$$B = 6 + 40X + 2XY - 3X^2 - Y^2$$

   in which X is the cost of the new CAD workstation and Y is that of the new printer, all in thousands of dollars. The three best candidates for the workstations cost \$10,000, \$12,000, and \$15,000, and printers are available at \$6,000, \$9,000, and \$11,000. Which machines should be selected if the total money available to be spent is \$40,000? What will be the expected annual benefit?

7. A highway bridge needs to span a gap of 2,000 m over a river gorge. Preliminary studies show that a series of equal-span trusses is most suitable. The two end piers will cost about \$40,000 each. The intervening piers, which will be located in the river itself, can be constructed anywhere within the crossing at a cost of \$80,000 each. The trusses will cost $8L^2$ dollars, where L is the length in meters.

   Determine the optimum number of the trusses, and give their length. What is the total cost of the bridge?

8. STRUCTEX, PA, are an engineering firm which specializes in the analysis and design of commercial buildings. They have studied their operations for a period of years, and have developed the equation

$$P = a + b\,x - c\,x^2 + d\,x\,y - e\,y^2$$

for the annual gross earnings of the firm, in which x is the number of steel structures projects and y is the number of reinforced concrete buildings in a given year. What is the optimum mix of projects to maximize the earnings?

What restrictions apply to the constants a, b, c, d, and e? Since x and y need to be integers, investigate the effect of rounding up or down upon the outcome for P.

9. A building supply firm stocks two major kinds of paints. The manager notes that, on an annual basis, the firm's profits are given by the equation

$$P = 2.31\,x_1\,x_2$$

in which x_1 and x_2 are the quantities of paint sold, in 1000s of gallons, of the two types, respectively.

However, the firm is limited by constraints $x_2 + 0.13\,x_1^2 \leq 12$ and $x_1 \leq 6.79$. Also, x_1, $x_2 \geq 0$, of course.

Determine the quantities of paint of each type which, when sold, will optimize the firm's earnings. Confirm your solution with a graphical answer.

CHAPTER 10

LAGRANGE MULTIPLIER METHODS

10.1 INTRODUCTION

Nearly one hundred years after Sir Isaac Newton and his calculus came another genius whose methods were to revolutionize the fields of mathematics, mechanics, and systems theory, although the latter was not recognized by that name at that time. The man was Joseph Louis comte Lagrange, the methods (for systems theory) are called ***Lagrange multipliers***. Lagrange's magnus opus, the *Mécanique analytique*, was published in 1789.

One of the salient aspects of Lagrange's work was the almost mystical manner in which certain key ingredients were introduced. In today's parlance, it might be said that "he did it with mirrors (or with smoke)." Perhaps a better form might be, in agreement with his national tongue, "C'est un mystère!" Without a doubt, there are elements of both the Lagrangean mechanics and the Lagrangean optimization methods which are metaphysical, or perhaps one should say, in this case, metamechanical.

One may also reasonably question whether Lagrange realized the full extent of the applicability of his new techniques. When he said, in effect, that he had included all of mechanics in one formula, and added "the world will be grateful to me for having done so," it was, as the saying goes, no brag—just fact. Even so, that one formula has been made to unify not only mechanics, but thermodynamics, electromagnetics, chemical reactivity, transport phenomena, and other branches of engineering science as well.

The Lagrange multiplier technique has also insinuated its way into several branches of engineering science. It is most unlikely that Lagrange would have identified it with all of its present day uses, but he may not have been surprised at its great utility in connection with the central business of optimization.

10.2 LAGRANGE MULTIPLIERS

Return once again to Black Ankle Three, the reservoir-park problem with equality constraint. Recall that the net benefit P is to be maximized subject to three constraints as follows:

$$P = 1.36 + 5.04\,X + 0.88\,X\,Y - 1.46\,X^2 - 1.03\,Y^2 \tag{10.1}$$

$$15\,X + Y - 20 = 0 \qquad 0 \le X \le 4.72 \qquad 0 \le Y \le 3.87 \tag{10.2}$$

Now, the function P will not be altered by adding to it the quantity 0.

$$P = 1.36 + 5.04\,X + 0.88\,X\,Y - 1.46\,X^2 - 1.03\,Y^2 + 0 \tag{10.3}$$

Let the 0 that is added to P be the specific quantity $15\,X + Y - 20 = 0$;

$$P = 1.36 + 5.04\,X + 0.88\,X\,Y - 1.46\,X^2 - 1.03\,Y^2 + 15\,X + Y - 20 \tag{10.4}$$

and, since it is zero anyway, P will not be changed if the 0 is multiplied by a parameter λ.

$$P = 1.36 + 5.04\,X + 0.88\,X\,Y - 1.46\,X^2 - 1.03\,Y^2 + \lambda(15\,X + Y - 20) \tag{10.5}$$

Once introduced into P, even by this back-door method, the parameter λ assumes a status equal-standing with that of X and Y, so that in the remainder of the optimization process it is treated as a variable of the problem. Then to determine the maximum, proceed in the usual way by taking derivatives of P with respect to each of the variables in turn, set each equal to zero and solve them simultaneously to obtain the three values of X, Y, and λ at the stationary point.

$$\frac{\partial P}{\partial X} = 5.04 + 0.88\,Y - 2(1.46)\,X + 15\,\lambda = 0 \tag{10.6}$$

$$\frac{\partial P}{\partial Y} = 0.88\,X - 2(1.03)\,Y + \lambda = 0 \tag{10.7}$$

$$\frac{\partial P}{\partial \lambda} = +(15\,X + Y - 20) = 0 \tag{10.8}$$

From Equation 10.7, $\lambda = 0.88\,X - 2(1.03)\,Y$, and using this in Equation 10.6 yields

$$\begin{aligned} 5.04 &+ 0.88\,Y - 2(1.46)\,X + 15\,[0.88\,X - 2(1.03)\,Y] \\ &= 5.04 - 16.12\,X + 31.78\,Y = 0 \end{aligned} \tag{10.9}$$

But $Y = -15\,X + 20$ from Equation 10.8, so

$$5.04 - 16.12\,X + 31.78(-15\,X + 20) = 0 \tag{10.10}$$

$$640.64 - 492.82\,X = 0 \qquad X = 640.64/492.82 = 1.30 \tag{10.11}$$

$$Y = -15(1.30) + 20 = 0.50 \tag{10.12}$$

$$\lambda = 2(1.03)(0.50) - 0.88(1.30) = -\,0.114 \tag{10.13}$$

and then

$$\begin{aligned} P = {} & 1.36 + 5.04(1.30) + (1.30)(0.50) - 1.46(1.30)^2 - (0.50)^2 \\ & + (-0.114)[(15)(1.30) + 0.50 - 20] = \$5.76 \times 10^6 \end{aligned} \tag{10.14}$$

as with the former method.

The parameter λ is called a *Lagrange multiplier*. It is in some ways similar to the virtual displacement of mechanics, with which it was of contemporary origin during the eighteenth century, in that it does no violence to the constraints and is otherwise arbitrary. However, its principal function is that of catalyst, or, more crudely, that of a prybar. It serves to "open up" the process of solution for the optimum values of the decision variables in constrained optimization problems. It seems to have been magically provided—the serendipitous flyby gadget with "handle toward the hand" just as the mind's hand reaches out for that very tool. A passerby stops to read a bulletin board and there, before the very eyes, is the discovery most wanted and most needed.

The Lagrange multiplier is a mathematical lubricant which eases the solution process by displacing the variables from their central role for a time, and replacing them with a new variable or variables unconstrained by the relationships between the other variables. The product of the Lagrange multiplier and the constraint equation is both seen and unseen, real and unreal, and above all, transient. It is an interloper, not there at the start, not there in the end, like a Good Samaritan by the road who helps out in a time of difficulty and then goes away, not to be heard from again.

As a second example, take the following function which is to be minimized, subject to two equality constraints:

$$\text{Minimize} \quad f = \tfrac{1}{2}(X_1^2 + X_2^2 + X_3^2) \tag{10.15}$$

$$\text{Subject to} \quad X_1 + 2\,X_2 + 3\,X_3 = 1 \tag{10.16}$$

$$3\,X_1 + 2\,X_2 + X_3 = 2 \tag{10.17}$$

Notice that this is a considerable escalation from the previous problem because there are three decision variables and two equality constraints, instead of the former two variables and one equality constraint.

Based upon what was learned in the previous case, there is now formed a Lagrangean function using the criterion function, f, and two Lagrange multipliers, one for each constraint. Thus

$$L(X_1, X_2, X_3, \lambda_1, \lambda_2) = \tfrac{1}{2}(X_1^2 + X_2^2 + X_3^2)$$

$$-\lambda_1(X_1 + 2\,X_2 + 3\,X_3 - 1) - \lambda_2(3\,X_1 + 2\,X_2 + X_3 - 2) \tag{10.18}$$

is the Lagrangean function and λ_1, λ_2 are the multipliers.

To determine the location of the stationary point, take the derivatives of L with respect to the three variables plus the two parameters, and set each equal to zero in turn.

$$\frac{\partial L}{\partial X_1} = X_1 - \lambda_1 - 3\,\lambda_2 = 0 \tag{10.19}$$

$$\frac{\partial L}{\partial X_2} = X_2 - 2\,\lambda_1 - 2\,\lambda_2 = 0 \tag{10.20}$$

$$\frac{\partial L}{\partial X_3} = X_3 - 3\,\lambda_1 - \lambda_2 = 0 \tag{10.21}$$

$$\frac{\partial L}{\partial \lambda_1} = X_1 + 2\,X_2 + 3\,X_3 - 1 = 0 \tag{10.22}$$

$$\frac{\partial L}{\partial \lambda_2} = 3\,X_1 + 2\,X_2 + X_3 - 2 = 0 \tag{10.23}$$

A suggested solution procedure is to solve the first three of these simultaneous equations for the X-variables in terms of λ_1 and λ_2 and then insert these into the last two equations of the set. Equations 10.22 and 10.23 will then become two equations in two unknowns.

Then from Equations 10.20, 10.21, and 10.22,

$$X_1 = \lambda_1 + 3\,\lambda_2 \tag{10.24}$$
$$X_2 = 2\,\lambda_1 + 2\,\lambda_2 \tag{10.25}$$
$$X_3 = 3\,\lambda_1 + \lambda_2 \tag{10.26}$$

so that Equations 10.22 and 10.23 are

$$(\lambda_1 + 3\,\lambda_2) + 2(2\,\lambda_1 + 2\,\lambda_2) + 3(3\,\lambda_1 + \lambda_2) - 1 = 0 \tag{10.27}$$

$$3(\lambda_1 + 3\,\lambda_2) + 2(2\,\lambda_1 + 2\,\lambda_2) + (3\,\lambda_1 + \lambda_2) - 2 = 0 \tag{10.28}$$

or

$$14\,\lambda_1 + 10\,\lambda_2 = 1 \tag{10.29}$$

$$10\,\lambda_1 + 14\,\lambda_2 = 2 \tag{10.30}$$

which have the solution $\lambda_1 = -1/16$, $\lambda_2 = 3/16$.

Once the multipliers have been found, they can be put into Equations 10.24, 10.25, and 10.26 to determine the values of the variables at the stationary point.

$$X_1 = -\tfrac{1}{16} + 3(\tfrac{3}{16}) = \tfrac{1}{2} \tag{10.31}$$
$$X_2 = 2(-\tfrac{1}{16}) + 2(\tfrac{3}{16}) = \tfrac{1}{4} \tag{10.32}$$
$$X_3 = 3(-\tfrac{1}{16}) + \tfrac{3}{16} = 0 \tag{10.33}$$

The value of the criterion function at the stationary point is

$$f = \tfrac{1}{2}[(\tfrac{1}{2})^2 + (\tfrac{1}{4})^2 + (0)^2] = \tfrac{5}{32} \tag{10.34}$$

10.3 OPTIMIZATION WITH SEVERAL VARIABLES

At this juncture it becomes possible to generalize the procedure which has been followed in the particular examples above. The procedure will start with the generalization based upon equality constraints first, then move to incorporate inequality constraints as well.

10.3.1 THE CASE OF EQUALITY CONSTRAINTS

Let there be an objective function $f(x, y)$ of the two variables x, y, and let there be a constraint function $g(x, y) = b$, where b is some constant. Then the Lagrangean function L is

$$L(x, y, \lambda) = f(x, y) - \lambda[g(x, y) - b] \tag{10.35}$$

The optimization technique now treats L as it would if there were no constraint. That is, by taking the partial derivatives of L with respect to x, y, and λ, and thereafter setting the results equal to zero. The simultaneous solution of the three equations produces the optimum.

$$\frac{\partial L}{\partial x} = \frac{\partial f}{\partial x} - \lambda\frac{\partial g}{\partial x} = 0 \tag{10.36}$$

$$\frac{\partial L}{\partial y} = \frac{\partial f}{\partial y} - \lambda\frac{\partial g}{\partial y} = 0 \tag{10.37}$$

$$\frac{\partial L}{\partial \lambda} = g(x, y) - b = 0 \tag{10.38}$$

The first two of these show that

$$\frac{\partial f}{\partial x} = \lambda\frac{\partial g}{\partial x} \qquad \text{and} \qquad \frac{\partial f}{\partial y} = \lambda\frac{\partial g}{\partial y} \tag{10.39}$$

at the stationary point. By taking the total derivative of f with respect to b, the result

$$\begin{aligned} \frac{df}{db} = \frac{dL}{db} &= \frac{\partial L}{\partial b} + \frac{\partial L}{\partial x}\frac{dx}{db} + \frac{\partial L}{\partial y}\frac{dy}{db} + \frac{\partial L}{\partial \lambda}\frac{d\lambda}{db} \\ &= \lambda + (0)\frac{dx}{db} + (0)\frac{dy}{db} + (0)\frac{d\lambda}{db} \equiv \lambda \end{aligned} \tag{10.40}$$

is obtained, so that

$$\lambda = \frac{\text{change in the objective function}}{\text{change in the right–hand side of the constraint}} \tag{10.41}$$

The significance of λ, the Lagrange multiplier, is that it serves as a measure of the degree of change in the objective function for a given change in the right-hand side of the constraint. It is a ***sensitivity factor***, by means of which it is possible to gauge the effect of changing the constant b on the objective function.

10.3.2 THE CASE OF INEQUALITY CONSTRAINTS

Move now to the case in which there is some general objective function $f(x, y)$ to be maximized subject to the constraint $g(x, y) \leq b$. The first step in the optimization process consists of converting the inequality constraint into an equality by inserting a quantity S, called the ***slack***, into the constraint equation. S is a parameter which "makes up the slack" between the left-hand side of the equation and the constant b on the right. Then

$$g(x, y) + S^2 = b \tag{10.42}$$

is the new form of the constraint, wherein the quantity S has been squared so that the slack is taken up whether S is positive or negative.

The Lagrangean function L becomes

$$L(x, y, \lambda, S) = f(x, y) - \lambda[g(x, y) + S^2 - b] \tag{10.43}$$

and it is now in a form suitable for processing by the usual technique. That is, derivatives are taken with respect to each of the parameters in L, the results are set equal to zero, and a simultaneous solution of these produces the collection of values for x, y, λ, and S which pertain to the optimum point in a space of rapidly increasing dimension.

$$\frac{\partial L}{\partial x} = 0, \qquad \frac{\partial L}{\partial y} = 0, \qquad \frac{\partial L}{\partial \lambda} = 0, \qquad \frac{\partial L}{\partial S} = 2\,\lambda\,S = 0 \tag{10.44}$$

The last equation is obviously different than the others of the set, and it will soon be seen to be of highest importance in the overall development of the method. For the moment, take an example of the procedure as follows.

Let the objective function (criterion function) be given as $f(x) = 9 + 4\,x - x^2$, and let the constraint be $x \leq 1.5$. The Lagrangean function is

$$L(x, \lambda, S) = 9 + 4\,x - x^2 - \lambda(x + S^2 - 1.5) \tag{10.45}$$

and the derivatives are

$$\frac{\partial L}{\partial x} = 4 - 2\,x - \lambda = 0 \tag{10.46}$$

$$\frac{\partial L}{\partial \lambda} = -(x + S^2 - 1.5) = 0 \tag{10.47}$$

$$\frac{\partial L}{\partial S} = -\,2\,\lambda\,S = 0 \tag{10.48}$$

Now it is clear that there are exactly two conditions under which a solution exists for the set above; namely,

$$S = 0 \qquad x = 1.5 \qquad \lambda = 1 \qquad f(x) = 12.75 \tag{10.49}$$

or

$$\lambda = 0 \qquad x = 2.0 \qquad S^2 = -\,0.5 \tag{10.50}$$

Of course, both S and λ might vanish simultaneously.

On the face of it, the latter solution is not suitable because it requires that S shall have an imaginary value, a condition which has no physical significance. In systems work, the language used for such a result is ***infeasible***. In other words, there is no feasible solution corresponding to these values.

The other one, however, is acceptable, and is therefore said to be feasible.

This case furnishes an easy piece in which to show the effect of varying the right-hand side of the constraint inequality. Suppose that b is 2.0 instead of the 1.5 previously given. The new solution is

$$S = 0 \qquad x = 2.0 \qquad \lambda = 0 \qquad f(x) = 13.00 \tag{10.51}$$

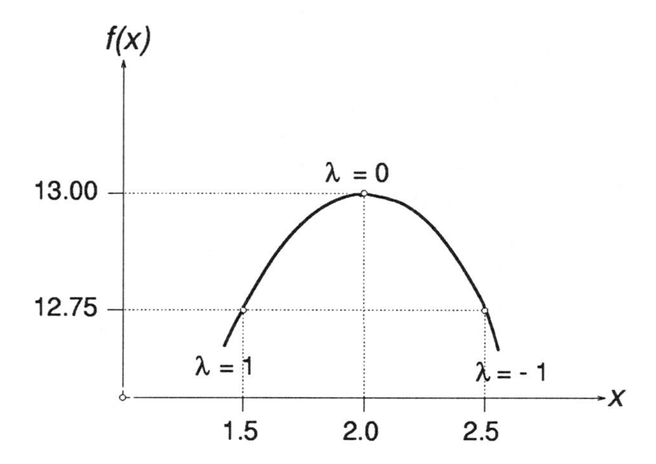

Figure 10.1: Sensitivity Index

which is still feasible. To complete the picture, take the case of b = 2.5, which gives

$$S = 0 \qquad x = 2.5 \qquad \lambda = -1 \qquad f(x) = 12.75 \tag{10.52}$$

The physical significance of these results is as shown in Figure 10.1. Clearly, the optimum (maximum) value of the criterion function, f, is located at $x = 2.0$. This is a particular case of the general criterion:

- If $\lambda \geq 0$, the constraint is not binding, and the value of increasing the right-hand side of the constraint inequality is greater than zero (i.e., there is an unused portion of the "resource" the utilization of which the constraint limits);
- If $\lambda = 0$, the value of increasing the right-hand side of the constraint inequality is zero;
- If $\lambda \leq 0$, the optimum point has been passed, although there is still a feasible solution.

With two decision variables, the objective function might be, say,

$$f = 1 + 4x + xy - x^2 - y^2 \tag{10.53}$$

with the constraint $1.4x + y \leq 3$. Convert the constraint inequality to an equality by introducing the slack, S, as

$$1.4x + y + S^2 = 3 \tag{10.54}$$

and create the Lagrangean function L as

$$L = 1 + 4x + xy - x^2 - y^2 - \lambda(1.4x + y + S^2 - 3) \tag{10.55}$$

Now take the partial derivatives with respect to x, y, λ, and S and set these equal to zero.

$$\frac{\partial L}{\partial x} = 4 + y - 2x - 1.4y = 0 \tag{10.56}$$

$$\frac{\partial L}{\partial y} = x - 2y - \lambda = 0 \tag{10.57}$$

$$\frac{\partial L}{\partial \lambda} = -(1.4x + y + S^2 - 3) = 0 \tag{10.58}$$

$$\frac{\partial L}{\partial S} = -2\lambda S = 0 \tag{10.59}$$

For the case in which $\lambda = 0$, to satisfy Equation 10.59, note that the problem degenerates to the solution of

$$4 + y - 2x - 1.4y = 0 \tag{10.60}$$

$$x - 2y = 0 \tag{10.61}$$

which are the very same equations that would be obtained if there were no constraint. The solution is $x = 2\ 2/3$, $y = 1\ 1/3$, and for S there is

$$S^2 = -1.4 \times (2\tfrac{2}{3}) - (1\tfrac{1}{3}) + 3 = -2.06 \tag{10.62}$$

Because this is negative, so that S is imaginary, there is no feasible solution.

For the case where $S = 0$ to satisfy Equation 10.59, the simultaneous solution of Equations 10.56, 10.57, and 10.58 gives $x = 1.77$, $y = 0.528$, $\lambda = 0.710$, and the objective function is $f = 5.6 \times 10^3$.

Consider next the case with an objective function $f(x, y)$ to be optimized in the presence of a constraint of the greater-than type, say $g(x, y) \geq b$. Here, the constraint inequality can be converted into an equality by subtracting a quantity S^2, which in this instance is called a ***surplus***, because it represents a surfeit of the resource represented by the constraint. Otherwise, the procedure remains the same. That is, construct the Lagrangean function L as

$$L = f(x, y) - \lambda[g(x, y) - S^2 - b] \tag{10.63}$$

then take its derivatives, set them equal to zero, and solve simultaneously.

$$\frac{\partial L}{\partial x} = \frac{\partial f}{\partial x} - \lambda\frac{\partial g}{\partial x} = 0 \tag{10.64}$$

$$\frac{\partial L}{\partial y} = \frac{\partial f}{\partial y} - \lambda\frac{\partial g}{\partial y} = 0 \tag{10.65}$$

$$\frac{\partial L}{\partial \lambda} = -[g(x, y) - S^2 - b] = 0 \tag{10.66}$$

$$\frac{\partial L}{\partial S} = -2\lambda S = 0 \tag{10.67}$$

The process is obviously the same as for the former instance.

Two special constraint situations need to be mentioned. These are the inequalities $x \leq 0$ and $x \geq 0$. They would be altered to show

$$x \leq 0 \qquad x + S^2 = 0 \tag{10.68}$$

$$x \geq 0 \qquad x - S^2 = 0 \tag{10.69}$$

to reflect the slack and surplus associated with the two cases. Each would merit its own Lagrange multiplier when it is entered into the formation of the Lagrangean function for the problems where it arises.

Building upon the examples cited above, a general procedure for addressing problems by using Lagrange multipliers can be enunciated as follows.

10.4 LAGRANGE MULTIPLIER METHOD

If the method of solving the Lagrangean equations is analytic (which is possible in many cases) then the Lagrange multiplier method can be used to find all local optima, and from these the global optimum can be had.

A formal statement of the method and a procedure for carrying it out for the general situation is given below. This declaration states whether there is to be a maximization or a minimization of the objective function, then states the constraints to which the optimization is subject, including the universal condition that all variables are to be positive quantities.

$$\text{MAX or MIN} \qquad Z = f(x_1, x_2, \ldots, x_n) \tag{10.70}$$

$$\text{ST} \qquad C_1(x_1, x_2, \ldots, x_n) \begin{array}{c} > \\ = \\ < \end{array} b_1 \tag{10.71}$$

$$\ldots\ldots\ldots\ldots\ldots\ldots \quad \ldots \quad \ldots$$

$$C_m(x_1, x_2, \ldots, x_n) \begin{array}{c} > \\ = \\ < \end{array} b_m \tag{10.72}$$

$$x_j \geq 0 \quad (j = 1, 2, \ldots, m) \tag{10.73}$$

Convert each inequation to an equation by adding slacks and surplus variables, S_i^2. There will be, say, l of these. Now multiply each constraint by a constant, λ_i, called the Lagrange multiplier. Add each constraint (multiplied by its λ_i) into the objective function to form

$$Z = f(x_1, x_2, \ldots, x_n; S_1, S_2, \ldots, S_l; \lambda_1, \lambda_2, \ldots, \lambda_m) \tag{10.74}$$

Take partial derivatives of Z with respect to each of its variables and set these equal to zero. Solve the simultaneous $n+l+m$ equations for each of these variables. The set will usually be nonlinear, so numerical techniques will probably be needed to solve them.

10.5 THE KUHN-TUCKER THEOREM

From the developments presented above, it will be appreciated that there is a special significance attached to those derivatives of the Lagrangean function with the form $-2\lambda S$, whether these result from slacks or surplusses. There will, naturally, be as many of these terms as there are inequalities, each with its own suffixes to indicate which λ and which S is involved.

The role played by these terms in the optimization process was illuminated in 1950 by studies of Kuhn and Tucker.[1] Their presentation at the Symposium and the paper which followed in the Proceedings constitute a dividing line which separates the limited theory and procedures of the Newton-Lagrange eras from modern practices, and are accordingly a watershed in the growth of contemporary methods. This progress has now been solidified into a theorem, bearing the name *Kuhn-Tucker Theorem*, which is the basis for the present state of knowledge in systems optimization.

The theorem specifies conditions that are necessary and sufficient to ensure that a global optimum exists. If there are a differentiable objective function

$$Z = Z(x_1, x_2, \ldots, x_n) \tag{10.75}$$

and constraints

$$C_i(x_1, x_2, \ldots, x_n) \leq b_i \quad i = 1, 2, \ldots, m \tag{10.76}$$

$$x_j \geq 0 \quad j = 1, 2, \cdots, n \tag{10.77}$$

Kuhn-Tucker states that an optimum solution $\overset{*}{x}_1$, $\overset{*}{x}_2$, ..., $\overset{*}{x}_n$ exists only if there are constants λ_1, λ_2, ... , λ_m such that:

$$\text{If } \overset{*}{x}_j \geq 0, \text{ then } \frac{dZ}{dx_j} + \sum_{i=1}^{m} \lambda_i \frac{dC_i}{dx_j}\bigg|_{\overset{*}{x}_j} = 0, \quad j = 1, 2, \ldots, n \tag{10.78}$$

$$\text{If } \overset{*}{x}_j = 0, \text{ then } \frac{dZ}{dx_j} + \sum_{i=1}^{n} \lambda_i \frac{dC_i}{dx_j}\bigg|_{\overset{*}{x}_j} \geq 0, \quad j = 1, 2, \ldots, n \tag{10.79}$$

$$\text{If } \lambda_i \geq 0, \text{ then } C_i(\overset{*}{x}_1, \overset{*}{x}_2, \ldots, \overset{*}{x}_n) = b_i, \quad i = 1, 2, \ldots, m \tag{10.80}$$

$$\text{If } \lambda_i = 0, \text{ then } C_i(\overset{*}{x}_1, \overset{*}{x}_2, \ldots, \overset{*}{x}_n) \leq b_i, \quad i = 1, 2, \ldots, m \tag{10.81}$$

$$\overset{*}{x}_j \geq 0, \quad j = 1, 2, \ldots, n \tag{10.82}$$

$$\lambda_i \geq 0, \quad i = 1, 2, \ldots, m \tag{10.83}$$

The constants are, of course, Lagrange multipliers.

The first requirement, Equation 10.78, is that at the site of the optimum, the total derivative of the objective function vanishes. In addition, if the optimum

[1]Kuhn, H. W., and Tucker, A. W., "Nonlinear Programming," *Proceedings of the Second Berkeley Symposium on Mathematical Statistics and Probability*, Jerzy Neyman, Editor, University of California Press, Berkeley, 1951, pp. 481-492.

occurs at the limit of non-negative value of a particular decision variable, the case is governed by the second requirement, Equation 10.79.

Equation 10.80, a condition on the sensitivity index λ_i, is for a constraint that is working up to its limit (i.e., is ***binding***), while Equation 10.81 applies if the constraint in question is slack (***loose***). Equations 10.82 and 10.83 provide the customary non-negative requirement for decision variables and Lagrange multipliers.

10.6 PROJECT #8

1. Maximize the function $Z = X_1 X_2$ subject to the conditions

$$0.5\,X_1 + X_2 \leq 9$$
$$X_1 \leq 13$$
$$X_1, X_2 \geq 0$$

   using a Lagrange multiplier method, confirmed by a graphical solution.

2. The owner of a family seafood restaurant (Calabash-style) plans to renovate his establishment during the off-season. He wishes to seat 120 customers at all times to keep the waiting line as short as posssible.

   Based upon his previous experience, he plans to install two sizes of tables; one to sit parties of four, another to handle groups of eight. His business records indicate that the net benefit from the operation is

$$B = 1{,}000 + 8\,X - 4\,X^2 + 2\,X\,Y - 6\,Y^2$$

   where X is the number of 8-person tables, Y the number of 4-person tables.

   Assume that each table is filled with customers all of the time. Use a Lagrange multiplier method to determine X and Y to maximize the owner's benefit.

3. The plan of a proposed racetrack is shown below. To minimize the cost of the paving, the centerline length of the lap is set at 500 m. Determine the radius of the ends, r, and the length of the straight portions, ℓ, so that the cross-sectional area within the loop is a maximum.

   What is the area?

   Confirm the computations using a Lagrange multiplier method.

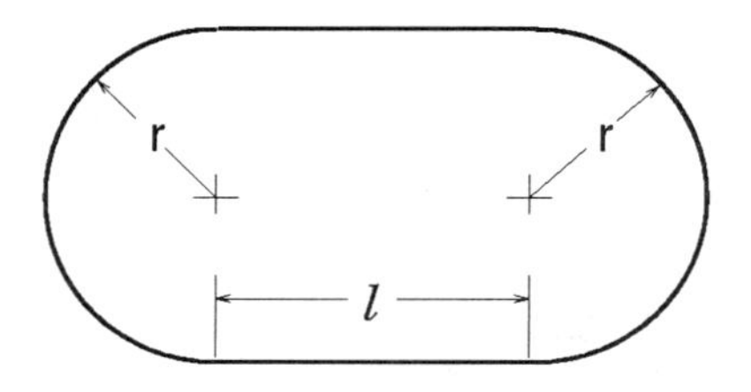

4. DOT has estimated that its cost for new interstate highway construction in mountainous country is

$$C = \tfrac{1}{4}(X - 4)^2 + 1$$

in which X is in kilometers and C is in tens of millions of dollars. They have also estimated that the minimum size of a construction project for economic feasibility is 5.2 kilometers. What is the optimum length of the highway project which DOT should let to contract? What is the minimum cost? Confirm this result using a Lagrange multiplier method. Show also a graphical solution.

5. Maximize the function $Z = X_1 X_2$ subject to the conditions

$$0.5\, X_1 + X_2 \le 9$$

$$X_1 \le 13$$

$$X_1,\, X_2 \ge 0$$

using a Lagrange multiplier method, confirmed by a graphical solution.

6. O'Connor and Goldberg is a consulting engineering company which specializes in environmental and soil testing jobs. Company officers have studied the firm's performance and have concluded that their income can be represented by the equation

$$I = 3 - 27\, X + 2\, X\, Y + 0.8\, X^3$$

in which I is in thousands of dollars, X is the number of environmental projects and Y is the number of soil testing projects. In addition, they have recognized that the company's resources limit the total number of projects which can be undertaken to 14, and the number of environmental projects at any time to 8.

Determine the number of projects in each category which will maximize the company's income. What is this maximum income.

Use a Lagrange multiplier method to confirm your solution to the problem.

7. The Senior Partner in an engineering firm is reviewing the firm's operations in preparation for establishing next year's budget. He notes that the productivity of the technical staff can be represented by the function

$$P = (X - 2.1)^2 + 1.1$$

in which X is the number of workstations available to them. The equation is based upon data taken from the few existing workstations in the office. He estimates, however, that the firm can only afford the money to purchase a maximum of 5 workstations.

How many workstations should the firm have?

8. The engineer in charge of planning notes that over a year's time the operations of the Materials Department can be described by the following equations:

$$Z = x_1 \, x_2$$
$$0.5 \, x_1 + x_2 \leq 9$$
$$x_1 \leq 15.0$$
$$x_1, x_2 \geq 0$$

Z is the total benefit of the Department's operations, x_1 and x_2 are the amounts of asphalt and concrete employed, and the constraints express the requirements of codes and good practice.

She asks you to maximize the benefit using a Lagrange multiplier method, and to confirm the computations with a graphical solution.

9. Given the objective function

$$f = x_1^2 + 2\,x_2^2 - 3\,x_1\,x_2 - 4\,x_1$$

and the constraint

$$5\,x_1 + x_2 = 4$$

determine the minimum of the function. Find the values of x_1 and x_2 at the minimum. Show the solution by the method of Lagrange multipliers and confirm the result using a direct substitution method.

CHAPTER 11

LINEAR PROGRAMMING

11.1 INTRODUCTION

The technique called Linear Programming, or LP, was born sometime after the end of World War II and just before the beginning of the so-called Cold War. It is a methodology drawn from the field known as Operations Research, which is a branch of learning that crosses disciplinary lines between mathematics or applied mathematics, and Industrial Engineering and closely-related technological specialties. OR is the science of logistics—of planning, optimizing, analyzing, controlling, and directing large-scale and far-flung enterprises.

LP is largely the result of one man's endeavors. He is George Dantzig, a legend in his own time. Dantzig was a member of the Rand Corporation "think tank," and he also taught classes at University of California, Berkeley. He managed to perfect his own method of solving optimization problems similar to those encountered in Chapter 4, and other related industrial and governmental oriented problems. Apparently he, and others like him who labored in this field, were driven to seek and refine the methodology because of the increasing complexity of the systems with which they worked.

Dantzig's toy is called the Simplex Method, not because it is simple (although it isn't *that* complex, either) but in recognition of the problem which he chose to illustrate it at the time of its annunciation. There was a company making and distributing a product for sale, and Dantzig analyzed the most efficient scheme for their doing business.

In emulation of Dantzig, the approach herein to the general procedure is through the route of a specific example which follows in course.

11.2 AN OPTIMIZATION EXAMPLE

Imagine that there is the following problem to be dealt with. It presents several decision variables and several inequality-type constraints. The problem is to maximize an objective function subject to these constraints. Note that there are two less-than type inequalities, and two greater-than type inequalities—the usual and customary requirements placed on the decision variables.

$$\text{MAX} \quad Z = 4\,x_1 + 5\,x_2 \tag{11.1}$$
$$\text{ST} \quad x_1 + 2\,x_2 \leq 8 \tag{11.2}$$
$$3\,x_1 + 2\,x_2 \leq 15 \tag{11.3}$$
$$x_1 \geq 0 \tag{11.4}$$
$$x_2 \geq 0 \tag{11.5}$$

The next step will be to formulate the Lagrangean function, L.

$$L = 4\,x_1 + 5\,x_2 - \lambda_1(x_1 + 2\,x_2 + S_1^2 - 8) - \lambda_2(3\,x_1 + 2\,x_2 + S_2^2 - 15)$$
$$-\lambda_3(x_1 - S_3^2) - \lambda_4(x_2 - S_4^2) \tag{11.6}$$

And after this has been done, as is shown above, derivatives of it with respect to each of the parameters in it will be taken and set equal to zero. Note that there are ten of these in the present instance; two decision variables, four λs, four Ss.

$$\frac{\partial L}{\partial x_1} = 4 - \lambda_1 - 3\,\lambda_2 - \lambda_3 = 0 \tag{11.7}$$

$$\frac{\partial L}{\partial x_2} = 5 - 2\,\lambda_1 - 2\,\lambda_2 - \lambda_4 = 0 \tag{11.8}$$

$$\frac{\partial L}{\partial \lambda_1} = x_1 + 2\,x_2 + S_1^2 - 8 = 0 \tag{11.9}$$

$$\frac{\partial L}{\partial \lambda_2} = 3\,x_1 + 2\,x_2 + S_2^2 - 15 = 0 \tag{11.10}$$

$$\frac{\partial L}{\partial \lambda_3} = x_1 - S_3^2 = 0 \tag{11.11}$$

$$\frac{\partial L}{\partial \lambda_4} = x_2 - S_4^2 = 0 \tag{11.12}$$

$$\frac{\partial L}{\partial S_1} = -\,2\,\lambda_1\,S_1 = 0 \tag{11.13}$$

$$\frac{\partial L}{\partial S_2} = -\,2\,\lambda_2\,S_2 = 0 \tag{11.14}$$

$$\frac{\partial L}{\partial S_3} = -\,2\,\lambda_3\,S_3 = 0 \tag{11.15}$$

$$\frac{\partial L}{\partial S_4} = -\,2\,\lambda_4\,S_4 = 0 \tag{11.16}$$

Equations 11.13, 11.14, 11.15, and 11.16 are referred to as the *Kuhn-Tucker conditions*, and it will be recalled that these are special to the solution of the set. In fact, it will presently be seen that they are critical to the solution (in accordance with the Kuhn-Tucker Theorem) precisely because they establish certain critical points in the feasible solution space. It is therefore with these equations that the solution process must begin.

Note first that since there are four of these equations and two variables in each, there are 2^4 or 16 possible combinations of the parameters to be investigated. Let these be taken seriatim in a systematic scheme as follows.

1. $\lambda_1 = 0 \quad \lambda_2 = 0 \quad \lambda_3 = 0 \quad \lambda_4 = 0$

   When these four values are inserted into Equations 11.7 through 11.12, it is seen that Equation 11.7 requires $4 = 0$ and Equation 11.8 calls for $5 = 0$, both of which are contradictions. Therefore this case is not possible.

2. $\lambda_1 = 0 \quad \lambda_2 = 0 \quad \lambda_3 = 0 \quad S_4 = 0$

   Substitution of these values in the first six equations calls for $4 = 0$ for Equation 11.7. This is impossible.

3. $\lambda_1 = 0 \quad \lambda_2 = 0 \quad S_3 = 0 \quad \lambda_4 = 0$

   This requires $5 = 0$ from Equation 11.8, and so is not possible.

4. $\lambda_1 = 0 \quad \lambda_2 = 0 \quad S_3 = 0 \quad S_4 = 0$

   Substituting these values in the first six equations produce these results: From Equation 11.7, $\lambda_3 = 4$; from Equation 11.8, $\lambda_4 = 5$; in Equation11.11 $x_1 = 0$; and from Equation 11.12, $x_2 = 0$. There is a solution to the set. The point in the feasible solution space corresponding to this condition is $x_1 = 0$, $x_2 = 0$, and of course the objective function has the value zero as well. Label this *Point 1*.

5. $\lambda_1 = 0 \quad S_2 = 0 \quad \lambda_3 = 0 \quad \lambda_4 = 0$

   From Equation 11.7, $\lambda_2 = 4/3$, and from Equation 11.8, $\lambda_2 = 5/2$. Since λ_2 cannot be equal to both of these numbers, no solution is permissible in this case.

6. $\lambda_1 = 0 \quad S_2 = 0 \quad \lambda_3 = 0 \quad S_4 = 0$

   From Equation 11.12, $x_2 = 0$, and from Equation 11.10, $x_1 = 5$. This produces the second feasible solution. Call it *Point 2*.

7. $\lambda_1 = 0 \quad S_2 = 0 \quad S_3 = 0 \quad \lambda_4 = 0$

   From Equation 11.10 there is $3\,x_1 + 2\,x_2 - 15 = 0$ and Equation 11.11 shows that $x_1 = 0$, while Equation 11.12 requires $x_2 = S_4^2$. Then Equation 11.9 gives $S_1^2 = 8 - 2\,S_4^2$. From Equation 11.10 $S_4^2 = 15/2$, hence $S_1^2 = -7$. S_1 would therefore need to be imaginary, so this case is infeasible.

8. $\lambda_1 = 0 \quad S_2 = 0 \quad S_3 = 0 \quad S_4 = 0$

   Equation 11.12 requires that $x_1 = 0$, while Equation 11.11 dictates that $x_2 = 0$, and with these two values Equation 11.10 shows $-15 = 0$, which is contradictory, so the case is also not possible.

9. $S_1 = 0 \quad \lambda_2 = 0 \quad \lambda_3 = 0 \quad \lambda_4 = 0$

   Equation 11.7 requires $\lambda_1 = 4$ and Equation 11.8 requires $\lambda_1 = 5/2$. This is not possible.

10. $S_1 = 0 \quad \lambda_2 = 0 \quad \lambda_3 = 0 \quad S_4 = 0$

    From Equation 11.12, $x_2 = 0$, then from Equation 11.9 $x_1 = 8$, so Equation 11.10 shows $S_2^2 = -9$. S_2 would need to be imaginary here, so the case is not feasible.

11. $S_1 = 0 \quad \lambda_2 = 0 \quad S_3 = 0 \quad S_4 = 0$

    Equation 11.11 gives $x_1 = 0$ and Equation 11.9 gives $x_2 = 4$. This is *Point 3*, a solution.

12. $S_1 = 0 \quad \lambda_2 = 0 \quad S_3 = 0 \quad \lambda_4 = 0$

    Equation 11.11 calls for $x_1 = 0$, and Equation 11.12 requires $x_2 = S_4^2$, so that Equation 11.9 would have $S_4^2 = 4$, and then $x_2 = 4$. Solution *Point 3* is also returned by this case.

13. $S_1 = 0 \quad S_2 = 0 \quad \lambda_3 = 0 \quad \lambda_4 = 0$

    Equation 11.9 gives $x_1 + 2\,x_2 = 8$, Equation 11.10 yields $3\,x_1 + 2\,x_2 = 15$; whereas Equation 11.7 has $4 - \lambda_1 - 3\lambda_2 = 0$, and from Equation 11.8 there is $5 - 2\lambda_1 - 2\lambda_2 = 0$. The simultaneous solution of these two sets of resulting equations yields $x_1 = 3.5$, $x_2 = 2.25$, $\lambda_1 = 13/4$, and $\lambda_2 = 3/4$. There is now *Point 4*.

14. $S_1 = 0 \quad S_2 = 0 \quad \lambda_3 = 0 \quad S_4 = 0$

    Equation 11.12 requires $x_2 = 0$, Equation 11.9 calls for $x_1 = 8$, and then Equation 11.10 needs $x_1 = 5$ to be satisfied. x_1 cannot be equal to both 5 and 8 at the same time, so the condition is not possible.

15. $S_1 = 0 \quad S_2 = 0 \quad S_3 = 0 \quad \lambda_4 = 0$

    Equation 11.11 calls for $x_1 = 0$, Equation 11.9 indicates $x_2 = 4$, then Equation 11.10 requires $x_2 = 15/2$. This is not possible.

16. $S_1 = 0 \quad S_2 = 0 \quad S_3 = 0 \quad S_4 = 0$

    From Equation 11.11 $x_1 = 0$ and from Equation 11.12 $x_2 = 0$. This gives from Equation 11.9, $x_1 + 2\,x_2 = 8$, and from (4), $3\,x_1 + 2\,x_2 = 15$. The simultaneous solution of these two gives $x_1 = 23/4$, $x_2 = -9/4$. This point falls outside the feasible solution space, and so is rejected.

Four possible solution points have been discovered which fulfill all of the requirements imposed upon the set of constraints for the above-mentioned problem. Obviously, they are very special.

A graphical solution of the problem is displayed in Figure 11.1. Note first that the coordinate axes themselves are two lines which are part of the constraints of the problem. That is, the solution must lie somewhere above the x_1-axis and somewhere to the right of the x_2-axis. When they are converted into equalities,

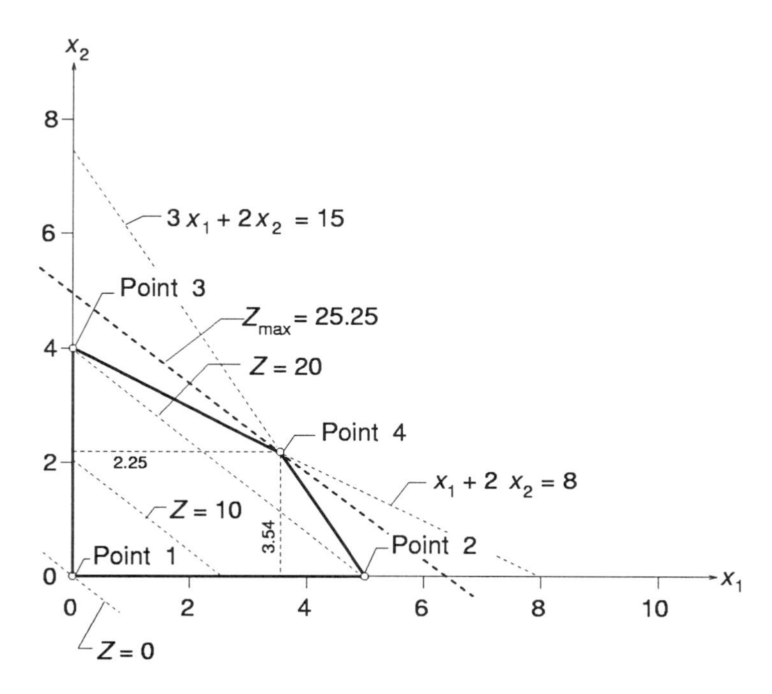

Figure 11.1: Graphical Solution for Linear Programming Problem

the other two constraints also furnish bounds to the solution space. These become the straight lines plotted in the graph as shown.

Simple tests show that any solution of the problem which satisfies these two inequality constraints must lie below these lines (that is, above the x_1 axis) and to the right of the x_2 axis also. The axes and the two constraint lines therefore bound a quadrilateral area within which must be found all solutions to the optimization problem.

The four corners of this quadrilateral are the four special points found in the analysis above.

The solution region is named the ***feasible solution space***, and every acceptable solution of the problem lies on or within these boundaries. The remaining question is, which of the infinitude of places within the area produces the maximum value of the objective function.

The objective function itself is a straight line within the space, with Z a parameter of the line. Let Z take on several values to determine a trend which may indicate where the maximum may be located. First, take $Z = 10$. The two intercepts are $x_1 = 0$, $x_2 = 2$; and $x_2 = 0$, $x_1 = 2.5$. This line is shown plotted on the graph in Figure 11.1. Next, let the objective function be $Z = 20$, with the intercepts $x_1 = 0$, $x_2 = 4$; $x_2 = 0$, $x_1 = 5$. This line is also plotted on the Figure. Clearly, increasing Z indicates that the line moves farther away from the origin.

Clearly, also, the several lines with different Z parameters all have exactly the same slope, hence the line can simply be moved parallel to itself as far away from the origin as possible and still remain within the feasible solution space.

That would be until it reaches Point 4, of course, and there would be found the maximum condition. This point can be solved for from the intersection of the two lines

$$3\,x_1 + 2\,x_2 = 15 \tag{11.17}$$
$$x_1 + 2\,x_2 = 8 \tag{11.18}$$

From the Equation 11.18, $2\,x_2 = 8 - x_1$, and when this is inserted into the Equation 11.17, there is

$$3\,x_1 + 8 - x_1 = 15 \tag{11.19}$$
$$2\,x_1 = 7 \tag{11.20}$$
$$x_1 = 7/2 = 3.5 \tag{11.21}$$
$$2\,x_2 = 8 - x_1 = 8 - 3.5 = 4.5 \tag{11.22}$$
$$x_2 = 2.25 \tag{11.23}$$

Thereafter, Z becomes

$$Z = 4\,x_1 + 5\,x_2 = 4(3.5) + 5(2.25) = 25.25 \tag{11.24}$$

The line representing this optimum is also included in Figure 11.1.

What was learned in the example above was that all of the extrema of the objective function are located at apexes of the quadrilateral. At the origin there is a local and a global minimum, with $Z = 0$. At $x_1 = 0$, $x_2 = 4$; at $x_1 = 2$, $x_2 = 0$; and at $x_1 = 3.5$, $x_2 = 2.25$ there are local maxima, with that one at $x_1 = 3.5$, $x_2 = 2.25$ the global maximum. And all feasible solutions lie on or within the area bounded by the constraint lines. This "in-the-paint" region is therefore called the feasible solution space.

In this case, drawn from Linear Programming theory, the point of global maximum was seen to be that one farthest removed from the origin. It must be emphasized at once that this is a condition which holds only for the linear case, and will in general be incorrect for nonlinear programming problems. Moreover, it will sometimes happen that the slope of the objective function line will equal that of one of the lines for the constraint equations which have been converted to equalities. In that instance, all points along those coincident lines will qualify as extrema, and, indeed, all will be global maxima. There will be an infinitude of solutions to the problem, all of which are equally valid.

Procedurally speaking, what was learned from the example was that all possible extrema are the result of that condition which prevails for each of the Kuhn-Tucker conditions, and these are therefore the only places which need to be investigated. From a practical point of view, this requires that each apex of the polygonal figure shall be visited and studied for effect. In the two-dimensional problem this is rather simple, the more so because a figure can be drawn to follow visually as the process unfolds. If the problem has more than three decision variables it becomes a very complex agenda, and no visual aid can be had because of the impossibility of rendering the hyperplanes in a two-dimensional space.

A systematic procedure of algebraic proportions is what is called for in the larger problems, and this is exactly what Dantzig contributed. And that is just what follows here.

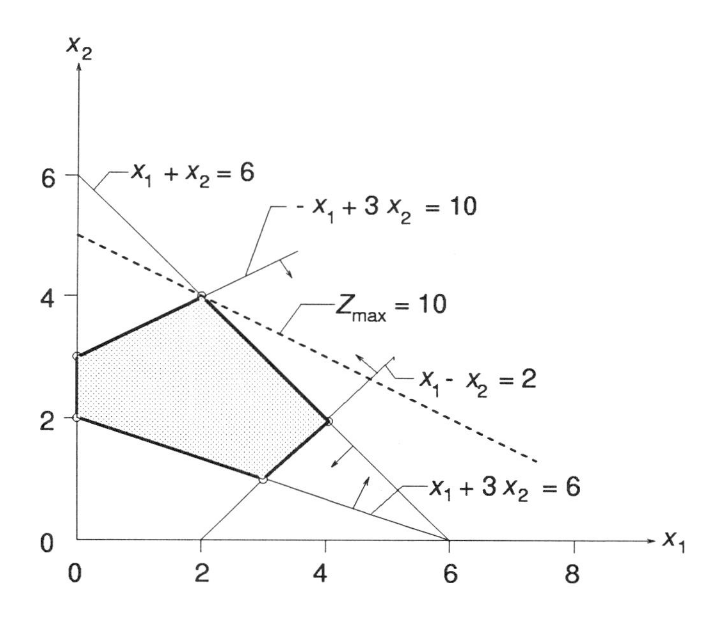

Figure 11.2: Simplex Solution Space

11.3 THE SIMPLEX METHOD

To display the full range of features involved in the Simplex Method, consider the following problem.

$$\text{MAX} \quad Z = x_1 + 2\,x_2 \tag{11.25}$$

$$\text{ST} \quad -x_1 + 3\,x_2 \leq 10 \tag{11.26}$$

$$x_1 + x_2 \leq 6 \tag{11.27}$$

$$x_1 - x_2 \leq 2 \tag{11.28}$$

$$x_1 + 3\,x_2 \geq 6 \tag{11.29}$$

$$x_1,\, x_2 \geq 0 \tag{11.30}$$

A two-dimensional problem is intentionally retained, as this will permit the graphical solution to be compared with the Simplex procedure. Note that there is now an explicit constraint equation of the greater-than type in addition to those usual and customary ones x_1, $x_2 \geq 0$.

When the constraints are converted to equalities by simply changing the inequality signs to equal marks, the graph shown in Figure 11.2 results. By performing some simple tests again as in the previous case, it can be learned that there is a five-sided figure bounded by the constraint lines (including $x_1 \geq 0$) within which any point claiming to solve the problem must reside.

To finish the graphical solution, which is already at hand, plot the equation of the objective function, and since it must have the same slope irrespective of the value of the parameter Z, it is clear that the optimum (maximum) Z is for the case of the line passing through the point $x_1 = 2$, $x_2 = 4$, because it is furthest from the origin of coordinates. At that point, $Z = 2 + 2(4) = 10$. This is the solution.

The Simplex method begins by reformulating the statement of the problem in a standard form suitable to the process. First note that there are two decision variables, and these are named x_1 and x_2. There are three constraints of the less-than type, so there can be expected to be three slack variables. Let these be named x_3, x_4, x_5. Additionally, there is one greater-than type constraint, and it will merit the inclusion of a surplus variable, called x_6. One more parameter, called an artificial variable, is added for each greater-than type constraint. In the present problem, say it is x_7.

The new construction of the system to be solved is as shown below.

$$
\begin{array}{lllllllll}
\text{MAX } Z & -x_1 & -2x_2 & -0x_3 & -0x_4 & -0x_5 & -0x_6 & +Mx_7 & = 0 \quad (A1) \\
\text{ST} & -x_1 & +3x_2 & +x_3 & & & & & = 10 \quad (B1) \\
 & x_1 & +x_2 & & +x_4 & & & & = 6 \quad (C1) \\
 & x_1 & -x_2 & & & +x_5 & & & = 2 \quad (D1) \\
 & x_1 & +3x_2 & & & & -x_6 & +x_7 & = 6 \quad (E1)
\end{array}
$$

The tabular-style array coincides with the systematic form of solution procedure which is expected. Note that the objective function has been organized in the same way as have the constraints, and see also that the variables have all been moved to the left-hand side of their respective equations. Only the right-hand sides of the constraint equations with their numerical values have been retained as in the constraints themselves.

A comment must be made concerning the coefficient M in the objective function line. The purpose of including the artificial variable x_7 in the first place is to drive that constraint condition out of the problem as quickly as possible, so that variable should be given as large a coefficient as is practical in order to make its effect in the first round of computations a maximum. It will therefore do its work and be gone. Rather than give it a numerical value in the algebraic operations that are to come, it is customary to show it as some number M, which is conceived to be a large number, but is never actually specified.

The Simplex scheme will proceed to investigate the value of the objective function at each vertex (apex) of the polygon that is the feasible solution space. In this problem, the origin is not one of the vertices. However, it is a very convenient starting point, the more so in those problems where the number of decision variables precludes the graphical alternative. Accordingly, it is a usual practice to begin with the origin, so that the first step in the solution process then becomes one of getting "in the paint" as an opening phase.

If the origin is chosen as the beginning, the consequence is

$$x_1 = 0 \quad x_2 = 0 \quad x_3 = 10 \quad x_4 = 6 \quad x_5 = 2 \quad x_6 = 0 \quad x_7 = 6 \quad Z = -6M$$

Look at the equation for Z. In its present tabular form, it is seen that the value of Z can be increased (which is the name of this game) by increasing that variable which has the largest negative coefficient. It is x_2, because its coefficient is -2, a number more negative that any other number on the line containing Z. So the agenda should be to make x_2 as large as possible in this round of computations.

Let the next round begin. Examine each constraint equation to find the upper limit on x_2 in the round. It will be, of course, the ***smallest*** positive value of x_2 from any of the equations.

$$x_1 = 0 \quad x_3 = 0 \quad x_2 = \tfrac{10}{3} \qquad (B1)$$

$$x_1 = 0 \quad x_4 = 0 \quad x_2 = 6 \qquad (C1)$$

$$x_1 = 0 \quad x_5 = 0 \quad x_2 = -2 \qquad (D1)$$

$$x_1 = 0 \quad x_6 = 0 \quad x_7 = 0 \qquad x_2 = 2 \quad (E1)$$

The upper bound on x_2 is 2, from Equation (E1), obviously.

From equation ($E1$): $\quad x_2 = 2 + \frac{1}{3}\,x_6 - \frac{1}{3}\,x_1 - \frac{1}{3}\,x_7$

Substitute this into the equation for Z and the other equations also to obtain the following set.

$$Z - \tfrac{1}{3}\,x_1 - \tfrac{2}{3}\,x_6 + (M + \tfrac{2}{3})\,x_7 = 4 \qquad (A2)$$

$$-2\,x_1 + x_3 + x_6 - x_7 = 4 \qquad (B2)$$

$$\tfrac{2}{3}\,x_1 + x_4 + \tfrac{1}{3}\,x_6 - \tfrac{1}{3}\,x_7 = 4 \qquad (C2)$$

$$\tfrac{4}{3}\,x_1 + x_5 - \tfrac{1}{3}\,x_6 + \tfrac{1}{3}\,x_7 = 4 \qquad (D2)$$

$$x_1 + 3\,x_2 - x_6 + x_7 = 6 \qquad (E2)$$

At this stage

$$x_2 = 2 \quad x_1 = 0 \quad x_6 = 0 \quad x_7 = 0 \quad x_5 = 4 \quad x_4 = 4 \quad x_3 = 4 \quad Z = 4$$

and an inspection of the equation for Z reveals that the variable with the largest negative coefficient (and therefore the greatest potential for increasing Z) is x_6. So it will be the purpose of the next round to increase x_6 as much as possible. The next round commences.

$$x_1 = 0 \quad x_3 = 0 \quad x_7 = 0 \quad x_6 = 4 \qquad (B2)$$

$$x_1 = 0 \quad x_4 = 0 \quad x_7 = 0 \quad x_6 = 12 \qquad (C2)$$

$$x_1 = 0 \quad x_2 = 2 \quad x_7 = 0 \quad x_6 = -12 \quad (D2)$$

$$x_1 = 0 \quad x_2 = 2 \quad x_7 = 0 \quad x_6 = 0 \qquad (E2)$$

From ($B2$): $\qquad x_6 = 4 + 2\,x_1 - x_3 + x_7$

$$Z - \tfrac{5}{3}x_1 + \tfrac{2}{3}x_3 + M\,x_7 = \tfrac{20}{3} \qquad (A3)$$

$$-2x_1 + x_3 + x_6 - x_7 = 4 \qquad (B3)$$

$$\tfrac{4}{3}x_1 - \tfrac{1}{3}x_3 + x_4 + \tfrac{1}{3}x_7 = \tfrac{8}{3} \qquad (C3)$$

$$\tfrac{2}{3}x_1 + \tfrac{1}{3}x_3 + x_5 + \tfrac{1}{3}x_7 = \tfrac{16}{3} \qquad (D3)$$

$$-x_1 + 3x_2 + x_3 + x_7 = 10 \qquad (E3)$$

The condition now is

$$x_1 = 0 \quad x_6 = 4 \quad x_3 = 0 \quad x_7 = 0 \quad x_4 = \tfrac{8}{3} \quad x_5 = \tfrac{16}{3} \quad Z = \tfrac{20}{3}$$

and the scan of the Z-line reveals that there is only one remaining negative coefficient which could allow an increase in Z. That one multiplies x_1, so it will be the purpose in the next round to increase x_1 as much as possible.

$$x_3 = 0 \quad x_6 = 4 \quad x_7 = 0 \quad x_1 = 0 \quad (B3)$$

$$x_3 = 0 \quad x_4 = 0 \quad x_7 = 0 \quad x_1 = 2 \quad (C3)$$

$$x_3 = 0 \quad x_5 = 0 \quad x_7 = 0 \quad x_1 = 8 \quad (D3)$$

$$x_3 = 0 \quad x_2 = \tfrac{10}{3} \quad x_7 = 0 \quad x_1 = 0 \quad (E3)$$

From $(C3)$: $\qquad x_1 = \tfrac{1}{4}x_3 - \tfrac{3}{4}x_4 - \tfrac{1}{4}x_7 + 2$

Substitute this into the other equations to obtain the next set.

$$Z + \tfrac{1}{4}x_3 + \tfrac{5}{4}x_4 + (\tfrac{5}{12} + M)\,x_7 = 10 \qquad (A4)$$

$$-2x_1 + x_3 + x_6 - x_7 = 4 \qquad (B4)$$

$$\tfrac{4}{5}x_1 - \tfrac{1}{3}x_3 + x_4 = \tfrac{8}{3} \qquad (C4)$$

$$\tfrac{1}{2}x_3 - \tfrac{1}{2}x_4 + x_5 + \tfrac{1}{6}x_7 = 4 \qquad (D4)$$

$$3x_2 + \tfrac{3}{4}x_3 + \tfrac{3}{4}x_4 + \tfrac{3}{2}x_7 = 12 \qquad (E4)$$

Now the condition is

$$x_1 = 2 \quad x_3 = 0 \quad x_4 = 0 \quad x_5 = 4 \quad x_6 = 8 \quad x_2 = 4 \quad x_7 = 0 \quad Z = 10$$

and it is clear that no further improvement in Z can be expected since there is no one of the coefficients in its equation that is negative. The procedure has come to the end of the line. The situation there is $x_1 = 2$, $x_2 = 4$, $Z = 10$. This is the global maximum. The solution has been obtained.

In reviewing what the process above displayed, it can be seen that the following steps were taken:

1. The problem was set up in a standard form. The objective function was written with all terms on the left-hand side of the equal marks, and the right-hand side was zero. One new variable—a slack—was added for each less-than type equalilty, and the inequalities were converted thereby to equalities. One new variable—a surplus—was added for each constraint of the greater-than type inequality. In addition, for each such greater-than type inequality a new variable called "an artificial variable" was introduced into the inequality, which was thus converted to an equality.
2. The objective function was modified to include each of the new slack and surplus variables with coefficients equal to zero, and to include the artificial variables with coefficients equal to M, a very large number.
3. A set of starting values is selected with x_1 and x_2 equal to zero and the other variables solved for from the constraint equations under these conditions.
4. The Z-line is scanned for the largest negative coefficient, and the variable which has this coefficient is made the "entering variable" in the next round.
5. The entering variable is solved for from each of the constraint equations, and the smallest positive value is obtained by scanning the column of its values associated with each constraint. The variable corresponding to this smallest positive value is designated "the leaving variable."
6. The constraint equation corresponding to the leaving variable is solved for the variable in terms of the others. The new value of this variable is then substituted in all of the other constraints and the objective function to create a new set.
7. New values of the variables are then computed for this condition, including a new value of the objective function.
8. The new objective function line is scanned for the largest negative coefficient, which is now made to be the new "entering variable."
9. The process of steps 5, 6, and 7 is repeated in a new round of computation or iteration. This process is continued until there is no longer a negative coefficient in the Z-line. This identifies the solution to the problem, because there is no possibility of further increasing the value of the objective function.

With respect to the graphical procedure, what took place in the example above was the following:

- The procedure started at the origin, $x_1 = 0$, $x_2 = 0$.
- It moved up the x_2-axis from $x_2 = 0$ to $x_2 = 2$, a local maximum.
- It moved further up the x_2-axis from $x_2 = 2$ to $x_2 = 4$, another local maximum.
- It moved along the constraint line $-x_1 + 3x_2 = 10$ to the point with the values $x_1 = 2$, $x_2 = 4$, where the global maximum occurs.
- Three iterations were required to converge upon the global maximum.

11.4 THE TABLEAUX METHOD

On the face of it, there needs to be a short-hand method for doing the computations indicated in the previous section. There were only two decision variables in that example, and only six constraints, counting the two customary ones which require the decision variables to be positive. Yet the solution process, with only three iterations to convergence, was rather tedious, to say the least.

The character of the solution was seen to contain a very repetitive process, with a number of steps which could have been made in an automatic manner. A computational algorithm could have been created to accomplish this in an automated way. This would be a literal necessity in a problem with many decision variables and many constraints.

Actually, the basis for just such an algorithm has been stated in the summary of the operations given at the end of the section. And, actually, Dantzig gave the program 40-odd years ago, albeit in the state-of-the-art format suitable to that time. There were not, of course, more than a handful of electronic computers in existence then, and they ran only a very few hours per day, because the galaxy of vacuum tubes of which they were composed was subject to failures at a high rate, and even when they were operational, the speeds were terribly slow compared to what was needed to solve the Simplex program.

The standard treatment extant in engineering and applied science at the time was a tabular form of computation. The French word for table is ***tableau***, and the plural is ***tableaux***, so these were the nomenclature Dantzig and others used to describe the iterations appropriate to the Simplex pattern.

Each round of computation is entered in a table or tableau, which therefore contains one iteration in the sequence of such iterations or tableaux until convergence occurs.

The tableaux merely mechanize the process given in the previous section, so that it proceeds without the extra baggage of the running commentary. The four such tables that are necessary for that problem are given below, one after the other, just as they would be constructed. See also Appendix F.

The Initial Tableau contains, as its first row, the column headings for the remaining rows. The first column identifies the row name, and the second column identifies the basic variable (BV) for that row. The number of entries in this column is equal to one (for Z) plus the number of basic variables, is this case four. The decision variables and the artificial variables are omitted, leaving Z, x_3, x_4, x_5, x_6.

There is then a column for Z, the objective function, and columns for each variable, including the decision variables and the arbitrary variable. After this comes a column labelled RHS, which stands for "RIGHT-HAND SIDE", obviously. A last column is for showing the upper bound on the entering variable ($UBEV$).

Entries within the cells of the matrix created by these rows and columns are then made. The actual numbers are those taken from the coefficients in the canonical form of the statement of the problem. In the row with Z, for example, the number 1 in the Z-column and Z-row represents the coefficient of Z in the equation of the objective function. The number -1 in the x_1-column at the Z-row represents the coefficient of x_1 in the objective function equation, and -2 in the cell of the x_2-column in the Z-row is the coefficient of x_2 in the objective function equation.

Zeros in the x_3, x_4, x_5, and x_6 columns of the Z-row are the coefficients of those variables, respectively, in the objective function equation. x_7 gets the "big M" coefficient, and the right-hand side of the objective function is 0 in the canonical format.

In the second row—that of the basic variable x_3—the 0 under Z in the third column says that Z is not involved in the equation ($B1$), which is a constraint. -1 in the x_1-column of the x_3-row tells that the coefficient of x_1 in the equation ($B1$) is -1, and the 3 in the x_2-column of the x_3-row is there to indicate that the coefficient of x_2 in the ($B1$) equation is 3. The number 1 under the x_3-column in the x_3-row means that the coefficient of x_3 in equation ($B1$) is 1.

Zeros in the x_4, x_5, x_6, and x_7 columns of the x_3-row say that none of these variables enter equation ($B1$). However, in the RHS-column of x_3-row, the number 10 is that of the right-hand side of equation ($B1$); namely, 10.

The numerical entries for each column and row for the remainder of the Initial Tableau are likewise taken directly from the defining equations of the problem statement.

Step One Algorithm for the solution process now comes to the front. Place a scanner on the Z-line and determine the *largest negative number*, which is -2 in the present instance. The variable above this column, x_2, is declared the *entering variable*. The column containing x_2 is highlighted to indicate its premier status.

Now it is time to construct the upper bound of the entering variable column entries. This is **Step Two Algorithm** in the solution process. The Z-row entry in this column is vacant, of course. For the x_3-row, take the coefficient of x_2, the entering variable, in this row and divide it into the RHS value at this row; i.e., 10/3. Enter this in the upper bound column in this row. In the x_4-row, divide the x_2-column cell number (which is 1) into the RHS-column number (here it is 6) to get $6/1 = 6$ as the entry in the upper bound column. Continue with -1 in the x_5-row at the x_2-column divided into 2, the RHS-column item in x_5-row, to get $2/-1 = -2$ for the upper bound number in x_5-row. Similarly, in x_6-row, divide 3, the x_2-column number, into 6, the RHS-column number of x_6-row to get $6/3 = 2$ for the upper bound on this row.

Here comes the **Step Three Algorithm** of the solution process. Put a scanner on the Upper-Bound Entering Variable Column and find the *smallest positive number* there. In the case at hand, it is 2, the number in the x_6-row. That variable, x_6, now becomes the *leaving variable*, and its row is highlighted to emphasize its exalted role.

At the intersection of the entering variable column and the leaving variable row is the cell with the number 3 in it. This cell is called the *pivot*, and the number in the cell is the *pivot element*. The location and the value are both critical to the next steps in the solution.

What happens next is the mechanization of the procedure employed in the Simplex Method when equation ($E1$) and variable x_2 were identified. x_2 was solved for from ($E1$) and substituted into the other equations to eliminate x_2 at this stage. A new tableau—Second Tableau—will be created from the Initial Tableau and the identification of the pivot.

In the column under x_2, there will need to be the number 1 at the ($E2$) equation row, now named x_2 , because the entering variable, x_2, has entered. All other entries in this column will need to be 0s, because x_2 does not appear on the right-hand side of the equation ($E2$). To accomplish these new entries, proceed with **Step Four Algorithm**. Divide the pivot element by itself to produce the required number 1 at that place. Divide each of the entries in the old x_2-column by the pivot element, change signs, and add to the original elements in the row in question. This makes the new numbers there 0. Now divide each of the elements of the leaving variable row by the pivot element and record the results in the entering variable row of the Second Tableau. For example, divide the 1 in the x_2-column at the x_6 row by the pivot element which is 3 to get 1/3, the entry under the x_2-column at the x_2 row in the Second Tableau. Do all of the other elements in this row the same (except, of course, the upper-bound on the entering variable).

Step Five Algorithm applies to the creation of the new Second Tableau entries which are not on the pivot row or in the pivot column. For a given row, divide the pivot column element in that row by the pivot element, change the sign, then go to the pivot row element under the cell where the computation is being done and multiply by the element found there and add the result to the existing element in the cell where the computation is being done. Record this as the element of that cell in the Second Tableau.

For example, the cell in the x_1-column at the Z-row has the element -1 in the Initial Tableau. Divide 3 (the pivot element) into -2 (the pivot column element corresponding to the Z-row), get $-2/3$, change the sign to $+2/3$, go to the x_6-row under the x_1-column and find 1, multiply 2/3 by that 1, add to -1, get $-1/3$, the entry at that cell in the Second Tableau. Or, for another example, in the Z-row under the x_7-column is the cell value M. Divide the pivot element, 3, into the pivot column element, -2, at that row, to get $-2/3$, change sign, $+2/3$, go to x_6-row under x_7-column, find 1 at that cell, multiply by that 1, get $+2/3$, add to M in the Z-row, x_7-column, get $M + 2/3$, record this in the same cell of the new Second Tableau.

This same algorithm holds for computing the RHS values in the Second Tableau. These elements are treated as if they were in some other column.

Next, revert to the scanning algorithms as applied to the partially-completed Second Tableau. Locate the largest negative cell element on the Z-line, which is $-2/3$, under the x_6-column. x_6 has thus been identified as the entering variable of the next round of computation. Highlight this column. Then divide each of the RHS-column numbers of the Second Tableau by the corresponding row-element of the new pivot column. For the x_3-row, the element is 1, divide it into 4, the RHS element on that row to get 4, the new Upper-Bound Column element in the Second Tableau at x_3-row. Divide 1/3 into 4 to get 12, the entry in x_4-row of the Upper-Bound Column. $-1/3$ divided by 4 gives -12, the element on x_5-row in the Upper-Bound column. Divide 2 by $-1/3$ to get -6 as the x_2-row Upper-Bound column element.

The Second Tableau is now complete.

Scan the Upper-Bound Column for the smallest positive element, which is 4, in the x_3-row. x_3 is thereby identified as the leaving variable in this iteration.

Highlight this row, and note that it intersects the entering variable column, x_6, at the pivot element 1. The entire procedure repeats itself from here, to generate the Third Tableau.

Enter x_6 in place of x_3 in Row $B3$, column BV, then divide each element on this row by the pivot element, 1, from the Second Tableau. Make each element of the x_6-column equal zero (except, of course, the pivot, which is 1).

Invoke the **Step Five Algorithm** to create the new entries for Tableau 3 elements which are not on the pivot row or in the pivot column, including the RHS elements.

Divide each of the RHS elements by the corresponding row-element of the pivot column, x_6, of the Second Tableau. This generates the $UBEV$ numbers 4, 2, 8, and −10 for the Third Tableau.

When the Third Tableau has been finished, a scan of its Z-row shows the largest negative value to be −5/3 in the x_1-column, which therefore becomes the new entering variable. The scan of the Upper-Bound Column finds the smallest positive number to be 2 in the x_4-row. Accordingly, the intersection of the column with the largest negative coefficient and the row with the smallest positive $UBEV$ element shows that 4/3, in the x_1-column at the $C3$ (x_4) row, is the new pivot element.

The procedure again repeats itself to produce the Fourth Tableau. Divide each element of Row $C3$ by 4/3 to produce Row $C4$. Make all elements of Column x_1 equal 0, except the pivot, which is 1. Compute all the other elements, including the RHS column ones, with **Step Five Algorithm**. At the completion of the Fourth Tableau, the scan of the Z-line shows no negative coefficient, which is evidence that the process is at an end.

To identify the values of the decision variables and the objective function at the end, note that in the x_1-row under the x_1-column is the number 1. Under the x_2-column at x_2-row is the number 1. On the Z-line under the Z-column is the number 1. These define the answers. Go to the RHS column in each of these rows and read off the values: $Z = 10$, $x_1 = 2$, $x_2 = 4$.

INITIAL TABLEAU

		Coefficients of									
Row	BV	Z	x_1	x_2	x_3	x_4	x_5	x_6	x_7	RHS	$UBEV$
$A1$	Z	1	−1	−2	0	0	0	0	M	0	
$B1$	x_3	0	−1	3	1	0	0	0	0	10	$\frac{10}{3}$
$C1$	x_4	0	1	1	0	1	0	0	0	6	6
$D1$	x_5	0	1	−1	0	0	1	0	0	2	−2
$E1$	x_6	0	1	$\boxed{3}$	0	0	0	−1	1	6	2

SECOND TABLEAU

Row	BV	Coefficients of Z	x_1	x_2	x_3	x_4	x_5	x_6	x_7	RHS	UBEV
A2	Z	1	$-\frac{1}{3}$	0	0	0	0	$-\frac{2}{3}$	$M+\frac{2}{3}$	4	
B2	x_3	0	-2	0	1	0	0	$\boxed{1}$	-1	4	4
C2	x_4	0	$\frac{2}{3}$	0	0	1	0	$\frac{1}{3}$	$-\frac{1}{3}$	4	12
D2	x_5	0	$\frac{4}{3}$	0	0	0	1	$-\frac{1}{3}$	$\frac{1}{3}$	4	-12
E2	x_2	0	$\frac{1}{3}$	1	0	0	0	$-\frac{1}{3}$	$\frac{1}{3}$	2	-6

THIRD TABLEAU

Row	BV	Coefficients of Z	x_1	x_2	x_3	x_4	x_5	x_6	x_7	RHS	UBEV
A3	Z	1	$-\frac{5}{3}$	0	$\frac{2}{3}$	0	0	0	M	$\frac{20}{3}$	
B3	x_6	0	-2	0	1	0	0	1	-1	4	4
C3	x_4	0	$\boxed{\frac{4}{3}}$	0	$-\frac{1}{3}$	1	0	0	0	$\frac{8}{3}$	2
D3	x_5	0	$\frac{2}{3}$	0	$\frac{1}{3}$	0	1	0	0	$\frac{16}{3}$	8
E3	x_2	0	$-\frac{1}{3}$	1	$\frac{1}{3}$	0	0	0	0	$\frac{10}{3}$	-10

FOURTH TABLEAU

Row	BV	Coefficients of Z	x_1	x_2	x_3	x_4	x_5	x_6	x_7	RHS	UBEV
A4	Z	1	0	0	$\frac{1}{4}$	$\frac{5}{4}$	0	0	M	10	
B4	x_6	0	0	0	$\frac{1}{2}$	$\frac{3}{2}$	0	1	-1	8	
C4	x_1	0	1	0	$-\frac{1}{4}$	$\frac{3}{4}$	0	0	0	2	
D4	x_5	0	0	0	$\frac{1}{2}$	$-\frac{1}{2}$	1	0	0	4	
E4	x_2	0	0	1	$\frac{1}{4}$	$\frac{1}{4}$	0	0	0	4	

11.5 THE LINEAR PROGRAMMING PROCEDURE

Based upon the previous theory and examples, a formal general statement of the Linear Programming Algorithm can now be given. There are a number of versions of this statement extant in the business at the present time, and all may be entirely correct. The following, however, applies to the analysis which has been developed in this exposition.

Find the values of the decision variables

$$x_1,\ x_2,\ x_3, \ldots,\ x_n \tag{11.31}$$

which optimize (maximize or minimize) the linear function (the criterion or objective function)

$$Z = c_1\, x_1 + c_2\, x_2 + \cdots + c_n\, x_n \tag{11.32}$$

subject to the constraints

$$a_{11}\, x_1 + a_{12}\, x_2 + \cdots + a_{1n}\, x_n \leq b_1 \tag{11.33}$$

$$\cdots\cdots\cdots\cdots\cdots\cdots\cdots\cdots \quad \cdot\cdot\ \cdot\cdot$$

$$a_{r1}\, x_1 + a_{r2}\, x_2 + \cdots + a_{rn}\, x_n \leq b_r \tag{11.34}$$

$$a_{s1}\, x_1 + a_{s2}\, x_2 + \cdots + a_{sn}\, x_n = b_s \tag{11.35}$$

$$\cdots\cdots\cdots\cdots\cdots\cdots\cdots\cdots \quad \cdot\cdot\ \cdot\cdot$$

$$a_{t1}\, x_1 + a_{t2}\, x_2 + \cdots + a_{tn}\, x_n = b_t \tag{11.36}$$

$$a_{u1}\, x_1 + a_{u2}\, x_2 + \cdots + a_{un}\, x_n \geq b_u \tag{11.37}$$

$$\cdots\cdots\cdots\cdots\cdots\cdots\cdots\cdots \quad \cdot\cdot\ \cdot\cdot$$

$$a_{v1}\, x_1 + a_{v2}\, x_2 + \cdots + a_{vn}\, x_n \geq b_v \tag{11.38}$$

$$\text{all} \quad x_j \geq 0 \tag{11.39}$$

where there are m constraints of the less-than type, p constraints of the equality type, and q constraints of the greater-than type. The as, bs, and cs are constants.

For the sake of brevity, the formal statement is sometimes shown as

$$\text{Minimize} \quad \mathbf{cx} \tag{11.40}$$

$$\text{Subject to} \quad \mathbf{Ax} = \mathbf{b} \tag{11.41}$$

$$\mathbf{x} \geq \mathbf{0} \tag{11.42}$$

in which matrix quantities have been employed. In this format all of the constraint inequalities are reduced to equalities by the introduction of slacks, surplusses, and artificial variables.

When the problem is stated in the canonical form given above, the next step in the solution algorithm is to introduce the slacks, surplusses and artificial variables. A slack will be named for each less-than type inequality, a surplus variable for each greater-than type inequality, and an additional variable—an artificial one—will be supplied for each equality and each greater-than type inequality.

Let n be the number of decision variables, k be the total number of constraint equations, and h be the number of the slack and the surplus variables. There are accordingly $n+h$ variables to be solved for from k constraint equations. Since the total number of variables is greater than the total number of constraint equations, $n+h-k$ variables may be selected to be zero, and the remaining k can then be solved for. The number of ways in which the $n+h-k$ variables might be chosen from among the k possibilities is given by combinatorics as

$$\frac{(n+h)!}{(n+h-k)!(k)!} \tag{11.43}$$

which is a very large number of potential arrangements, given any size of system at all. For instance, in the case of the simple system of the previous sections, there were $n = 2$ decision variables, $k = 4$ constraints and $h = 4$ slacks and surplus variables. Therefore

$$\frac{(n+h)!}{(n+h-k)!(m)!} = \frac{(2+4)!}{(2+4-4)!(4)!} = \frac{6!}{2!4!} = 15 \tag{11.44}$$

a rather sizeable number of ways in which the starting condition for the solution might be chosen.

The k variables which are selected and solved for are called the ***basic solution***. A basic solution which falls within the feasible solution space (paint area) is called a ***basic feasible solution***, and a basic feasible solution with all k variables positive is called a ***basis***.

The starting point of the Linear Programming algorithm is to be a basic feasible solution. This was the case in the illustration previously used, where x_1 and x_2 were taken to be zero, and the other xs were solved for. This same starting point (i.e., decision variables set equal to zero) is employed in many, if not most, procedures. It is convenient, especially when using computers to process the technique, because the registers containing these variables can easily be initialized to those values.

It should be noted and emphasized that the optimization process by the Simplex and Linear Programming procedures is valid for both maximizing and minimizing. The understructure as developed grew out of the fundamentals of the calculus, so the end results are first of all optima, and thereafter they are identified as either maxima or minima by applying the usual tests. In practice, for minimization, one simply changes the sign of the objective function to $-Z$, and then proceeds to treat the problem as one of maximization. That is,

$$\text{MAX }(-Z) = \text{MIN } Z. \tag{11.45}$$

A simple demonstration for a single decision variable is that furnished by the function (a, b, and c are constants)

$$Z(x) = a\,x^2 + b\,x + c \tag{11.46}$$

with which

$$\frac{dZ}{dx} = 2\,a\,x + b \tag{11.47}$$

$$\frac{d^2 Z}{dx^2} = 2\,a \tag{11.48}$$

and

$$-Z(x) = -a\,x^2 - b\,x - c \tag{11.49}$$

with

$$\frac{d(-Z)}{dx} = -2\,a\,x - b \tag{11.50}$$

$$\frac{d^2(-Z)}{dx^2} = -2\,a \tag{11.51}$$

In the first case, a positive second derivative indicates a minimum of the function Z, whereas, in the second, a negative second derivative shows there is a maximum of the function $-Z$.

Graphically, the minimization procedure requires that the critical point *nearest the origin of coordinates* be taken as the place where the optimum (minimum) objective function line must pass. *This is valid only for linear programming*, not for other cases. The procedure for maximizing the objective function requires the selection of the critical point *farthest removed from the origin of coordinates* as the place where the optimum (maximum) objective function line shall pass. *This, too, is valid only for linear programming*, not for other examples.

11.5.1 A MINIMIZATION EXAMPLE

Watauga Builders, Inc., have been awarded the contract to construct a large vacation resort atop Puppy Foot Mountain in the Great Smokies. The location is remote from populated areas, but two sources of tradespersons are identified in reasonably accessible valleys to the northeast and southwest. The project manager estimates that 43 carpenters, 58 plumbers, and 39 electricians will be needed for an estimated 2,600 hours.

The established wage scales in the two sites are somewhat different: \$22/hour in Cherokee, and \$28/hour in Sylvania. The composition of the available workforce is also somewhat different, as evidenced in the table below.

	CHEROKEE (%)	SYLVANIA (%)
Carpenters	28	31
Plumbers	26	49
Electricians	46	20

How many workers should the manager employ from each area to minimize the total cost?[1]

The total cost is formulated as

$$C = 22 \times 2,600\,x_1 + 28 \times 2,600\,x_2 \tag{11.52}$$

in which x_1 is the number of workers to be recruited in Cherokee, and x_2 is the corresponding number for Sylvania.

[1]This is actually an integer programming problem. Please see Section 11.6.

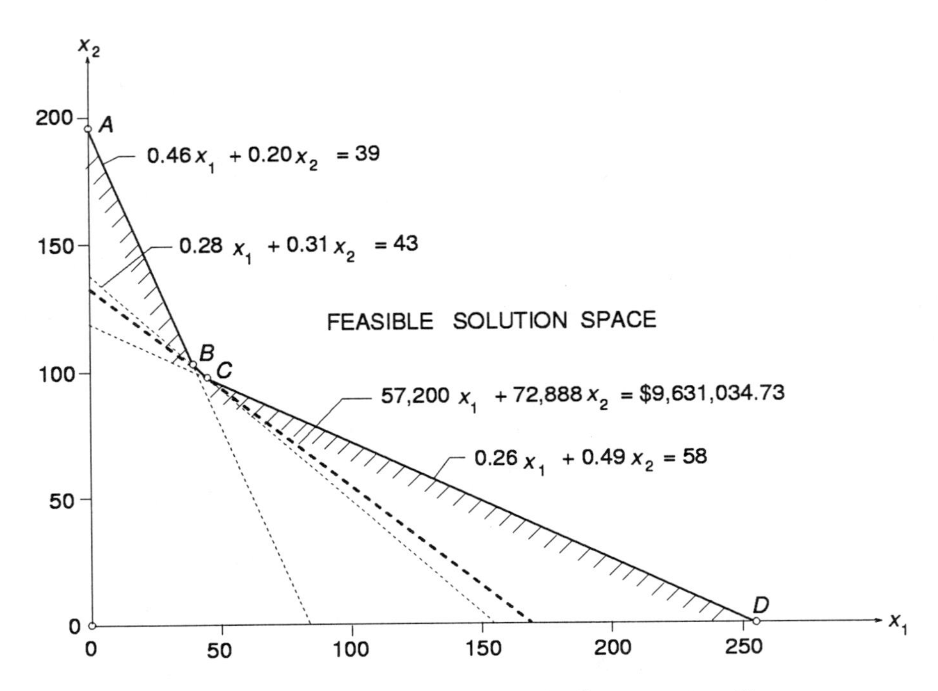

Figure 11.3: Minimizing Watauga's Construction Costs

Of course, there must be enough workers in each trade to accomplish the job, so there are three constraints.

$$0.28\, x_1 + 0.31\, x_2 \geq 43 \tag{11.53}$$

$$0.26\, x_1 + 0.49\, x_2 \geq 58 \tag{11.54}$$

$$0.46\, x_1 + 0.20\, x_2 \geq 39 \tag{11.55}$$

A graphical solution can be constructed by converting each constraint inequality to an equation, then plotting the resulting lines as shown in Figure 11.3. The feasible solution space is bounded by the coordinate axes and the line segments $\overline{AB}$, $\overline{BC}$, and $\overline{CD}$, and includes the area above and to the right of these. Obviously, the point C is the closest place to the origin of coordinates, and is therefore the site of the minimum cost. The intersection of lines $0.28\, x_1 + 0.31\, x_2 = 43$ and $0.26\, x_1 + 0.49\, x_2 = 58$ determines the location of C. From $0.28\, x_1 + 0.31\, x_2 = 43$,

$$0.28\, x_1 = 43 - 0.31\, x_2$$

$$x_1 = \tfrac{43}{0.28} - \tfrac{0.31}{0.28} x_2 = 153.57 - 1.107\, x_2 \tag{11.56}$$

into $0.26\, x_1 + 0.49\, x_2 = 58$,

$$0.26(153.57 - 1.107\, x_2) + 0.49\, x_2 = 58$$

$$39.929 - 0.2879\, x_2 + 0.49\, x_2 = 58$$

$$0.202\,x_2 = 18.071$$

$$x_2 = 89.399 \tag{11.57}$$

and

$$x_1 = 153.57 - 1.107(89.399) = 153.57 - 98.98 = 55.59 \tag{11.58}$$

The minimum cost is

$$C = 22 \times 2600(55.59) + 28 \times 2600(89.399) = \$9,631,034.73 \tag{11.59}$$

with the objective function the line that passes through point C

$$57,200\,x_1 + 72,800\,x_2 = \$9,631,034.73 \tag{11.60}$$

The criterion that the objective function is a minimum for that line which passes through the point closest to the origin is valid only for a linear program, naturally.

11.5.2 SOME CLOSING REMARKS

In the case of both maximizing and minimizing, one can imagine in the mind's eye the operation of a butterfly triangle from the draftsman's kit, set so its adjustable angle is that of the objective function line, and translated until it passes through the critical point. There the line is drawn, to establish the optimum.

The local tactic in each iteration of the simplex method (and in Linear Programming as well) is to make the next move in accordance with a simple criterion: improve the value of the objective function. It can happen that at some point there may be no improvement in the objective function, even though there is a negative coefficient on the Z-line. This situation takes place in the event that two critical points coalesce—a purely coincidental numerical circumstance. The solution obtained thereby is referred to as a *degenerate* solution. These occur infrequently in practice.

In case of a degenerate solution, it is theoretically possible for the procedure to cycle back and forth from one of the vertices to the other (using graphical language), a phenomenon named *cycling*, naturally, but there are apparently no instances of this having actually happened. A large amount of theoretical effort has been expended in the investigation of this possibility.

One of the principal limitations of Linear Programming is the difficulty of interpreting the solution once it has been obtained. This is a consequence of the enormous variety of particular results which may be had, and of the non-standard individual forms taken. The situation is aggravated as the number of decision variables and the number of constraint relationships grows. Large-scale operations which are being modeled by LP techniques normally present thousands of constraints and variables, and it is no small matter to sort out the meaning of a given set of numbers in a sea of the same.

A practical consideration in this connection is the matter of the printout from the computer. When there are some thousands of decision variables and possibly a comparable number of constraints, the format of the output from a computer program becomes a formidable undertaking. No paper is wide enough to list all of this in a meaningful manner, and it is certainly out of the question to display anything like the tableau arrangement of the data, hence it is unreasonable to expect to use the previous methods to identify solution variables and their values.

Therefore, it cannot be said with assurance that some highly improbable answer may not someday present itself to the problem solver in very unfamiliar garb, as if to knock on the door and say, when it is opened, "Here I am, Dad, don't you recognize me?"

11.6 INTEGER PROGRAMMING

It is a not infrequent occurrence that a programming situation will require an integer answer for one or more of the decision variables. If all of the variables are to be integers, the process is called *integer programming*, whereas if only some are to be integers while the remainder may be real numbers, the description is *mixed programming*.

Integer outcomes are clearly indicated in those models which are posed in terms of numbers of "units" of the variables in question. Loads of gravel, boxes of nails, bays or floors of buildings, persons—all of these are generally taken to be indivisible, and are therefore enumerated by integer counts. It would be patently ridiculous, for example, to accept a result which calls for 10.2 workers, or 3.6 dumptrucks.

Integer programming is developed to a very highly specialized procedure. Unfortunately, it is not so simple as linear programming with continuous outcomes expected. The algorithms are more esoteric and definitely more computer intensive, although most moderate-to-large computing centers will possess routines for this purpose.

Most of the references to Linear Programming will contain a chapter or sections on integer programming, and there are those[2] which specialize mainly to integer problems.

While the rigorous treatment of integer programming is somewhat beyond the scope of this writing, a more heuristic approach can be made to serve for many practical situations. Basically, this involves proceeding to treat the integer programming problem as if it were continuous—i.e., linear programming; then after the answer is obtained, it is rounded off to integer amounts.

Although it is not impossible that such a tactic will occasionally produce an integer solution that does in fact satisfy the linear programming model, it is a rare event. Problems in which the right-hand side of the constraints as well as the coefficients of the objective function and constraints are integers seem to furnish more such integer solutions than do those where the numbers are reals.

It should immediately be mentioned that rounding off reals to integers as linear programming results is fraught with certain dangers. The rounding can produce numbers outside of the feasible solution space, and even if the answer is "nearly" feasible, this may still not be acceptable. Only if the constraints are "somewhat flexible" can this practice be of use.

---

[2]Garfinkel, R. S., and Nemhauser, G. L., *Integer Programming*, John Wiley and Sons, New York, 1972.

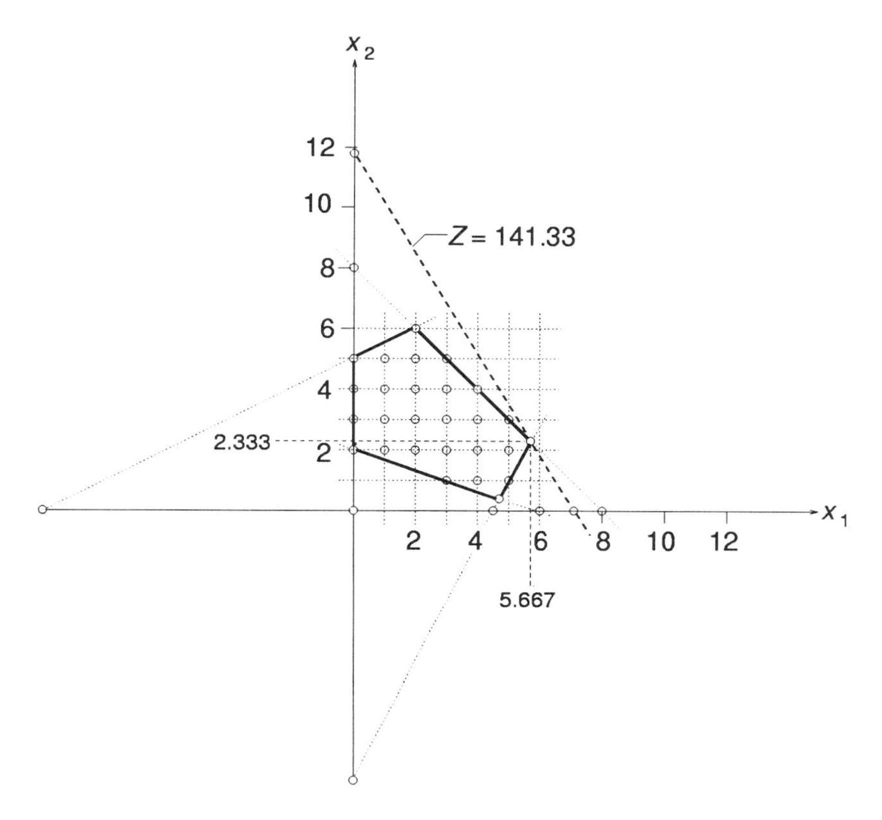

Figure 11.4: Integer Programming Example

Another danger is that, when rounded down, a decision variable, or several such, may mask the discovery of other, more optimum, integer variables which are not necessarily proximate to the continuous variable optimum solution. Therefore, it becomes necessary to exercise great care to examine all potential candidates among the available integers in such problems.

Consider the following example.

$$\text{MAX} \quad Z = 20\,x_1 + 12\,x_2 \tag{11.61}$$

$$\text{ST} \quad -x_1 + 2\,x_2 \leq 10 \tag{11.62}$$

$$2\,x_1 + 6\,x_2 \geq 12 \tag{11.63}$$

$$x_1 + x_2 \leq 8 \tag{11.64}$$

$$4\,x_1 - 2\,x_2 \geq 18 \tag{11.65}$$

$$x_1,\, x_2 \geq 0 \tag{11.66}$$

From the illustration, which, because there are only two decision variables, is a planar figure, Figure 11.4, the optimum ***continuous*** solution is $Z = 141.33$ at $x_1 = 5.667$, $x_2 = 2.333$.

Suppose that x_1 is rounded down to 5, while x_2 is at 2. The value of Z is 124. However, the points $x_1 = 5$, $x_2 = 3$ and $x_1 = 4$, $x_2 = 4$, which are on the constraint line $x_1 + x_2 \leq 8$, give Z values of 136 and 128, which are *larger* than the Z for $x_1 = 5$, $x_2 = 2$.

Probably the most conservative process will be to overlay the feasible solution space with a lattice having its grid points at all the possible integer combinations within the feasible region. Then compute the value of the Z function at most of those locations near the constraint lines, even those which are removed from the continuous linear program optimum.

While such a process might be called "crude," it is nevertheless effective at rendering the integer solutions to many programming situations.

11.7 SENSITIVITY ANALYSIS

Optimization is the stated objective of the linear programming agenda, and, as has been shown in the preceding sections, the simplex algorithm and tableaux method achieve this objective in a definitive manner. Therefore, given the specific numbers for the a_{ij}, b_i, and c_i in Equations 11.32–11.42, the best values of the x_i in this model can be obtained.

There is still a higher level of optimization which can be attained by ***sensitivity analysis***. The origin of this procedure was established in Chapter 10, where the ***sensitivity index*** was described. There it was shown that λ, the Lagrange multiplier or sensitivity factor, is a measure of the change in an objective function caused by a change in the right-hand side of a constraint relationship.

Sensitivity analysis is a generalized method for investigating a system model as formulated in the linear programming format to obtain valuable information concerning the effect of varying the system parameters upon the optimum values of the decision variables. The procedure contemplates determining the best values of the Z-function for changes in the "cost" coefficients, c_i, as well as for the right-hand side "resource" numbers, b_i.

As a matter of fact, even the a_{ij} coefficients, which represent the resource allocations associated with the decision variables, x_i, can also be allowed to assume perturbed values, although it must be understood that this optimization will need to be carried out piecemeal; i.e., while holding all the other a_is constant.

To begin the study of sensitivity analysis, take the system model to be

$$\text{MAX} \quad Z = 3x_1 + 4x_2 \tag{11.67}$$

$$\text{ST} \quad x_1 + 2x_2 \leq 8 \tag{11.68}$$

$$2x_1 + x_2 \leq 8 \tag{11.69}$$

$$x_1, x_2 \geq 0 \tag{11.70}$$

as shown in Figure 11.5. Set up the initial tableau and generate the second and third tableaux from it as usual. Note that the iteration ends at the third tableau with $x_1 = 8/3$, $x_2 = 8/3$, and $Z = 56/3$, which confirms the graphical solution.

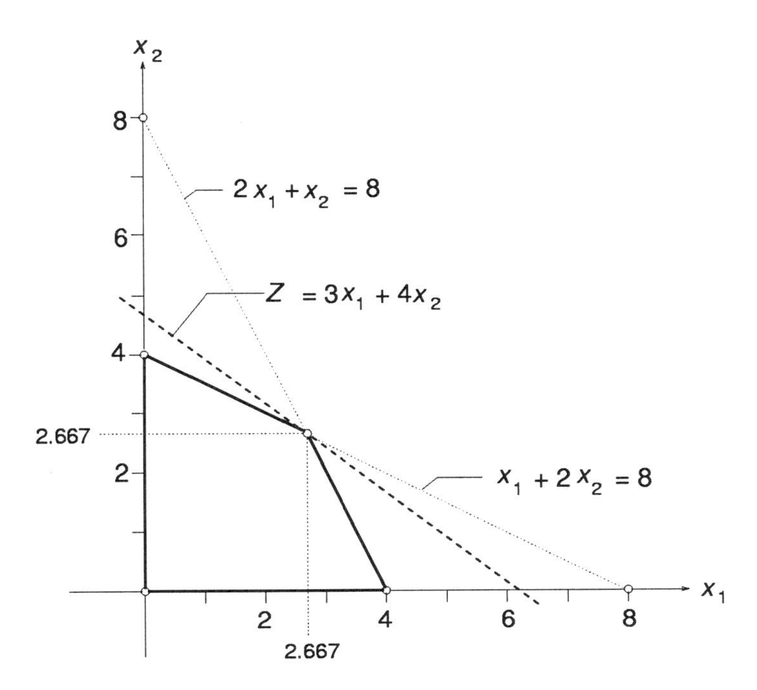

Figure 11.5: System Model

INITIAL TABLEAU

Row	BV	Z	x_1	x_2	x_3	x_4	RHS	$UBEV$
		Coefficients of						
$A1$	Z	1	−3	−4	0	0	0	
$B1$	x_3	0	1	$\boxed{2}$	1	0	8	4
$C1$	x_4	0	2	1	0	1	8	8

SECOND TABLEAU

Row	BV	Z	x_1	x_2	x_3	x_4	RHS	$UBEV$
		Coefficients of						
$A2$	Z	1	−1	0	2	0	16	
$B2$	x_2	0	$\frac{1}{2}$	1	$\frac{1}{2}$	0	4	8
$C2$	x_4	0	$\boxed{\frac{3}{2}}$	0	$-\frac{1}{2}$	1	4	$\frac{8}{3}$

THIRD TABLEAU

		Coefficients of						
Row	BV	Z	x_1	x_2	x_3	x_4	RHS	$UBEV$
$A3$	Z	1	0	0	$\frac{5}{3}$	$\frac{2}{3}$	$\frac{56}{3}$	
$B3$	x_2	0	0	1	$\frac{2}{3}$	$-\frac{1}{3}$	$\frac{8}{3}$	
$C3$	x_1	0	1	0	$-\frac{1}{3}$	$\frac{2}{3}$	$\frac{8}{3}$	

11.7.1 CHANGES IN COEFFICIENTS c_i

To investigate possible changes in the cost coefficients, c_i, assume that the objective function had been

$$Z = c_1\, x_1 + c_2\, x_2 \tag{11.71}$$

where c_1 and c_2 are some other constants than -3 and -4. The third tableau expression for Z would have been

$$\begin{aligned} Z &= c_1\,(\tfrac{8}{3} - \tfrac{1}{3}\,x_3 + \tfrac{2}{3}\,x_4) + c_2\,(\tfrac{8}{3} + \tfrac{2}{3}\,x_3 - \tfrac{1}{3}\,x_4) & (11.72)\\ &= \tfrac{8}{3}(c_1 + c_2) + (-\tfrac{1}{3}\,c_1 + \tfrac{2}{3}\,c_2)\,x_3 + (\tfrac{2}{3}\,c_1 - \tfrac{1}{3}\,c_2)\,x_4 & (11.73)\end{aligned}$$

Now, the third tableau must remain the final one if the optimum solution is to be preserved. That is, there cannot be another iteration beyond the third tableau, or, in other words, the coefficients of x_3 and x_4 in Equation 11.73 cannot be permitted to go negative, otherwise there would be the possibility of another round of iteration. This places the following requirements upon c_1 and c_2:

$$\begin{aligned} -\tfrac{1}{3}\,c_1 + \tfrac{2}{3}\,c_2 &\geq 0\\ 2\,c_2 - c_1 &\geq 0\\ 2\,c_2 &\geq c_1 \end{aligned} \tag{11.74}$$

$$\begin{aligned} \tfrac{2}{3}\,c_1 - \tfrac{1}{3}\,c_2 &\geq 0\\ \tfrac{2}{3}\,c_1 - \tfrac{1}{3}\,c_2 &\geq 0\\ 2\,c_1 &\geq c_2 \end{aligned} \tag{11.75}$$

Obviously, for $c_1 = 3$, $c_2 = 4$, the relationships above are satisfied, and it is clear that c_1 and c_2 can be allowed to take on other values, so long as that is the case. The value of the objective function will change, of course, as c_1 and c_2 are altered, even though the optimal values $\overset{\star}{x}_1$ and $\overset{\star}{x}_2$ are the same.

A reference to the graph, Figure 11.5, adds to an understanding of the situation. The slope of the objective function line can be made to pass through the optimum point so long as the slope is such that $x_1 = 8/3$, $x_2 = 8/3$ is still that point. When the slope changes enough that one or the other vertices of the quadrilateral becomes the optimum, then there is a new system model to be treated.

Suppose $Z = 2x_1 + x_2$, so that $c_1 = 2$, $c_1 = 1$. There would then need to be $2c_2 = 2(1) = 2 = c_1$, $2c_1 = 2(2) = 4 > c_2 = 1$. This is thus a limiting line beyond which the negative slope could not be allowed to increase without causing the optimum point to shift to the lower right apex of the feasible solution space on the x_1-axis.

Similarly, if $Z = x_1 + 2x_2$, with $c_1 = 1$, $c_2 = 2$, there would then need to be $2c_2 = 2(2) = 4 > c_1$, $2c_1 = 2(1) = 2 = c_2$. Again, there is a limit line with negative slope which cannot be permitted to decrease without shifting the optimum point to the upper left apex of the feasible solution space on the x_2-axis.

These limit lines are the constraint lines, of course, and the objective function has the value 8 for both cases. This is less than the optimum, $Z = 56/3$. However, multiples of the c_i values which maintain the same slopes, could be made to yield larger Zs, which would still be optima.

11.7.2 CHANGES IN CONSTRAINTS, b_i

Each constraint relationship is of the form

$$\sum_{j=1}^{n} a_{ij}\, x_j \begin{pmatrix} \geq \\ = \\ \leq \end{pmatrix} b_i \tag{11.76}$$

or

$$\sum_{j=1}^{n} a_{ij}\, x_j = b_i \tag{11.77}$$

when slacks, surplusses, and artificial variables have been introduced. To use the language appropriate to the system model proposed above, each constraint is a straight line which serves as one boundary of the feasible solution space. Then each such constraint line has specific a_{ij} coefficients, and these determine the slope of the line in question.

When the "resource" parameter b_i is changed, the slope of that constraint line is not altered; rather, the line is displaced parallel to itself to accomodate the new b_i. For example, if the constraint $2x_1 + x_2 = 8$ is changed into $2x_1 + x_2 = 6$, the line is translated parallel to itself to the position shown in the dashed line in Figure 11.6. On the other hand, if it is changed into $2x_1 + x_2 = 10$, the dotted line shows its new location. In the first case, the optimum shifts to $x_1 = 4/3$, $x_2 = 10/3$, where $Z = 52/3$. In the second, the values are $x_1 = 4$, $x_2 = 2$, $Z = 60/3$.

Similarly, if the constraint $x_1 + 2x_2 = 8$ is altered to $x_1 + 2x_2 = 6$, the new line is as shown dashed in Figure 11.7, and when it is $x_1 + 2x_2 = 10$, the line is as seen dotted in Figure 11.7. In the first instance, the optimum is relocated to $x_1 = 10/3$, $x_2 = 4/3$, which gives $Z = 46/3$. In the latter case, the numbers are $x_1 = 2$, $x_2 = 4$, $Z = 66/3$.

What these special cases show is that the system analyst can determine that an increment in the "resource" quantity b_i associated with the constraint can cause the objective function to increase or decrease, depending upon the sense of the increment.

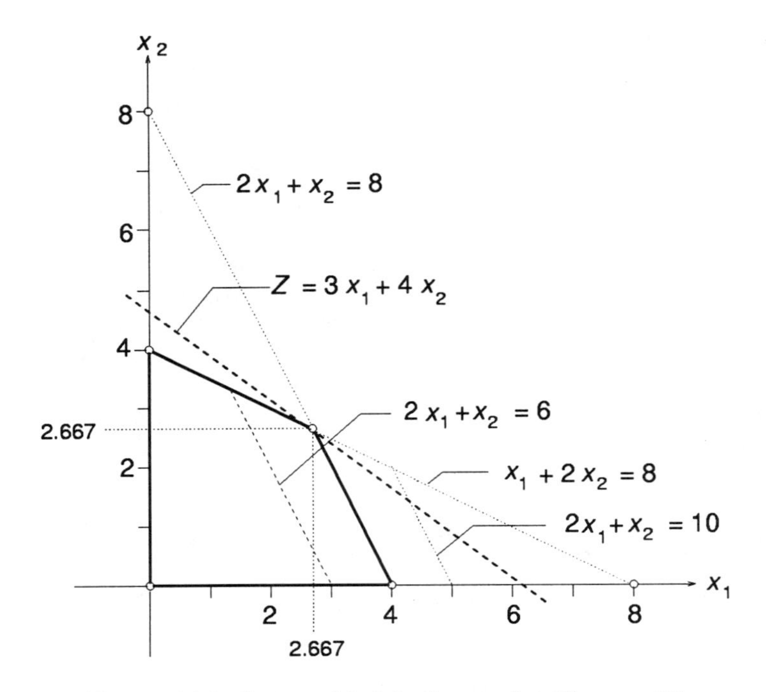

Figure 11.6: System Model, Constraint Change #1

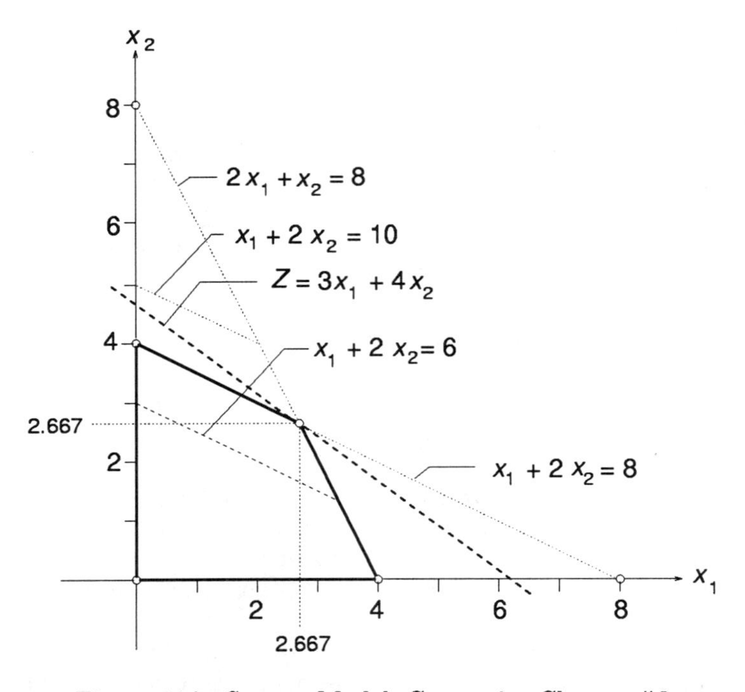

Figure 11.7: System Model, Constraint Change #2

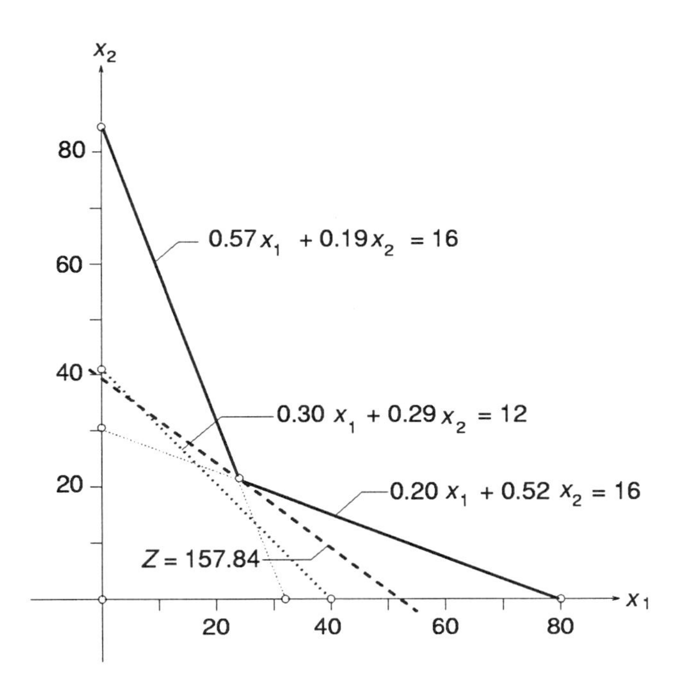

Figure 11.8: Minimization System Model

11.7.3 LOOSE CONSTRAINTS

All of the changes discussed in the specific problem above pertain to what are described as "tight" constraints; that is, the resource parameters b_i are those associated with a full utilization of the quantity in question. There is no slack remaining to be taken up by the system in attaining the optimum. This is due to the simplified nature of the problem as posed, in which there are only the two constraint relationships, and they have both been converted to equality constraints—always "tight."

In the more common system models, there are at least several constraints, and inequalities of both types are included. Take, for instance, the minimization

$$\text{MIN} \quad Z = 3.00\,x_1 + 4.00\,x_2 \tag{11.78}$$
$$\text{ST} \quad 0.20\,x_1 + 0.52\,x_2 \geq 16 \tag{11.79}$$
$$0.30\,x_1 + 0.29\,x_2 \geq 12 \tag{11.80}$$
$$0.50\,x_1 + 0.19\,x_2 \geq 16 \tag{11.81}$$
$$x_1,\, x_2 \geq 0 \tag{11.82}$$

As is shown in Figure 11.8, the constraint $0.30\,x_1 + 0.29\,x_2 \geq 12$ is "slack" or "loose," and does not play a determinative role in the solution for the optimum. Whatever resource may be described by this constraint is evidently not working up to its fullest utilization.

The minimum, $Z = 157.838$, is located at $x_1 = 23.784$, $x_2 = 21.622$, as shown.

The question immediately arises whether or not an increment in the b_i value, 12, of the second constraint, will change the outcome, and by how much. How sensitive is the system to alteration of this parameter? Clearly, the constraints $0.20\,x_1 + 0.52\,x_2 \geq 16$ and $0.50\,x_1 + 0.19\,x_2 \geq 16$ are "tight," whereas constraint $0.30\,x_1 + 0.29\,x_2 \geq 12$ can be permitted to be displaced parallel to itself without affecting the mimimum, and is therefore "loose."

In preparation for a discussion of the sensitivity of an outcome to variation in the right-hand side of a constraint, the tableau method solution should be obtained, as shown below. Because the constraints are all greater-than types, surplus and artificial variables are created for each, and these are named x_3 and x_4, x_5 and x_6, and x_7 and x_8, respectively. Construction of the tableaux proceeds in the usual manner.

Note that in the final tableau, the variable x_5 remains (Row 4). It is the surplus variable associated with the second constraint, Equation 11.80. The fact that it remains in the final tableau along with the decision variables x_1, and x_2 is evidence that the constraint with which it is identified is loose. Notice also that there is a right-hand side value of 1.405 linked to the x_5 variable.

If the right-hand side constant $b_2 = 12$ of this second constraint is increased by an amount Δ, the right-hand side value of Row 4 in the final tableau will be changed to $1.405 - \Delta$, since from the analog of Equation 11.80 there is

$$0.30\,x_1 + 0.29\,x_2 - x_5 = 12 + \Delta \tag{11.83}$$

or

$$\begin{aligned} x_5 &= 0.30\,x_1 + 0.29\,x_2 - 12 - \Delta \\ &= 0.30(23.784) + 0.29(21.622) - 12 - \Delta \\ &= 1.405 - \Delta \end{aligned} \tag{11.84}$$

INITIAL TABLEAU

		Coefficients of										
Row	*BV*	Z	x_1	x_2	x_3	x_4	x_5	x_6	x_7	x_8	*RHS*	*UBEV*
$A1$	Z	1	−3	−4	0	M_1	0	M_2	0	M_3	0	
$B1$	x_3	0	0.20	[0.52]	−1	1	0	0	0	0	16	30.769
$C1$	x_4	0	0.30	0.29	0	0	−1	1	0	0	12	41.379
$D1$	x_5	0	0.50	0.19	0	0	0	0	−1	1	16	84.211

SECOND TABLEAU

Row	BV	Z	x_1	x_2	x_3	x_4	x_5	x_6	x_7	x_8	RHS	UBEV
		Coefficients of										
A2	Z	1	−1.462	0	7.692	M_1-	0	M_2	0	M_3	123.077	
B2	x_2	0	0.385	1	−1.923	1.923	0	0	0	0	30.769	80.00
C2	x_4	0	[0.188]	0	0.558	−0.558	−1	1	0	0	3.077	16.327
D2	x_5	0	0.427	0	0.365	−0.365	0	0	−1	1	10.154	23.784

THIRD TABLEAU

Row	BV	Z	x_1	x_2	x_3	x_4	x_5	x_6	x_7	x_8	RHS	UBEV
		Coefficients of										
A3	Z	1	0	0	12.046	M_1+	−7.755	M_2+	0	M_3	146.939	
B3	x_2	0	0	1	−3.061	3.061	2.041	−2.041	0	0	24.490	12.00
C3	x_1	0	1	0	2.959	−2.959	−5.306	5.306	0	0	16.327	−3.007
D3	x_5	0	0	0	−0.898	0.898	[2.265]	−2.265	−1	1	3.184	1.405

FOURTH TABLEAU

Row	BV	Z	x_1	x_2	x_3	x_4	x_5	x_6	x_7	x_8	RHS	UBEV
		Coefficients of										
A4	Z	1	0	0	8.952	M_1	0	M_2	3.436	M_3+	157.837	
B4	x_2	0	0	1	−2.252	2.252	0	0	0.901	−0.901	21.622	
C4	x_1	0	1	0	0.856	−0.856	0	0	−2.342	2.342	23.784	
D4	x_5	0	0	0	−0.396	0.396	1	−1	−0.441	0.441	1.405	

Because x_5, like all the variables, must be positive, it can be seen that Δ can take on any negative amount, but when $\Delta = 1.405$, $x_5 = 0$, and thereafter any larger positive increment Δ will drive x_5 negative, and the solution becomes infeasible. $\Delta = 1.405$, $x_5 = 0$, corresponds to a parallel-displaced line for the second constraint, one which passes through the solution point $x_1 = 23.784$, $x_2 = 21.622$.

In a more general system model, each constraint relationship can be supplied with its own Δ_i sensitivity increment, and the range of values of Δ_i can be determined as above for each in turn, using the final tableau right-hand side value in question. For less-than type constraints, the relationship is

$$x_i = (RHS)_i + \Delta_i$$

of course, rather than $x_i = (RHS)_i - \Delta_i$, as in the numerical example. Equality-type constraints are always "tight," so there is no possibility of increments without altering the optimal solution.

To summarize, when the constraints are "tight," any proposed alteration of the right-hand sides amounts to a new solution of the problem, with altered optimum; whereas, if the constraints are "loose," increments can be discovered for the right-hand sides which, within the limits as determined, will retain the system optimum.

Information contained in analyses such as this may be of considerable utility to the civil engineer who is faced with a system design. Some constraints can be modeled more easily than can others, because of the difficulty of obtaining valid data or other information. When such constraints are slack, or the sensitivity to right-hand sides is not strong, the analyst's attention can be directed toward those which are more decisive. Also, expensive resource parameters b_i can be exchanged for less costly ones, if there is a sufficiently wide tolerance, in a given case.

11.8 COMPUTER METHODS

The numerical complexity and repetitive nature of the Linear Programming scheme figuratively cries out for adaptation to machine computation. If ever there was a problem area in which computers could be said to be a natural aide, this would surely be it. The point is, as the lawyers might say, a case of *res ipsa loqitur*—"the thing speaks for itself"; i.e., it is self-evident.

There are a number of programs available for running LP on computers, from supercomputers to mainframes, to minis, through workstations to desktop PCs to programmable hand-helds. Historically, only mainframes were of large enough capacity and sufficient speed to justify programming for this purpose, and, in addition, the other machines were nonexistent at that stage. Thus the code which was written was in FORTRAN, using punched cards for input.

As other machines have come into being and have evolved into more powerful instruments, the small-scale software appropriate to early models has evolved also to the point that problems of modest size can be accomodated.

Two particular computer programs will be mentioned as representative of those available for student work in a first exposure to Linear Programming. The first is SAS LP/OR, a program originally for mainframes, and now available for machines ranging from PCs to supercomputers. The other is *LP.go*, a locally-prepared program available on the *eos* network.

SAS LP/OR

SAS is an acronym for *Statistical Analysis Systems*, a company which produces and sells software for computers. It is a company which shares the rags-to-riches syndrome common to a number of high-tech enterprises in the last four decades. SAS was founded by students at North Carolina State University in the mid-1960s who were engaged in studies in Experimental Statistics. While they were still students they wrote programs for doing statistical models, and later sold these under license to various universities and research groups.

SAS is now a global corporation with a large campus in the Research Triangle area of North Carolina. While their product still features statistical and related computations, it is highly diversified in terms of technical specialties and geographical distribution. One of their most important packages contains the LP/OR routines.

LP/OR, as the name implies, is a program for doing Linear Programming and Operations Research. It is in the library of *eos* and many other networks as well. As a software, it was originally based on FORTRAN source code, but is now mainly a C language product. To use it, one logs onto the net, with the system in command mode, then toggles to collect mode. With the system in collect mode, the following commands are entered.

```
PROC LP DATA=PROBLEM;
 VAR     VAR1     VAR2 ··· VARn
 TYPE RELAT;
 RHS VALUE;
```

Exit the collect mode and return to command mode and enter the command

```
SUBMIT SAS;
```

and any other system commands as desired or appropriate. Then give the additional commands DATA PROBLEM;

```
 INPUT    VAR1    VAR2 ··· VARn    RELAT $ VALUE;
 CARDS;

 c1     c2 ··· cn     MAX

 a11    a12 ··· a1m    LE    b1

 ··· ··· ··· ··· ···

 ap1    ap2 ··· apm    EQ    bp

 ··· ··· ··· ··· ···

 as1    as2 ··· asm    GE    bs

 ;
```

From the literal nature of these commands it is apparent that in the collect mode a file is being created that contains the LP PROCEDURE in a problem format, after which, in command mode, the data to supply the entries for this model are input. Obviously, the *cs* are the coefficients of the objective function, the *as* are coefficients of the variables in the constraints, which are listed by type (LE for less-than-or-equal, EQ for equality constraints, GE for greater-than-or-equal), and the *bs* are the right-hand-side numbers for the constraints. MAX tells the program that the user wishes to maximize the objective function. The final semi-colon tells the machine that data input is at an end. The problem is then run and printed out. It's just that simple.

The most recent version of SAS on *eos* net is probably even simpler to use.

LP.go

This program was prepared by the CE 375 staff especially for students to use in connection with the course. It is FORTRAN source code which has been compiled and is ready to run. It is resident in the locker *mcdonald_src* on the *eos* network. It is interactive and user-friendly. When the program runs, the user is prompted on the screen for keyboard input.

The user will be informed of the features of the program and asked to supply the data for the *as*, *bs*, and *cs*, and for the number and types of the constraints. A query will also prompt the user for a choice of maximizing or minimizing.

When all data have been entered, the program will run. It will display (echo) the Initial Tableau entries and pause while the user confirms that the data is correct. Thereafter, the program will stop at each tableau for inspection, and proceed to the Final Tableau. There is a printout to a file named *LP.dat*.

11.9 PROJECT # 9

1. A structural engineer is designing a high-rise building. The estimates are that at least 10,000 #10, 15,000 #8, and 12,000 #12 screws will be needed for the interior finishing. Previous experience indicates that there are two major suppliers of fasteners who can furnish this requirement. SCREW-ZLOOZ offers a package which contains 20% #10s, 18% #8s, 25% #12s and 37% other sizes for $8 each. For $16 each, FASTENZ has a package which contains 30% #10s, 50% #8s, and 20% #12s.

   How many packages (of 1000 screws each) from each supplier should be specified to minimize the cost? What is the cost?

2. YURA GOODE, CE, is doing a job which requires large amounts of brick. She estimates that the job will require: Regular Brick, 290,000; Half-brick, 100,000; Cap Brick, 20,000. The brick will come from two makers who supply brick in lots of 1000 each. Analysis shows the following composition of the two makers' lots:

	CHUCKACHEE	BURREN BRICK
	(%)	(%)
Regular Brick	42	49
Half Brick	27	31
Cap Brick	20	20
Half Cap Brick	11	0

It costs $500 per lot of 1000 brick from CHUCKACHEE and $800 from BURREN BRICK per lot of 1000 brick. How much should be bought from each to minimize the total job cost? What is this total cost?

3. Donaldson and James is a family-owned construction company which specializes in residential and light business or industrial jobs. Company officers have studied the firm's performance and have concluded that their net income can be represented by the equation

$$P = 14X + 9Y$$

in which P is in tens of thousands of dollars, X is the number of residential projects and Y is the number of business/industrial projects. In addition, they have recognized that the company's resources are sufficient to undertake a maximum of 14 projects, and, further, that there should be at least 5 business/industrial projects to fully utilize their crews which are specialized to this work.

Moreover, the banks have established the following maximum amount of credit against which the company can draw (all items in tens of thousands of dollars)

$$10X + 5Y \leq 100$$

Determine the number of projects in each category which will maximize the company's income. What is this maximum income?

Use a tableaux method to obtain the answer requested above, and confirm your solution using both a graphical method and the program *LP.go* in the locker *mcdonald_src* on *eos* net.

4. Lickens Steel Company sells rolled steel beams and steel reinforcing bars. The beams are employed by builders who erect steel frame structures, while the rebars are used by contractors in the reinforced concrete structures business. Lickins estimates that its annual net profit can be represented by the equation

$$P = 7X + 5Y$$

in which P is in hundreds of thousands of dollars, X is the number of steel projects and Y is the number of reinforced concrete projects built in the area annually. They also notice that the total number of projects is less than 16 per year, and the number of reinforced concrete building projects is less than 7.

Use a tableaux method to determine the number of projects which optimizes Lickins' earnings. What is the earnings?

Solve the above problem by a graphical technique.

5. Mr. Grey Slate operates ROCK CITY QUARRIES, LOADED. The firm has received a substantial order from Mr. Coarse Rubble, a local contractor. Mr. Slate has decided that the material will be quarried from two pits by his operator, F. Flintstone, and subject to constraints $-x_1 + 2x_2 \leq 6$ and $4x_1 - 3x_2 \leq 8$, in which x_1 and x_2 are the amounts mined from pits A and B, respectively. These constraints represent the limitations imposed by time, labor, machinery, and supplies.

   If the selling price of rubble from *PitA* is \$3 per m^3, and that from *PitB* is \$4, what amounts should be taken from each pit in order to maximize Mr. Slate's earnings?

   Solve this problem by the tableaux method and confirm the solution graphically.

6. IMA GOODE, CE, is doing a job which requires large amounts of sand and gravel. She estimates that it will need: coarse gravel, 19,500 m^3; fine gravel, 31,000 m^3; sand, 20,500 m^3. The material will come from two pits, and will be screened at the job site.

   Analysis shows the following material composition of the two pits:

Material	**Pit A** (%)	**Pit B** (%)
Coarse gravel	20	30
Fine gravel	14	50
Sand	25	20
Waste	41	0

   It costs \$8 per cubic meter for material and haulage from Pit A and \$10 per cubic meter from Pit B. How much should be hauled from each pit to minimize the total job cost? What is this cost?

   Solve this problem using the tableaux method, a graphical method, and a computer method.

7. A builder is beginning a job which will require large amounts of framing lumber. The estimate is that the need is for: 8-ft 2×4s, 10,000; 12-ft 2×4s, 5000; 16-ft 2×4s, 10,000. The lumber will come from two lumberyards which furnish timbers in lots of 100 each. Analysis shows the following composition of the lots obtained from the two lumberyards:

	BILDEX (%)	ALL-STUDS (%)
8-ft 2×4s	20	31
10-ft 2×4s	44	0
12-ft 2×4s	15	52
16-ft 2×4s	21	17

   It costs \$225 per lot from BILDEX and \$300 per lot from ALL-STUDS. Any 10-ft 2×4s purchased will be used on other jobs or resold.

How many lots should be purchased from each yard to minimize the builder's total lumber cost? What is the total cost?

Solve this problem using the tableaux method, a graphical method, and a computer method.

8. Concrete Canoe Bunch, Collaborators, makes two kinds of concrete canoes for sale. The standard model, "BOB," requires 2 hours to make the hull, 3 hours to install flotation and trim and to paint and decorate. Times required for the deluxe model, "BABE," are 3 hours, hull; 2 hours for flotation and trim and paint and decorating.

   A work crew specializes in making hulls, another crew does the flotation, trim, painting, and decorating. The sales prices are $130 and $170, respectively.

   If there are 600 hours allocated in the next production period for each work crew, determine the number of each model which should be produced to optimize the profits. What is the gross income?

9. Use a linear programming software of your choice to determine the capacity of the network illustrated below.

   Let your report include a print-out of the solution, and also show that you have checked the answer by primal and dual net analyses.

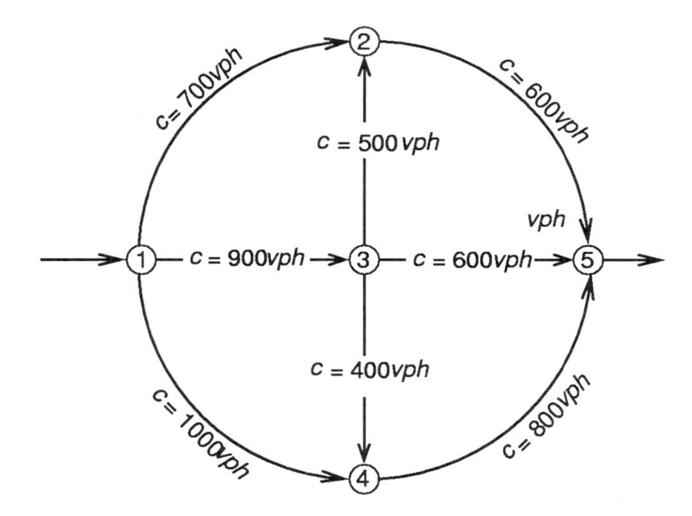

CHAPTER 12

NONLINEAR PROGRAMMING

12.1 INTRODUCTION

The previous chapter treated the problem of optimizing a criterion function which was linear in the decision variables and the constraints. The methods developed therein are strictly appropriate to linear problems, and do not allow themselves to be applied when either the objective function or the constraints become nonlinear.

This situation is typical of that throughout the world of engineering science. Some have even said that there is a fundamental and irreducible discontinuity between the two realms, a difficulty not broached by extension of the methodology of linear systems to the nonlinear domain. As a matter of fact, linearizations and quasi-linearizations do not come close to representing some of the most important features of nonlinear systems, and are therefore not suitable in most applications.

The most important first attribute of nonlinear systems is that it is not even known whether a solution to the given problem exists. In a great many cases, it is simply not possible to demonstrate that there is an answer, let alone find what it is. And even in those cases where an answer is found, it is usually uncertain that it is the only one or that it is anything more than a local optimum. So it is necessary to devote a considerable effort to the business of trying to demonstrate existence of solutions and to find upper and lower bounds on the range of whatever solutions there may be.

There are a few good procedures, most of which are the result of the particular acts of genius of certain individual persons. Of course, this has been the case with all of the general methods to date, as calling to mind the names of Newton, Lagrange, Kuhn, Tucker, Dantzig, etc., will confirm. But in the case of things nonlinear, the particular solutions and methods do not seem to lend themselves to easy generalization. The serendipity involved in each special case does not appear to extend much beyond the limits of the speciality which loosed it in the first place.

There is nothing like the Linear Programming algorithm or its accompanying computer programs for the nonlinear type of problem. Nothing, that is, which can provide for various classes and degrees of nonlinearity. Only within certain grades of nonlinearity can methods of solution be applied to a range of problems. It would be better to say, probably, that dealing with nonlinear problems is an ad hoc business. Mixtures of methods, half-methods, and heuristic processes are commonly seen. Some methods of high sophistry and great complexity may be feeders to the roughest or crudest approximations.

Some of the methods and names associated with them that are utilized in nonlinear optimization are ***Half-Interval*** or ***Bisection*** algorithm, ***Newton-Raphson*** method, the ***Minimax Principle***, ***Fibonacci Search***, ***Golden Section Search***, ***Lattice Search***, ***Powell's Method***, ***Steepest Gradient***, ***Pattern Search***, ***Method of Approximate Programming (MAP)***, and ***Successive Quadratic Programming***.

Typical of the computer programs employed in nonlinear programming is MINOS, a software developed in Systems Optimization Laboratory at Stanford University, Stanford, California. It is a FORTRAN-based optimization system designed to efficiently solve large-scale linear or nonlinear programming problems. When the objective function is nonlinear, the user must supply a FORTRAN subroutine to compute the function and its gradient. MINOS then uses a reduced-gradient algorithm combined with a quasi-Newton procedure to compute the solution. When some or all of the constraints are nonlinear, the user must furnish a second subroutine to compute the constraint functions and their gradients. MINOS invokes a projected augmented Lagrangean algorithm to obtain a solution. Obviously, this software is intended for use with a mainframe, and results[1] from problems run on an IBM 3033 were impressive for certain classes of constraints.

Whatever works in nonlinear theory should be acclaimed, and that which is unable to promise existence, set limits or purchase numbers, however exalted otherwise, may be allowed to seek its place in the arcane lore of the unopened tomes.

There is a saying attributed to "the old army": "When a soldier leaves his company, all debts and friendships are cancelled." In programming, when a problem leaves the linear world—when it goes nonlinear—all bets are off!

In recognition of these circumstances, it has seemed desirable to limit the nonlinear encounter in a first course to some necessary topological concepts and a proven method for the numerical determination of optima in two-dimensional problems using a pattern-search technique which has application to Civil Engineering practice. The student is encouraged to pursue this subject, not only to more advanced levels, but to make a career of it, for there is room enough for many and rewards and satisfaction yet undreamed of for some who do.

12.2 NONLINEAR SYSTEMS TOPOLOGY

Many of the concepts and methods utilized in nonlinear optimization work employ the idea of gradient or slope to guide the direction of progress. Or again, language such as "climbing techniques" suggests a feasible solution space as a surface with hills, valleys, ridges, and contours. Therefore both physical and mathematical topology undergird the subject.

---

[1]Simon, J. D., and Azma, H. M., "Exxon Experience with Large Scale Linear and Nonlinear Programming Applications," ***Computers and Chemical Engineering***, Vol. 7, No. 5, 1983, p. 605.

Figure 12.1 shows the graph of a function of a single variable, x, with a single maximum. Figure 12.2 illustrates another function of a single variable, but this one has two maxima, so they are called ***local maxima***, and that one with the largest value of the function will earn the additional name, ***global maximum***. Of course, there are similar pictures that can be drawn if the discussion is about minima, as in Figures 12.3 and 12.4.

Procedures for locating such local optima, and for identifying whether or not they are minima or maxima, were discussed in Chapter 9. Unfortunately, there is no known method for systematically progressing from one local maximum to the global maximum. So while it is relatively easy to find a local optimum, finding the global optimum in the case of nonlinear problems is fortuitous if it happens at all.

It therefore becomes important to know if there is only one local optimum, since this would guarantee that it is also the global one.

A general topological concept called ***convexity*** can be invoked to assist with this task. In Figure 12.1, for example, there is only one local maximum, and from the figure it is seen that the shape of the function is such that it always has a curvature of the same sign. That is, it is ***unimodal*** with respect to curvature. The unimodality is said to be positive in this case. On the other hand, in Figure 12.2 the function goes through several changes of sign of the curvature, which reflect the torture of its topology and accompany its multiple maxima.

There is the companion concept called ***concavity***, a term applied to those shapes with negative curvatures. Figure 12.3 shows such a function, and Figure 12.4 illustrates a function with multiple minima and negative unimodality.

There is an analytical test which can be applied to a function to determine if it is convex. Take any two points on the graph with the arguments x_1 and x_2. Then if for all k such that $0 \leq k \leq 1$

$$f[k\,x_2 + (1-k)\,x_1] \leq k\,f(x_2) + (1-k)\,f(x_1) \tag{12.1}$$

the function is convex. Correspondingly, if the $\leq$ is changed to a $\geq$ sign, the function is concave.

The nature of the test can be seen graphically in Figures 12.1 and 12.3, where the chord drawn between the values of the function at x_1 and x_2 lies entirely below the values of the function between x_1 and x_2, for the convex curve, and entirely above the values on the curve between x_1 and x_2 for the concave function.

Of course, the test requires that it is to be fulfilled for all pairs of points on the curves. Making this test numerically will then be comparable to simply computing the function itself and plotting the curve in the usual straightforward way to see whether or not it has multiple optima. Otherwise, two points such as x_1 and x_2 in Figure 12.2 or Figure 12.4 might be taken, the test would be passed, and a wrong conclusion would be drawn about the nature of the function.

The true merit of the label convex (concave) as applied to a function undergoing optimization is that it does guarantee a single global minimum (maximum).

When the function to be optimized becomes one with many decision variables, the same principles of convexity apply. Their application to the multidimensional case is simply more complex.

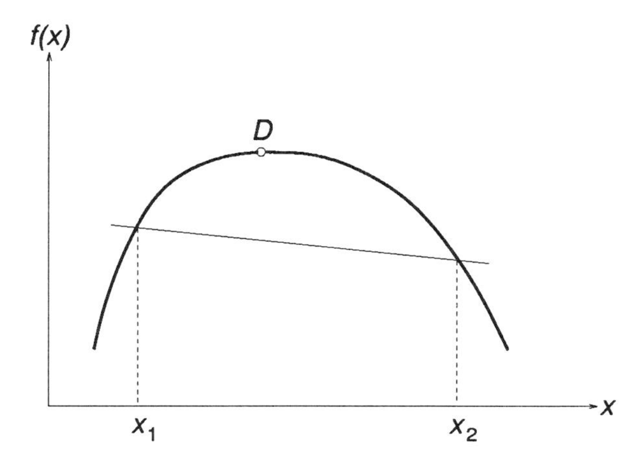

Figure 12.1: Local Maximum

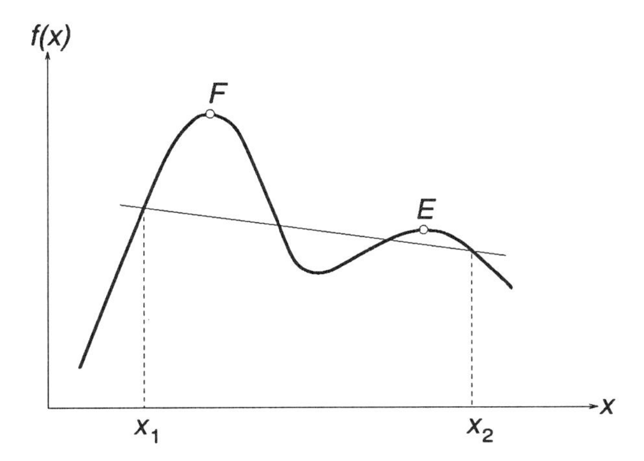

Figure 12.2: Global Maximum

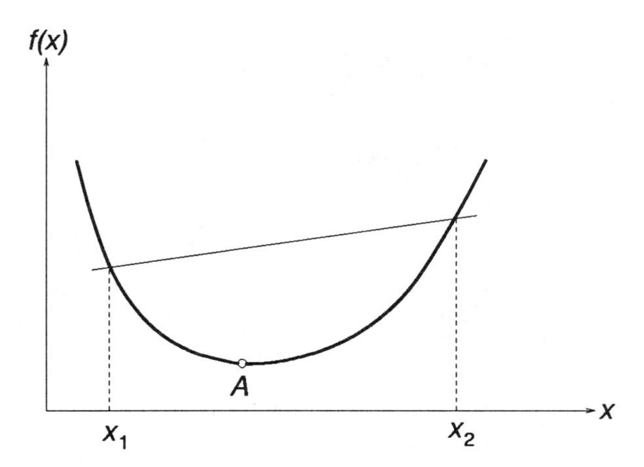

Figure 12.3: Local Minimum

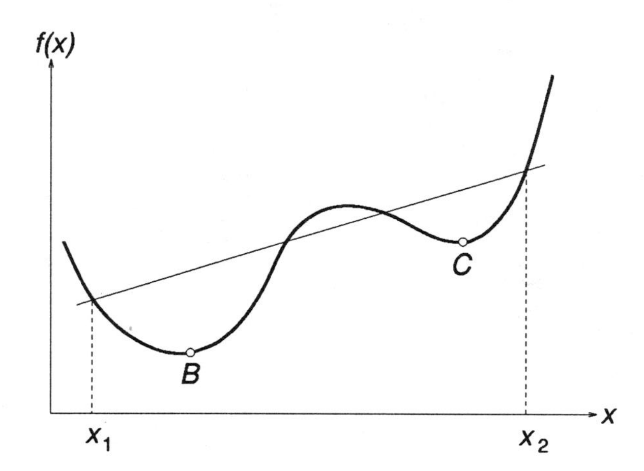

Figure 12.4: Global Minimum

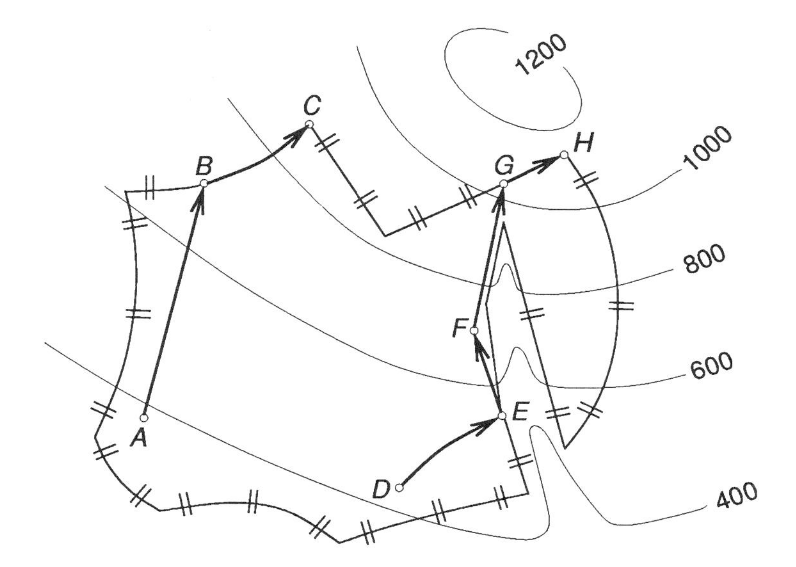

Figure 12.5: Mountainside Meadow

A more important consideration is presented when the constraints are added to the problem. The nature of the difficulty can be brought out by the following analogy. One of the world's most beautiful scenic routes is that called the *Blue Ridge Parkway* in western North Carolina and southwestern Virginia. This road, now 50 years old, was constructed along and near the crest of the Blue Ridge chain of mountains. It was built during the Great Depression, a time when labor was cheap and when public works were created to furnish jobs for people who had none. It was also a time when land was easy to acquire, when design services and park consultants were inexpensive. These circumstances conspired to permit a project of incomparable beauty and facility.

The vistas one sees as one drives along the Parkway are formed from mountainside fields and meadows, framed by split rail fences, rock outcrops and forests. Imagine that Figure 12.5 is the plot of one such mountainside pasture which is limited by fences, stone walls, forests, and ravines. These are its constraints. The contours of the land are suggested on the plot, so that the highest point in view—a mountaintop of elevation 1200 m—is revealed.

A Scout Troop of two squads plans a hiking venture to the top of the mountain, from the vicinity of which the view is said to be spectacular. One squad will start from location A, the other from its camp at D. The cleared area allows for line-of-sight visual guidance in an uphill climb for both groups on the first leg of the trip from A to B and from D to E. At B the one squad encounters a fence, and decides to follow this to point C, which is obviously still higher up in the direction they wish to go. Here they are stopped by a stone wall. The other group comes to a sharp crevice at E, and elects to follow the rim of this obstruction to F, also visibly uphill, in the right direction.

When they reach F, the second group can again see, in line of sighting, an additional stretch of terrain leading uphill, and so take the straight route up to point G. There they find a rock wall, and follow it, still climbing as they go, to H. At H a dense stand of spruce pine inhibits their further progress.

With the benefit of the topological map, an analyst can easily inform the second squad of the troop that they have attained a higher standing (and presumably a better view) than have the other group whose advance was halted at C. Both groups might survey the scene and report that they had reached the highest possible point. Indeed, both have achieved local maxima, but one is higher than the other one. Neither group can see the other group, yet can logically claim the highest point, and will probably say that the others were lost. Hopefully, both will obtain the benefit of good clean air and wholesome exercise and a glorious backward overlook.

This is the nature of the constrained nonlinear problem of optimization. Except that it is usually more difficult than the little adventure detailed above. For one thing, there is frequently a fog over the entire landscape to prevent any visual guidance to the climbing programme. As in real life, the view from the bottom of the mountain is screened by the hill mass itself to exclude the peak.

Nevertheless, the topological analogy is most useful in connection with nonlinear programming devices.

12.3 CLIMBING TECHNIQUES

Mountain climbing techniques have, in general, two principal objectives (other than the customary "because it's there"); these are

- To achieve a better value of the objective function by using a climbing procedure.
- To generate important information which will be helpful in deciding upon the merits of future excursions.

Generally, three distinct phases are observed in climbing procedures. These are

- The original game plan. This generally amounts to an exploratory operation, testing the waters, learning the lay of the land or other metaphor for the first steps.
- Now the tactic changes into a deliberate struggle to gain height, step-by-upward-step, one increment after another.
- A final gambit which strives to reach the ultimate goal in the neighborhood of a near place. This stage is somewhat like the starting phase in that it is exploratory.

A number of different procedures by different names are available which utilize climbing techniques. In the main, they differ from each other by the tactics employed in the second phase, above.

12.4 PATTERN SEARCH

An important member of the climbing techniques family of algorithms for nonlinear programming is called *Pattern Search*. It is quite simple to program for both hand computation and machines. It offers the possibility of accelerated pursuit when conditions are favorable, and it has ridge-following properties which allow it to strike out along the crest of a ridge for the top of the hill.

Pattern Search is based upon the conjecture that a set of moves which was successful in first trials is worth trying again. It is in fact the same premise as that behind the coach's maxim, "We're gonna run that play again until they stop it!"

Possibly the purest version of the Pattern Search algorithm is that offered by Hooke and Jeeves.[2] It has two parts. One is a "local" exploratory scheme, the other is the acceleration part. The local explorations of the first part take place about a "base camp" (point) by perturbing each of the decision variables in turn, one at a time. As each takes its increment and a better value of the objective function is found, that place becomes the base rather than the original camp. Hooke and Jeeves gave the name *temporary heads* to these vectors.

When each decision variable has been perturbed and the objective function evaluated and compared, an accelerated move is made based upon this local pattern to a new location from which the pattern of local exploration takes place again. The entire process is continually repeated until no improvement can be had. Hooke and Jeeves listed a dozen detailed steps for carrying out the procedure. They are enumerated below.

1. Begin at a base point whose location is given by the vector $\mathbf{b}_1$. This location is arbitrary, although a judicious choice of it can both speed up convergence to the optimum and lessen potential computational difficulties in process.

2. Choose a step size δ_i for each decision variable, x_i. This then creates the step vector $\boldsymbol{\delta}_i = (0 \cdots \delta_i \cdots 0)$.

3. Compute the objective function at $\mathbf{b}_1$ and then at $\mathbf{b}_1 + \boldsymbol{\delta}_1$. If the new value of the objective function is better than the one at the base, then the temporary head is $\mathbf{t}_{11} = \mathbf{b}_1 + \boldsymbol{\delta}_i$. If the objective function here is not as good as that at the base, try $\mathbf{b}_1 - \boldsymbol{\delta}_1$, and if this gives a better result, then name it $\mathbf{t}_{11}$. The literal scheme to be implemented is given herewith.

FOR MAXIMIZING

$$\text{if } Z(\mathbf{b}_i + \boldsymbol{\delta}_i) \leq Z(\mathbf{b}_i) \qquad \mathbf{t}_{ii} = \mathbf{b}_i + \boldsymbol{\delta}_i \tag{12.2}$$

$$\text{if } Z(\mathbf{b}_i - \boldsymbol{\delta}_i) \leq Z(\mathbf{b}_i) \leq Z(\mathbf{b}_i + \boldsymbol{\delta}_i) \qquad \mathbf{t}_{ii} = \mathbf{b}_i - \boldsymbol{\delta}_i \tag{12.3}$$

$$\text{if } Z(\mathbf{b}_i) \geq \max\left[Z(\mathbf{b}_i + \boldsymbol{\delta}_i), Z(\mathbf{b}_i - \boldsymbol{\delta}_i)\right] \quad \mathbf{t}_{ii} = \mathbf{b}_i \tag{12.4}$$

[2]Hooke, R. and Jeeves, T. A., "Direct Search Solution of Numerical and Statistical Problems," *Journal of the Association of Computational Machines*, Vol. VIII, No. 2, April, 1961, pp. 212-229.

FOR MINIMIZING

$$\text{if } Z(\mathbf{b}_i + \boldsymbol{\delta}_i) \le Z(\mathbf{b}_i) \qquad \mathbf{t}_{ii} = \mathbf{b}_i + \boldsymbol{\delta}_i \tag{12.5}$$

$$\text{if } Z(\mathbf{b}_i - \boldsymbol{\delta}_i) \le Z(\mathbf{b}_i) \le Z(\mathbf{b}_i + \boldsymbol{\delta}_i) \qquad \mathbf{t}_{ii} = \mathbf{b}_i - \boldsymbol{\delta}_i \tag{12.6}$$

$$\text{if } Z(\mathbf{b}_i) \le \min\,[Z(\mathbf{b}_i + \boldsymbol{\delta}_i), Z(\mathbf{b}_i - \boldsymbol{\delta}_i)] \qquad \mathbf{t}_{ii} = \mathbf{b}_i \tag{12.7}$$

4. Next perturb x_2 about the temporary head $\mathbf{t}_{11}$. The recurrence formula for the temporary head is

$$\text{if } Z(\mathbf{t}_{1j} + \boldsymbol{\delta}_{1+j}) \le Z(\mathbf{t}_{1j}) \qquad \mathbf{t}_{1,j+1} = \mathbf{t}_{1j} + \boldsymbol{\delta}_{1+j} \tag{12.8}$$

$$\text{if } Z(\mathbf{t}_{1+j} - \boldsymbol{\delta}_{1+j}) \le Z(\mathbf{t}_{1j}) \le Z(\mathbf{t}_{1j} + \boldsymbol{\delta}_{1+j}) \quad \mathbf{t}_{1,j+1} = \mathbf{t}_{1j} - \boldsymbol{\delta}_{1+j} \tag{12.9}$$

$$\text{if } Z(\mathbf{t}_{1j}) \le \min\,[Z(\mathbf{t}_{1j} + \boldsymbol{\delta}_{1+j}), Z(\mathbf{t}_{1j} - \boldsymbol{\delta}_{1+j})] \quad \mathbf{t}_{1,j+1} = \mathbf{t}_{1j} \tag{12.10}$$

5. Perturb all of the decision variables, following which the last temporary head is then designated the second base point, $\mathbf{b}_2$. $\mathbf{b}_1$ and $\mathbf{b}_2$ establish the first pattern.

6. Assume that a similar pattern holds for the steps ahead. Skip the local excursions and double the length of the next pattern. Get a new temporary head $\mathbf{t}_{20}$ for the second pattern (based at $\mathbf{b}_2$).

$$\mathbf{t}_{20} = \mathbf{b}_1 + 2(\mathbf{b}_2 - \mathbf{b}_1) = 2\,\mathbf{b}_2 - \mathbf{b}_1 \tag{12.11}$$

7. Make a local exploration about $\mathbf{t}_{20}$. Use the logic (recurrence formulas) given at step 4 with new temporary heads $\mathbf{t}_{21}$, $\mathbf{t}_{22}$, ... , $\mathbf{t}_{2n}$. When all variables have been perturbed, the last temporary head $\mathbf{t}_{2n}$ makes the new base point $\mathbf{b}_3$.

8. Establish a new temporary head $\mathbf{t}_{30}$.

$$\mathbf{t}_{30} = \mathbf{b}_2 + 2(\mathbf{b}_3 - \mathbf{b}_2) = 2\,\mathbf{b}_3 - \mathbf{b}_2 \tag{12.12}$$

9. Continue with the third and following patterns. Should the ith pattern fail to improve, but $Z(\mathbf{t}_{i0}) \le Z(\mathbf{b}_i)$, the pattern maintains its direction and length, but there is no growth.

10. For pattern k, if $\mathbf{t}_{k0}$ does not improve $\mathbf{b}_k$, then $\mathbf{b}_{k+1} = \mathbf{b}_k$, and the pattern is broken. The process has possibly come to the peak of a ridge or may be crossing a valley.

11. At this stage a new pattern designated $\mathbf{b}_{k+1}$ based at $\mathbf{b}_k$ should be attempted. Make $\mathbf{t}_{k+1,0} = \mathbf{b}_{k+1}$. If this locates a better point, a new pattern begins. Otherwise, shorten the steps and try again.

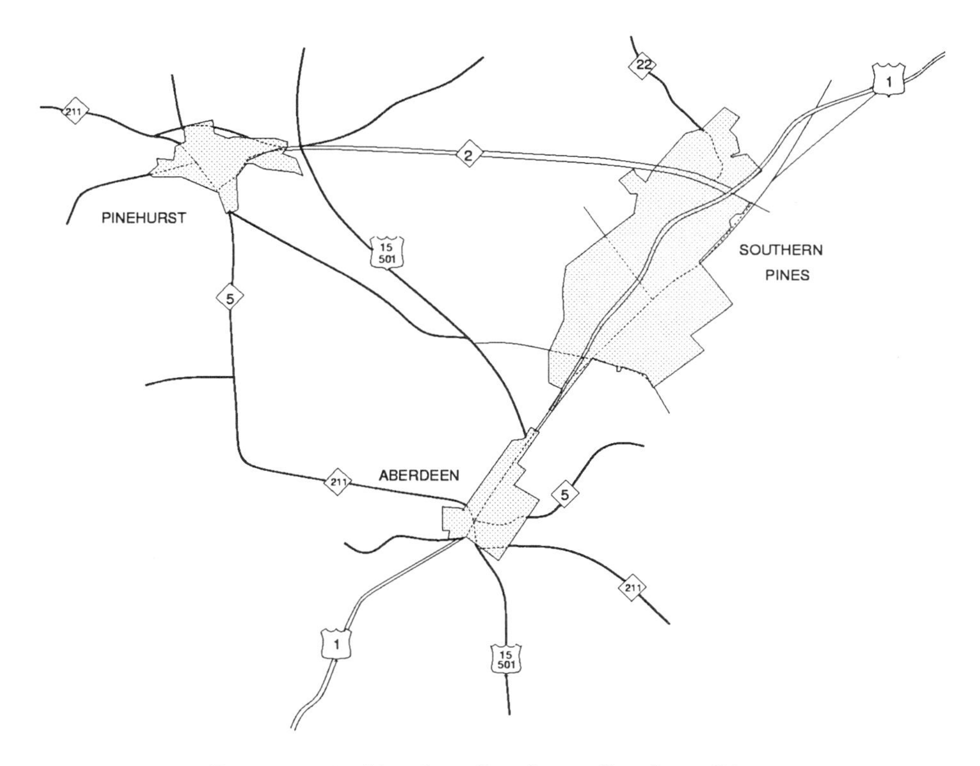

Figure 12.6: Aberdeen-Pinehurst-Southern Pines

12. A local optimum is easy to find, while a global optimum may be ignored if the search procedure converges on a local optimum with a better value of the objective function. Make a policy to conduct at least two searches from randomly selected, distinct points. If both converge to the same optimum, confidence in the result increases. Other possibilities include a search by regions.

12.5 A PATTERN SEARCH EXAMPLE

An example of the Hooke-Jeeves Pattern Search procedure applied to a problem drawn from civil engineering practice is furnished by the following.

Aberdeen, Pinehurst, and Southern Pines have decided to pool their resources and construct a new waste treatment plant somewhere within their environs. As a guide to the location of the new facility, they wish to determine the point at which the cost of the project will be a minimum. Since the plant cost will be the same wherever it is located, the total cost will be least at that location where the piping expense is smallest.

A convenient point within each town has been located for the start of a pipe run to the facility. The plan of this geography is shown in Figure 12.6. The place where the plant will be located is designated P in Figure 12.7.

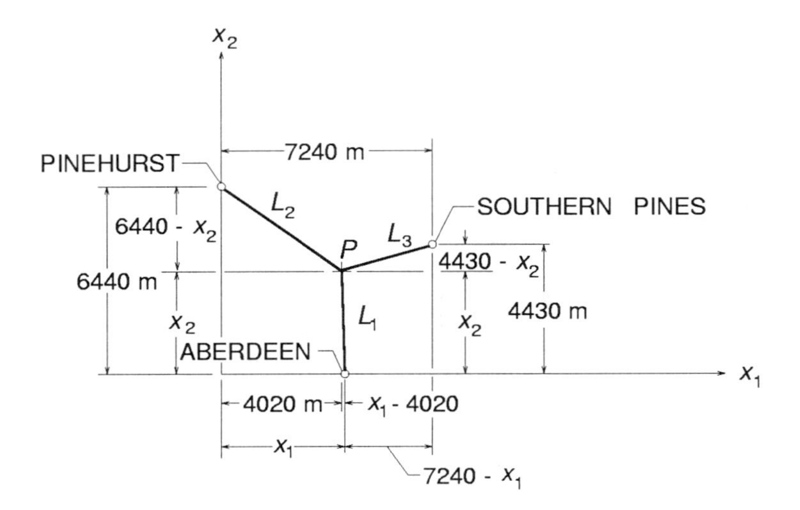

Figure 12.7: Aberdeen-Pinehurst-Southern Pines

Pinehurst has the smallest population of the three towns, and the estimate is that a pipe 0.3 m in diameter will be needed, and that its cost will be $12 per meter of length.

Aberdeen's population is about 1.5 times as large as that of Pinehurst, and the pipe is thought to be about 0.40 m in diameter to serve its requirements. The cost will amount to $15 per meter.

Southern Pines is five times as large as Pinehurst, the pipe diameter is believed to be 0.5 m in diameter to accomodate its needs, and the cost will be $18 per meter of running length.

From the given dimensions, the lengths of pipe from the three towns to the plant are

$$L_1 = [(x_1 - 4,020)^2 + x_2^2]^{1/2} \tag{12.13}$$

$$L_2 = [x_1^2 + (6,440 - x_2)^2]^{1/2} \tag{12.14}$$

$$L_3 = [(7,240 - x_1)^2 + (4,430 - x_2)^2]^{1/2} \tag{12.15}$$

so that the total piping cost is

$$Z = 15\,L_1 + 12\,L_2 + 18\,L_3 \tag{12.16}$$

This is the objective function which is to be minimized by a pattern search procedure. It is obviously very nonlinear, but it will be easy to compute, given x_1 and x_2 values.

Let the base camp be Pinehurst, $\mathbf{b}_1 = (0, 6400)$, and take the initial step size as $\boldsymbol{\delta}_1 = (100, 100)$. The following numbers show how the Hooke-Jeeves Algorithm unfolds with an accelerating factor of 2, and a deceleration of 0.5.

The cost function difference at convergence can be taken to be one cent.

Z(0000,6440) = 249,124.60		Z(5600.000,3640.00000) = 167,420.156714
Z(0100,6440) = 247,803.94		Z(5400.000,3440.00000) = 167,335.412827
Z(0100,6540) = 250,083.92		Z(5000.000,3040.00000) = 167,920.804313
Z(0100,6340) = 246,545.94		Z(5200.000,3240.00000) = 167,502.517664
Z(0300,6140) = 241,432.31		Z(5300.000,3440.00000) = 167,342.272798
Z(0700,5740) = 231,402.59		Z(5400.000,3540.00000) = 167,336.311582
Z(1500,4940) = 212,367.30		Z(5400.000,3340.00000) = 167,421.784282
Z(3100,3340) = 181,634.15		Z(5450.000,3440.00000) = 167,353.959220
Z(6300,0140) = 220,230.93		Z(5350.000,3440.00000) = 167,331.563713
Z(4700,1740) = 174,378.36		Z(5350.000,3390.00000) = 167,353.801311
Z(4800,1740) = 174,588.81		Z(5350.000,3490.00000) = 167,330.919945
Z(4600,1740) = 174,272.30		Z(5250.000,3590.00000) = 167,487.698034
Z(4600,1840) = 173,573.39		Z(5300.000,3540.00000) = 167,382.228333
Z(4400,2040) = 172,609.82		Z(5375.000,3490.00000) = 167,326.021363
Z(4000,2440) = 172,925.39		Z(5375.000,3515.00000) = 167,331.512838
Z(4200,2240) = 172,425.12		Z(5375.000,3465.00000) = 167,326.093431
Z(4300,2240) = 171,979.72		Z(5387.500,3490.00000) = 167,324.902521
Z(4300,2340) = 171,576.07		Z(5387.500,3502.50000) = 167,326.323659
Z(4500,2540) = 170,147.44		Z(5387.500,3477.50000) = 167,324.876129
Z(4900,2940) = 168,226.70		Z(5393.750,3477.50000) = 167,324.967061
Z(5700,3740) = 167,560.90		Z(5381.250,3477.50000) = 167,325.009282
Z(7300,5340) = 199,007.96		Z(5387.500,3483.62500) = 167,324.715335
Z(6500,4540) = 172,328.37		Z(5393.625,3483.62500) = 167,324.649874
Z(5800,3740) = 167,643.58		Z(5393.625,3486.68750) = 167,324.618572
Z(5600,3740) = 167,530.03		Z(5395.160,3486.68750) = 167,324.616686
Z(5600,3840) = 167,747.10		Z(5395.160,3588.21875) = 167,324.622866

The first pattern is broken at $x_1 = 4,000$, $x_2 = 2,440$, and a second one fails at $x_1 = 4,200$, $x_2 = 2,240$. A third ends with $x_1 = 5,700$, $x_2 = 3,740$, while a fourth extends through $x_1 = 5,400$, $x_2 = 3,440$. The final gambit stops at a value of $167,324.62 for the cost function.

Figure 12.8 shows the overall development of the Hooke-Jeeves pattern as computed above. Figures 12.9, 12.10, and 12.11 provide magnified portions of the patterns in sequence, and the final gambit is displayed in Figure 12.12, which is also a magnified section of the pattern.

It is interesting to note that the search overshoots several times in the course of hunting for the minimum. Figure 12.9 displays two of the more obvious of these, as well as the automatic corrections that are made to rectify the excesses.

The overshoot phenomenon can perhaps be better appreciated if it is realized that the contours in this problem create a "bowl-shaped" topology, with the minimum as the lowest place in the bowl. The pattern search could be compared to the trace left by a marble rolling down into the bottom of the bowl when released from a point on the rim. The marble would acquire a motional energy as it rolls down the slope, and would overshoot the lowest point on its trajectory, only to roll backwards until a new start could be made in a second pattern tending further down the slope. Too much should not be made of such an analogy, because, of course, a real marble would roll downward in a spiral motion, rather than in straight lines as required in Hooke-Jeeves patterns.

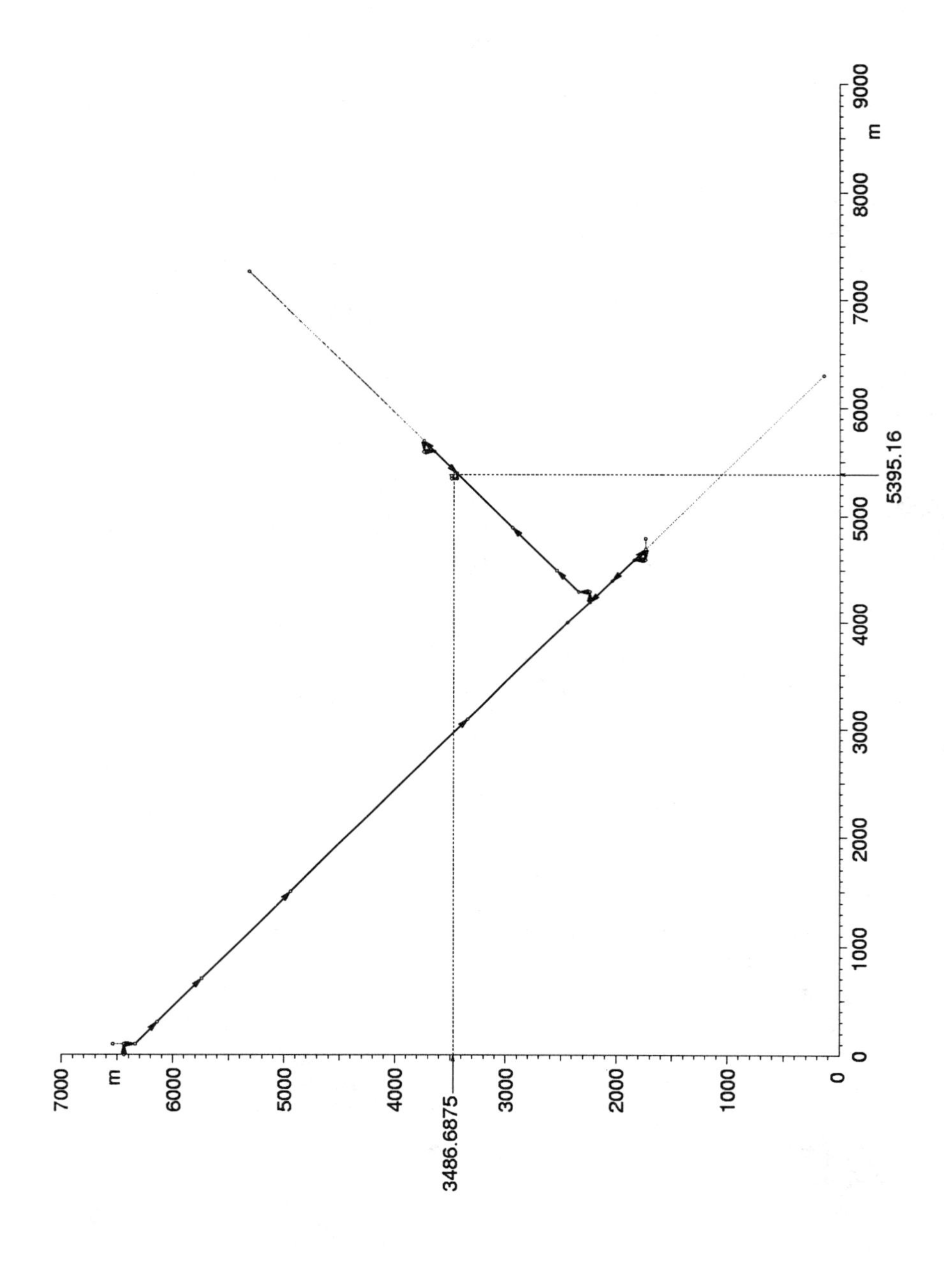

Figure 12.8: Aberdeen-Pinehurst-Southern Pines

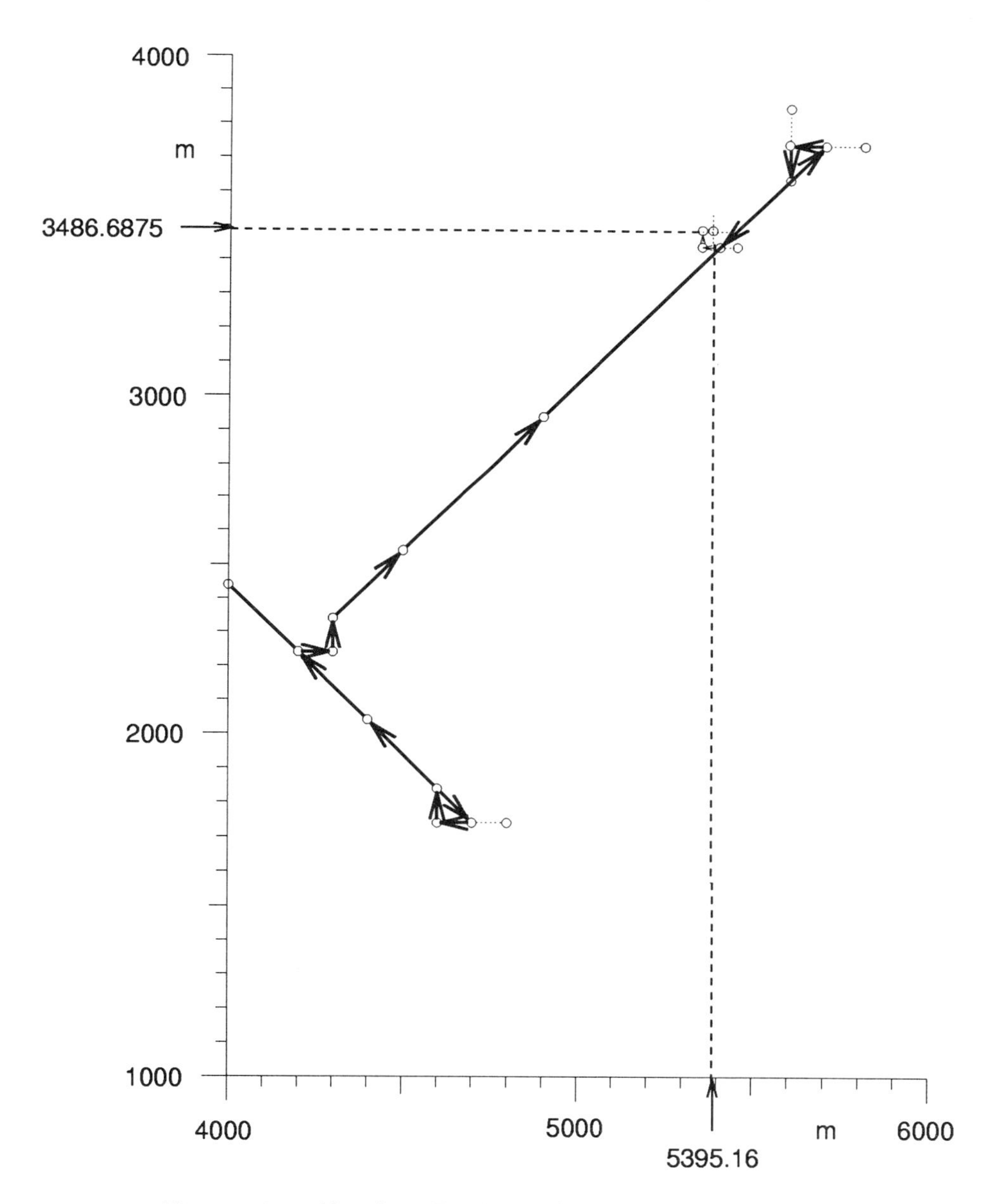

Figure 12.9: Aberdeen-Pinehurst-Southern Pines, Zoom 1

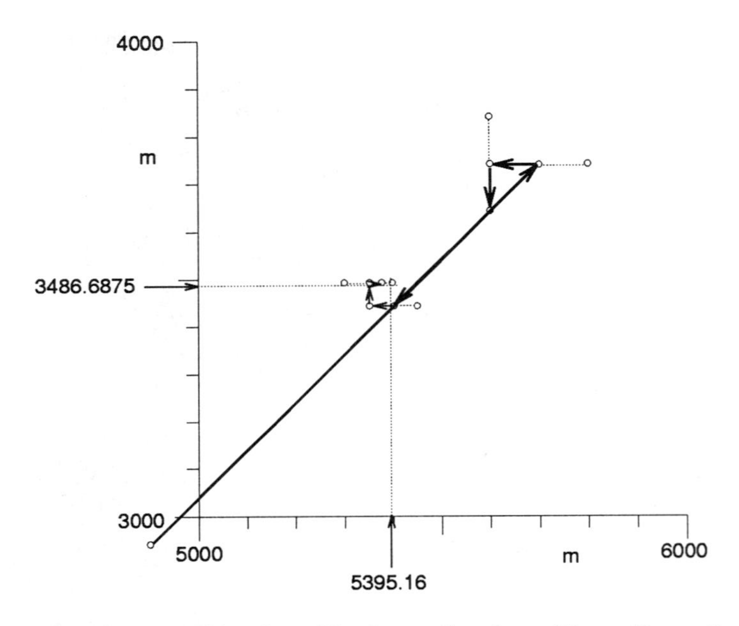

Figure 12.10: Aberdeen-Pinehurst-Southern Pines, Zoom 2

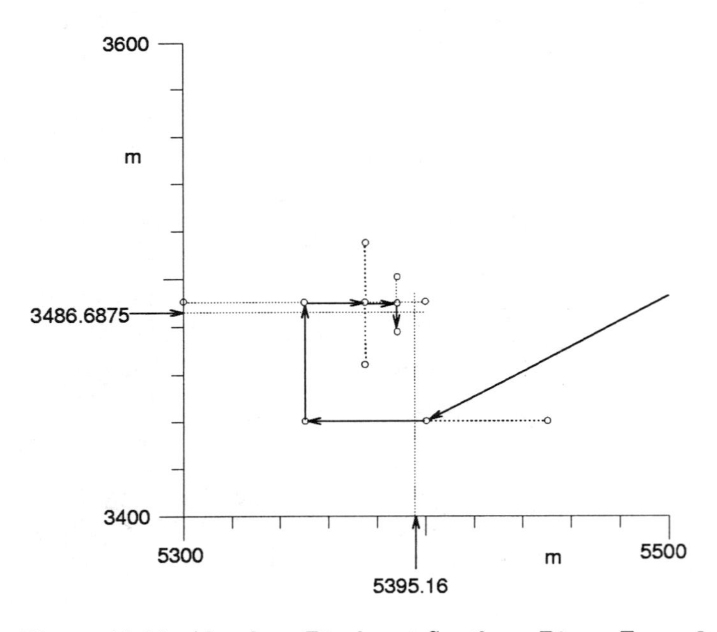

Figure 12.11: Aberdeen-Pinehurst-Southern Pines, Zoom 3

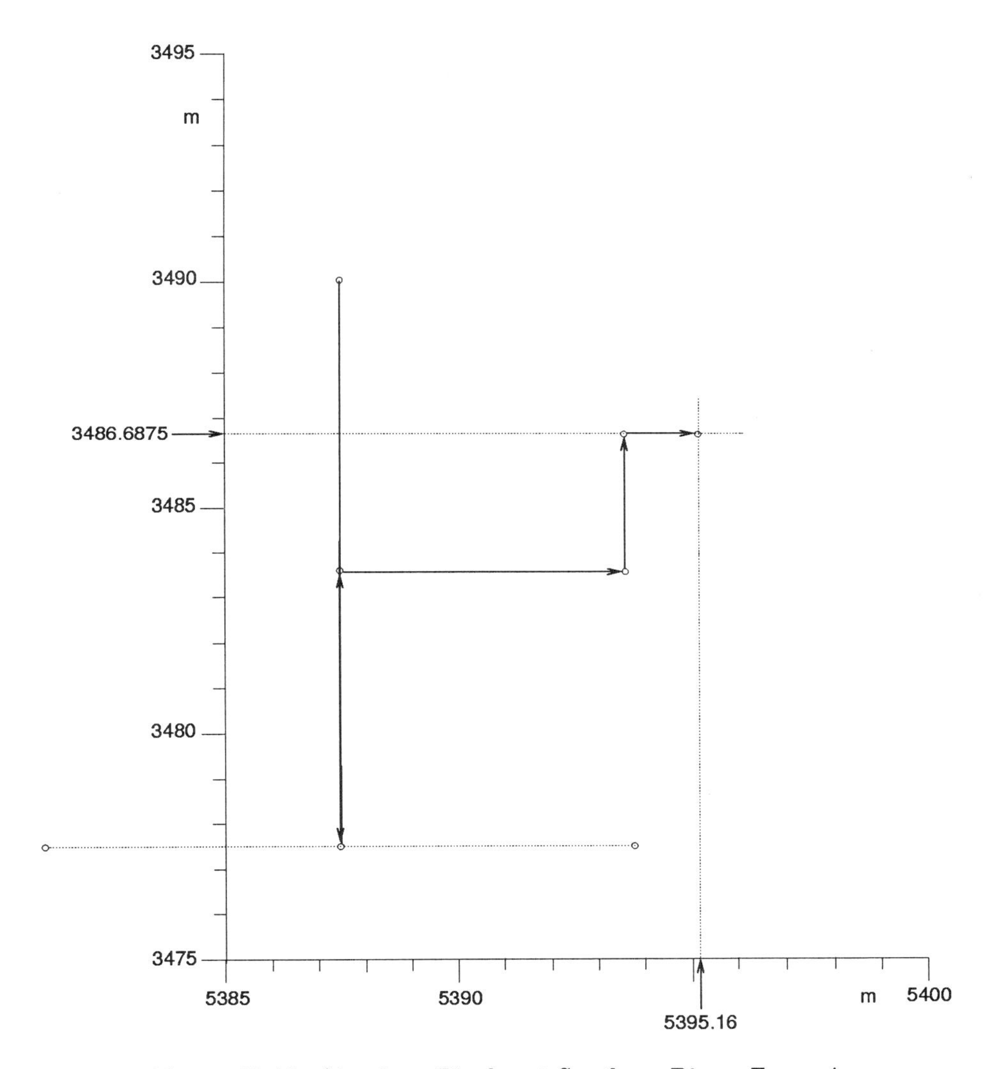

Figure 12.12: Aberdeen-Pinehurst-Southern Pines, Zoom 4

12.6 HOOKE-JEEVES PROGRAMS

Programs to do maximizing and minimizing using the Hooke-Jeeves algorithm have been prepared by the CE 375 staff for students to use with their projects. These are called HJMAX.f and HJMIN.f, and both are written in FORTRAN source code which is resident in the locker *mcdonald_src* on the *eos* net.

HJMAX.f is obviously for maximizing an objective function, and HJMIN.f does minimizing. The programs are main programs which call subroutines OFMAX.f and OFMIN.f at each stage of the Hooke-Jeeves procedure when the objective function is to be compared with its value at some other location. In other words, the main programs are prepared so as to be independent of the particular form of the objective function, and the subroutines are specific to the problem in question.

The user will therefore need to write the subprogram to compute the objective function. The main and subprograms will then be compiled and linked to produce run programs HJMAX.go and HJMIN.go.

The programs are interactive and user-friendly. When they run, the user will be prompted for inputs directly from the keyboard. The first query will be to give the number of decision variables. Provision is made for up to 8. The user will be asked to furnish a starting point, to give the initial step size, a final closure differential, the maximum number of step size reductions, the maximum number of patterns, and factors for acceleration and deceleration. These declarations furnish the limits within which the program will search for the optimum.

Because the programs are written in FORTRAN, a format has been assumed for each response from the user when prompted. An integer response for inputs which can only be integers is assumed. All other numerical entries are expected to be reals format. Alphabetic material is to be in capital letters.

The program has the option of showing the conditions at the end of each pattern search. This allows the operator to estimate the progress toward convergence upon the optimum.

The subroutine for the Aberdeen-Pinehurst-Southern Pines waste treatment plant optimization problem described in the previous section is given below. In order to preserve the proper communication between the main program and the subroutine it is necessary to adhere strictly to the format shown. Each problem is made specific by rewriting lines 8 through 11, which are the FORTRAN codes for the pipeline lengths L_1, L_2, and L_3. SUM is the name of the objective function, and 12.0, 15.0, and 18.0 are parameters which represent the costs per meter of each of the pipelines. That is, \$12/m, \$15/m, and \$18/m. The numbers actually entered at the prompts are 12., 15., and 18., of course, because this is the format employed in FORTRAN for real numbers.

RK(8) is the dimension statement for the array used to name the decision variables found in the main program; i.e., RK(1) is the first decision variable, RK(2) the second, and so on. In the subroutine RK(1) is renamed x_1, RK(2) becomes x_2. These are the names of the coordinate axes in the plot of the Aberdeen-Pinehurst-Southern Pines geography.

It is not necessary to limit the number of lines of code in the subroutine to exactly those shown below. For example, between line 8 and line 11 lines may be added to represent constraints. Had there, for instance, been a boundary placed upon the region in which the optimization is to be carried out, this bounded region might be described in standard FORTRAN statements at this place in the program. As the pattern search proceeds, when a particular heading would require the process to intercept a boundary (read constraint), the command can be made to require the variables to assume the constraint coordinates at that point. The processing will then stop at that point with the objective function having attained its optimum value on the constraint boundary.

```
      SUBROUTINE OFMIN(SUM,RK,N)
C
      DIMENSION RK(8)
      DOUBLE PRECISION SUM,X1,X2,L1,L2,L3
C
      X1 = RK(1)
      X2 = RK(2)
      L1 = SQRT((X1-4020.0)*(X1-4020.0)+X2*X2)
      L2 = SQRT(X1*X1+(6440.0-X2)*(6440.0-X2))
      L3 = SQRT((7240.0-X1)*(7240.0-X1)+(4430.0-X2)*(4430.0-X2))
      SUM = 15.0*L1+12.0*L2+18.0*L3
C
      RETURN
      END
```

Listed below is a printout from HJMIN which shows a typical run for an Aberdeen-Pinehurst-Southern Pines-type project. It will be seen that the input parameters have been echoed so as to form thereby a statement of the problem which is being solved. The printout contains the minimized value of the objective function, the values of the decision variables corresponding to the minimum, the final step sizes and a statement of the number of times the objective function was evaluated.

```
                 HOOKE-JEEVES PATTERN SEARCH

 ALPHA = 2.00        BETA = .50        ITMAX = 500        REDNS 20
      NUMBER OF DECISION VARIABLES = 2
      STARTING VALUE OF X1 = .0000E+00
      INITIAL STEP IN X1 = .1000E+03
      STARTING VALUE OF X2 = 0.6440E+04
      INITIAL STEP IN X2 = .1000E+03

      DIFFERENCE IN OBJECTIVE FUNCTION
      VALUES AT CONVERGENCE = .99999998E-02

      END OF FINAL PATTERN
      THE OBJECTIVE FUNCTION MINIMUM = .16732461E+06
      X1 = .5395E+04
      STEP1 = .9766E-01
      X2 = .3487E+04
      STEP2 = .9766E-01
      OBJECTIVE FUNCTION EVALUATED 193 TIMES
```

The printout above shows that the starting point (base camp) was at Pinehurst, obviously. In keeping with the policy of making runs that start at different locations, the exercise can be repeated with the starting points chosen to be at Aberdeen and Southern Pines, for instance. In these cases the printouts would be:

```
                 HOOKE-JEEVES PATTERN SEARCH
ALPHA = 2.00       BETA = .50       ITMAX = 500       REDNS 20
     NUMBER OF DECISION VARIABLES = 2
     STARTING VALUE OF X1 = .4020E+04
     INITIAL STEP IN X1 = .1000E+03
     STARTING VALUE OF X2 = 0.0000E+00
     INITIAL STEP IN X2 = .1000E+03

     DIFFERENCE IN OBJECTIVE FUNCTION
     VALUES AT CONVERGENCE = .99999998E-02

     END OF FINAL PATTERN
     THE OBJECTIVE FUNCTION MINIMUM = .16732462E+06
     X1 = .5393E+04
     STEP1 = .3906E+00
     X2 = .3486E+04
     STEP2 = .3906E+00
     OBJECTIVE FUNCTION EVALUATED 107 TIMES

                 HOOKE-JEEVES PATTERN SEARCH
ALPHA = 2.00       BETA = .50       ITMAX = 500       REDNS 20
     NUMBER OF DECISION VARIABLES = 2
     STARTING VALUE OF X1 = .7240E+04
     INITIAL STEP IN X1 = .1000E+03
     STARTING VALUE OF X2 = 0.4430E+04
     INITIAL STEP IN X2 = .1000E+03

     DIFFERENCE IN OBJECTIVE FUNCTION
     VALUES AT CONVERGENCE = .99999998E-02

     END OF FINAL PATTERN
     THE OBJECTIVE FUNCTION MINIMUM = .16732461E+06
     X1 = .5395E+04
     STEP1 = .3906E+00
     X2 = .3486E+04
     STEP2 = .3906E+00
     OBJECTIVE FUNCTION EVALUATED 95 TIMES
```

with a difference of only a penny or two in the outcomes.

Because of the wide range of problems to which the software might be applied, it will be noted that the exponential format has been employed so as to be able to accomodate a broad spectrum of magnitudes.

12.7 PROJECT #10

1. As the City Engineer for a North Carolina municipality you are planning the layout of a pumping station on a lot in a suburban area. The plan details are illustrated in the sketch below.

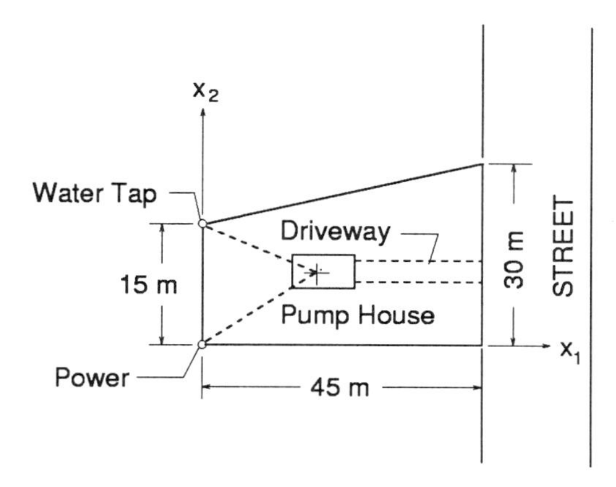

You expect to locate the station on the plat so that the cost of the infrastructure is minimized. To accomplish this, you will use a Hooke-Jeeves pattern search technique, with the objective function which is to be minimized equal to the total cost of these services.

The driveway will cost \$18/m, the water line \$14/m, and the buried power line \$10/m.

It is recommended that you use the HJMIN.f program available on the *eos* network. First, copy the source code for HJMIN.f from the locker **mcdonald_src** to your own directory by using the command **attach mcdonald_src** followed by **cp /ncsu/mcdonald_src/HJMIN.f␣.**. Then check to see that it appears in your directory.

You will, of course, need to write your own OFMIN.f source-code program for the cost function, and then you will compile and link HJMIN.f with OFMIN.f to produce the run-file HJMIN.go. On *eos* net, which runs on an *UNIX* platform, the command to accomplish this is **f77 -o HJMIN.go HJMIN.f OFMIN.f**.

The output is written to a file named HJMIN.dat, and you will need to print this file in order to have a hard copy of the result. The program provides that multiple runs may be written to the same file, but when the program is closed, HJMIN.dat will need to be removed before a new program call is made.

The computer monitor will display the result of each run in the fashion shown in the Aberdeen-Pinehurst-Southern Pines example above.

In your report give the x_1 and x_2 coordinates of the optimum location of the pumping station, and state the minimum cost and percentage saving over the cost had the station been located at $x_1 = 30$ m, $x_2 = 10$ m.

2. *O. U. Mudcat, Inc.*, contractors, are building three large projects as shown on the location plot below. Water must be carried by truck to each of the projects to wet down the streets and driveways prior to paving. It is estimated that the trucks must make one trip per day to *Ancient Oaks*, six per day to *Berry Hills*, and three per day to *Club Dues*. Where should the water tap be located so that the total cost will be minimized?

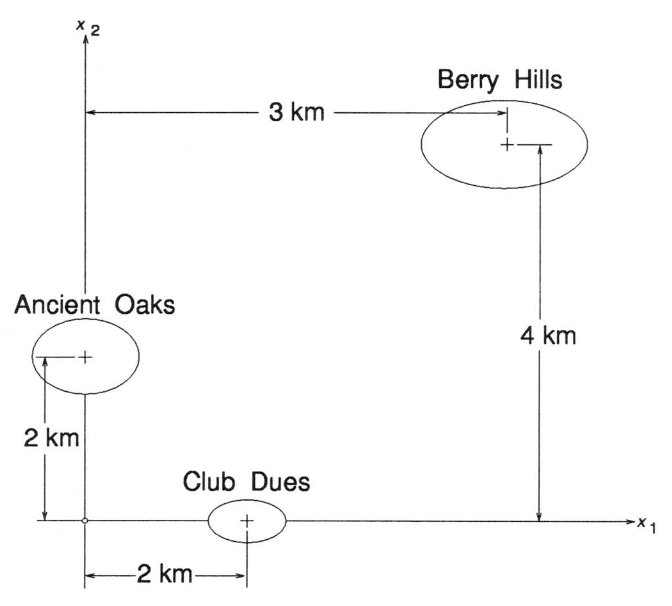

- Write equations for the distances from each project to the water tap, T, and for the total distance for all three, in terms of the coordinates x_1 and x_2.
- Employ the Hooke-Jeeves pattern search algorithm to locate the best place for T. Begin your search at Ancient Oaks, using a step size of 100 *m* for both x_1 and x_2. Confirm your answers by repeating the search beginning at Berry Hills and then at Club Dues.

3. SMELLS Sulphur Mining and Refinery Company has a loading area on a point of land near the river. Sulphur will be pumped to three loading docks as shown on the plat below. The proposal is to so locate the pumping point, P, that the total length of piping is minimized.

- Write equations for the lengths ℓ_1, ℓ_2, and ℓ_3 in terms of the coordinates x_1 and x_2. Then make a formula for the total piping length.
- Use a Hooke-Jeeves pattern search procedure to locate the optimum place for P. Start the search at the point (1000,500) and use a step size of 100 m for both x_1 and x_2. Draw a sketch showing the vectors of the search.
- Repeat the search from some other starting point with an appropriate step size to confirm the first answer. Work to an accuracy of plus-or-minus 10 m.

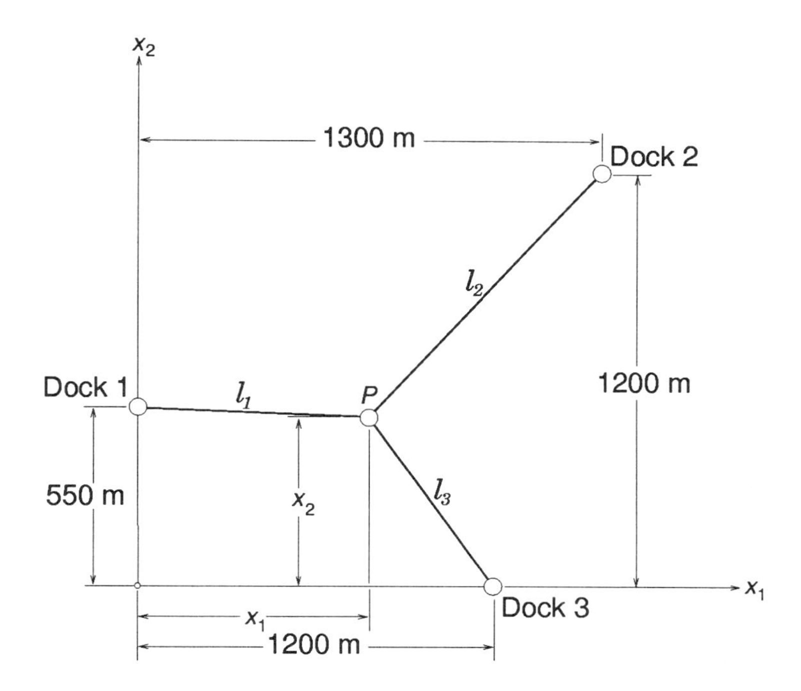

4. The community of ***Dry Gulch*** needs to establish a water system for its inhabitants. The Town Fathers, prodded by the Town Mothers and Bessie Blossom (owner/manager of Dry Gulch Inn), have concluded that a system of wells must be developed to tap the Hydralode aquifer as a source for the water system. They have consulted Redd Hedd, a well-known dowser, who has investigated the local terrain and selected three sites where wells might be drilled with a good probability of striking water.

   Based upon Hedd's judgement, it has been concluded that the three sites (shown on the plat below) which will provide different flow rates, will all three be needed. Because the flow rates are different, the pipe sizes may also be different, and the costs will likewise vary. However, only one water tank is required, and the decision that must be made is where to put it so as to minimize the overall cost of the project.

   Locate the optimum water tank site, P, for Dry Gulch. Use the software *HJMIN.go* in the locker *mcdonald_src* on the *eos* net, and compare this with a Hooke-Jeeves pattern search done by hand calculations. Show a plot of the several steps in the search to convergence in your report.

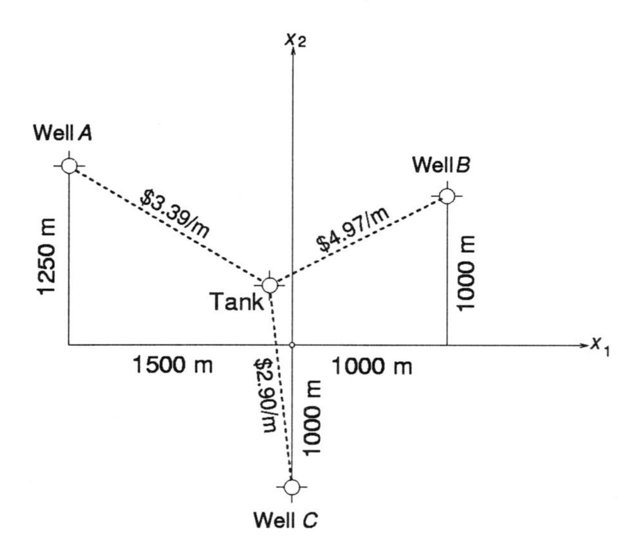

5. A central pumping station is to supply water for three high-pressure hydraulic mining sites as shown in the layout below. Different flow rates are required at the sites, therefore different piping sizes (and correspondingly different costs) are indicated.

   Use a Hooke-Jeeves pattern search procedure to determine the location of the pumping station which will minimize the piping cost.

   Begin the search at site A with a step size of 100 m for both x_1 and x_2, after having written equations for the pipe lengths from each site to the pumping station. Work to an accuracy of 10 m at closure.

   Confirm the answer by repeating the search beginning at sites B and C.

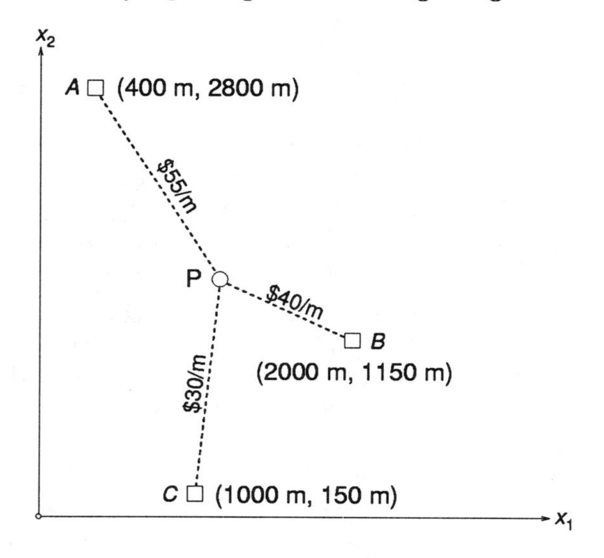

CHAPTER 13

SYSTEMS SIMULATION

13.1 INTRODUCTION

An ideal circumstance for the design engineer would be to possess a model of the system upon which analysis and synthesis is being carried out. The model would have the capacity of being altered by the designer at will without the necessity of resorting to new hardware representing the altered model, and the performance of the system in its new configuration would be readily observed and compared. In this manner the system model could be optimized without its having ever been realized in terms of hardware.

Systems simulation is the business of creating a synthetic version of a real system which will nevertheless possess a faithful rendition of the actual system's properties and behaviors. The simulated system will probably reside in a computer, and will therefore be as accessible as would any program for data inputs and modeled output. The key concept is necessarily contained in the phrase "faithful rendition."

To be useful the simulated system must perform reasonably well in conformity to that of the "real" thing, otherwise it is merely a theoretical device absent of any utility in the optimization role insofar as the approach to reality is concerned. "Reasonably well" is certainly open to interpretation, but a simulated system which has high reproducibility in terms of predicting real world performance of the prototype surely qualifies as desirable.

Perhaps the most difficult aspect of the simulated modeling enterprise is the capability of reflecting the variability of natural phenomena through all of the vagaries that must be encountered. There is a randomness in natural processes which defies representation by the laws of ordinary mechanics. The words stochastic and statistical tell the nature of the phenomena which must be dealt with.

To say that something varies in a random manner means that there is no physical or mathematical law which governs whether it behaves thus and so in a particular instance. In other words, there is the law of chance operative in the situation. However, this is not to say that there is no law governing a series of events of the type in question. Indeed, it is a fundamental assumption of statistical theory that behind every phenomenon there is a law of frequency for the occurrences under observation. There may be no law of motion and initial

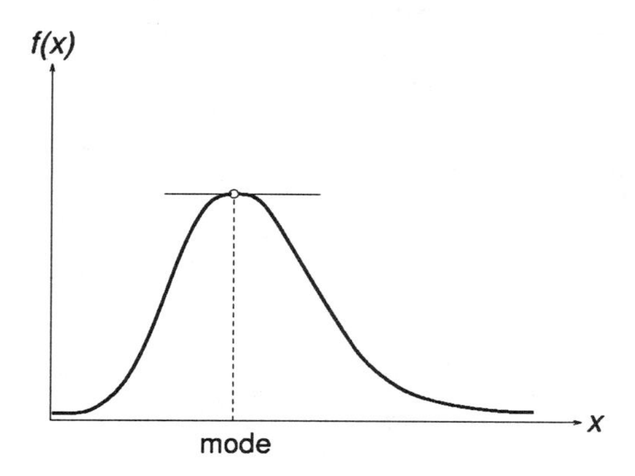

Figure 13.1: Frequency Distribution

conditions to govern a stochastic event, but there is a collective law hanging over the head of the family of many events of the type.

A typical example of the randomness involved in simulation is that of weather-related conditions. This was previously encountered in Chapters 3 and 8.

13.2 ELEMENTS OF STATISTICS

The fundamental assumption of statistics is that there exists a law which presides over a given phenomenon to say with what frequency it may have certain outcomes. This law is embodied in an equation of frequency distribution. This frequency is a continuous mathematical function of an argument x, which stands for the measure of the outcome. The relationship is between "how many" and "size."

Figure 13.1 shows a frequency distribution, f, distributed over a range represented by the argument, x. There are indeed many shapes which are known in the theory of statistics for the frequency distribution, but a large fraction of those which occur in systems modeling and simulation possess certain features in common.

Among these common attributes are a zero slope at three places: on the far left, on the far right, and somewhere in between, occasionally even near the middle, and, sometimes, symmetrically disposed even as between the right and the left. In other words, there is a common slope condition for the frequency curve at these three places. There is also a "bulge" or hump somewhere between the left and right, and just underneath the place where the slope is zero in between the two extremes.

If this frequency distribution is representative of a phenomenon of nature, then it shows that Mother Nature is trying to tell us something. It suggests that the measure x which is being represented is a random distribution of chance occurrences spread about a central place—the bulge—where it occurs most frequently. This message comes through the fog and smoke and filthy air: there is a preferred place, a favored value of x, of greater significance than other places and xs.

To be sure, the shape of the frequency distribution also has significance. Whether it is a low smooth bump in the mid-range, or whether it is mounded up high; whether there is more of the bump near the middle of it or whether it is broader; whether it is symmetrical or tilted to one side; these are also features that are meaningful. These are also the basis of the statistics of the distribution.

In Chapter 8 probability was defined as an area (under a curve). Or, the probability of an event is the area associated with the occurrence of that event when compared with the area possessed by the entire distribution. The area under the frequency distribution is normalized to 1.0, or certainty that the event is somewhere within the reach of the frequency curve. Therefore

$$1.0 = \int_0^\infty f(x)\,dx \tag{13.1}$$

is a property of the frequency distribution.

Other properties which relate directly to the observations recounted above are:

The first moment of the distribution:

$$1.0\,\overline{x} = \int_0^\infty x\,f(x)\,dx \tag{13.2}$$

This serves to define the quantity $\overline{x}$, which is called the *mean*. If the distribution is symmetrical, $\overline{x}$ is the average of the quantity x. Its significance is similar to that of the centroid or center of gravity in mechanics. It is a "representative" place of the distribution—a measure of the centrality or a spot where the whole weight of the distribution can be hung at balance.

The *mode* is the most frequent value of x. For symmetrical frequency distributions, the mode, mean, and average all have the same value.

The second moment:

$$\sigma^2 = \int_0^\infty (x - \overline{x})^2\,f(x)\,dx \tag{13.3}$$

This integral furnishes the definition of the standard deviation, a quantity which measures the degree to which the distribution spreads broadly or is alternatively clustered about the mean. It is analogous to the second moment of area in mechanics, or to the moment of inertia.

The third moment:

$$\mu = \int_0^\infty (x - \overline{x})^3\,f(x)\,dx \tag{13.4}$$

The skewness is μ, a measure of how skewed or tilted or slanted away from the mean is the distribution. It is a central indicator of how unsymmetrical is the spread.

Higher moments are also defined, but seldom used. The fourth, for example, defines the peakedness, or how sharply focussed (peaked) is the bulge at the mean.

Taken together, the three measures $\overline{x}$, σ, and μ plus the parameter obtained by the normalizing integral constitute the statistics which identify a frequency distribution and permit it to be used to establish probabilities for a variable subject to it.

13.3 SAMPLES

The frequency distribution discussed above is a continuous mathematical function. It exists in a mathematical universe of points or sets. Compared to the "real" world, it is a fiction—a convenient myth image. The chance occurrences or random events observed in nature are discrete, although large samples may be said to approach the continuum distribution as an asymptotic limit.

When a phenomenon is measured or sampled, only a few of the infinitude of possible places on the frequency distribution curve are tested. The idea and hope is that by taking a sample, the properties of the continuous variable may be approached. Thus, a sample is the litmus of statistics. Discrete paired readings of frequency (f) and size (x) are made, with the assumption that the sample obtained thereby does in fact represent in microcosm the attributes of the universe from which it has been drawn.

The size of the sample is n, the number of paired values of f and x. The larger it is, the more accurately it represents the universe behind it. Many authorities adjust the statistics by some factors which depend upon the size, especially if the size is small. Chauvenet's Criterion[1] is available to reject spurious data, particularly those which might alter the results in small-sized samples.

The fundamental premise, then, is that one can take a sample of data, theoretically drawn from a universe of values, and with the sample compute the statistics by the same methods as were employed for the universe, and obtain results which are as well unique for the sample as they were for the universe. Naturally, the integrals will have to be replaced by discrete sums, but the work will otherwise be the same.

The data from which a sample derives is divided into classes, an organization which represents a discrete subdivision of the whole range into convenient bins or boxes whose walls are the class limits. All data which accumulate in any one class are identified with that class in much the same way as, for example, students who earn an A on a course of study are all known as A-Students, etc. When the grades are all in, no notice is taken thereafter whether one student who got a B is any different (better or worse) than any other B-Student.

The graph of the sampled data, now classified, is plotted as in Figure 13.2, where the vertical bars constructed upon the base of each class width represent the frequency of sample data falling within that class. The whole diagram is called the *histogram*.

The equivalent of the normalization integral is

$$N = \sum_{i=1}^{n} f_i \tag{13.5}$$

and the other moments follow.

First moment:

$$\overline{x} = \frac{1}{N} \sum_{i=1}^{n} f_i \, x_i, \text{ the } \textit{mean} \tag{13.6}$$

[1]Worthing, A. G., and Geffner, J., *Treatment of Experimental Data*, John Wiley & Sons, Inc., New York, 1943, pp. 171, 172.

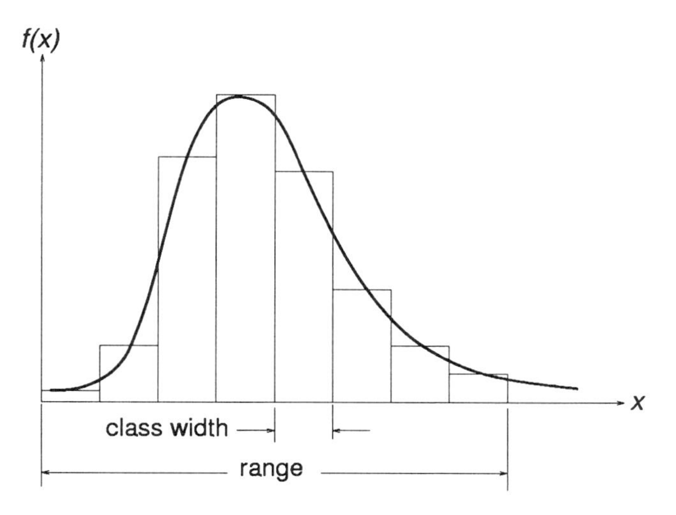

Figure 13.2: Histogram

Second moment:

$$s_x^2 = \frac{1}{N}\sum_{i=1}^{n} f_i\, x_i^2 - \overline{x}^2, \text{ the } \textit{variance} \tag{13.7}$$

Third moment:

$$m_3 = \frac{1}{N}\sum_{i=1}^{n} f_i\,(x_i - \overline{x})^3, \text{ the } \textit{skewness} \tag{13.8}$$

Several other measures of the third moment are used in various circumstances.

$$a_3 = \frac{m_3}{s_x^3} \qquad A = \frac{2}{a_3} \qquad M = -\frac{a_3}{2} \tag{13.9}$$

It is customary to define a standard variable which normalizes the frequency distribution in the form

$$t = \frac{x - \overline{x}}{s_x} \tag{13.10}$$

Thus t, the ***standard variable***, is here measured with respect to the mean as an origin. Again, there is an analogy with mechanics. The standard variable is similar to coordinates in mechanics which are taken relative to the centroid or mass center, whereas the variable x is with respect to some origin of coordinates far removed from the mean.

So, there is a sample $x_i,\ f_i,\quad i = 1, 2, \ldots, n$ of size n, with standard variable t and statistics $\overline{x}$, s_x, and m_3 that display its mean, its variance, and its skewness. Taken together these present a picture of a phenomenon modeled by that sample.

13.4 RANDOM NUMBERS

The device which makes it possible to "run" a simulated system is the random number. As the name implies, the random number is a number produced at random; i.e., purely by chance. Since it is obtained by a random process, the number replicates, in theory, the operations which produced the data from which are drawn the statistics of a frequency distribution in the first place, and therefore the simulated data set is as likely to occur as is the prototypical data.

A crude but valid method of obtaining random numbers is to draw them out of a hat in the old fashioned manner. There are, naturally, machine-generated random numbers, and almost all technically-oriented hand-held programmable computers and virtually all desk-top machines have simple programs to produce a set upon call. Certainly, work stations, minicomputers, mainframes, and supercomputers have the potential for this job. Usually the numbers produced range from 0.0000000000 to 1.0000000000, so that, thus normalized, they may be used for all purposes.

Therefore a string of random numbers is produced and recorded. The range of cumulative probability is divided into classes whose class width corresponds to that of the data set from the prototype system, and since the cumulative probabilty varies from 0.0000000000 to 1.0000000000, the probability range associated with each class boundary is at hand. When each random number is drawn (produced), it will therefore "belong" within one of these classes. That is, all random numbers falling between the cumulative probabilities at the class width boundaries will generate one unit of frequency at that class. When all have been drawn, the total frequency in each class is the sum of the individual numbers in that "bucket." The simulated histogram can next be drawn.

13.5 THE NORMAL DISTRIBUTION

The most famous frequency distribution is that called *normal* or *Gaussian* or *bell-shaped*. It is very common in nature, and in fact represents the case where there is a random opportunity for the event or trial to be greater than or less than the mean. In other words, there is a fifty-fifty chance of being over or under the mean or average.

Biological measurements are often found to conform to a normal frequency distribution. The heights of ninth-grade boys follow this distribution, and "normal height" can be taken literally as a description for those boys whose measure is within one standard deviation of the mean. The normal distribution is typical of many manufactured products, because the manufacturer will take all reasonable pains to have the finished product meet certain standards, and quality control plans will ensure that variations at the end of the final stage of manufacture are solely of a random nature.

The normal distribution equation is[2]

$$f(x) = c e^{-\frac{1}{2}\left(\frac{x-a}{b}\right)^2} \tag{13.11}$$

in which a, b, and c are parameters such that the required properties of a frequency distribution are fulfilled. The normalizing constant is c, adjusted so that the total

[2]Hoel, P. G., ***Introduction to Mathematical Statistics***, John Wiley & Sons, Inc., New York, 1947, pp. 28-35.

probability (area under the curve) is unity. The constant a turns out to be the mean, $\overline{x}$, and b is the standard deviation σ, so $c = \frac{1}{\sigma\sqrt{2\pi}}$. Therefore

$$f(x) = \frac{1}{\sigma\sqrt{2\pi}} e^{-\frac{1}{2}\left(\frac{x-\overline{x}}{\sigma}\right)^2}, \tag{13.12}$$

and the normal distribution is seen to be fully determined when the mean and standard deviation are supplied. The skewness is zero, of course, and the normalized peakedness is

$$\alpha_4 = \frac{\mu_4}{\mu_2^2} = \frac{\mu_4}{\sigma^2} = 3 \tag{13.13}$$

Using the definition

$$t = \frac{x - \overline{x}}{\sigma} \tag{13.14}$$

for the standard variable, the *standard normal curve* is

$$f(t) = \frac{1}{\sqrt{2\pi}} e^{-\frac{1}{2}t^2} \tag{13.15}$$

13.5.1 CUMULATIVE PROBABILITY

In order to employ the normal distribution as a model for purposes of simulation, it is necesary to provide for the cumulative probability of the variable x. That is,

$$P(x) = \int_0^x f(x)\,dx = \int_0^x \frac{1}{\sigma\sqrt{2\pi}} e^{-\frac{1}{2}\left(\frac{x-\overline{x}}{\sigma}\right)^2} dx = \frac{1}{\sqrt{2\pi}} \int_{-\frac{\overline{x}}{\sigma}}^{\frac{x-\overline{x}}{\sigma}} e^{-\frac{1}{2}t^2}\,dt \tag{13.16}$$

This indefinite integral represents the area under the normal curve to the left of the value x. There is not an elementary closed-form function for the integral, so it becomes necessary to resort to a numerical method for accomplishing the task of evaluating it. One method uses a series expansion of the exponential in the integrand, followed by term-by-term integration, and subsequent substitution of the upper and lower limits.

Appendix G furnishes the details of such a method. It should be noted that the computation of $P(x)$ is not simple. The series for $e^{-\frac{1}{2}t^2}$ is oscillatory (i.e., the signs change from plus to minus from one term to the next), therefore a large number of terms will be needed for good accuracy. Also, there is a factor $n!$ in the denominator of each term, and when the value of n becomes large, $n!$ may easily exceed the floating decimal point capacity of the computer being used to determine the result. Accordingly, care must be exercised to formulate the computer program in such a way that a direct computation is avoided in favor of a process by which each succeeding term in the series is calculated from its predecessor.

One method of accomplishing this is to compute each new term recursively; that is, from the immediately preceeding term, by dividing by the new n. Of course, this requires that the preceeding term must be stored independently of the series sum to that point, so that the necessary operations can be performed.

13.5.2 EXAMPLE

The systems manager for a local area network (LAN) in a large civil engineering design office wishes to determine the mean length of time associated with calls for a particular software by the engineers in the office who have workstations on the net. Data are available for the 1,000 most recent calls on this software as shown in the table below.

The data have been conveniently collected in classes of width of 100 seconds each.

Compute the mean, standard deviation, skewness, and peakedness for this sample, and plot the normal frequency distribution superposed upon the histogram of the data.

x seconds	f
0-100	6
100-200	28
200-300	88
300-400	180
400-500	247
500-600	231
600-700	162
700-800	42
800-900	11
900-1000	5
Total	1000

Computation of the statistics begins with the mean, followed by the variance, skewness, and peakedness.

$$
\begin{aligned}
\overline{x} &= \frac{1}{N}\sum_{i=1}^{n} f_i\,x_i \\
&= \frac{1}{1,000}[6(50)+28(150)+88(250)+180(350)+247(450) \\
&\quad +231(550)+162(650)+42(750)+11(850)+5(950)] \\
&= \frac{478,600}{1,000} = 478.6\,\text{seconds} = 7.977\,\text{minutes}
\end{aligned}
\tag{13.17}
$$

$$
\begin{aligned}
s_x^2 &= \frac{1}{N}\sum_{i=1}^{n} f_i\,x_i^2 - \overline{x}^2 \\
&= \frac{1}{1000}[6(50)^2+28(150)^2+88(250)^2+180(350)^2+247(450)^2+231(550)^2 \\
&\quad +162(650)^2+42(750)^2+11(850)^2+5(950)^2] - (478.6)^2 \\
&= \frac{252,620,000}{1,000} - (478.6)^2 = 252,620 - 229,057.96 = 23,562.04
\end{aligned}
$$

Therefore

$$s_x = \sqrt{23,562.04} = 153.499 \tag{13.18}$$

$$
\begin{aligned}
m_3 &= \frac{1}{N} \sum_{i=1}^{n} f_i \left(x_i - \overline{x}\right)^3 \\
&= \frac{1}{1,000}[6(50 - 478.6)^3 + 28(150 - 478.6)^3 + 88(250 - 478.6)^3 \\
&\quad +180(350 - 478.6)^3 + 247(450 - 478.6)^3 + 231(550 - 478.6)^3 \\
&\quad +162(650 - 478.6)^3 + 42(750 - 478.6)^3 + 11(850 - 478.6)^3 \\
&\quad +5(950 - 478.6)^3] \\
&= -\frac{79,016,688}{1,000} = -79,016.688
\end{aligned}
$$

This yields

$$
a_3 = \frac{m_3}{s_x^3} = \frac{-79,016.688}{(153.499)^3} = -0.021847386 \qquad (13.19)
$$

$$
\begin{aligned}
m_4 &= \frac{1}{N} \sum_{i=1}^{n} f_i \left(x_i - \overline{x}\right)^4 \\
&= \frac{1}{1,000}[6(50 - 478.6)^4 + 28(150 - 478.6)^4 + 88(250 - 478.6)^4 \\
&\quad +180(350 - 478.6)^4 + 247(450 - 478.6)^4 + 231(550 - 478.6)^4 \\
&\quad +162(650 - 478.6)^4 + 42(750 - 478.6)^4 + 11(850 - 478.6)^4 \\
&\quad +5(950 - 478.6)^4] \\
&= \frac{1.648533613 \times 10^{12}}{1000} = 1,648,533,613 \qquad (13.20)
\end{aligned}
$$

Then

$$
a_4 = \frac{m_4}{s_x^4} = \frac{1,648,533,613}{(153.499)^4} = 2.969422731 \qquad (13.21)
$$

It may be noted that for this sample the skewness is very small, and the peakedness is quite close to 3. These numbers suggest that the model is essentially a purely normal distribution.

Using the statistics in the generic form of the frequency distribution allows the values of the frequency to be computed for comparison with those of the sample. For instance,

$$
f(50) = \frac{1}{153.499\sqrt{2\pi}} e^{-\frac{1}{2}\left(\frac{50-478.6}{153.499}\right)^2} \times 1000 \times 100 = 5.2704 \qquad (13.22)
$$

The other frequencies and sample values are displayed below.

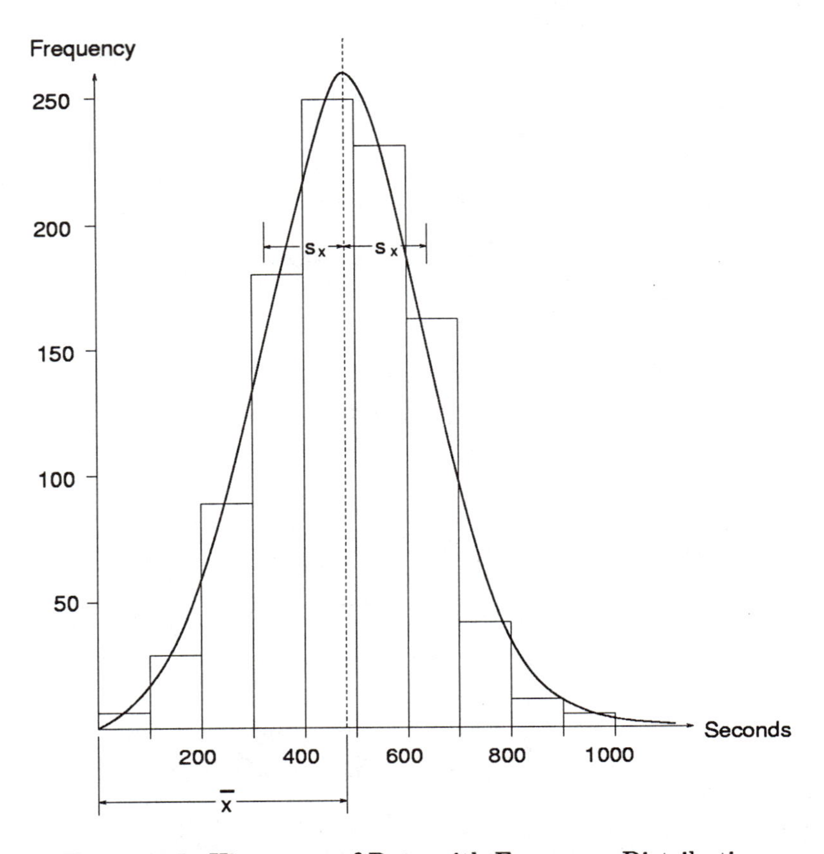

Figure 13.3: Histogram of Data with Frequency Distribution

x	$f(x)$	Sample
50	5.2704	6
150	26.2833	28
250	85.7424	88
350	182.9743	180
450	255.4261	247
550	233.2496	231
650	139.3334	162
750	54.4469	42
850	13.9178	11
950	2.3273	5

The histogram of the sample with normal frequency distribution superposed is shown in Figure 13.3.

The network manager could use the statistics in a program for computing cumulative probabilities, together with a random number generator to simulate the calls for the software in question over whatever length of time is desired.

13.6 SKEWED DISTRIBUTIONS

There are very many excellent treatises which deal with symmetrical frequency distributions. These same books also consider distributions such as the *Poisson*, *Student's t*, the *binomial*, several exponential types and other common elementary spreads.

There are few good sources of information concerning the skewed frequency distributions which are so prevalent in much of engineering science, and are particularly evident in systems simulation. It has been mentioned several times before that weather-related phenomena are highly skewed. Most queuing activities display a skewed graph. In general, those situations in which there is some form of bias, constraint, channelization, rectification, or other form of control are productive of skewed distributions. Obviously, this includes most projects in civil engineering practice.

Accordingly, the remainder of the present chapter will be devoted to a study of especially those types, since they are so ubiquitous.

The individual who may be said to have been the progenitor of much of statistical science as it is known today is Karl Pearson, an English statistician whose contributions came largely in the first decades of the Twentieth Century. For the present purposes, Pearson codified the commonly found frequency distributions according to a certain parameter which foretells their general shape. In particular, there are two main skewed types, one of which is widely known and used. It is the one called *Pearson Type III*. The other, *Pearson Type I*, is also useful in simulation work.

13.6.1 PEARSON TYPE III

The details of how this equation is derived are included in Appendix C, which the reader is encouraged to visit and digest. The material is not difficult from a theoretical point of view, but is sometimes complex algebra and integration, and is a diversion from the context. The frequency distribution is

$$f = y(t) = K\,(A+t)^{A^2-1}\,e^{-At} \tag{13.23}$$

in which K is a constant to be determined from the normalizing condition that the integral of the function shall equal 1.0. A is the measure of the skewness, and since the equation is presented in terms of the standard variable, t, both the mean, $\overline{x}$, and the variance, s_x, are also involved.

Given a sample of data which is supposed to conform to Pearson Type III, the sample statistics would be run to obtain the mean, variance and skewness. After that the area under the curve would be computed and set equal to 1.0. The unknown in this procedure will be K, so it is obtained from this analysis. Complete details of the work are displayed in Appendix C, and the results are:

$$K = \frac{1}{\sqrt{2\pi}\,A^{A^2-1}\,Q(A)} \tag{13.24}$$

$$Q(A) = 1 + \frac{1}{12\,A^2} + \frac{1}{288\,A^4} - \frac{139}{51840\,A^6} - \frac{571}{2488320\,A^8} + O(\frac{1}{A^{10}}) \tag{13.25}$$

This famous asymptotic series expansion involved in the Gamma Function—a higher transcendental—is described by Whittaker and Watson.[3] It is highly convergent, so that it is necessary to compute only a few terms in order to determine $Q(A)$ and subsequently K. Moreover, it is very easy to program for machine computation.

Determination of K completes the arithmetic required to have the Pearson Type III frequency distribution. One would customarily plot the curve superposed upon the histogram of the sample data so as to visually inspect the degree to which they are congruent. If the fit is acceptable, the conclusion would be drawn that the Pearson Type III curve with those particular statistics represents the phenomenon which was measured by the data. This constitutes a model of the system in question, one which is now analytical, and it can be employed to generate mathematically as many operations of the system as are desired.

13.6.2 CUMULATIVE PROBABILITY

Even when one knows the frequency distribution of the system model as given by the Pearson Type III statistics, in order to use this to simulate the operation of the system it is necessary to establish the cumulative probability $P(x)$ for the distribution. In the case of Pearson Type III there is not known any neat closed-form determination of this quantity. Pearson himself, in association with a staff of several statisticians on-and-off over a period of a quarter of a century, used hand computations of the most laborious sort to prepare a table of values for $P(x)$. This table is very difficult to use, and because of the nature of the animal, there is considerable doubt about its overall accuracy.

Unfortunately, there were no computers of the type in common use today which were available to Pearson. Unfortunately, also, even today there are few individuals with the means to render the necessary computations. Appendix C shows the details of a series expansion which does the job of obtaining $P(x)$ when the statistics of the distribution are provided. The series is one of the worst of all possible types for numerical work, because it is oscillatory (that is, it changes sign from one term to the next) and it contains $n!$ in the denominator of each term, where n is the index of the term in question. When it is seen that because of the slow convergence due to the oscillation of signs, it can be appreciated that the $n!$ term becomes a problem.

In practice, the expansion needs to be carried out to 50 or 60 terms to achieve a satisfactory overall percentage error in $P(x)$, and 50! is already a very large number which exceeds the floating decimal point capacity of many computers and programs. Such are the obstacles to determining $P(x)$, especially for higher values of x. The difficulty can be worked through by a technique of recursion formulas for the $(n + 1)$th term in terms of the nth term and by using math coprocessors and quadruple accuracy.

In other words, the function $P(x)$ has been made practical for use by these techniques.

---

[3]Whittaker, E. T., and Watson, G. N., "A Course of Modern Analysis," Fourth Edition, Cambridge University Press, 1940, p. 253.

13.6.3 EXAMPLE

The Chief Engineer asks you to plan the purchase of a new server for the workstations using *CAD*[4] software with the new office computer. You consult the log for the existing system and obtain the following data.

User 1		User 2		User 3		User 4		User 5	
On	Off	On	Off	On	Off	On	Off	On	Off
8:15	9:30	8:30	10:00	9:20	9:50	10:30	11:00	10:00	12:00
10:00	10:30	10:45	12:00	11:00	11:30				
11:00	11:20	1:00	1:45	1:30	3:00				
1:15	2:00								
2:20	4:00								
4:15	4:45								

Construct the histogram of number of users for each of the working hours in the intervals 8:00-12:00 and 1:00-5:00. Use a random number generator to produce the simulated histogram of usage for a total of 100 events.

The table of frequencies versus hours is constructed as:

8:00-9:00	2
9:00-10:00	3
10:00-11:00	4
11:00-12:00	4
1:00-2:00	3
2:00-3:00	2
3:00-4:00	1
4:00-5:00	1

where it is obvious that the class-width has been taken as one hour. The histogram based upon this data is shown in Figure 13.4. The total area under the frequency distribution curve is 20. The first moment computation then shows

$$\begin{aligned} 20\,\overline{x} &= 2\times 1 + 3\times 2 + 4\times 3 + 4\times 4 + 3\times 5 + 2\times 6 + 1\times 7 + 1\times 8 \\ &= 2 + 6 + 12 + 16 + 15 + 12 + 7 + 8 = 78 \end{aligned}$$

Accordingly,

$$\overline{x} = \frac{78}{20} = 3.90 \tag{13.26}$$

or about 11:24 a.m.

The other statistics can be developed as follows:

$$\begin{aligned} s_x^2 &= \frac{1}{N}\sum_{i=1}^{n} f_i\, x_i^2 - \overline{x}^2 \\ &= \frac{1}{20}[2(1)^2 + 3(2)^2 + 4(3)^2 + 4(4)^2 + 3(5)^2 + 2(6)^2 + 1(7)^2 + 1(8)^2] - (3.9)^2 \end{aligned}$$

[4]Computer Aided Design

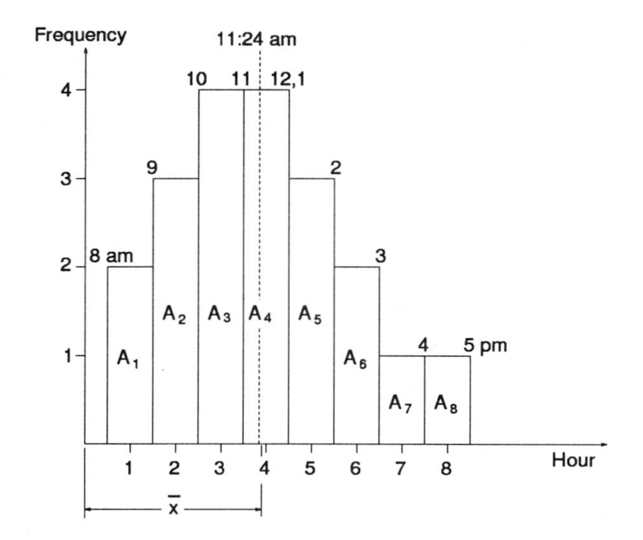

Figure 13.4: Histogram of Data

$$
\begin{aligned}
s_x^2 &= \frac{1}{20}[2+12+36+64+75+72+49+64]-(3.9)^2 \\
&= \frac{1}{20}(374)-(3.9)^2 = 18.7-15.21 = 3.49 \\
s_x &= \sqrt{3.49} = 1.868154169 && (13.27)
\end{aligned}
$$

$$
\begin{aligned}
m_3 &= \frac{1}{N}\sum_{i=1}^{n} f_i\,(x_i-\overline{x})^3 \\
&= \frac{1}{20}[2(1-3.9)^3+3(2-3.9)^3+4(3-3.9)^3+4(4-3.9)^3 \\
&\qquad +3(5-3.9)^3+2(6-3.9)^3+1(7-3.9)^3+1(8-3.9)^3] \\
&= \frac{1}{20}[48.96] = 2.448 \\
a_3 &= \frac{m_3}{s_x^3} = \frac{2.448}{(1.868)^3} = 0.375468297 && (13.28)
\end{aligned}
$$

$$
\begin{aligned}
m_4 &= \frac{1}{N}\sum_{i=1}^{n} f_i\,(x_i-\overline{x})^4 \\
&= \frac{1}{20}[2(1-3.9)^4+3(2-3.9)^4+4(3-3.9)^4+4(4-3.9)^4 \\
&\qquad +3(5-3.9)^4+2(6-3.9)^4+1(7-3.9)^4+1(8-3.9)^4] \\
a_4 &= \frac{1}{20}[601.3939999999] = 30.0696999999 \\
a_4 &= \frac{m_4}{s_x^4} = \frac{30.0696999999}{(3.49)^2} = 2.468756414 && (13.29)
\end{aligned}
$$

Cumulative probabilities associated with the class width boundaries are established as follows, using the areas.

Class	Class Probability	Class Boundaries
1.0	$\frac{A_1}{A} = \frac{2}{20} = 0.100000$	0.000000 – 0.100000
2.0	$\frac{A_2}{A} = \frac{3}{20} = 0.150000$	0.100001 – 0.250000
3.0	$\frac{A_3}{A} = \frac{4}{20} = 0.200000$	0.250001 – 0.450000
4.0	$\frac{A_4}{A} = \frac{4}{20} = 0.200000$	0.450001 – 0.650000
5.0	$\frac{A_5}{A} = \frac{3}{20} = 0.150000$	0.650001 – 0.800000
6.0	$\frac{A_6}{A} = \frac{2}{20} = 0.100000$	0.800001 – 0.900000
7.0	$\frac{A_7}{A} = \frac{1}{20} = 0.050000$	0.900001 – 0.950000
8.0	$\frac{A_8}{A} = \frac{1}{20} = 0.055556$	0.950001 – 1.000000

One hundred random numbers are drawn.

0.258164	0.356517	0.328117	0.913235	0.490798	0.366460
0.547250	0.136699	0.966490	0.473198	0.941463	0.196284
0.279794	0.919290	0.792363	0.835986	0.642372	0.176396
0.676883	0.910170	0.748793	0.010233	0.849870	0.057200
0.056700	0.454833	0.487669	0.061361	0.093443	0.342842
0.339846	0.725920	0.922713	0.367781	0.560052	0.054743
0.036783	0.350587	0.530000	0.204205	0.437881	0.624326
0.220480	0.101159	0.176412	0.223866	0.668517	0.609543
0.321406	0.606294	0.057296	0.341698	0.006566	0.653122
0.926274	0.633650	0.343418	0.047882	0.039384	0.914302
0.551342	0.797598	0.058191	0.286994	0.236066	0.479593
0.340312	0.780166	0.348781	0.720032	0.414815	0.874310
0.823825	0.675692	0.090331	0.315302	0.486088	0.239915
0.937351	0.049565	0.541400	0.889689	0.741722	0.445296
0.617730	0.297081	0.244733	0.332082	0.248744	0.061141
0.702187	0.780488	0.466759	0.990251	0.778968	0.602271
	0.208356	0.677624	0.797398	0.934772	

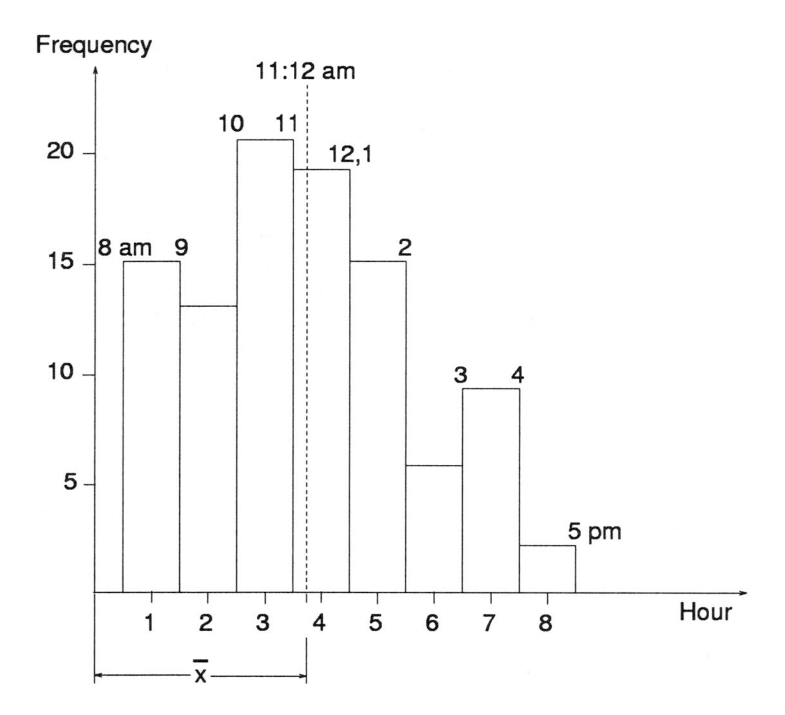

Figure 13.5: Simulated Histogram

Now it is possible to associate each random number with a particular class (hour) by noting the class limits within which it falls. The following table is then constructed.

Class	Class Boundaries	Random Number Tick Mark	Class Total
1.0	0.000000-0.100000	||||| ||||| |||||	15
2.0	0.100001-0.250000	||||| ||||| |||	13
3.0	0.250001-0.450000	||||| ||||| ||||| |||||		21
4.0	0.450001-0.650000	||||| ||||| ||||| ||||	19
5.0	0.650001-0.800000	||||| ||||| |||||	15
6.0	0.800001-0.900000	|||||		6
7.0	0.900001-0.950000	||||| ||||	9
8.0	0.950001-1.000000	||	2

The simulated histogram for 100 events can next be built from these latter data as shown in Figure 13.5.

It is of interest to determine how well the simulation may compare with the data, although it is already apparent visually that the object is generally congruent. A simple calculation will establish the mean of the simulated model as shown below. The total area under the distribution is, of course, equal to $100 \times 1 = 100$.

$$100\,\overline{x} = 1 \times 15 + 2 \times 13 + 3 \times 21 + 4 \times 19 + 5 \times 15 + 6 \times 6 + 7 \times 9 + 8 \times 2 = 370$$

$$\overline{x} = 3.70 \tag{13.30}$$

or about 11:12 a.m.; "close enough," as they say, "for systems work."

It may be noted that the period of most frequent usage is the hour from 10:00 to 11:00, when the frequency is 21. If the staff using CAD software is 100 individuals, then provision should be made for 21 of them to be on-line at the mid-morning hour.

The remaining statistics for the simulated histogram can be computed as follows.

$$\begin{aligned} s_x^2 &= \frac{1}{100}[15(1)^2 + 13(2)^2 + 21(3)^2 + 19(4)^2 + 15(5)^2 \\ &= \quad +6(6)^2 + 9(7)^2 + 2(8)^2] - (3.7)^2 \\ &= \frac{1}{100}(1720) - (3.7)^2 = 17.20 - 13.69 = 3.57 \\ s_x &= \sqrt{3.57} = 1.873499399 \end{aligned} \tag{13.31}$$

$$\begin{aligned} m_3 &= \frac{1}{100}[15(1-3.7)^3 + 13(2-3.7)^3 + 21(3-3.7)^3 + 19(4-3.7)^3 \\ &= \quad +15(5-3.7)^3 + 6(6-3.7)^3 + 9(7-3.7)^3 + 2(8-3.7)^3] \\ &= \frac{1}{100}[222.5999999999] = 2.225999999 \\ a_3 &= \frac{m_3}{s_x^3} = \frac{2.225999999}{(1.873499399)^3} = 0.338504530 \end{aligned} \tag{13.32}$$

$$\begin{aligned} m_4 &= \frac{1}{100}[15(1-3.7)^4 + 13(2-3.7)^4 + 21(3-3.7)^4 + 19(4-3.7)^4 \\ &\quad +15(5-3.7)^4 + 6(6-3.7)^4 + 9(7-3.7)^4 + 2(8-3.7)^4] \\ &= \frac{1}{100}[2872.77] = 28.7277 \\ a_4 &= \frac{m_4}{s_x^4} = \frac{28.7277}{(1.873499399)^4} = 2.331774907 \end{aligned} \tag{13.33}$$

These can be compared with the corresponding ones for the sample histogram. Additional comparisons may be obtained from the machine computations using *pearson3.go* on *eos* network. The latter are given below.

	Sample	Simulation
$\overline{x}$	3.9000	3.7000
s_x	1.8682	1.8735
A	5.3267	5.9083

These might be considered good simulations, particularly in light of the small number of data in the sample. Quantities determined from higher-order moments (here, the As) are generally less accurate than the means and variances.

13.6.4 PEARSON TYPE I

The Pearson-I frequency distribution is given by the equation[5]

$$y = y_0 \left(1 + \frac{x}{a_1}\right)^{m_1} \left(1 - \frac{x}{a_2}\right)^{m_2} \qquad -a_1 < x < a_2 \tag{13.34}$$

in which $m_1/a_1 = m_2/a_2$, and the origin of x is at the mode. The parameters m_1, m_2, a_1, and a_2 are related to the mean, variance, skewness, and peakedness of the distribution, and y_0 is a normalizing constant so chosen that the area under the frequency distribution curve equals unity. That is, the cumulative probability adds up to certainty that given data will be found within the range of the distribution.

13.6.5 THE NORMALIZING CONSTANT

The usual assumption is that the parameters or "statistics" of the continuous distribution can be equated to those of the sample. Thus $\bar{x}$, s_x, a_3, and a_4 are determined from the first, second, third, and fourth moments of the data, respectively. There is, therefore, one remaining parameter—namely, y_0—to be ascertained. Its value is set by the normalizing condition that the area under the frequency distribution curve must be equal to unity. Thus, with

$$z = (a_1 + x)/(a_1 + a_2) \tag{13.35}$$

$$\begin{aligned}
N &= \int_{-a_1}^{a_2} y_0 \left(1 + \frac{x}{a_1}\right)^{m_1} \left(1 - \frac{x}{a_2}\right)^{m_2} dx = \int_{-a_1}^{a_2} \frac{y_0}{a_1^{m_1} a_2^{m_2}} (a_1 + x)^{m_1} (a_2 - x)^{m_2}\, dx \\
&= \int_0^1 \frac{y_o}{a_1^{m_1} a_2^{m_2}} [z(a_1 + a_2)]^{m_1} [(1 - z)(a_1 + a_2)]^{m_2} (a_1 + a_2)\, dz \\
&= \int_0^1 \frac{y_0 (a_1 + a_2)^{m_1 + m_2 + 1}}{a_1^{m_1} a_2^{m_2}} z^{m_1} (1 - z)^{m_2}\, dz \\
&= \frac{y_0 (m_1 + m_2)^{m_1 + m_2} (a_1 + a_2)}{m_1^{m_1} m_2^{m_2}} \frac{\Gamma(m_1 + 1)\Gamma(m_2 + 1)}{\Gamma(m_1 + m_2 + 2)}
\end{aligned} \tag{13.36}$$

A similar method can be employed to obtain the moments. For the nth one, taken with respect to an origin on the vertical line through $-a_1$,

$$\begin{aligned}
N\, \mu_n &= \int_{-a_1}^{a_2} \frac{y_o}{a_1^{m_1} a_2^{m_2}} (a_1 + x)^n (a_1 + x)^{m_1} (a_2 - x)^{m_2}\, dx \\
&= \frac{y_o (m_1 + m_2)^{m_1 + m_2} (a_1 + a_2)^{n+1}}{m_1^{m_1} m_2^{m_2}} \frac{\Gamma(m_1 + n + 1)\Gamma(m_2 + 1)}{\Gamma(m_1 + m_2 + n + 2)}
\end{aligned} \tag{13.37}$$

There is the recurrence relationship $\Gamma(x) = (x - 1)\Gamma(x - 1)$ for the Gamma function, therefore

$$\mu_1 = \frac{(a_1 + a_2)(m_1 + 1)}{m_1 + m_2 + 2} \tag{13.38}$$

[5]Elderton, W. P., and Johnson, N. L., *Systems of Frequency Curves*, Cambridge University Press, 1969, p. 51.

This locates the mean, and then the origin can be shifted to the mean for the higher moments. Use $m_1' = m_1 + 1$, $m_2' = m_2 + 1$, $b = a_1 + a_2$ and $r = m_1' + m_2'$ for convenience so as to have

$$\mu_2 = \frac{b^2 m_1' m_2'}{r^2(r+1)} \qquad \mu_3 = \frac{2b^3 m_1' m_2'(m_2' - m_1')}{r^3(r+1)(r+2)} \tag{13.39}$$

$$\mu_4 = \frac{3b^4 m_1' m_2'[m_1' m_2'(r-6) + 2r^2]}{r^4(r+1)(r+2)(r+3)} \tag{13.40}$$

For a further convenience in the calculations, create the parameters

$$\varepsilon = m_1' m_2' \qquad \beta_1 = \frac{\mu_3^2}{\mu_2^2} \qquad \beta_2 = \frac{\mu_4}{\mu_2^2} \tag{13.41}$$

with which

$$r = \frac{6(\beta_2 - \beta_1 - 1)}{3\beta_1 - 2\beta_2 + 6} \qquad \varepsilon = \frac{r^2}{4 + \frac{1}{4}\beta_1 \frac{(r+2)^2}{r+1}} \tag{13.42}$$

$$b^2 = (a_1 + a_2)^2 = \frac{\mu_2(r+1)r^2}{\varepsilon} \tag{13.43}$$

Also,

$$m_1, m_2 = \frac{1}{2}\left[r - 2 \pm r(r+2)\sqrt{\frac{\beta_1}{\beta_1(r+2)^2 + 16(r+1)}}\right] \tag{13.44}$$

$$y_0 = \frac{N}{a_1 + a_2} \frac{m_1^{m_1} m_2^{m_2}}{(m_1 + m_2)^{m_1+m_2}} \frac{\Gamma(m_1 + m_2 + 2)}{\Gamma(m_1 + 1)\Gamma(m_2 + 1)} \tag{13.45}$$

and the mode can be located with respect to the mean by the formula

$$x_{\max} = \overline{x} - \frac{1}{2}\frac{\mu_3}{\mu_2}\frac{r+2}{r-2} \tag{13.46}$$

With these parameters determined, the frequency distribution can be displayed. Whittaker and Watson[6] gave an asymptotic series expansion for the Gamma function which can be used to shorten the labor of computing the functions which occur in the formulas above. The series is very highly convergent, so that only a few terms are needed.

13.6.6 CUMULATIVE PROBABILITY

Most of the applications of Pearson I employ the frequency distribution to compute the area under the curve up to a particular value of x; that is, the *cumulative probability* of some event or value. This means that indefinite Gamma function integrals must be evaluated. This is a formidable task, given the fact that there are no closed-form representations of such integrals, and given the nature of the function's behavior.

---

[6]Whittaker, E. T., and Watson, G., *Modern Analysis*, Cambridge University Press, 1952, p. 253.

As supplied above, the frequency distribution is valid on the range

$$-a_1 \leq x \leq a_2 \tag{13.47}$$

Therefore the cumulative probability up to an indefinite upper limit x is given by the integral

$$P(x) = \int_{-a_1}^{x} y\, d\xi = \int_{-a_1}^{x} y_0 \left(1 + \frac{\xi}{a_1}\right)^{m_1} \left(1 - \frac{\xi}{a_2}\right)^{m_2} d\xi \tag{13.48}$$

It may be expected that this integral is a form of the indefinite Gamma function with the argument x. Up to a point, a procedure similar to that invoked to determine y_0 can be followed. Thus, make the change of variable

$$\zeta = (a_1 + \xi)/(a_1 + a_2) \tag{13.49}$$

so that the integral becomes

$$P(x) = \int_0^z \frac{y_0(a_1 + a_2)^{m_1+m_2+1}}{a_1^{m_1} a_2^{m_2}} \zeta^{m_1}(1 - \zeta)^{m_2} d\zeta \tag{13.50}$$

When this is integrated, the result is

$$P(x) = \frac{y_0(a_1 + a_2)(m_1 + m_2)^{m_1+m_2}}{m_1^{m_1} m_2^{m_2}} \frac{\Gamma_z(m_1 + 1)\Gamma_z(m_2 + 1)}{\Gamma_z(m_1 + m_2 + 2)} \tag{13.51}$$

in which the Γ_zs are indefinite Gamma functions of their respective arguments. Direct numerical computation of the indefinite integral can be performed by first substituting a series expansion of the integrand. This procedure was elaborated in a previous publication.[7]

13.6.7 THE PROGRAMS

Computation of Pearson I statistics is accomplished in a program which is named *pearson1.p*, with a compiled or execute file called *pearson1.go*. It is written in Pascal source code. The program accepts as input up to 100 pairs of data as given below. It is quite easy to enlarge this number if needed. The program output is the $\bar{x}$, σ_x, μ_3, and μ_4 set of statistics for the data, under the assumption that it is a Pearson I Type distribution. Data are also computed to enable the plotting of the frequency distribution. The output is written to a file named *pearson1.dat*.

A companion program is presented for computing the cumulative probability for the data above. It is called *pearson1p.p*, for the source code, also written in Pascal, and the execute file is named *pearson1p.go*. As in the previous case, the output is written to a file named *pearson1p.dat*. The details of these programs are presented in a recent publication.[8]

---

[7]McDonald, P. H., "Computing Pearson-III Statistics," *Developments in Civil and Construction Engineering Computing*, CIVIL-COMP Press, Edinburgh, 1993, pp. 65-71.

[8]McDonald, P. H., "Computing Pearson I Statistics," *Developments in Computer Aided Design and Modeling for Civil Engineering Computing*, CIVIL-COMP Press, Edinburgh, 1995, pp. 263-269.

It might appear that the computation of cumulative probability ought to be incorporated in the program above, so as to minimize re-entry of data and other inefficiencies. In a given location with specific hardware and software capabilities it may indeed be possible to incorporate the two programs in one operation, probably called from a single shell. However, there is a need to use the maximum capability of the user's computer for *pearson1p.go*. Some systems will provide only double precision, whereas others will allow quadruple precision for the data. Some hand-held computers may actually give a better accuracy with the program than some mainframes.

For this reason the routines are kept separate, so that the user may decide how to obtain the better result. In any case, the maximum allowable precision should be called for.

As input, *pearson1p.go* accepts $\overline{x}$, σ_x, μ_3, and μ_4 from *pearson1.go*. It then asks the user to select a value of x, the variable which sets the cumulative probability. The program provides for the user to select a number of xs in sequence, so as to furnish adequate output for plotting a complete "S-curve" if desired.

13.6.8 EXAMPLE

PARSONS and WHITTED, PA are consulting engineers who employ a large staff in their home office. Their computing services manager is considering purchasing the new version of CAD software used by the people in the office. In order to negotiate the optimum site license, the manager wishes to determine as accurately as possible the frequency of calls for the software as a function of the daily work schedule. From the log, the following sample of number of users for each half-hour of the work day is obtained for one week's time.

8-8:30	8:30-9	9-9:30	9:30-10	10-10:30	10:30-11
34	145	156	145	123	103
11-11:30	11:30-12	12-12:30	12:30-1	1-1:30	1:30-2
86	71	55	37	21	13
2-2:30	2:30-3	3-3:30	3:30-4	4-4:30	4:30-5
7	3	1	0	0	0

1. Assuming that the data from the sample is fitted with a Pearson-I Type distribution, determine the mean, variance, skewness, and peakedness.

2. Plot the fit of the distribution to the data.

3. Using a random number generator, produce a simulated operation for three months duration. Plot the simulated operation on the same graph as that of the sample data and the Pearson-I distribution.

The program *pearson1.go* is called and run. The following is a printout of *pearson1.dat* from the run.

xbar	=	155.2500	sx	=	2.7681	d	=	- 0.2647
M2	=	7.6624	M3	=	15.1069	M4	=	172.3253
A	=	15.4858	A1	=	4.2006	A2	=	11.2851
B1	=	0.5073	B2	=	2.9351	r	=	5.1867
e1	=	0.4069	e2	=	2.7798	K	=	0.2440
G12	=	14.0128	G1	=	0.7476	G2	=	2.3568
tmax	=	- 2.2232	ye	=	125.3118			
xmax	=	88.5550	ymax	=	152.0288	P	=	1000.0000

x	=	30.0000	f	=	34.0000	y	=	37.7411
x	=	60.0000	f	=	145.0000	y	=	140.6391
x	=	90.0000	f	=	156.0000	y	=	152.0080
x	=	120.0000	f	=	145.0000	y	=	144.4042
x	=	150.0000	f	=	123.0000	y	=	128.5431
x	=	180.0000	f	=	103.0000	y	=	109.1527
x	=	210.0000	f	=	86.0000	y	=	88.8785
x	=	240.0000	f	=	71.0000	y	=	69.3518
x	=	270.0000	f	=	55.0000	y	=	51.6079
x	=	300.0000	f	=	37.0000	y	=	36.2844
x	=	330.0000	f	=	21.0000	y	=	23.7280
x	=	360.0000	f	=	13.0000	y	=	14.0556
x	=	390.0000	f	=	7.0000	y	=	7.11897
x	=	420.0000	f	=	3.0000	y	=	2.8777
x	=	450.0000	f	=	1.0000	y	=	0.6955
x	=	480.0000	f	=	0.0000	y	=	0.0289
x	=	510.0000	f	=	0.0000	y	=	0.0000
x	=	540.0000	f	=	0.0000	y	=	0.0000

Next, *pearson1p.go* is run, with the following output:

$\overline{x}$ = 155.250000 sx = 2.7681 M3 = 15.1069 M4 = 172.3253

x	P
x = 0.500000	P = 0.000000
x = 1.500000	P = 0.042273
x = 2.500000	P = 0.313566
x = 3.500000	P = 0.615207
x = 4.500000	P = 0.814263
x = 5.500000	P = 0.918575
x = 6.500000	P = 0.966595
x = 7.500000	P = 0.986942
x = 8.500000	P = 0.995080
x = 9.500000	P = 0.998199
x = 10.500000	P = 0.999357
x = 11.500000	P = 0.999776
x = 12.500000	P = 0.999924
x = 13.500000	P = 0.999976
x = 14.500000	P = 0.999993
x = 15.500000	P = 1.000000

With the information above, the computation of statistics for the model is complete. The simulation process can now begin.

Since the number of calls for a three-month period is estimated to total to 12000, that amount of random numbers should be generated. Listed below is a representative sample of the random numbers drawn.

0.258164	0.547250	0.279794	0.676883	0.056700
0.339846	0.036783	0.220480	0.321406	0.926274
0.551342	0.304312	0.823825	0.937351	0.617730
0.702187	0.208356	0.248744	0.490798	0.941463
0.642372	0.849870	0.093443	0.560052	0.356517
0.136699	0.819290	0.910170	0.454833	0.725920
0.350587	0.101159	0.606294	0.633650	0.797598
0.780166	0.675692	0.049565	0.297081	0.780488
0.677624	0.061141	0.366460	0.196284	0.176396
0.057200	0.342842	0.054743	0.328117	0.966490
0.792363	0.748793	0.487669	0.922713	0.530000
0.176412	0.057296	0.343418	0.058191	0.348781
0.090331	0.541400	0.244733	0.466759	0.797396
0.778968	0.668517	0.006566	0.039383	0.236066
0.414815	0.486088	0.913235	0.473198	0.835986
0.010233	0.061361	0.367781	0.204205	0.223866
0.341698	0.047882	0.286994	0.720032	0.315302
0.889689	0.332082	0.990251	0.934772	0.602271
0.609543	0.653122	0.914302	0.479593	0.874310
0.239915	0.437881	0.624326	0.741722	0.445296
0.668767	0.008061	0.048345	0.289772	0.736682
...	...	...	...	...
0.557661	0.342185	0.050801	0.304492	0.824904
0.943814	0.656465	0.934335	0.599657	0.593857
0.559225	0.351556	0.106967	0.641104	0.842276
0.047928	0.287273	0.721707	0.325338	0.949838
0.692568	0.150708	0.903251	0.413364	0.477391
0.861112	0.160820	0.963853	0.776563	0.654104
0.920183	0.514838	0.085541	0.959712	0.751744
0.505360	0.028735	0.172246	0.032328	0.193777
0.161368	0.967134	0.796227	0.771953	0.626475
0.754598	0.522461	0.131225	0.786484	0.713559
0.276511	0.657207	0.938779	0.626291	0.753498
0.515872	0.091735	0.549817	0.295173	0.769050
0.609073	0.650304	0.284925	0.707634	0.241001
0.444391	0.663342	0.975552	0.846672	0.074279
0.445197	0.668169	0.004479	0.268876	0.161107
0.965570	0.786849	0.715749	0.289632	0.735843
0.410058	0.457579	0.742379	0.449229	0.692334
0.149304	0.894838	0.362940	0.175190	0.049977
0.299551	0.062825	0.376550	0.256756	0.538814
0.229231	0.373848	0.240567	0.441791	0.647755

Using the class limits defined by the cumulative probability data above, the following simulated histogram data can be constructed:

8:00-8:30	am	435	calls
8:30-9:00	am	1726	calls
9:00-9:30	am	1894	calls
9:30-10:00	am	1738	calls
10:00-10:30	am	1452	calls
10:30-11:00	am	1258	calls
11:00-11:30	am	1006	calls
11:30-12:00	am	839	calls
12:00-12:30	pm	671	calls
12:30-1:00	pm	443	calls
1:00-1:30	pm	263	calls
1:30-2:00	pm	143	calls
2:00-2:30	pm	84	calls
2:30-3:00	pm	36	calls
3:00-3:30	pm	12	calls
3:30-4:00	pm	0	calls
4:00-4:30	pm	0	calls
4:30-5:00	pm	0	calls

If desired, the simulated data can also be processed through the ***pearson1.go*** program to compare sample and simulated statistics. Thus

Sample Statistics

$\overline{x} = 155.2500$ $\quad s_x = 2.7681$ $\quad \mu_3 = 15.1069$

$\mu_4 = 172.3253$

Simulated Statistics

$\overline{x} = 154.8425$ $\quad s_x = 2.7705$ $\quad \mu_3 = 15.1168$

$\mu_4 = 172.5075$

As usual, there is excellent agreement on the lower moments, and in this case for the higher moments as well, possibly because of the large sample size.

The manager of computing services can use the simulated data to assist in negotiating the site license.

Figure 13.6 shows the histograms and frequency distributions and Figure 13.7 displays the cumulative probability "*S*-curve."

13.7 QUEUES

A queue is a line of persons or vehicles waiting to receive some service or attention, as in a cafeteria, in a jobshop, at a stoplight, at a ticket office, or in a Post Office. There are a number of queue-type systems in civil engineering practice, especially in traffic studies and construction operations.

The customary frequency distribution used to model a queue is that called the ***Poisson equation***, after the French engineer S. D. Poisson. It is

$$P(x) = \frac{e^{-\mu}\,\mu^x}{x!} \tag{13.52}$$

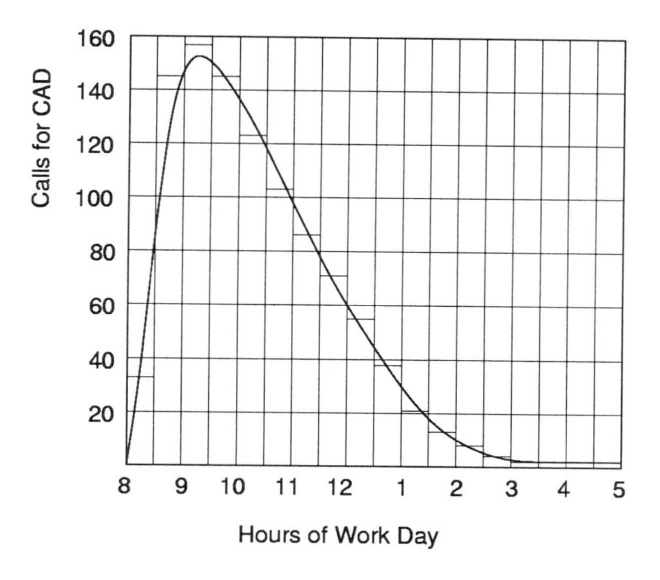

Figure 13.6: Simulated Histogram

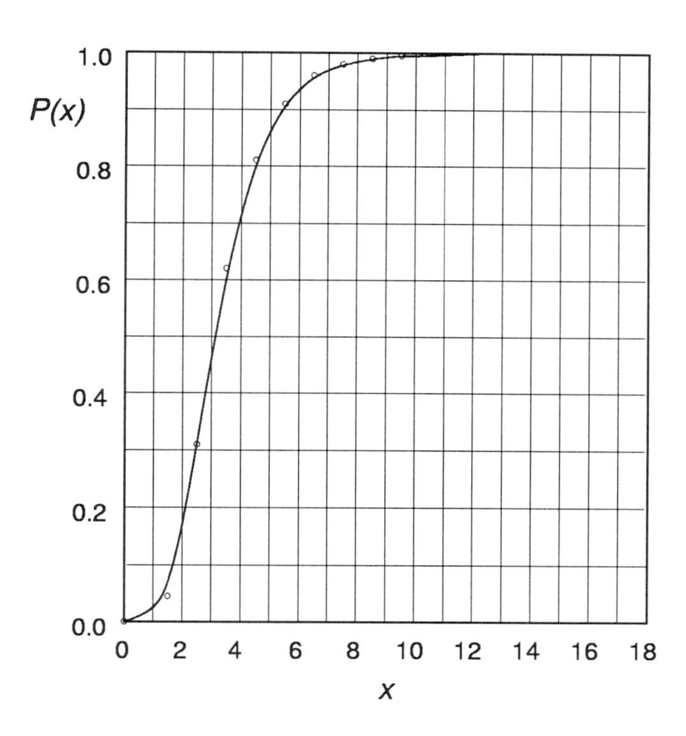

Figure 13.7: Cumulative Probability

in which $P(x)$ is the probability of x arrivals, and μ is the mean arrival rate. This formula is employed in treating the occurrence of what are called "rare events" in a smooth-flowing river of time. As mentioned, μ is the mean, and the standard deviation for this distribution is $\sqrt{\mu}$.

Queues are found in drive-thru or "minute" service plazas in a variety of contemporary enterprises such as bank tellers, fast-food emporia, car wash stations and the like. In a typical situation, the client or customer arrives at the tail end of the queue at a variable rate, so that the number waiting in line is a variable. Upon taking samples of the arrival rate, it is then possible to establish the average number of arrivals per unit time. After waiting in the queue the clients are serviced at a rate μ per unit time, and then they depart at a constant rate.

The system analyst will wish to know the length of the queue in order to provide for the space which will be needed to accomodate the clients or their vehicles. Further, if the length becomes "too great," some clients will lose patience and leave the queue.

13.7.1 EXAMPLE

Vehicles arrive at the toll gate on a turnpike at an average rate of one every 50 seconds, or $\mu = \frac{60}{50} = 1.2$ vehicles per minute. Suppose that the number of vehicles which are processed through the turnstiles per minute is a variable, with a schedule of probability given in the table below.

Number of Vehicles Processed per Minute	Probability
0	0.12
1	0.58
2	0.30

To simulate the queue, construct the following tables of probabilities and class limits for random numbers, using Equation 13.52.

Arrivals

x	$P(x)$	Class Boundaries
0	0.301194211	0.000000000-0.301194211
1	0.361433054	0.301194212-0.662627266
2	0.216859832	0.662627267-0.879487098
3	0.086743933	0.879487099-0.966231031
4	0.026023179	0.966231032-0.992254210
5	0.006245563	0.992254211-0.998499773
6	0.001249112	0.998499774-0.999748885
6	0.000214133	0.999748886-0.999963018
8	0.000032120	0.999963019-0.999995138
9	0.000004282	0.999995139-0.999999420
10	0.000000513	0.999999421-1.000000000

Processing

x	$P(x)$	Class Boundaries
0	0.12000	0.00000-0.12000
1	0.58000	0.12001-0.70000
2	0.10000	0.70001-1.00000

Next, draw some random numbers, say 50, for the arrivals, and 50 others for the processing, and enter these into the table below to construct the length of the waiting line.

Time Period	Random Number	Number of Arrivals	Random Number	Number Processed	Number Waiting
1	0.0347437386	0	0.4698637162	-	0
2	0.9774246762	4	0.4253323732	1	3
3	0.1676622766	0	0.3290797853	1	2
4	0.1256859926	0	0.0503729315	0	2
5	0.5559563564	1	0.3162430990	1	2
6	0.1622779439	0	0.1737932378	1	1
7	0.8442175331	2	0.7704744767	2	1
8	0.6301637859	2	0.9810507175	2	1
9	0.3321123429	1	0.5242074438	1	1
10	0.5760863244	1	0.4917460962	1	1
11	0.1818079246	0	0.7983861962	2	0
12	0.2662389775	0	0.8311463224	2	0
13	0.2342406474	0	0.0745321408	0	0
14	0.5236281995	1	0.0056763138	0	1
15	0.3785943512	1	0.4234079688	1	1
16	0.7020171213	2	0.1389510374	1	2
17	0.5662183735	1	0.9712259347	2	1
18	0.9949572277	5	0.1664361600	1	5
19	0.1608150472	0	0.4559346849	1	4
20	0.3116933243	1	0.2015370049	1	4
21	0.6834301370	2	0.4422788426	1	5
22	0.7457256576	2	0.7191386754	2	5
23	0.2742378653	0	0.9657693610	2	3
24	0.0039682961	0	0.7784570329	2	1
25	0.2994989424	0	0.5375919330	1	0
26	0.1690826659	0	0.6719007174	1	0
27	0.1127947506	0	0.0294373402	0	0
28	0.3524101620	1	0.7978450491	2	0
29	0.3823168638	1	0.8775668141	2	0
30	0.3196259147	1	0.3486825391	1	0
31	0.6667406714	2	0.1105650968	0	2
32	0.1490844511	0	0.5227411486	1	1
33	0.6805511800	2	0.0760629355	0	3
34	0.2046787390	0	0.0402333108	0	3
35	0.6419589779	1	0.0190107506	0	4

Time Period	Random Number	Number of Arrivals	Random Number	Number Processed	Number Waiting
36	0.0526937060	0	0.9203515977	2	2
37	0.0797108823	0	0.6171629915	1	1
38	0.6871868585	2	0.7332081112	2	1
39	0.2699616553	0	0.4210506742	1	0
40	0.1465526875	0	0.2678630655	1	0
41	0.3696473661	1	0.3326168045	1	0
42	0.4281145720	1	0.2707360751	1	0
43	0.5650267107	1	0.1355385859	1	0
44	0.9696682731	4	0.5712101421	1	3
45	0.3854824622	1	0.0618443253	0	4
46	0.4954435482	1	0.8735209643	2	3
47	0.5724550688	1	0.2176335025	1	3
48	0.1695556719	0	0.1286735807	1	2
49	0.7864560782	2	0.1551001342	1	3
50	0.0947279654	0	0.9052847727	2	1

Evidently, the queue reaches its maximum length of 5 vehicles at about 18 minutes after startup, but by 22 minutes it returns to a level of 3 or four or less, and is often vacant.

13.8 PROJECT #11

1. The manager of an airport (who is a Civil Engineer) has noted that a local airline is using the airport for a regional center of operations. In order to better plan the activities of the facility, he has taken the following sample data on the "cluster flight" departure intervals at certain hours:

0.5-1.5 minutes	2 flights
1.5-2.5 minutes	15 flights
2.5-3.5 minutes	7 flights
3.5-4.5 minutes	2 flights
4.5-5.5 minutes	1 flight

   a) Construct the histogram of this data

   b) Assuming that the model is Pearson III, what are $\overline{x}$, s_x, and A?

   c) Use a random number generator to produce a simulated cluster flight schedule (histogram) for a gaggle of 20 flights.

2. AJAX, LTD, are builders engaged in the construction of a large power plant structure. The plant is to be over an old landfill, so the foundation is being supported on piles driven to solid earth. They find that of the first 20 piles, 6 went 2 m, 9 went 3 m, 3 went 5 m, 1 went 6 m and 1 went 7 m. There are to be 150 piles altogether.

   a) Assuming that the data from the first 20 piles is fitted with a Pearson III-type distribution, determine the mode, variance, and skewness.

b) Plot the fit of the distribution to the histogram of the data.

c) Using a random number generator, produce a simulated operation for the total of 150 piles. Plot the simulated operation on the same graph as that of the sample data and the Pearson III distribution. Comment upon the results.

3. Wolfsden Road is a boulevard through an upscale residential neighborhood. The traffic engineers responsible for this division are planning some improvements in the road net to facilitate commuting in the area. They have taken a sample of the hourly traffic (in hundreds of vehicles) on Wolfden, and have made the assumption that it is representative of the other streets feeding an arterial thoroughfare leading into the city. The data are shown below.

Hour	Vehicles
7:00-8:00	9
8:00-9:00	12
9:00-10:00	7
10:00-11:00	3
11:00-12:00	1

Make the further assumption that the sample is drawn from a Pearson III distribution.

- Compute the mean, variance, and skewness for the sample.
- Determine the cumulative probability class limits, and then draw 100 random numbers (to represent the traffic on several streets).
- Construct the histogram of the simulated traffic flow, and compare it with the sample histogram.

4. A consulting engineering firm specializes in making tests of concrete core samples. The firm's president has reviewed the records of recent operations, and extracted the following sample showing the number of jobs which clients have requested during the sample period.

Number of Jobs	Frequency Requested
1	24
2	54
3	12
4	6
5	4

In order to make sure that the firm has enough equipment to serve the needs of his clients promptly, the president wishes to simulate the projected operations for an extended period.

- Assume that the data can be fitted by a Pearson III frequency distribution, and use a random number generator to produce a histogram based upon a run of fifty trials to help him decide the number of testing machines available.
- Display the sample statistics and the simulated histogram as well as the computer printouts in your report.

5. The manager of the documents room in a large engineering firm which specializes in site lay-out work wishes to improve the efficiency of her unit's operations. She has recorded the following sample data for the usage of the room by staff engineers during a representative period of time.

Number of Documents Requested	Number of Engineers Using Documents
1	4
2	17
3	9
4	2
5	1

- Construct the histogram of this sample data for the manager, and determine the mean number of documents requested.
- Produce random numbers to generate a simulated histogram for a run of fifty request events. Calculate the mean number of documents requested in the simulation and compare this with the sample mean.

6. The Chief Engineer asks you to plan the design of a new materials storage building. In preparation for this assignment, you review the usage log for the old hodge-podge arrangement of sheds, and obtain the following data on the times workers visit the material facilities to obtain supplies.

User 1		User 2		User 3		User 4	
In	Out	In	Out	In	Out	In	Out
8:15	8:30	8:45	9:05	9:20	9:30	9:45	9:55
9:15	9:30	9:50	10:05	10:30	10:40	11:05	11:15
10:20	10:35	10:50	11:10	11:40	11:55	12:20	12:30
11:05	11:20	11:45	12:00	12:45	1:05		
12:10	12:25						

User 5		User 6	
In	Out	In	Out
10:10	10:20	10:50	11:05
11:40	11:55	12:30	12:40
12:55	1:15		

Construct the histogram of number of users for each of the working hours 8:00-2:00. What is the time of most-frequent use?

Use a random number generator to produce the simulated histogram of usage for a total of 100 events.

7. The copier in your engineering office is one of the most important items of equipment. The Chief Engineer is considering replacing it with a newer and faster model, because it is being used more-and-more heavily. She asks you to study present usage and make a recommendation.

   You think it will be best to simulate its behavior, so a record is kept of a typical day's operation as shown in the table below. Now construct histograms showing (a) the inter-arrival times of users, (b) times to make copies, and (c) deviations of the copying times from the average time per copy.

   Also display a pie-chart giving the probability of number of copies. Using random numbers, generate a simulated model of the number of copies for one day.

User Number	Arrival Time	Interval From Last User	Number Copies	Copying Time Minutes	Time per Copy	Deviation from Average Time per Copy
1	8:15		60	2.00		
2	8:32		20	1.00		
3	9:03		50	1.00		
4	9:47		60	2.05		
5	10:00		50	2.30		
6	11:16		70	2.25		
7	1:03		140	4.25		
8	2:06		100	3.00		
9	2:18		40	1.50		
10	2:49		40	2.15		
11	4:06		30	1.25		

8. Workers on a construction project must go periodically to a stores area to pick up parts and supplies. Assume that their mean rate of arrival is one every two minutes, and assume also that a study of previous operations indicates that the following distribution of processing their needs prevails:

Number of Workers Processed per Minute	Probability of Occurrence
0	0.15
1	0.55
2	0.30

   If the arrivals are randomly distributed in time, the Poisson frequency distribution can be used to estimate the probability of a specific number of arrivals.

   Use a random number generator to simulate the operation of the stores center for twenty time periods. Give the number of arrivals during the time period in question, as well as the number still waiting at the end of the period.

Plot the graph of the frequency distribution superposed upon the histogram of the arrivals.

9. You have been engaged as a consultant to design the parking area for a suite of professional offices. You observe the traffic at a similar complex in another location as given in the table below.

Arrivals		Length of Visit, Minutes					
Time	Number of Cars	Car 1	Car 2	Car 3	Car 4	Average	Deviation from Average
8:00-8:45	2	64	69			67	− 3, 2
8:45-9:30	1	75					
9:30-10:15	3	73	77	70			
10:15-11:00	3	123	116	123			
11:45-12:30	2	105	104				
12:30-1:15	4	27	34	34	30		
1:15-2:00	1	48					
2:00-2:45	2	55	59				
2:45-3:30	1	83					
3:30-4:15	3	50	46	54			
4:15-5:00	2	61	65				

Plot histograms showing the inter-arrival times, the average length of visit, and the deviations in length of visit, and a pie-chart model of number of cars arriving. Use a random number generator to simulate the arrivals schedule for your parking area.

CHAPTER 14

DIFFERENTIAL EQUATIONS FOR SYSTEMS

14.1 INTRODUCTION

A most important class of systems is that in which time is a parameter. One might label this classification Dynamical Systems, although such a description usually connotes mechanical systems of the hardware type, whereas the present model is envisaged to apply primarily to systems in which there is little or no material subject to Newtonian laws of mechanics. Rather, it is imagined that the systems in question are those modeled by a description of the type studied extensively by Poincaré[1] and Volterra.[2]

That is to say, the fundamental description of the system is through differential equations, particularly those in which the differentiation is with respect to time. The system dynamic is, therefore, a trajectory of something in time and space, but the equations of the motion are first order differential equations, not the familiar second order ones from mechanics.

Such an organization of the problem is especially appropriate to the realm of infrastructural systems in civil engineering practice. As the analysis of the time value of money showed, in Chapter 5, things do change with the passage of time, and civil engineering works frequently either outlast the frame of reference in which they were produced or are forced to adapt to conditions vastly different than those envisaged in their design process.

Therefore it is important to be able to project a system's performance (under suitable assumptions, of course) as far forward in time as can be reasonably supported with an extrapolation of rational character. Our client—the public—expects this of us in increasing measure, and since the Civil Engineer's work is primarily and uniquely devoted to the infrastructure of society, it is there that the changes wrought with the passage of time are so important.

---

[1]Poincaré, H., "Mémoire sur les courbes définie par une équation différentielle", *Journal de Mathématique*, Vol. 7, No. 3, 1881, pp. 375-442; Vol. 8, No.3, 1882, pp. 251-296; Vol. 1, No. 4, 1885, pp. 167-244. Also, *Œuvres*, Vol. 1, Paris, 1928, pp. 3-84, 90-161.

[2]Volterra, V., *Leçons sur la theorie mathématique de la lutte pour la vie*, Paris, 1931, vi+214 pp.

14.2 THE ECOLOGY OF SYSTEMS

It is convenient to describe the performance of the systems in question as though their elements were living entities. The terminology preferred here is "protagonist," a label which suggests that the element wearing it plays a role, especially vis-a-vis other protagonists, and it is thus imagined that a collection of such protagonists is bound together in a web of relationships delimited by a system boundary, all of which might be thought to be similar to a theatrical performance on a stage or in an amphitheater.

A mathematical description of the manner in which a number of system variables (protagonists) are coupled is the set of equations

$$\frac{dN_i}{dt} = P_i(N_1, N_2, N_3, \ldots, N_k), \qquad i = 1, 2, \ldots, k \tag{14.1}$$

where the N_i are the k quantities which interact with each other, and the P_i are the specific functions that define the behavior.

Some other labels are applied to the character of the relationships between the protagonists. Consider, for the sake of simplicity, a two-variable model of the form

$$\frac{dN_1}{dt} = a_{11}\,N_1 + a_{12}\,N_2 \qquad \frac{dN_2}{dt} = a_{21}\,N_1 + a_{22}\,N_2 \tag{14.2}$$

in which the as are constants. Some nomenclature borrowed from ecological science can be applied to four cases as follows.

Competition

$$\frac{\partial(dN_2/dt)}{\partial N_1} = a_{21} \leq 0 \tag{14.3}$$

$$a_{11} \geq 0 \qquad a_{22} \geq 0 \qquad a_{12} \leq 0 \qquad a_{21} \leq 0 \tag{14.4}$$

Predation

$$\frac{\partial(dN_1/dt)}{\partial N_2} = a_{12} \leq 0 \tag{14.5}$$

$$a_{11} \geq 0 \qquad a_{22} \leq 0 \qquad a_{12} \leq 0 \qquad a_{21} \geq 0 \tag{14.6}$$

Saprophytism

$$\frac{\partial(dN_1/dt)}{\partial N_2} = a_{12} = 0 \tag{14.7}$$

$$a_{11} \geq 0 \qquad a_{22} \leq 0 \qquad a_{12} = 0 \qquad a_{21} \geq 0 \tag{14.8}$$

Symbiosis

$$\frac{\partial(dN_2/dt)}{\partial N_1} = a_{21} \geq 0 \tag{14.9}$$

$$a_{11} \leq 0 \qquad a_{22} \leq 0 \qquad a_{12} \geq 0 \qquad a_{21} \geq 0 \tag{14.10}$$

14.2.1 PREDATION EXAMPLE

In Chapter 3 Volterra's prey-predator problem was mentioned as an example of this type of system. The more interesting nonlinear version is

$$\frac{dN_1}{dt} = a_{11}\,N_1 - a_{12}\,N_1\,N_2 \qquad \frac{dN_2}{dt} = -\,a_{21}\,N_2 + a_{22}\,N_1\,N_2 \tag{14.11}$$

where all of the as are positive numbers.

The construction of this model is derived from a line of reasoning as follows. Assume that two species N_1 and N_2 are interacting within a bounded environment. N_1 is the prey, and if it were alone in the environment it would grow in accordance with the equation

$$\frac{dN_1}{dt} = a_{11}\,N_1 \tag{14.12}$$

which is the mathematical statement of the law "The rate of growth is proportional to the amount present." N_2 is the predator, and, alone in the environment, it would die off as given by the equation

$$\frac{dN_2}{dt} = -\,a_{21}\,N_2 \tag{14.13}$$

since there would be no food for it to eat.

Next, assume that when both species are present the rate of encounters between them will be proportional to the product $N_1\,N_2$ per unit of time. The rate of growth of the prey is diminished by $-a_{12}\,N_1\,N_2$, because of the encounters, while that of the predator is increased by $+a_{22}\,N_1\,N_2$, because in each encounter the number of prey is one less, and one predator lives on.

The normalization

$$N_1 = \frac{a_{21}}{a_{22}}\,x \qquad N_2 = \frac{a_{11}}{a_{12}}\,y \tag{14.14}$$

gives

$$x^{-a_{21}}\,e^{a_{21}x} = C\,y^{a_{11}}\,e^{-a_{11}y} \qquad C = e^K \tag{14.15}$$

for the phase equation and then the trajectories are found from

$$\xi = (x^{-1}\,e^{x})^{a_{21}} \qquad \xi = C\,\eta \qquad \eta = (y\,e^{-y})^{a_{11}} \tag{14.16}$$

Examples showing the phase plots and numerically determined waveforms for this system are given by Professor Davis.[3]

A representative graph of the nonlinear system of Equations 14.11 is shown in Figure 14.1, and the phase diagram is given in Figure 14.2. The latter shows the graphical method of solution employed by Volterra. Note that the constant C is determined by the initial values N_{10} and N_{20} of N_1 and N_2. The Volterra systems mentioned above are drawn from the predation category of systems ecology. They have been variously cast as the problem of the sharks and soles, or coyotes and jackrabbits or, generally, as "big fish eat little fish," but Volterra himself wrote

---

[3]Davis, H. T., ***Introduction to Nonlinear Differential and Integral Equations***, U.S. Atomic Energy Commission, 1960.

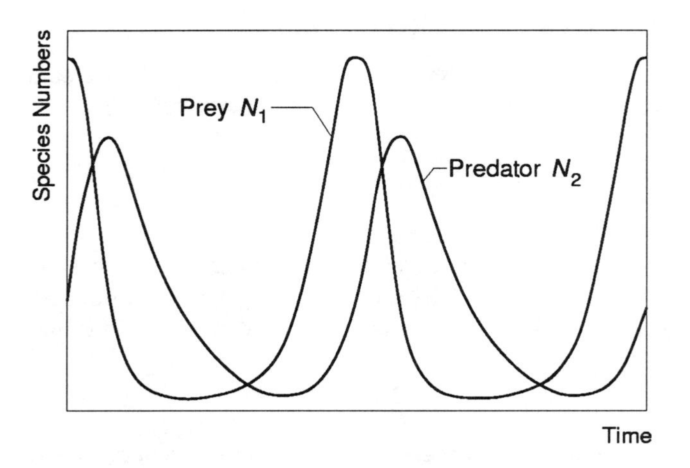

Figure 14.1: Nonlinear Prey-Predator Waveforms

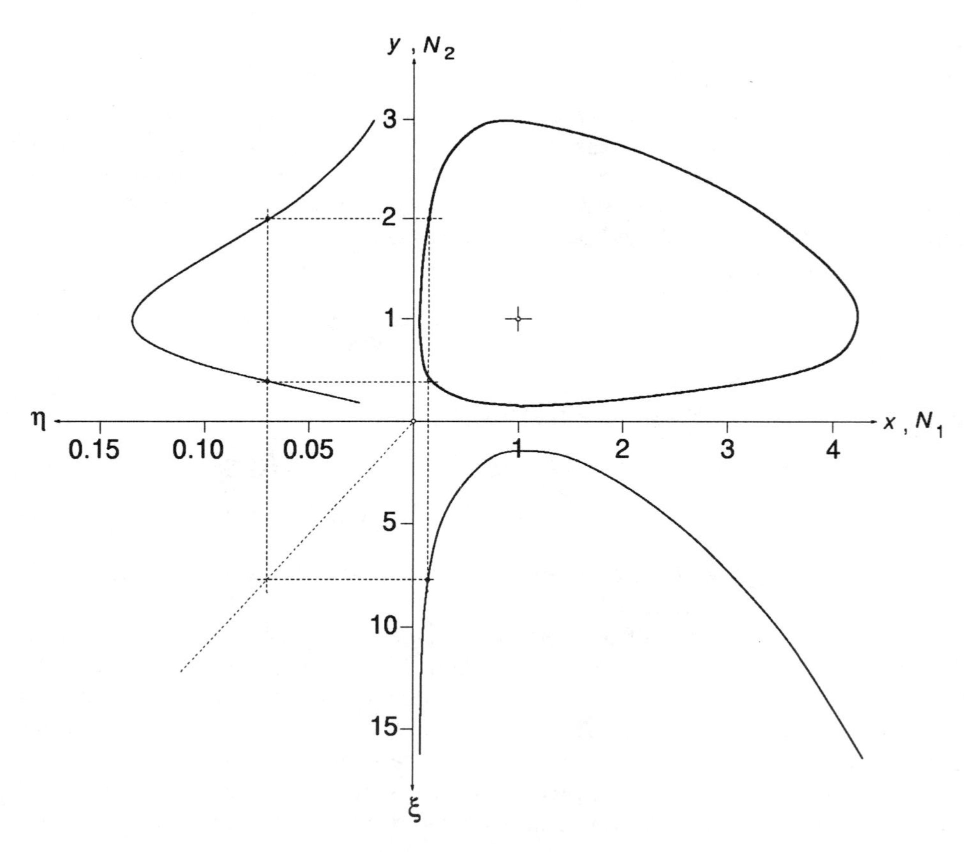

Figure 14.2: Nonlinear Prey-Predator Phase Space

"la lutte de la vie" (the struggle for life). As can be seen from either Figures 3.12 and 3.13 for the linear case or these of the present nonlinear model, there is a cyclic repetition of the magnitudes of the prey and the predator. Neither species ever dies out altogether; they rise again (like the South) out of the bones of their vanquished fellows, and the cycle repeats itself ad infinitum.

Examples drawn from the categories of competition and symbiosis are more directly relevant to the task of modeling the infrastructure. Attention turns immediately to this work.

14.3 DEMOGRAPHY

At the heart of all descriptions of society lies the enumeration of the socials themselves; in other words, demography. All other quantities, especially those which are the consequence of human technological activity, are directly coupled thereto. The ancient Romans initiated the census, primarily, we may suppose, as a basis for taxation and military conscription. Our own nation established it as a constitutional requirement on a decennial basis which began in 1790. Other governments in the modern world followed suit, so that there is available data on population growth for many major nations from the beginning of the nineteenth century to date.

The Reverend Thomas Malthus,[4] an English cleric, appears to be the progenitor of analytical population studies. His theory was combined with factors of economics, war, and pestilence, but basically assumed that in the absence of these latter, human numbers would grow unchecked in an exponential manner. His model corresponds to that of Equations 14.2 with $a_{12} = a_{21} = 0$, and is the same as that which is nowadays called the "Population Explosion."

The Malthusian theory created quite a stir in learned circles, and was soundly attacked by those who understood its implications by arguments of *reductio ad absurdam*. Quetelet[5] and Verhulst,[6] Belgian astronomer-statisticians, attempted to alter the theories by the addition of a factor which would cause the growth to conform to the reality of a saturating population. The equation they came to was the S-shaped logistic

$$\frac{dN}{dt} = a_{11}\,N - a_{12}\,N^2 \tag{14.17}$$

which has the solution

$$N = \frac{b}{e^{-at} + c} \tag{14.18}$$

with $a = a_{11}$, $b = a_{11}/a_{12}$, and c is an integration constant.

In 1924, biologist Raymond Pearl[7] made extensive calculations of the logistic curve fitted to the populations of many nations of the world, to cities, and to the global population itself. This highly important contribution was apparently

[4] Malthus, T. R., "An Essay on the Principle of Population, or a View of its Past and Present Effects on Human Happiness, with an Inquiry Into Our Prospects Respecting the Future Removal or Mitigation of the Evils Which It Occasions," Sixth Edition, 1826.

[5] Quetelet, A., *Sur l'Homme et le Developpement de Ses Facultes*, 1835.

[6] Verhulst, P. F., "Notice sur la Loi Que la Population Suit Dans Son Accroissement," Correspondence Mathematique et Physique, 1838.

[7] Pearl, R., *Studies in Human Biology*, Williams and Wilkins Company, 1924.

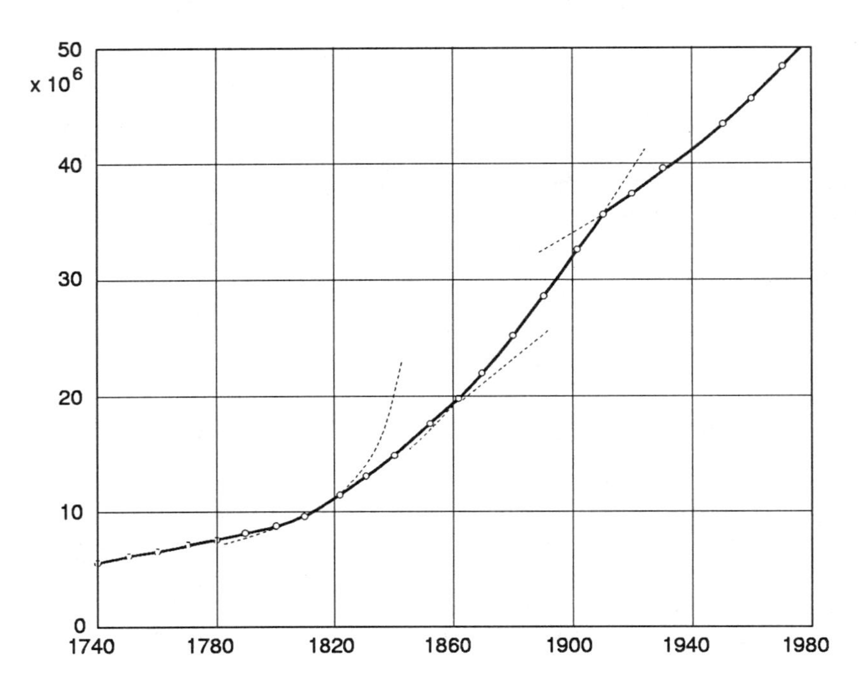

Figure 14.3: Population of England and Wales

discounted not long afterwards because data from the censuses of such major countries as the United States and the United Kingdom showed significant departures from the curve in the 1920s, 1930s, 1940s, and subsequent years.

However, a return to the logistic as a basically correct model of population growth follows from an updated analysis similar to Pearl's. Figure 14.3 shows the population of England and Wales from 1740 to 1980. The data are from Mitchel.[8] Figure 14.4 shows the structure of the English population in the Nineteenth Century and the fitting of logistics to the male, female, and total populations in the several regimes of this period.

It is most unnatural to find cusps in the graphs of natural phenomena, but in the proximity of the 1911 census there is definitely such a discontinuity of slope. A careful inspection of the graph reveals at least two other such breaks, of lesser severity, in the early 1820s and 1860s. Logistic curves can be fitted to the segments between these points with very great accuracy as Figure 14.3 shows.

If it is accepted that there are in fact "switching points" at which population growth changes from a regime with parameters a_1, b_1, c_1 to another with a_2, b_2, c_2, the system can be modeled as follows:

$$\frac{dN_+}{dt} = \alpha N_+ - \beta N_+ N^{\nearrow} \qquad \frac{dN^{\nearrow}}{dt} = \gamma N^{\nearrow} - \delta N_+ N^{\nearrow} \tag{14.19}$$

$$N = N_+ + N^{\nearrow} \tag{14.20}$$

[8] Mitchel, B. R., ***Abstract of British Historical Statistics***, Phyllis Deane, Collaborator, Cambridge Press, 1962, pp. 5, 6.

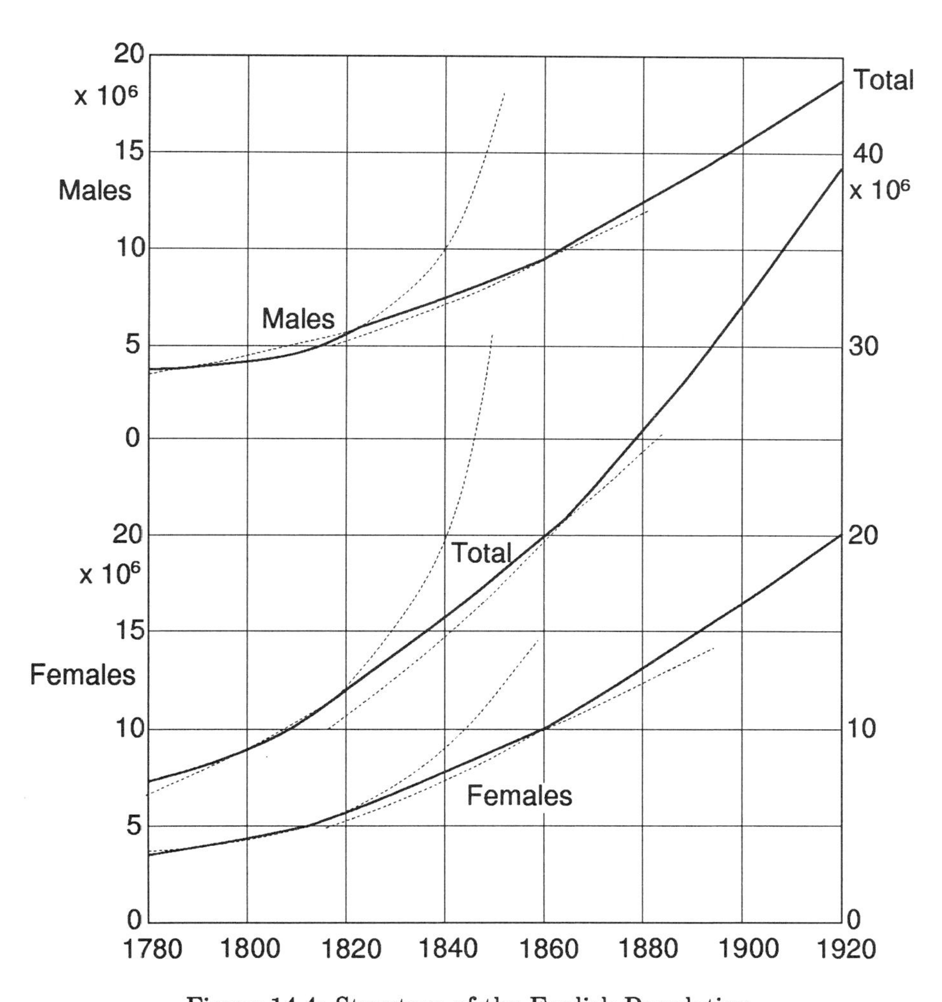

Figure 14.4: Structure of the English Population

Further, if the numbers of males and females in the total population are definite ratios, and if, as is actually the case, the ratios are each about 0.5, then the system is

$$N_+ = p\,N \qquad p+q=1 \qquad N^{\nearrow} = q\,N \tag{14.21}$$

$$\frac{dN}{dt} = \alpha\,N - \beta\,q\,N^2 \qquad \frac{dN}{dt} = \gamma\,N - \delta\,p\,N^2 \qquad \frac{dN}{dt} = a_{11}N - a_{12}\,N^2 \tag{14.22}$$

$$\alpha = \gamma = a_{11} \qquad \beta q = \delta p = a_{12} \tag{14.23}$$

and the curves for all three are logistics.

14.4 SWITCHING POINT THEORY

In the previous section it was shown that human population growth can be explained by logistic systems within regimes defined by sharp transitions in the parameters. A more adequate interpretation of this switching point behavior is desirable.

Assume that in each case there are protagonists N_1 and N_2 operative. (For the demographic problem, these are N_+ and $N^{\nearrow}$.) Let the basic system be

$$\frac{dN_1}{dt} = \alpha\,N_1 - \beta\,N_1\,N_2 \qquad \frac{dN_2}{dt} = \gamma\,N_2 - \delta\,N_1\,N_2 \tag{14.24}$$

$$N = N_1 + N_2 \tag{14.25}$$

and eliminate time as a parameter to obtain the phase trajectories

$$f(N_1) = \frac{(N_1)}{(N_1)_o}\,\gamma\,e^{-\delta[(N_1)-(N_1)_o]} = \frac{(N_2)}{(N_2)_o}\,\alpha\,e^{-\beta[(N_2)-(N_2)_o]} = f(N_2) \tag{14.26}$$

With the parameters $N_1 = (\gamma/\delta)x$, $N_2 = (\alpha/\beta)y$, differentiate Equations 14.24 above to find

$$x\,\frac{d^2x}{dt^2} = \left(\frac{dx}{dt}\right)^2 - \alpha\,\gamma\,x^2 + \gamma\,x\,(1-x)\frac{dx}{dt} + \alpha\,\gamma\,x^3 \tag{14.27}$$

$$y\,\frac{d^2y}{dt^2} = \left(\frac{dy}{dt}\right)^2 - \alpha\,\gamma\,y^2 + \alpha\,y\,(1-y)\frac{dy}{dt} + \alpha\,\gamma\,x^3 \tag{14.28}$$

These latter are highly nonlinear, and the most expeditious reduction is probably the numerical technique described in Professor Davis' book.[9] The phase equation, however, is directly informative.

For specific values of the parameters α, β, γ, and δ, and with specific initial conditions defined by N_1 and N_2, a continuous graph of x and y will be described by Equations 14.27 and Equation 14.28. Such a graph will be valid until one or more of the parameters change because of the action of some external agent not visible in the modeling of the system equations. It is appropriate to name this event a *switching point*. It is further appropriate to investigate the evidences of such a switching point in the system phase plane. Singular points in the phase plane are indicators of the existence of switching points.

[9]Davis, H. T., loc. cit., pp. 102-109.

14.4.1 SINGULAR POINTS IN THE PHASE PLANE

Consider the system of linear equations

$$\frac{dx}{dt} = C\,x + D\,y \qquad \frac{dy}{dt} = A\,x + B\,y \tag{14.29}$$

in which A, B, C, and D are constants.

Solve the first equation for dt, and insert it into the second one. Then cross-multiply to present the result in the form

$$\frac{dy}{dx} = \frac{A\,x + B\,y}{C\,x + D\,y} \qquad BC - AD \neq 0 \tag{14.30}$$

This is the linear version of the more general phase plane trajectories of Equation 14.26, above.

The waveform equations analogous to Equations 14.27 and 14.28 are easily obtained by differentiating both Equations 14.29 to get the identical equations for x and y

$$\ddot{x} - (B + C)\,\dot{x} + (BC - AD)\,x = 0 \tag{14.31}$$

$$\ddot{y} - (B + C)\,\dot{y} + (BC - AD)\,y = 0 \tag{14.32}$$

From the theory of linear differential equations, there is the *characteristic equation*

$$m^2 - (B + C)\,m + BC - AD = 0 \tag{14.33}$$

which has the roots m_1 and m_2, say, so that there are the two solution functions

$$x = a\,e^{m_1 t} + b\,e^{m_2 t} \qquad y = c\,e^{m_1 t} + d\,e^{m_2 t} \tag{14.34}$$

where a, b, c, and d are arbitrary constants to be found from the initial conditions.

Now solve these simultaneously for $e^{m_1 t}$ and $e^{m_2 t}$ and afterwards take the logarithms of both, which yields

$$m_1\,t = log\,k(d\,x - b\,y) \quad m_2\,t = log\,k(a\,y - c\,x) \quad 1/k = a\,d - b\,c = \Delta \tag{14.35}$$

Again, solve each equation for t, then eliminate t from the results to put the phase equation into the form

$$(d\,x - b\,y) = K(a\,y - c\,x)^{m_1/m_2} \qquad K = \Delta^{1-m_1/m_2} \tag{14.36}$$

For convenience, write Equation 14.36 in the form

$$w = K\,z^{m_1/m_2} \tag{14.37}$$

where, of course, $w = d\,x - b\,y$, $z = a\,y - c\,x$.

An investigation of the nature of Equation 14.37 as it relates to the roots of the characteristic equation will now be conducted.

Case 1. *Both roots are real, unequal, and negative.*

From Equations 14.34 it is seen that as t varies from $-\infty$ to $+\infty$, x and y continually decrease and finally approach zero as a limit. At the origin of the z coordinates the slope to the curves is also zero, so there is a family of curves all of which pass through the origin and are tangent to the line $a\,y = c\,x$. There is a ***stable node*** at $z = 0$.

Figure 14.5(a) illustrates such a family of curves for the condition $m_1 = -4$, $m_2 = -3$, for various values of K.

Case 2. *Both roots are real, unequal, and positive.*

From Equations 14.34 it is seen that when t varies from $-\infty$ to $+\infty$, x and y start at zero and increase indefinitely without limit. The slope of the curves at the origin is again zero, so the family of curves all pass through the origin and are tangent to the same line $a\,y = c\,x$ as in the previous case. However, there is now an ***unstable node*** at $z = 0$.

Figure 14.5(b) displays this family of curves for $m_1 = 4$, $m_2 = 3$, for various values of K.

Case 3. *Both roots are real, but differ in sign.*

From Equations 14.34 it is seen that both x and y will become infinite when t approaches either $+\infty$ or $-\infty$, so the condition is ***unstable***. Equation 14.37 for this case embodies a family of curves all of which are hyperbolas such that at $z = 0$, $w \to \infty$. The description for this condition is ***unstable saddle point***.

Figure 14.5(c) shows the picture for $m_1 = -4$, $m_2 = 3$, for various values of K.

Case 4. *The roots are pure imaginaries.*

For this case the solution of Equation 14.30 is a family of ellipses

$$A\,x^2 + 2\,B\,x\,y - D\,y^2 = K \qquad -A\,D \geq b^2 \tag{14.38}$$

The system performs endless cycles in a stable manner, so that the singular point at the origin is called a ***vortex point***. (Sometimes called a ***center***.)

Figure 14.5(d) shows, for $A = 1$, $B = -1$, $C = 1$, $D = 3$, and $m_1 = \sqrt{2}\,i$, $m_2 = -\sqrt{2}\,i$, this family of ellipses for various values of K. The origin is the vortex point.

Case 5. *The roots are complex conjugate numbers with the real part positive.*

For this case Equations 14.34 are converted to

$$x = a\,e^{m_1 t}\,cos(\mu t + b) \qquad y = c\,e^{m_2 t}\,cos(\mu t + d) \tag{14.39}$$

since $m_1 = \lambda + \mu\,i$, $m_2 = \lambda - \mu\,i$, say.

The solution of Equation 14.30 for this case is given by

$$\mu \log r = \lambda \tan^{-1}\frac{u}{v} + k \tag{14.40}$$

where $u = C\,x + D\,y - \lambda\,x$, $v = \mu\,x$, $r^2 = u^2 + v^2$. This equation describes a family of spirals. The amplitudes of x and y both increase without limit as $t \to +\infty$, so the case is an ***unstable focus***.

Figure 14.6(a) displays the curves for $\lambda = 1/2\pi$, $\mu = 1$, for several values of K. Here the equation is $r = K\, e^{\theta/2\pi}$, so that the spirals unwind from about the origin.

Case 6. *The roots are complex conjugate numbers with the real part negative.*

This case is like that of Case 5 except that as $t \to +\infty$, the amplitudes of x and y both decrease toward zero. There is again a family of spirals, but in this case the spirals wind inward upon the origin, so that there is a *stable focus*.

Figure 14.6(b) illustrates this case for $m_1 = -\frac{1}{2\pi} + i$, $m_2 = -\frac{1}{2\pi} - i$, or $r = K\, e^{-\theta/2\pi}$, for various values of K.

Case 7. *The roots are equal.*

In this case Equations 14.34 become, with $m_1 = m_2 = m$,

$$x = e^{m\,t}\,(a + b\,t) \qquad y = e^{m\,t}\,(c + d\,t) \tag{14.41}$$

Therefore, the system is *stable* if $m = \frac{1}{2}(B + C) \leq 0$, and *unstable* if $m \geq 0$. The curves may be straight lines, parabolas, or complex logarithmic figures.

Because the roots are equal, the constants in Equation 14.30 are required to satisfy $(B - C)^2 + 4\,A\,D = 0$.

Illustrations of this case are shown in Figures 14.6(c) and 14.6(d) for the conditions $A = -B$, $B = C$, and $D = 0$. The system trajectories are given by the equation

$$y = x\,(K - \ln\, x) \tag{14.42}$$

with these conditions.

The specific differential equations that govern the amplitudes of x and y as functions of time t are

$$\frac{dx}{dt} = A\,x \qquad \frac{d^2y}{dt^2} - 2\,B\,\frac{dy}{dt} + B^2\,y = 0 \tag{14.43}$$

Herein, both roots are equal to B, so that if B is negative, the system is *stable*, whereas if B is positive, it is *unstable*.

14.4.2 SEPARATRICES

Singular points in the phase plane dominate the character of the system trajectories. In fact, they often divide the plane into regions or theaters which are self-contained or exclusive. The saddle point, for instance, displays two asymptotic lines which are called *separatrices* because they literally separate the plane into four divisions. When the initial conditions are such that a phase trajectory starts in one of these quadrants, it cannot ever cross a separatrix and enter into another region. The importance of identifying separatrices cannot be overemphasized, obviously, but as these lines constitute singular solutions to the phase equations, it is often exceedingly difficult to obtain them.

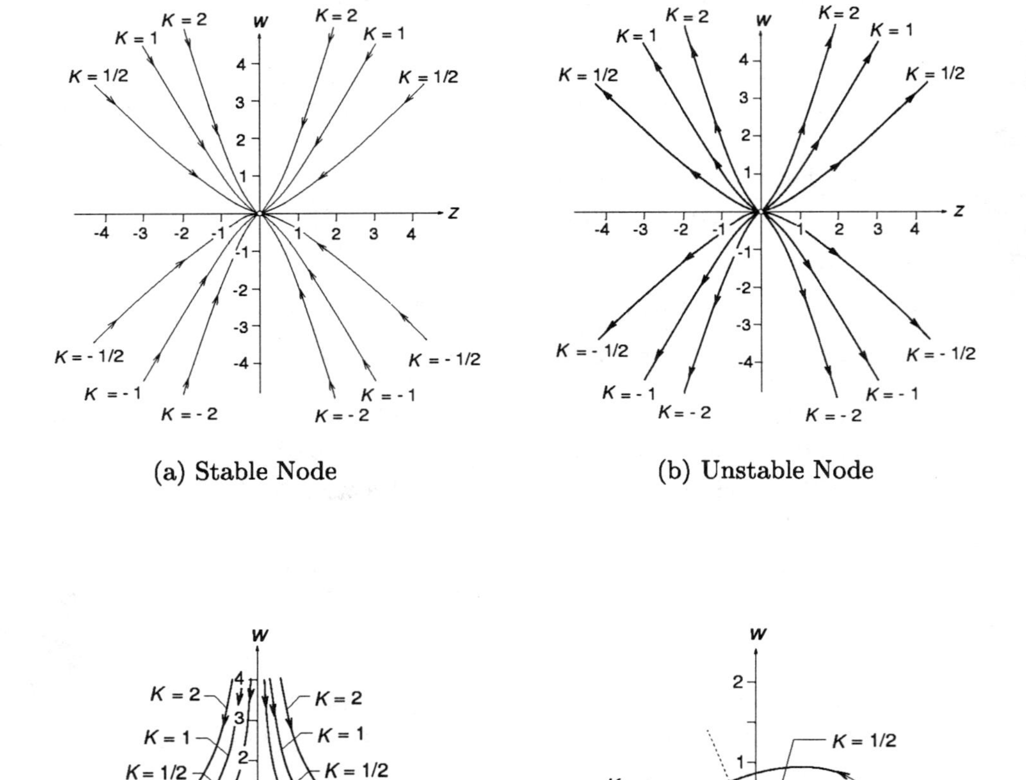

(a) Stable Node

(b) Unstable Node

(c) Saddle Point

(d) Vortex Point

Figure 14.5: Phase Plane Singular Points, I

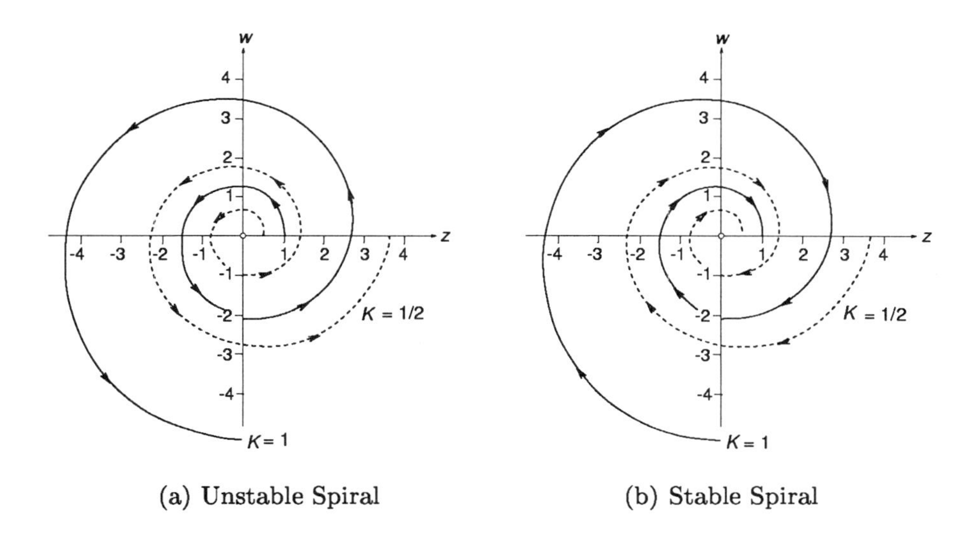

(a) Unstable Spiral (b) Stable Spiral

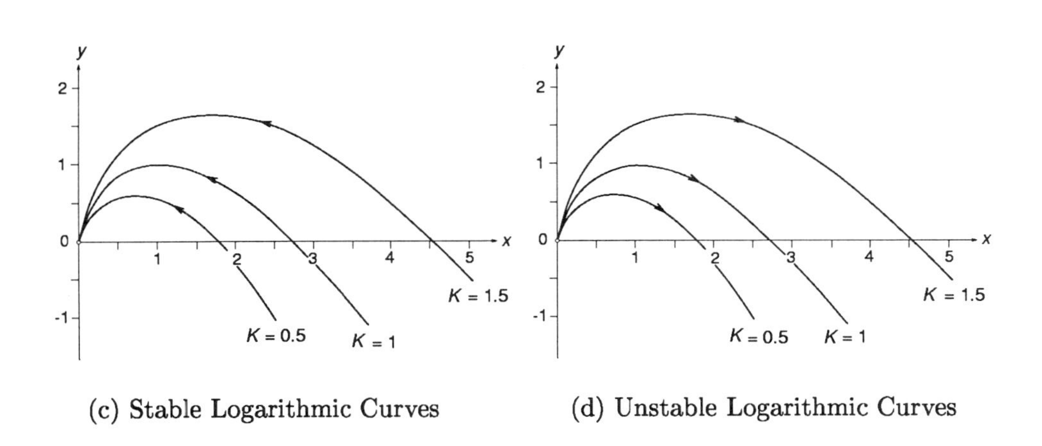

(c) Stable Logarithmic Curves (d) Unstable Logarithmic Curves

Figure 14.6: Phase Plane Singular Points, II

14.4.3 THE LIAPOUNOV STABILITY THEOREM

The example of the previous section in which the stability of a system of linear equations was considered can be extended to general nonlinear systems by a similar analysis. The principal contributors to this work were Poincaré[10] and Liapounov.[11] The general stability criterion is appropriately called the *Liapounov Theorem.*

The system under study can be written

$$\frac{dy}{dt} = P(x,y) \qquad \frac{dx}{dt} = Q(x,y) \tag{14.44}$$

in which $P(x,y)$, $Q(x,y)$ are suitable general functions of the variables x, y in the phase plane.

Since it is the purpose of the analysis to investigate the behavior of the system in the neighborhood of the singular points in the phase plane, it will be appropriate to make the transformation to new coordinates $x = X+\xi$, $y = Y+\eta$, where ξ and η are assumed to be variables which describe the small departures of the system point away from a singular point located at X, Y. Thus

$$\begin{aligned}\frac{d\eta}{dt} &= P(\xi+X,\eta+Y) = P(X,Y) + \frac{\partial P(X,Y)}{\partial x}\xi + \frac{\partial P(X,Y)}{\partial y}\eta \\ &+\frac{1}{2!}\left[\frac{\partial^2 P(X,Y)}{\partial x^2}\xi^2 + 2\frac{\partial^2 P(X,Y)}{\partial x \partial y}\xi\eta + \frac{\partial^2 P(X,Y)}{\partial y^2}\eta^2\right] + \cdots \\ &= P(X,Y) + \frac{\partial P(X,Y)}{\partial x}\xi + \frac{\partial P(X,Y)}{\partial y}\eta + \phi(\xi,\eta)\end{aligned} \tag{14.45}$$

$$\begin{aligned}\frac{d\xi}{dt} &= Q(\xi+X,\eta+Y) = Q(X,Y) + \frac{\partial Q(X,Y)}{\partial x}\xi + \frac{\partial Q(X,Y)}{\partial y}\eta \\ &+\frac{1}{2!}\left[\frac{\partial^2 Q(X,Y)}{\partial x^2}\xi^2 + 2\frac{\partial^2 Q(X,Y)}{\partial x \partial y}\xi\eta + \frac{\partial^2 Q(X,Y)}{\partial y^2}\eta^2\right] + \cdots \\ &= Q(X,Y) + \frac{\partial Q(X,Y)}{\partial x}\xi + \frac{\partial Q(X,Y)}{\partial y} + \psi(\xi,\eta)\end{aligned} \tag{14.46}$$

in which $\phi(\xi,\eta)$, $\psi(\xi,\eta)$ are functions of higher order in ξ and η. For small (infinitesimal) motions, these terms can be assumed to be negligible when compared to the remaining ones, hence

$$\frac{d\eta}{dt} = P(X,Y) + \frac{\partial P(X,Y)}{\partial x}\xi + \frac{\partial P(X,Y)}{\partial y}\eta \tag{14.47}$$

$$\frac{d\xi}{dt} = Q(X,Y) + \frac{\partial Q(X,Y)}{\partial x}\xi + \frac{\partial Q(X,Y)}{\partial y}\eta \tag{14.48}$$

The singular points of Equations 14.47 and 14.48 are the places where the curves $P(X,Y) = 0$ and $Q(X,Y) = 0$ intersect. At such singular points, X_i, Y_i,

---

[10]Poincaré, op. cit.

[11]Liapounov, A., "Problème général de la stabilité du mouvement," *Annales de Toulouse*, Volume 9, Number 2, 1907, pp. 203-474. Originally published in Russian in 1892.

Equations 14.47 and 14.48 become

$$\frac{d\eta}{dt} = A\,\xi + B\,\eta \tag{14.49}$$

$$\frac{d\xi}{dt} = C\,\xi + D\,\eta \tag{14.50}$$

where

$$A = \frac{\partial P(X_i, Y_i)}{\partial X} \quad B = \frac{\partial P(X_i, Y_i)}{\partial Y} \quad C = \frac{\partial Q(X_i, Y_i)}{\partial X} \quad D = \frac{\partial Q(X_i, Y_i)}{\partial Y} \tag{14.51}$$

Now the equations are like Equations 14.29; accordingly,

$$\ddot{\eta} - (B + C)\,\dot{\eta} + (BC - AD)\,\eta = 0 \tag{14.52}$$

$$\ddot{\xi} - (B + C)\,\dot{\xi} + (BC - AD)\,\xi = 0 \tag{14.53}$$

and the characteristic equation is

$$m^2 - (B + C)\,m + B\,C - A\,D = 0 \tag{14.54}$$

If the roots are m_1 and m_2, the solution functions are

$$\eta = a\,e^{m_1 t} + b\,e^{m_2 t} \qquad \xi = c\,e^{m_1 t} + d\,e^{m_2 t} \tag{14.55}$$

and the phase equation is

$$(d\eta - b\,\xi) = K(a\,\xi - c\,\eta)^{m_1/m_2} \qquad K = \Delta^{1-m_1/m_2} \tag{14.56}$$

Therefore, the stability criterion can be formulated as follows.

Liapounov Theorem

> ***In the neighborhood of a singular point of the system defined by Equations 14.44, the stability characteristics of the solution are determined by the characteristic roots of Equation 14.54 in the same sense as they characterize the solutions of the linear system as described in the several cases of the previous section.***

As in the former (linear) instance, there are five cases to be investigated. These are: I. *Both roots are real and positive.* II. *Both roots are real and negative.* III. *Both roots are real, but differ in sign.* IV. *The roots are complex conjugate numbers.* V. *Both roots are pure imaginaries.* The singular points associated with these cases are identified as *spirals*, (unstable), *nodes*, (stable), *saddle points*, (unstable), *focus*, (stable or unstable, according to sign), and *center*, (stable vortex motion).

For a more extended discussion of each of these cases, please again see Professor Davis' book.[12]

Illustrations of the singular points are shown in Figures 14.7(a), (b), (c), and (d).

---

[12]Davis, H. T., loc. cit., pp. 317-322.

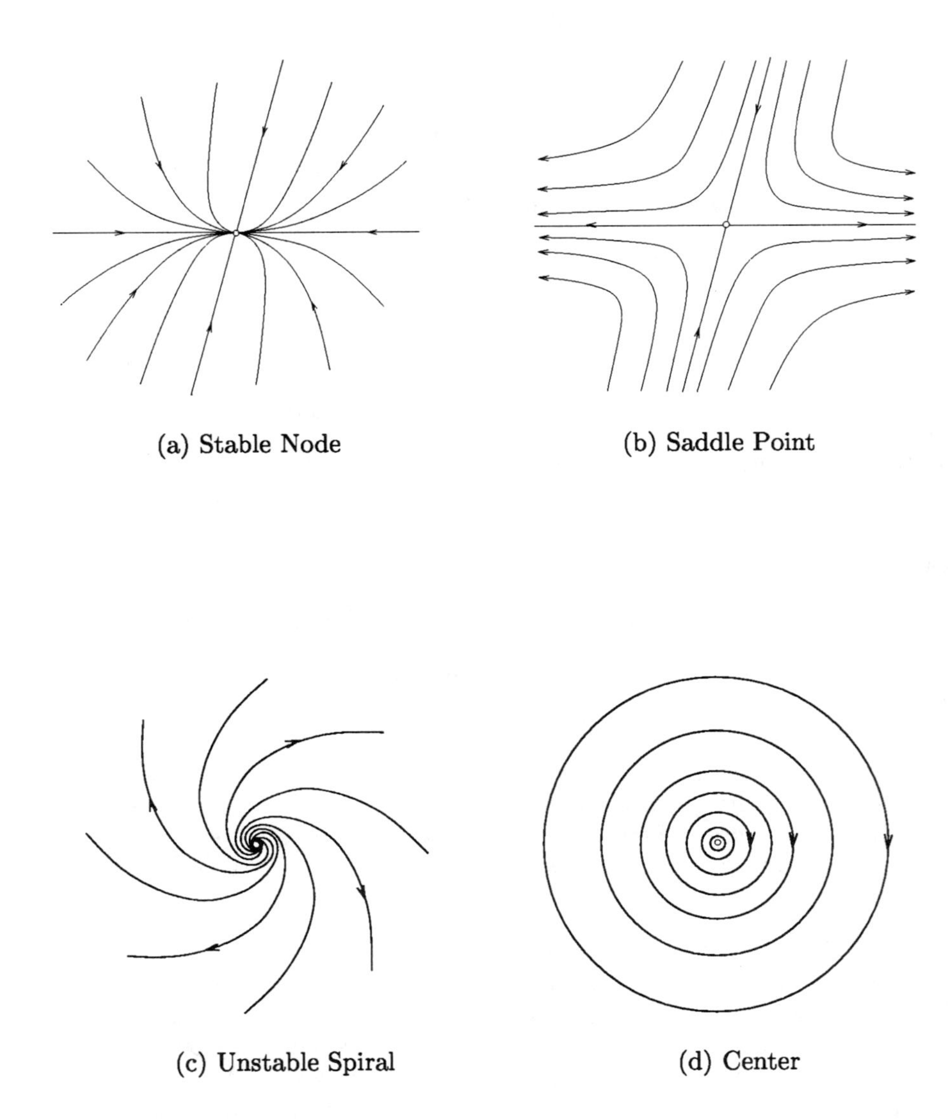

(a) Stable Node

(b) Saddle Point

(c) Unstable Spiral

(d) Center

Figure 14.7: Phase Plane Singular Points

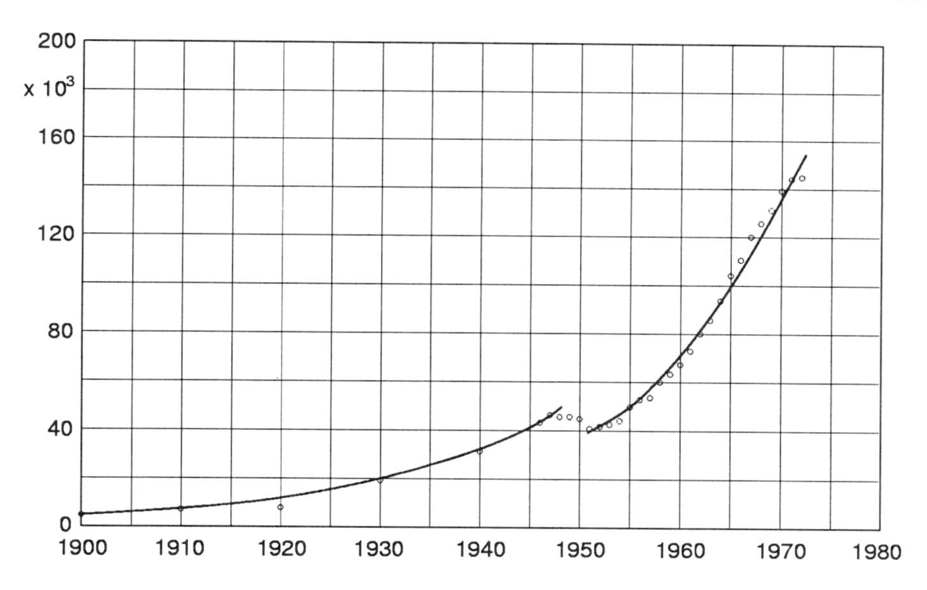

Figure 14.8: Student Enrollment in North Carolina Colleges

14.4.4 COLLEGE ENROLLMENT EXAMPLE

As an example, take the case of student enrollment in North Carolina Colleges as shown in Figure 14.8.[13] During the period 1900 through 1947, this enrollment followed very closely the logistic curve shown as fitted with the least squares technique. A general logistic curve is also shown for the period 1951 through 1972, but it is readily apparent that the fit is nowhere adequate by comparison with other fittings. Figure 14.9 gives the same quantity fitted by a number of logistic segments with appropriate switching points.

Let the protagonists for this example be N_1, the enrollment in private universities, and N_2 that for the public ones, and consider the logistic segment from 1951 to 1956. During this time the ratio of N_1 to N_2 remained virtually constant with N_1 at 47% and N_2 at 53% of the total. Figure 14.10 shows the phase space and trajectories calculated from the data for the period, and the waveforms are displayed in Figures 14.11 and 14.12. The phase plane is dominated by an unstable saddle point near $N_1 = 40 \times 10^3$, $N_2 = 40 \times 10^3$.

It is clear that N_1 reaches a maximum before N_2 does, and since both must follow the projected waveforms and N_2 cannot do this and maintain its prescribed relationship to N_1, a discontinuity or switching point must occur. For the given data, the calculated switching point is early autumn of 1956. In the actual event, the regime changed in midsummer, which, since college enrollment is "digitized" to the opening of the Fall Semester, is a very reasonable agreement.

[13]Davis, J. B., Statistical Abstract of Higher Education in North Carolina, *The University of North Carolina*, 1973.

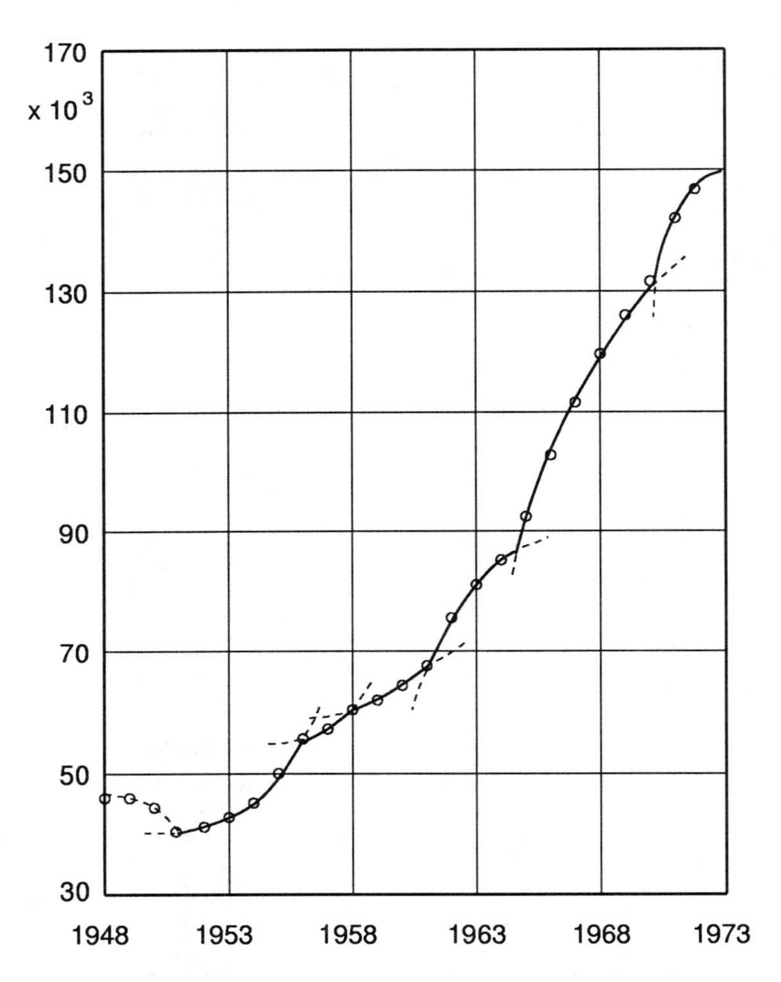

Figure 14.9: North Carolina College Student Enrollment

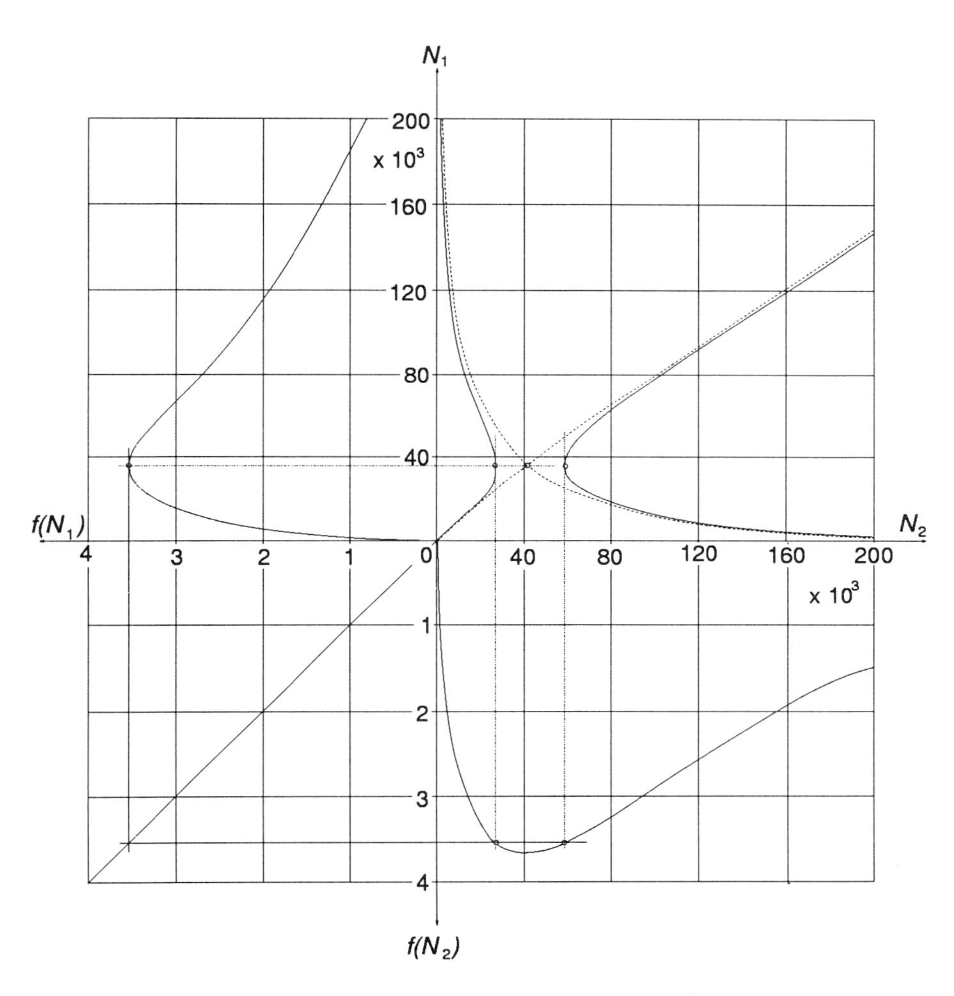

Figure 14.10: College Enrollment Phase Space

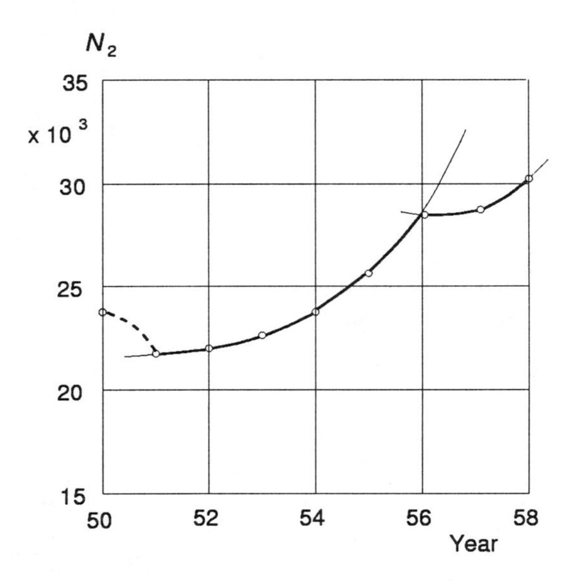

Figure 14.11: N_2 Enrollment Trajectory

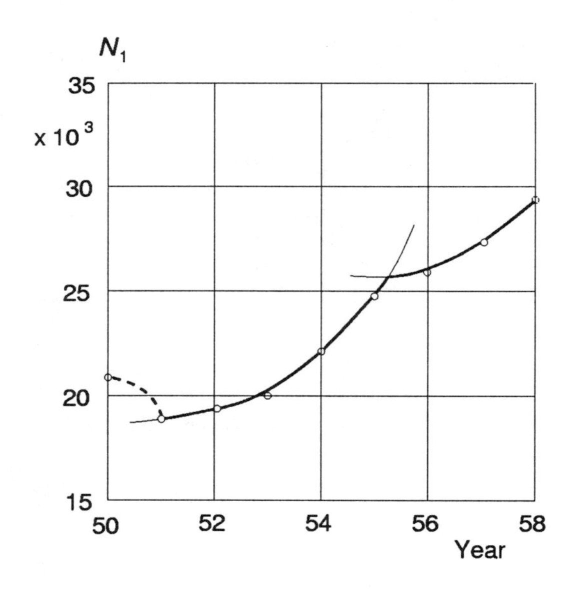

Figure 14.12: N_1 Enrollment Trajectory

14.5 FITTING LOGISTICS TO DATA

Professor Pearl[14] showed how to fit a logistic curve to data. If one has a collection of data, which consists of pairs of values of, say, population and time, one should try to find three pairs which span the spread of the data in such a way that one pair is at one end of the range, another pair is at the other end, and the third pair is near the center of the range. Select pairs which are equally-spaced in time (i.e., are symmetric about the center.)

The logistic will pass exactly through the three data points if the parameters a, b, and c are chosen so that

$$k = \frac{2\,N_0\,N_1\,N_2 - N_1^2(N_0 + N_2)}{N_0\,N_2 - N_1^2} \tag{14.57}$$

$$c = \frac{N_0}{k - N_0}, \qquad b = k\,c \tag{14.58}$$

$$-a = \frac{1}{t_1}\ln\frac{N_0(k - N_1)}{N_1(k - N_0)} \tag{14.59}$$

In these relationships N_0 is the value of N at $t = t_0$, N_1 is the value of N at t_1, the intermediate point, and N_2 corresponds to t_2 at the far end of the data range. As a matter of fact, it is not even necessary to use real data points in order to pass a logistic through a cloud of data. One can actually manufacture representative points within the field of the data, and compute synthetic values of a, b, and c to allow a curve to be drawn.

The tactics mentioned above, specious though they may be, serve as a starting point for an accurate fitting of the data in question. Thereafter, a least-squares technique (also perfected by Professor Pearl) furnishs a rational—and much better—fit. The so-called "three-point-fit" is a first-order approximation which allows the least-squares procedure to begin. See Appendix B.

Pearl's computations were carried out in a tabular form typical of the state-of-the-art in his time. The laborious task of reducing the numbers is today a modest effort on the machines available to everyone. A suitable program (written in Pascal source code) is installed on *eos* net. It is user-friendly, keyboard-interactive, with prompts for appropriate inputs as it runs. The outputs include the parameters of the three-point fit, the same parameters obtained by least-squares technique, and computed data points for a selected range of times within the field.

The program is *logcur.go* in the locker *mcdonald_src*.

14.5.1 DEMOGRAPHY EXAMPLE

Given below are the data for the population of England and Wales for the years 1911 through 1971. The British censuses are taken on the year following the decennial year. Note that there is no data point for 1941, when the country was under siege during World War II. See also Figure 14.3.

[14]Pearl, R., op. cit.

1911	36,070,000
1921	37,887,000
1931	39,952,000
1941	Not taken
1951	43,758,000
1961	46,072,000
1971	48,594,000

The three-point fit procedure would indicate that the figure for 1941 would be named N_1, but since there is no such number available, the average of the 1931 and 1951 years is used instead. This number is 41,855,000. Then

$$N_0 = 36,070,000 \tag{14.60}$$

$$N_1 = \tfrac{1}{2}(39,952,000 + 43,758,000) = 41,855,000 \tag{14.61}$$

$$N_2 = 48,594,000 \qquad t_1 = 30 \tag{14.62}$$

$$k = \frac{2\times 36.07\times 10^6\times 41.855\times 10^6\times 48.594\times 10^6 - (41.855\times 10^6)^2(36.07\times 10^6 + 48.59\times 10^6)}{36.07\times 10^6\times 48.59\times 10^6 - (41.855\times 10^6)^2}$$
$$= -1,685,648,415.2 \tag{14.63}$$

$$c = \frac{36,070,000}{-1,685,648,415.2 - 36,070,000} = -0.02095000 \tag{14.64}$$

$$b = kc = (-1,685,648,415.2)(-0.02095000) = 3,531,334 \tag{14.65}$$

$$-a = \tfrac{1}{30}\ln\frac{36,070,000(-1,685,648,415.2 - 41,855,000)}{41,855,000(-1,685,648,415.2 - 36,070,000)}$$
$$= \tfrac{1}{30}\ln 0.86468034 = -0.00484651 \tag{14.66}$$

Therefore

$$N = \frac{35,314,334}{e^{-0.00484651t} - 0.02095000} \tag{14.67}$$

Although this equation represents the data with very good accuracy, there is still the better least squares result which starts from the above as a first approximation. Thus

$$N = \frac{26,007,657}{e^{-0.0033173321t} - 0.280899108} \tag{14.68}$$

t	Year	N	% Error
0	1911	36,166,910	0.27
10	1921	37,885,982	− 0.003
20	1931	39,711,967	− 0.60
30	1941	41,654,054	NA
40	1951	43,722,507	− 0.08
50	1961	45,928,816	− 0.31
60	1971	48,285,899	− 0.63

is that better representation. (See Appendix B.)

14.5.2 METROPOLITAN AIRPORT EXAMPLE

The Raleigh-Durham International Airport furnishes a very good example for the application of logistic modeling analysis to a system which is an important infrastructural element that depends directly upon the demography of the Research Triangle Region. As such, it can be taken to be representative of the transportation needs of the society in this area.

The primary data needed to establish this model are the annual boardings of passengers through RDU terminals for as many years as are readily available. That is, from 1955 forward. Prior to that date, service to the Raleigh area was supplied through another airfield.

Table 14.1 displays the data,[15] which is also depicted in Figure 14.13. A first inspection of this graph reveals that there are evidently two distinct overall growth regimes: the 30 year period from 1955 through 1985, and the three years 1985 through 1988. The latter regime corresponds to the time since the opening of a new 10,000-ft runway, and since there are three data points available, the obvious thing to do is to pass a single logistic curve through these three points. The parameters associated with this equation are

$$a = 0.4917155869 \qquad b = 1680778 \qquad c = 0.0806427257 \tag{14.69}$$

and there is no error of representation because the curve is made to pass exactly through the three points.

TABLE 14.1

Enplaned Passengers,
Raleigh-Durham International Airport

Year	Number	Year	Number
1955	101,818	1972	571,385
1956	114,685	1973	604,079
1957	126,966	1974	633,205
1958	124,014	1975	597,281
1959	141,214	1976	635,107
1960	134,836	1977	711,657
1961	144,687	1978	809,917
1962	141,878	1979	927,873
1963	169,568	1980	886,848
1964	203,782	1981	868,216
1965	243,030	1982	942,205
1966	274,835	1983	1,147,571
1967	357,101	1984	1,319,680
1968	427,583	1985	1,381,688
1969	477,719	1986	1,555,350
1970	491,416	1987	2,428,101
1971	521,833	1988	3,696,712

[15]Courtesy Mr. John C. Brantley, Manager, RDU International Airport.

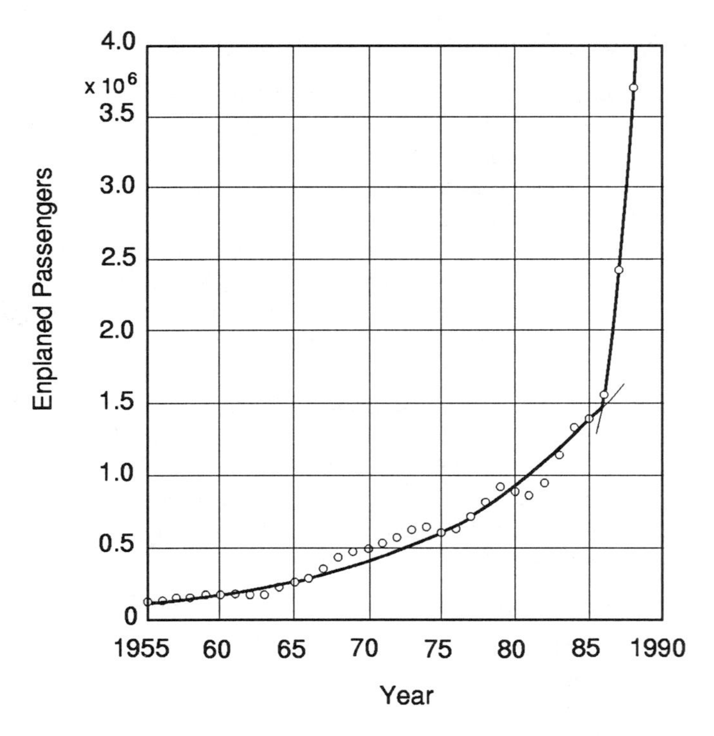

Figure 14.13: Raleigh-Durham International Airport

With respect to the three decades of the first regime, it would seem that, at least as a rough approximation, a single equation might be employed to describe the growth. The three-point fit provides the numbers

$$a = 0.1187641812 \qquad b = 106801 \qquad c = 0.0489418457 \tag{14.70}$$

which are utilized as the starting point for a least-squares fit with the parameters

$$a = -0.1247809305 \qquad b = 87854 \qquad c = 0.0470317377 \tag{14.71}$$

Although the single equation for this regime provides reasonable accuracy, a better depiction can be had by subdividing the epoch into a number of lesser regimes. A three-point fit is made for each of these, followed by a least-squares fitting, with the results shown in Figure 14.14 and Table 14.2.

Only three data points within this time-span fail by as much as 5% of accurate representation by the appropriate equation, and the data for most years lie within 2% of their model numbers.

Perhaps the most interesting phase is that of the 1981-1985 epoch. There are actually two regimes within this period, one of rapid increase and one of equally rapid saturation. The protagonists—in this case the major airline groups which are competing for the relevant shares of the growing market—are causing a 'hunting-like' phenomenon to be plainly evident. In fact, this suggests that a similar condition is probably operative throughout much of the growth period

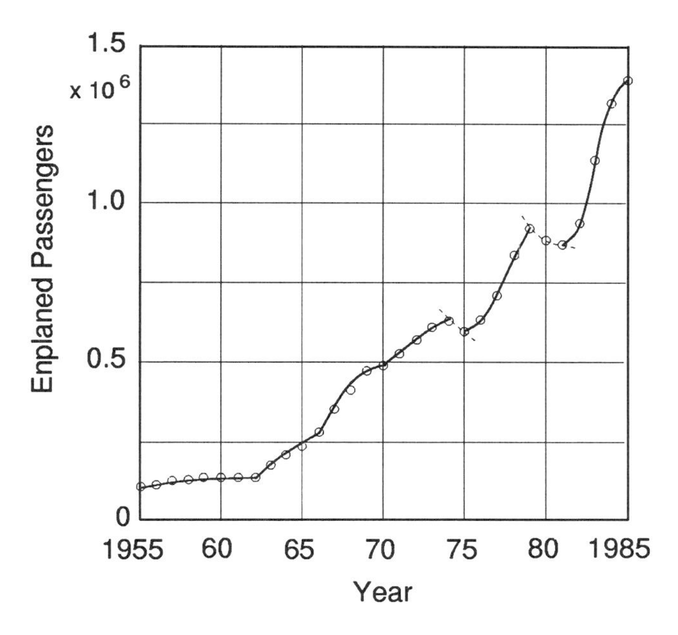

Figure 14.14: RDU Airport 1955-1985

from 1955 through 1985 (see Figure 14.14). There is a clear indication for further investigation into the fine-scale details of this regime. See Figure 14.15.

TABLE 14.2

RALEIGH-DURHAM AIRPORT EQUATIONS

	Logistic Curve Parameters		
Years	a	b	c
1955-1962	− 0.1838445891	248,159	1.3815686516
1962-1966	− 0.2642693140	189,285	0.3396557500
1966-1970	− 0.6903570042	572,092	1.0888160669
1971-1974	− 0.4337888498	2,228,052	3.2670422692
1975-1977	0.5297390952	− 7,004,784	− 12.7277867720
1977-1979	− 0.0826041694	465,074	− 0.3464910016
1981-1983	0.7419150243	− 25,538,167	− 31.5144849550
1983-1985	− 1.2065229863	6,166,184	4.3732480330

The most dramatic aspect of the airport model is the sharp cusp at the switching point in the 1985–1986 year. As was previously mentioned, a 10,000-ft runway was opened, on April 1, 1986, as a matter of fact, and this was followed soon after (June 15, 1987) by the establishment of American Airlines' North-South Hub. It is difficult to extrapolate the growth phenomenon with confidence in a time of such rapid change, but it is clear that a new order of growth of numbers is involved.

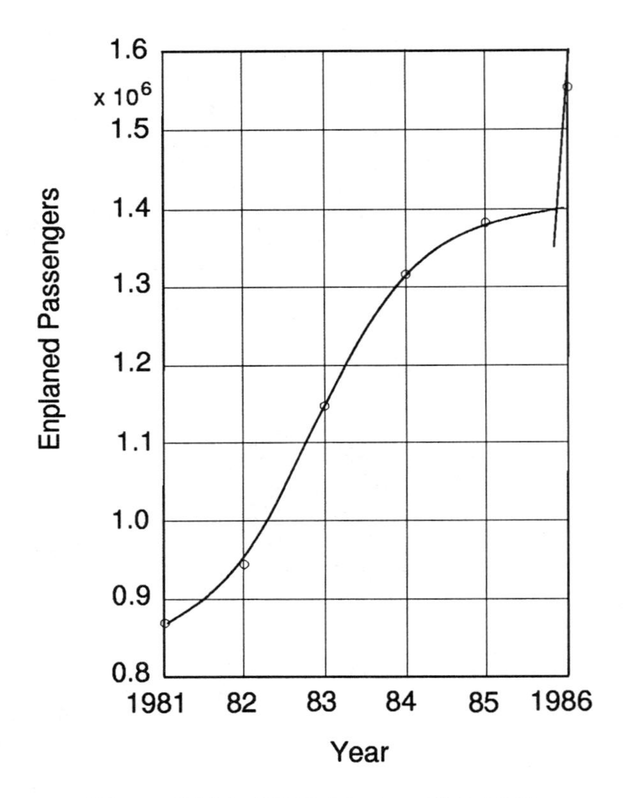

Figure 14.15: RDU Airport 1981-1986

14.5.3 COUNTY FINANCES EXAMPLE

Table 14.3 lists the revenues and expenditures of the County of Wake in North Carolina from 1973 to 1988. The data are furnished by Mr. Wally Hill of the Wake County Manager's Office. The graphical versions of these are shown in Figures 14.16 and 14.17. It is immediately seen that there is growth, and that there are cleanly-divided regimes within the overall pattern.

There is a clear indication of a distinct change in both the revenues and expenditures in the year 1984.

The county budgets data can be fitted with high accuracy by logistic curves in the two regimes defined above.

Wake County Revenues

1974-1984

$$a = -0.4181960961 \quad b = 235068000 \quad c = 1.0630523637 \tag{14.72}$$

1984-1988

$$a = -0.1517476874 \quad b = 38596000 \quad c = 0.1158040506 \tag{14.73}$$

Wake County Expenditures

1974-1984

$$a = -0.1349111712 \quad b = 35663000 \quad c = 0.0524508927 \tag{14.74}$$

1984-1988

$$a = -0.2289698455 \quad b = 165439000 \quad c = 0.4807617058 \tag{14.75}$$

The accuracy of fit for the revenues is within 3% throughout the range, and, except for two years at 5% and 7%, is within 4% for the expenditures. Both revenues and expenditures in the 1984-1988 years are described to within 1.5%!

The curves plotted with the above parameters are superposed on the data in Figures 14.16 and 14.17.

Protagonists can be identified in the case of revenues as General Revenues and Functional Revenues. Figures 14.18 and 14.19 show the division of the revenues into these two categories, and Figures 14.20 and 14.21 display the phase planes for the 1974-1984 and 1984-88 regimes. The switching points are clearly revealed in these two graphs.

This system has unstable singular points at the normalized points (0,0) and (1,1) in the phase plane. The origin is a node, and (1,1) is a saddle point. Therefore the system point grows away from the origin toward the saddle point, but is deflected away from the saddle. Because it is not permitted to cross a separatrix, the system trajectory encounters a mandatory switching point, where a new system, continuous with the former, comes into being with new parameters.

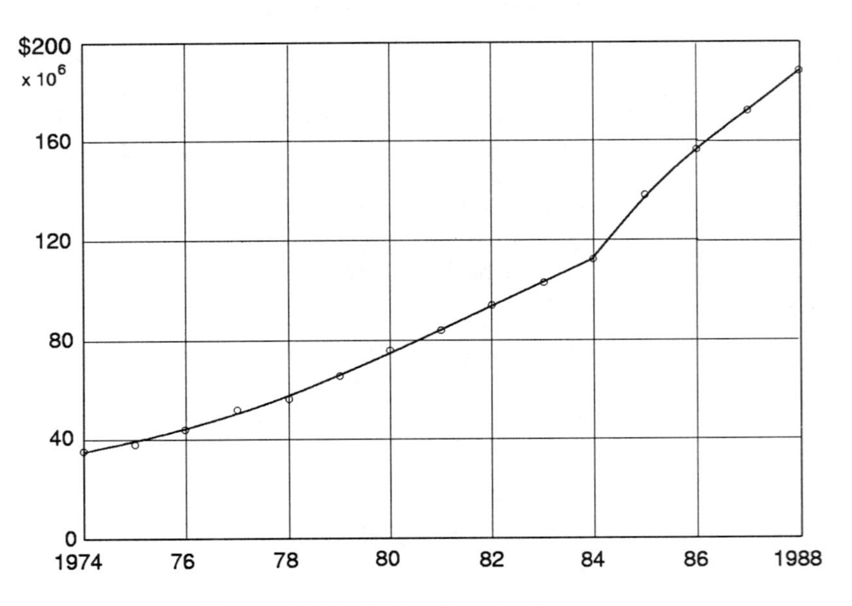

Figure 14.16: Wake County Revenues

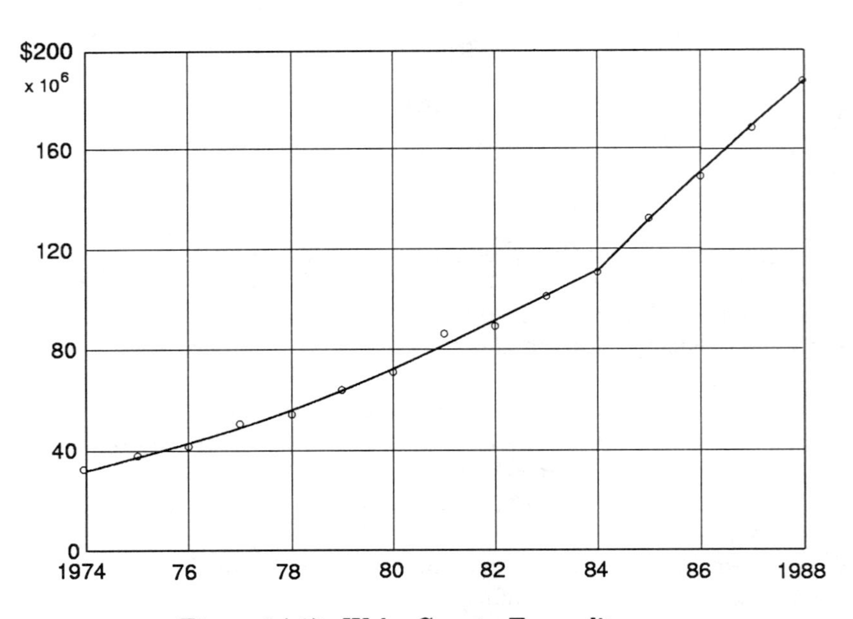

Figure 14.17: Wake County Expenditures

TABLE 14.3

WAKE COUNTY FINANCES

	Revenues ($)			Expenditures ($)	
Fiscal Year	General Revenues	Functional Revenues	Total Revenues	General Admin.	Human Services
73-74	31,157,911	3,872,402	35,030,316	3,021,848	6,710,829
74-75	33,902,659	5,816,343	38,719,002	3,137,431	9,500,411
75-76	37,648,610	6,587,314	44,235,924	3,841,862	10,926,829
76-77	43,778,583	8,601,156	52,379,939	3,633,170	14,529,662
77-78	47,228,900	9,216,095	56,444,995	4,077,049	16,097,346
78-79	55,164,940	10,564,725	65,729,665	5,201,528	17,707,623
79-80	61,870,702	13,986,665	75,857,367	6,154,955	20,729,042
80-81	68,960,939	15,927,562	84,888,501	6,548,744	24,110,631
81-82	77,469,075	17,253,026	94,722,101	7,468,342	26,563,895
82-83	83,509,946	20,437,846	103,947,792	7,933,381	31,336,345
83-84	89,458,082	23,798,575	113,256,657	7,929,383	34,114,554
84-85	112,225,936	26,596,673	138,822,609	9,664,031	37,491,754
85-86	126,468,520	29,488,606	155,957,126	10,689,140	41,622,951
86-87	142,605,417	30,827,275	173,432,692	11,141,103	47,280,043
87-88	155,330,285	33,658,542	188,988,827	12,692,601	53,684,930

	Expenditures ($)				
Fiscal Year	Education, Culture	Community Devel.	Public Safety	Other	Total Spending
73-74	13,148,725	73,443	911,193	10,879,486	34,745,524
74-75	13,453,502	400,718	1,049,981	11,284,383	38,826,426
75-76	16,285,305	321,590	997,142	9,053,030	41,425,758
76-77	19,560,217	393,464	1,597,492	10,865,049	50,579,054
77-78	23,631,773	600,675	2,237,047	7,553,915	54,197,805
78-79	27,805,775	742,504	2,705,566	10,431,837	64,594,833
79-80	30,793,180	835,569	3,451,777	9,357,117	71,321,640
80-81	35,137,948	986,037	3,998,592	15,900,159	86,682,111
81-82	35,722,426	1.034,269	4,451,842	14,088,748	89,329,522
82-83	40,372,112	1,209,348	5,154,024	16,719,481	102,724,691
83-84	45,318,249	1,592,917	5,559,490	16,483,661	110,998,254
84-85	48,963,658	2,113,662	6,632,193	26,670,495	131,535,793
85-86	54,706,808	3,083,584	7,536,714	29,481,222	147,120,419
86-87	64,775,278	4,106,760	9,372,537	32,728,466	169,404,187
87-88	72,833,867	5,530,771	11,277,560	31,435,968	187,455,697

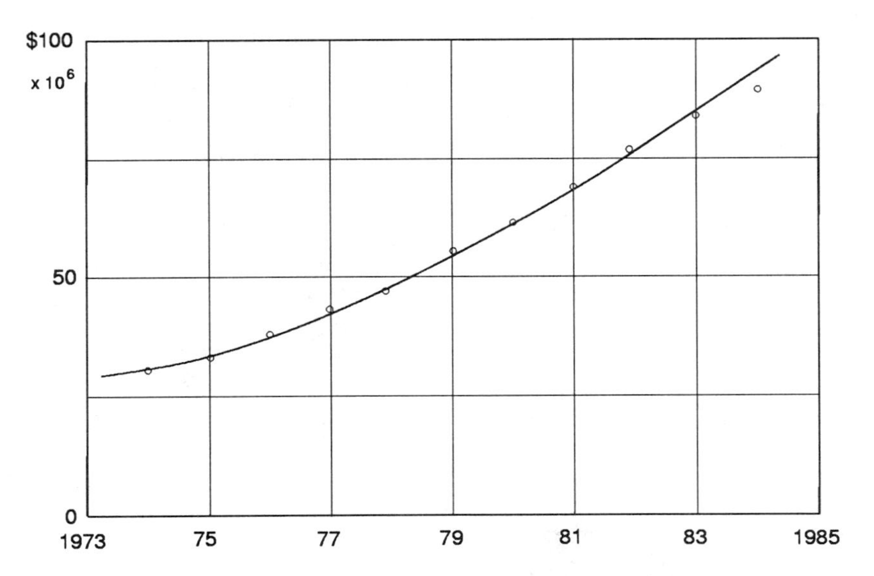

Figure 14.18: Wake County General Revenues

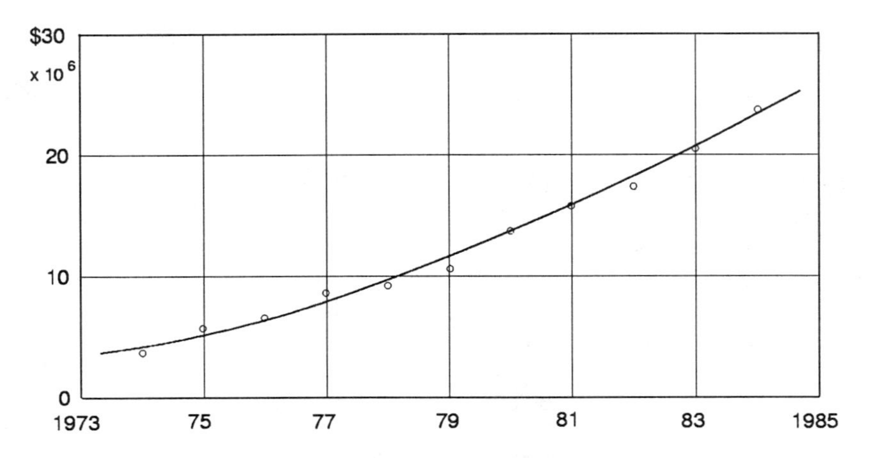

Figure 14.19: Wake County Functional Revenues

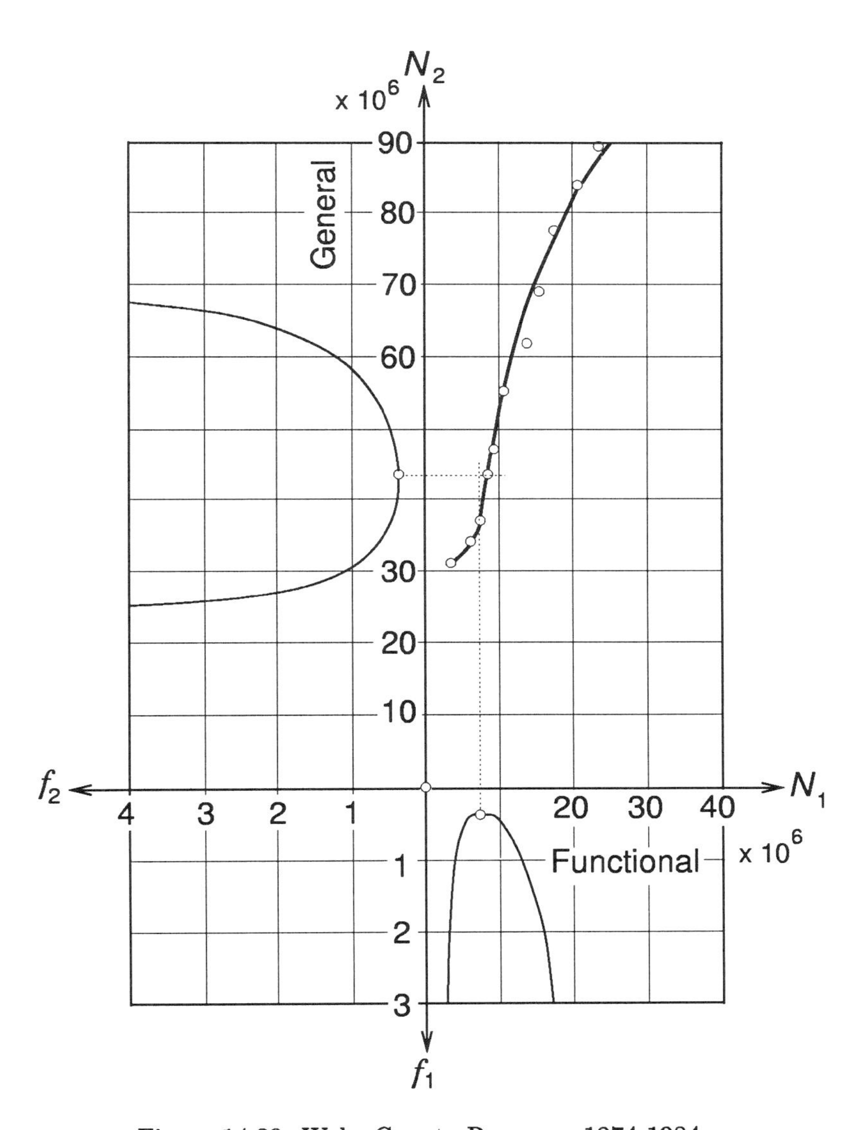

Figure 14.20: Wake County Revenues 1974-1984

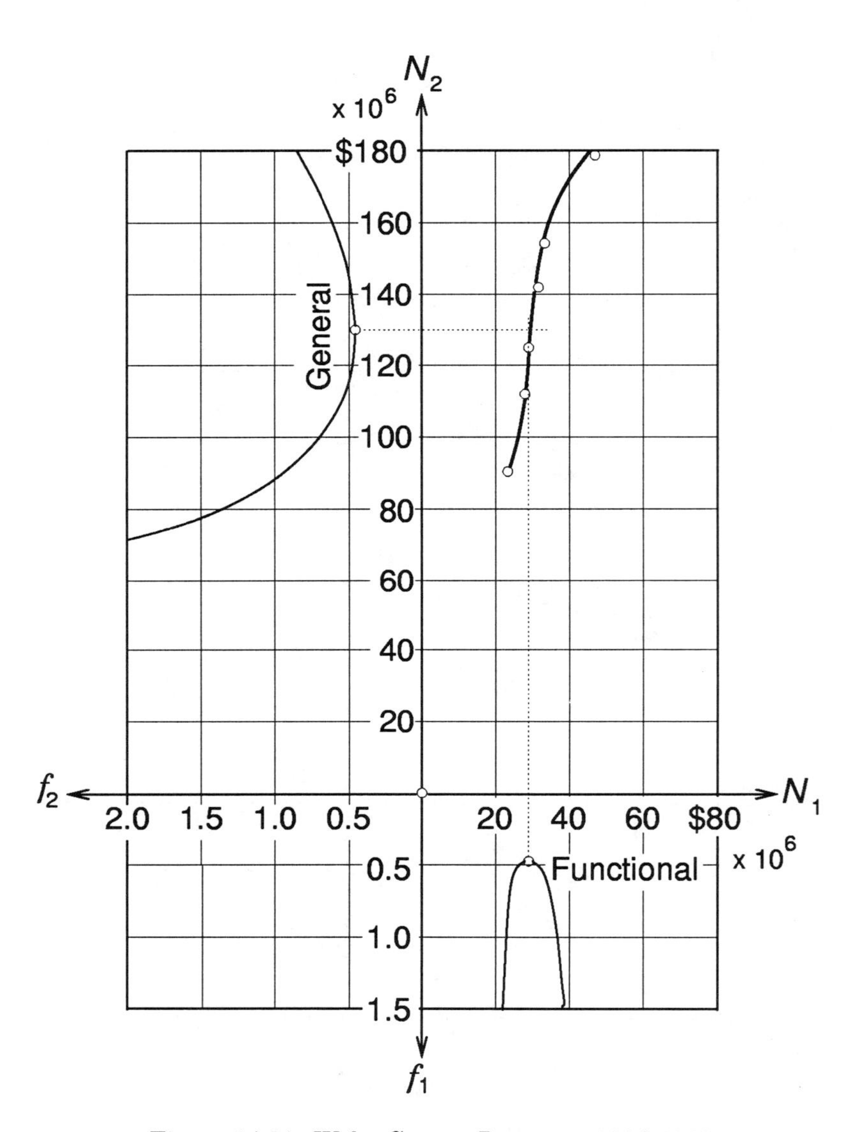

Figure 14.21: Wake County Revenues 1984-1988

14.5.4 RADIOACTIVE WASTE EXAMPLE

Safety and cost are the foremost aspects of nuclear power generation today. The production and management of the radioactive wastes are inextricably bound to safety and cost in a direct way, so that a very high premium is put upon the estimation of the accumulated amount of such wastes.

Moreover, public resistance to the locations and mechanisms for storing and treating wastes is increasing county-by-county as proposals are aired for these purposes. Indeed, it must be said that dealing with the wastes is definitely a hard constraint upon utilization of nuclear technology in our present civilization.

It is thus a matter of the highest importance to be able to calculate the amount of the wastes with the greatest possible accuracy.

In the case of LWR (Light Water Reactor) power generation, protagonists can be identified as BWR (Boiling Water Reactor) and PWR (Pressurized Water Reactor) systems, and the mode of interaction is that of competition. It is a question of "zero-sum-game" in an environment where the "sum" is growing logistically.

For LWR fuel, there are two switching points: in 1972 and in 1980. These divide the history of LWR waste generated in the 1969-1988 span into three epochs or regimes, each with its own set of parameters, and there are three corresponding regimes in the phase plane. These are shown in Figures 14.22, 14.23, 14.24, and 14.25. Using available data,[16] these graphs are plotted in terms of MTIHM (Metric Tons of Initially Heavy Metal). The protagonists: N_1 is BWR Waste, N_2 is PWR Waste. Figure 14.26 shows the 1972-1980 phase plane.

The Liapounov stability analysis of the system equations shows that there are two singular points in the phase plane: an unstable node at the origin, and an unstable saddle point at the normalized (1,1) point. Therefore, as soon as there are non-zero values of N_1 and N_2, the system point will grow outward from the origin (the unstable node) toward (1,1), the unstable saddle point. The saddle point acts to turn the system point away, and, in fact, the field is divided by separatrices so that a system point once started in a portion of the field is prohibited from crossing over a separatrix to enter another area. Thus the system is destroyed and a new system with different parameters but continuous with the old system is formed. The new system then pursues its own dynamic until it, too, must in turn also switch.

The following parameters have been determined.

Spent LWR Fuel 1970-1972

$$a = -1.5178985074 \quad b = 10.369160398 \quad c = 0.015932113302 \tag{14.76}$$

Spent LWR Fuel 1972-1980

$$a = -0.49787940049 \quad b = 91.5605017129 \quad c = 0.010251006054 \tag{14.77}$$

Spent LWR Fuel 1980-1988

$$a = -0.15639148553 \quad b = 1384.9639357 \quad c = 0.027440772392 \tag{14.78}$$

[16] U.S. Department of Energy, "Integrated Data Base for 1988: Spent Fuel and Radioactive Waste Inventories, Projections and Characteristics," DOE/RW-0006, Rev., 4, Washington, D.C., (1988).

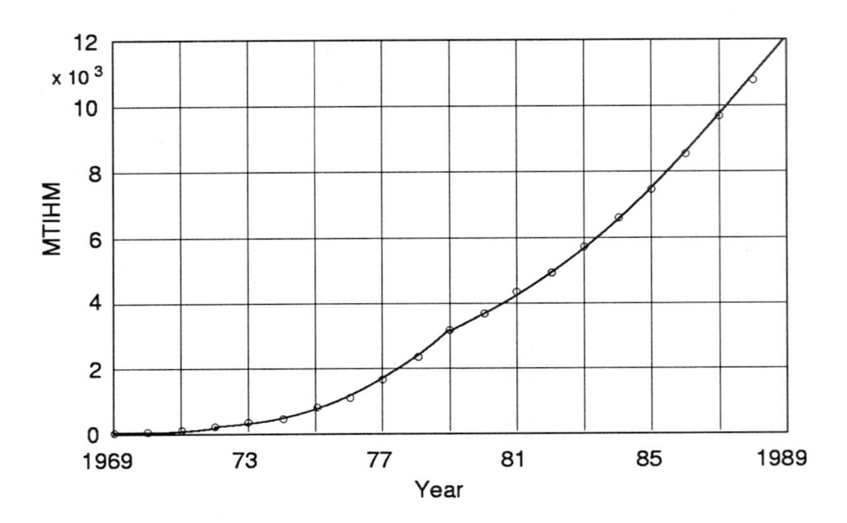

Figure 14.22: Mass of Spent Reactor Fuel, PWR

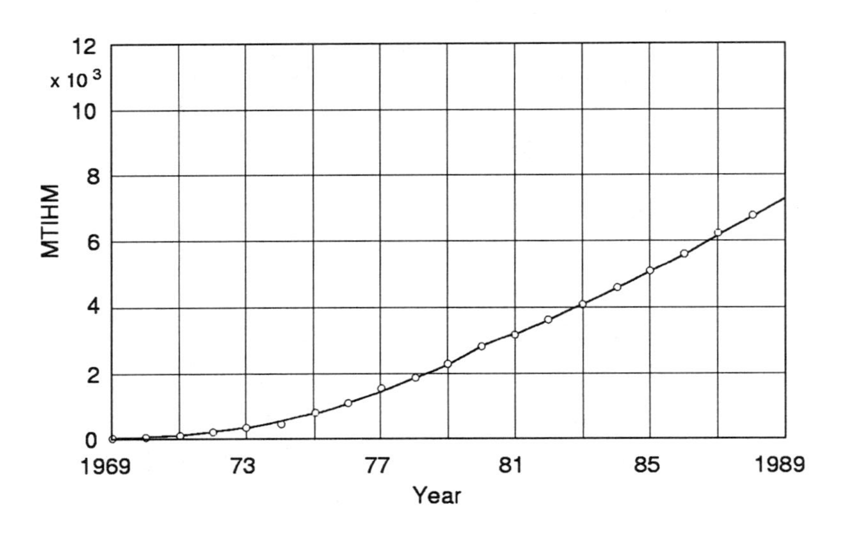

Figure 14.23: Mass of Spent Reactor Fuel, BWR

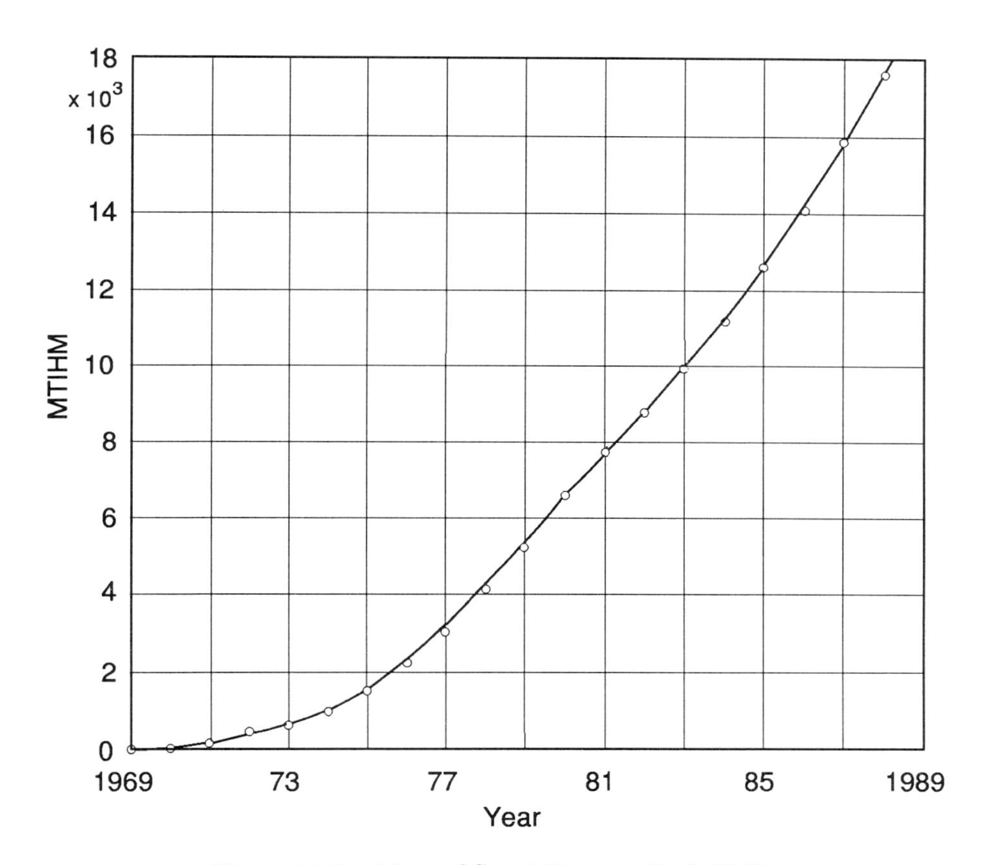

Figure 14.24: Mass of Spent Reactor Fuel, LWR

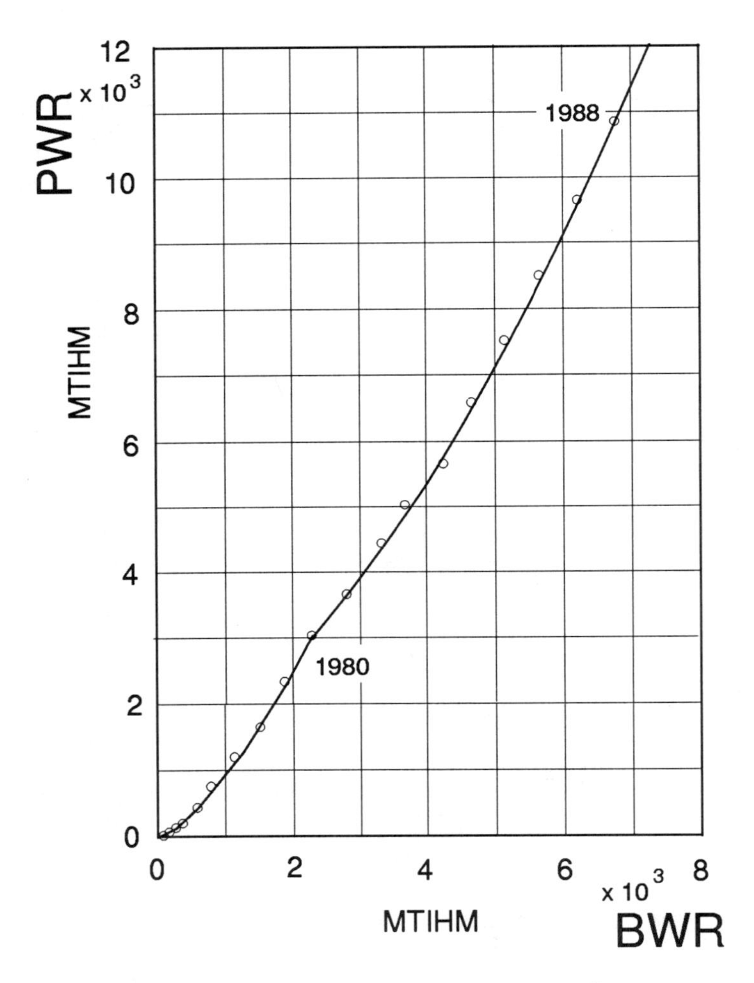

Figure 14.25: Reactor Protagonists Phase Space

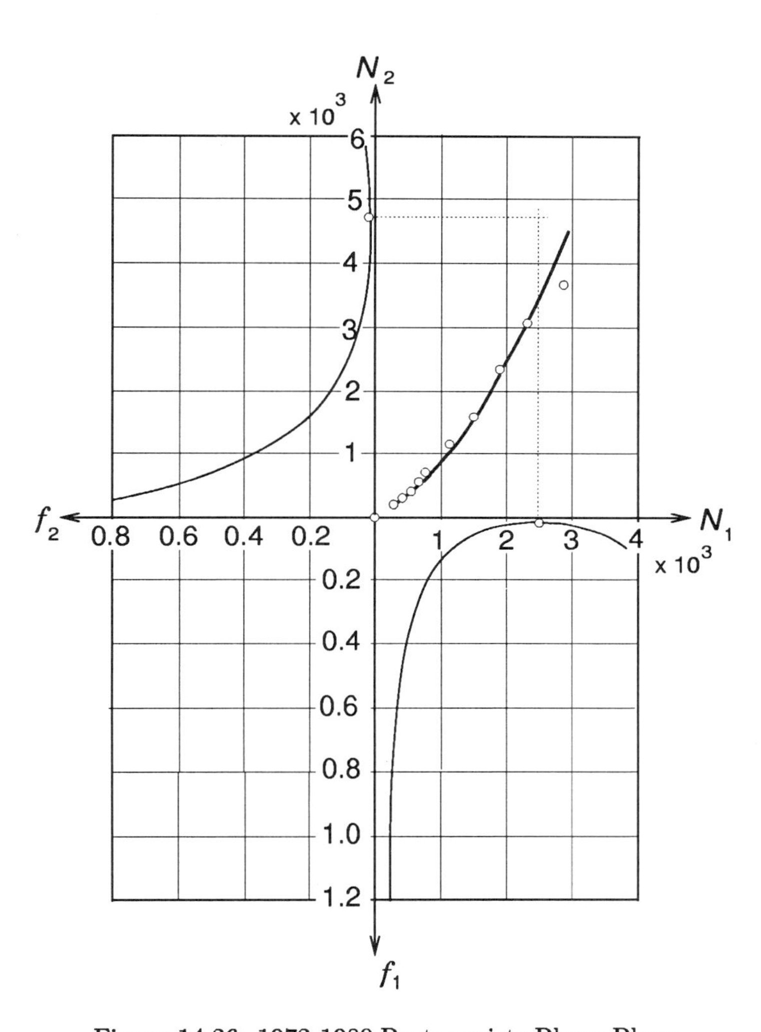

Figure 14.26: 1972-1980 Protagonists Phase Plane

14.6 PROJECT #12

1. A newspaper article describing the growth of the Research Triangle Park for a 30-year period cited the figures below for the number of employees in the Park from 1960 to 1987.

   Assume that your engineering firm has been retained to assist in the planning and design of the infrastructure in and near the Park for the immediate future. Obviously, a first task you will face is that of modeling the growth of the Park in the near-range future, so that the proper facilities can be realized. A further reasonable assumption to make is that the amount of these facilities will be proportional to the number of employees there.

   Accordingly, you are asked to produce the following:

   - A graph of the number of employees for each of the years from 1960 to 1987, using the data in the newspaper article.
   - An equation (or equations) of the logistic type which fits the data for the years 1960-1987.
   - Using the equation above, project the number of employees in the Park for the years 1988-1998. This assumes, of course, that the growth rate will be very similar to what it has been through the 1960-1987 period.

Year	Number of companies	Number of employees
1960	2	500
1965	6	908
1970	16	8,000
1975	16	10,400
1976	16	10,680
1977	21	11,000
1978	22	11,200
1979	26	12,639
1980	36	17,500
1982	43	20,000
1983	45	22,000
1984	46	23,000
1985	47	26,000
1986	54	27,000
1987	54	30,000

2. Maintenance costs (in dollars) for a pumping station show the following history.

Year	1981	1982	1983	1984	1985	1986	1987	1988	1989	1990	1991
Cost	$471	529	591	628	706	797	834	918	1012	1087	1156

   Assuming that the growth in costs will follow a similar pattern in the near-range future, what will be the costs in the years 1992-2000?

3. Your engineering firm has been retained by the County of Catawba to assist in the planning and design of infrastructures within its jurisdiction. Naturally, a first task you will face is that of modeling the growth of the County in the near-range future, so that the amount of these facilities will be proportional to the population there.

   Accordingly, you are asked to produce the following:

   - A graph of the population of the county from 1910 to date, using U.S. Census data.
   - An equation (or equations) of the logistic type which fits the data for the present century.
   - With the equation above, predict the population of the County for the years of the decade 1990-2000. This assumes, of course, that the growth rate will be very similar to that which has taken place in recent times. Show the projected growth on the graph above.

   The software ***logcur.go*** is available in the locker *mcdonald_src* on *eos* net to assist with this project.

4. An article in the Raleigh *News and Observer*, Page 4 C, January 29, 1987, described the drawdown of groundwater levels in Eastern North Carolina as the pumping by municipalities and industries has increased. The article stated that the levels have decreased by 30 m since 1900, and an accompanying graph displayed the consumption by decade during the intervening years. The data were supplied by the U.S. Geological Survey in a report authored by hydrologist M. D. Winner, Jr., of USGS. The figure below shows the picture.

   The following quote is from the newspaper account:

   > "The USGS began studying ground water in an area centered around the cities of Kinston and Greenville in 1985, he [M. D. Winner] said, after a consortium of Coastal Plain communities asked the federal agency for help in managing their water supplies. The federal government and the communities are splitting the $500,000 cost of the study, which is to be completed in 1988."

   Using the software *logcur.go* in the locker *mcdonald_src* on *eos* net, produce a model for the pumped water from these aquifers, and project the amounts which can be expected in the 1990, 2000, and 2010 years.

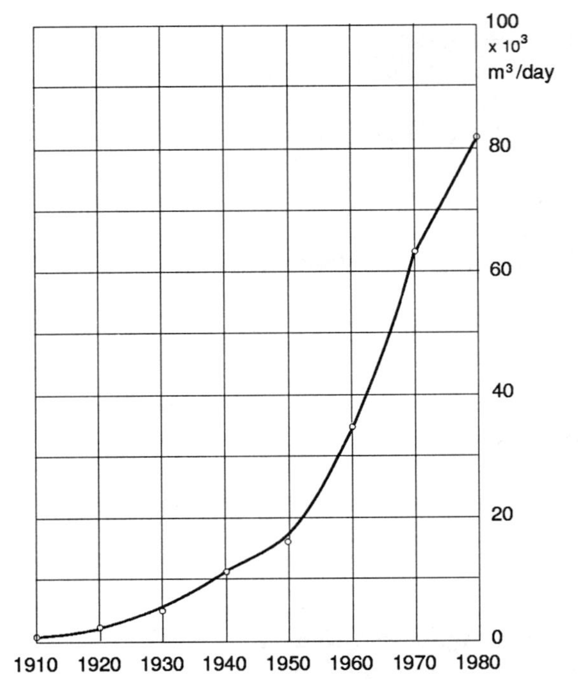

Water Pumped from Black Creek and Upper Cape Fear Aquifers

Year	m^3/day
1910	454.25
1920	2,309.10
1930	4,769.62
1940	11,431.94
1950	16,050.15
1960	34,787.93
1970	63,329.94
1980	81,121.37

CHAPTER 15

STATE SPACE SYSTEMS

15.1 INTRODUCTION

A fundamental understanding of the nature of systems is outlined in Chapter 1. There it is established that systems include *elements*, *relationships*, *boundaries*, and *totality*. These components are amenable to representation in the form of affine vectors and the vector spaces of which the vectors are elements.

When a system is modeled as a vector system, it is convenient to refer to the *state vector* of the system in the *state space*; to the agents which may cause a change in the system's state as *inputs* or *input vectors*; and to the resultant changes of the system as *outputs* or the *output vector*.

The *output* or the *output vector* of the system is therefore functionally related to the system *state* or *state vector* and the *input* or *input vector*. This functional relationship is thus a matter of the geometry, algebra, and calculus of vector spaces. The formal mathematics of such vector relationships is highly developed and readily available to the system engineer for purposes of analysis, design, and optimization.

15.2 VECTORS

Vectors are fundamental things—primitive elements which exist even before one attempts to describe them. Most engineers who are products of mechanics-based curricula have been introduced early to some elementary vectors such as displacement, velocity, force, and the like. These are usually presented as Cartesian vectors; that is, quantities tied to a Cartesian geometry.

However, it is not at all necessary to restrict vectors to such a frame of reference. Specifically, it is not necessary to have the same units of measurement on each of the components of a vector. The crude analogy of vector components as fruits and the vector itself as a fruit salad in Chapter 1 makes the point, if badly.

It is better to think of a vector as a primitive object. One can imagine that upon landing and exiting from a flying saucer an alien comes forward smiling to greet the earthling leader, and holding in outstretched hands a shining arrow as a gift. The arrow has a length (its magnitude or substance), an arrowhead (its direction or sense), and the inclinations relative to the world as its line of action.

Each of these properties would be the same anywhere in the universe *except* for the inclinations with respect to the viewer's references.

The vector itself would exist *before* anyone—either the alien or the earthling, or any other being—would attempt to describe it. In other words, as a primitive object, it is aboriginal. In this sense vectors are like pure numbers: they were always there, with their same values, before they were discovered.

15.2.1 BASE VECTORS

Think about a collection of vectors (perhaps like a quiver of arrows) each with its own properties, and each with one unit of length in its own measure. These are often described as unit vectors, but the better name is ***base vectors***. A common way of writing these independent entities is

$$\mathbf{e}_1 = \begin{bmatrix} 1 \\ 0 \\ \vdots \\ 0 \end{bmatrix} \quad \mathbf{e}_2 = \begin{bmatrix} 0 \\ 1 \\ \vdots \\ 0 \end{bmatrix} \quad \cdots, \quad \mathbf{e}_n = \begin{bmatrix} 0 \\ 0 \\ \vdots \\ 1 \end{bmatrix} \tag{15.1}$$

This employs a column matrix representation of the vectors. If the row matrix representation is employed, the form is:

$$\mathbf{e}_1 = [\,1\ 0\ \cdots\ 0\,] \quad \mathbf{e}_2 = [\,0\ 1\ \cdots\ 0\,] \quad \mathbf{e}_n = [\,0\ 0\ \cdots\ 1\,] \tag{15.2}$$

These vectors are for a space of n dimensions.

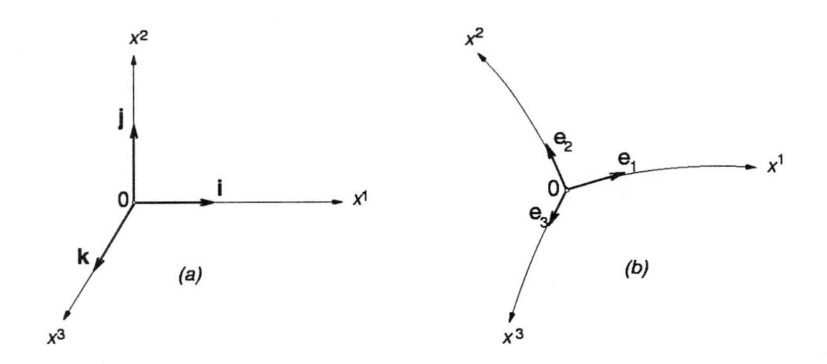

Figure 15.1: (a) Cartesian Base Vectors. (b) Euclidean Base Vectors.

The unit vectors of mechanics, $\mathbf{i}$, $\mathbf{j}$, and $\mathbf{k}$, are base vectors of equal length (unity) along the Cartesian coordinate axes as shown in Figure 15.1(a). They are thus special base vectors in that system. Figure 15.1(b) illustrates the more general Euclidean base vectors $\mathbf{e}_1$, $\mathbf{e}_2$, and $\mathbf{e}_3$ for a space of three dimensions.

When this collection of basis vectors is supplied (i.e, when the quiver of arrows is furnished), an n-dimensional vector space is defined by them. The unitary primitive basis vectors are members of the vector space, and the space is sometimes said to be *generated* or *spanned* by the basis vectors.

The properties of the vector space are derived from the basis vectors. In particular, other vectors of the space can be formulated in terms of the basis vectors, as for example

$$\mathbf{v} = v^1\mathbf{e}_1 + v^2\mathbf{e}_2 + v^3\mathbf{e}_3 + \cdots + v^n\mathbf{e}_n \tag{15.3}$$

where $\mathbf{v}$ is such a vector, and

$$\mathbf{u} = u^1\mathbf{e}_1 + u^2\mathbf{e}_2 + u^3\mathbf{e}_3 + \cdots + u^n\mathbf{e}_n \tag{15.4}$$

is some other vector of the space. The scalar numbers v^1, v^2, ..., v^n and u^1, u^2, ..., u^n are the components of $\mathbf{v}$ and $\mathbf{u}$ in the space.

15.2.2 VECTOR SPACES

When the basis vectors have been identified, the vector space is defined. When this basis is supplied, and the method of representing any other vector of the space is specified, the properties of the space can be established. These properties are generally sought to be such that the ordinary operations with vectors of the space are both possible and consistent. The operations are hopefully efficient and understandable to those who may invoke them.

Here are some commonly held requirements that govern the ordinary algebra relating to any two vectors of the space:

1. $\mathbf{v} = \mathbf{v}^1 + \mathbf{v}^2$ is also a member of the space
2. $\mathbf{v} = \mathrm{k}\,\mathbf{v}^1$ is also of the space, where k is a scalar number
3. $\mathbf{v}^1 + \mathbf{v}^2 = \mathbf{v}^2 + \mathbf{v}^1$ addition is commutative
4. $\mathbf{v} + (\mathbf{v}^1 + \mathbf{v}^2) = (\mathbf{v} + \mathbf{v}^1) + \mathbf{v}^2$ addition is associative
5. $\mathbf{v} + \mathbf{0} = \mathbf{v}$ this defines the vector of zero length
6. $\mathbf{v} + (-\,\mathbf{v}) = \mathbf{0}$ this defines a negative vector
7. $1\,(\mathbf{v}) = \mathbf{v}$
8. $\mathrm{p}\,(\mathrm{q}\,\mathbf{v}) = (\mathrm{p}\,\mathrm{q})\,\mathbf{v}$ where p and q are scalars
9. $(\mathrm{p} + \mathrm{q})\,\mathbf{v} = \mathrm{p}\,\mathbf{v} + \mathrm{q}\,\mathbf{v}$
10. $\mathrm{k}\,(\mathbf{v}^1 + \mathbf{v}^2) = \mathrm{k}\,\mathbf{v}^1 + \mathrm{k}\,\mathbf{v}^2$

When defined in accordance with these principles, the object that is outlined thereby is formally called a ***vector space*** or a ***linear space*** because any vector formed of it is a linear function of the basis vectors.

There is also the ***subspace*** of the vector space, which is defined such that for a set of elements of the space principles 1. and 2., above, are fulfilled for the set.

The abstract n-dimensional real vector space is sometimes denoted $R(n)$, and $C(n)$ identifies an n-dimensional complex vector space. The vector space coordinate vectors (n-ordered numbers) are written R^n and C^n, respectively.

The most general n-dimensional vector space is *Riemannian* or *affine*, and the geometry is often not known or specified. In many ordinary problems of mechanics and other engineering sciences, the space in question is *Euclidean*, where a *Cartesian* coordinate geometry is always available. For systems that are modeled in state space, no such simple geometries are indicated, in general.

15.2.3 VECTOR ALGEBRA

Suppose there are two vectors

$$\mathbf{a} = a^1\mathbf{e}_1 + a^2\mathbf{e}_2 + \cdots + a^n\mathbf{e}_n = \sum_{i=1}^{n} a^i\mathbf{e}_i \tag{15.5}$$

and

$$\mathbf{b} = b^1\mathbf{e}_1 + b^2\mathbf{e}_2 + \cdots + b^n\mathbf{e}_n = \sum_{i=1}^{n} b^i\mathbf{e}_i \tag{15.6}$$

Their sum can be formulated as

$$\mathbf{c} = \mathbf{a}+\mathbf{b} = (a^1+b^1)\,\mathbf{e}_1 + (a^2+b^2)\,\mathbf{e}_2 + \cdots + (a^n+b^n)\,\mathbf{e}_n = \sum_{i=1}^{n}(a^i+b^i)\,\mathbf{e}_i \tag{15.7}$$

in which

$$\mathbf{a} = [a^1\ a^2\ \ldots\ a^n] \qquad \mathbf{b} = [b^1\ b^2\ \ldots\ b^n] \tag{15.8}$$

$$\mathbf{c} = \mathbf{a}+\mathbf{b} = \sum_{i=1}^{n}(a^i+b^i)\,\mathbf{e}_i = [a^1+b^1\ \ a^2+b^2\ \ +\cdots+a^n+b^n] \tag{15.9}$$

Likewise, subtraction is formed as negative addition:

$$\mathbf{d} = \mathbf{a}-\mathbf{b} = \sum_{i=1}^{n}(a^i-b^i)\,\mathbf{e}_i = [a^1-b^1\ \ a^2-b^2\ \ \ldots\ \ a^n-b^n] \tag{15.10}$$

Multiplication of a vector by a scalar goes as follows:

$$k\mathbf{v} = k(v^1\mathbf{e}_1 + v^2\mathbf{e}_2 + \cdots + v^n\mathbf{e}_n) = kv^1\mathbf{e}_1 + kv^2\mathbf{e}_2 + \cdots + kv^n\mathbf{e}_n$$
$$= [kv^1\ \ kv^2\ \ \cdots\ \ kv^n] \tag{15.11}$$

The operation of division with vectors is not defined. Sometimes the functional equivalent can be obtained by multiplying by the inverse of a vector quantity, if the inverse is known. In general, it cannot be assumed that the inverse exists.

When the question becomes that of multiplying a vector by another vector, there are several possibilities. First, the expected answer may be a scalar. This is called the ***scalar product*** of vectors. A second product is that in which the outcome is a vector, called, appropriately, the ***vector product*** of the two vectors. A third type is the ***open*** or ***dyadic product***, which turns out to be of higher order than the elements of which it is composed.

A fourth type of vector product is that which transforms a vector constructed from one basis to the same vector in its form constructed upon another basis. A fifth type of vector product changes a vector into a new entity (frequently to the same basis) but with differing physical dimensions.

These several operations are best developed by referring to the general properties of the space in the following way.

Establish a basis $\mathbf{e}^i$ that is *dual* to $\mathbf{e}_i$ with the relationships

$$\mathbf{e}^i \cdot \mathbf{e}^j = g^{ij} \qquad \mathbf{e}^i \cdot \mathbf{e}_j = g^i_j \qquad \mathbf{e}_i \cdot \mathbf{e}_j = g_{ij} \tag{15.12}$$

wherein the g^{ij}, $g^i_{\ j}$, and g_{ij} are related to the transformation properties of the vectors in the dual space. The gs are the components of the *metric tensor*, and are functions of the specific geometry (if any) of the space.

Any vector of the space can be transformed by using the dual bases. The base vector $\mathbf{e}_k$ can be written in terms of the duals $\mathbf{e}^i$, and $\mathbf{e}^k$ can be obtained from duals $\mathbf{e}_i$ as follows:

$$\mathbf{e}_k = \sum_{i=1}^{n} g_{ki}\, \mathbf{e}^i \qquad \mathbf{e}^k = \sum_{i=1}^{n} g^{ki}\, \mathbf{e}_i \tag{15.13}$$

together with

$$\sum_{i=1}^{n} g^{ki}\, g_{i\ell} = \delta^k_\ell \tag{15.14}$$

In the last equation δ^k_ℓ is the *Kronecker's delta* symbol

$$\delta^k_\ell = 1 \qquad \text{if } k = \ell \tag{15.15}$$

$$= 0 \qquad \text{if } k \neq \ell \tag{15.16}$$

and there are the additional relationships

$$\delta^{kl} = 1 \qquad \text{if } k = \ell \tag{15.17}$$

$$= 0 \qquad \text{if } k \neq \ell \tag{15.18}$$

$$\delta_{kl} = 1 \qquad \text{if } k = \ell \tag{15.19}$$

$$= 0 \qquad \text{if } k \neq \ell \tag{15.20}$$

Then the transformation of some other vector of the space takes the forms

$$\mathbf{v} = \sum_{i=1}^{n} v^i\, \mathbf{e}_i = \sum_{i=1}^{n} v_i\, \mathbf{e}^i \tag{15.21}$$

The components of the vector transform according to

$$v_k = \sum_{i=1}^{n} g_{ki}\, v^i \qquad v^k = \sum_{i=1}^{n} g^{ki}\, v_i \tag{15.22}$$

In these formulas v_k are called *covariant* components of $\mathbf{v}$ because they transform as do the base vectors $\mathbf{e}_k$, and the v^k are called *contravariant* components of the vector because they contratransform as do the base vectors.

The dual nature of general vector spaces (affine geometries) is a direct consequence of the necessity to address the transformation properties of vectors in the space before a full and complete algebra and calculus of the vectors can be undertaken. Mathematical operations with such general vector spaces are in fact limited or bounded by the requirements of the dual transformation relationships.

In simpler geometries, such as the Cartesian, for example, the metric tensor components g_{ij}, $g^i_{\ j}$, and g^{ij} reduce to the Kronecker deltas δ_{ij}, $\delta^i_{\ j}$, and δ^{ij}, respectively. Thus, the Kronecker deltas become the agents of erecting an *ortho-normal* system of coordinate axes in the Cartesian vector space, because the axes are at right angles to each other and the base vectors are independent unit vectors.

For Euclidean spaces, there is usually a metric tensor which is a function of the coordinates, as, say, in cylindrical coordinates

$$r = x^1 = [(y^1)^2 + (y^2)^2]^{1/2} \qquad \theta = x^2 = tan^{-1}\left(\frac{y^2}{y^1}\right) \qquad z = x^3 = y^3$$

or

$$y^1 = x^1 \cos x^2 \qquad y^2 = x^1 \sin x^2 \qquad y^3 = x^3$$

Here, the metric tensor components are $g_{11} = 1$, $g_{22} = (x^1)^2$, $g_{33} = 1$, $g_{12} = 0$, $g_{13} = 0$, and $g_{23} = 0$. The system of axes is obviously still ortho-normal, but the unit of measure on the angular or x^2 sense is a function of the r or x^1 coordinate.

As the level of complexity of the vector space increases, the available operations with vectors in the space become more constrained. In the last analysis, only the more basic things can be accomplished, such as the linear form in its fundamental version, which is the only allowed form of multiplication between such vectors.

15.2.4 A SUMMATION CONVENTION

As the number of developments increased in research in vector space analysis, those who worked in this field found it desirable to utilize some notations which economized upon the evident labor of simply writing so many summations and other repetitive constructions. There came to be what is called the *Einstein running sum convention*. In this notation, when an index (whether it be a superscript or a subscript) is repeated, it is assumed that summation with respect to this index is intended, and the summation sign is omitted for the sake of brevity.

For a space of n dimensions, the so-called *running index* (the one which is repeated), is allowed to take on successively all possible values 1, 2, 3, ..., n, and the summation sign is abandoned or "understood" to be there.

Using the summation convention, the transformation formulas for vectors and bases developed above can be reformulated as follows.

$$\mathbf{v} = v^i\,\mathbf{e}_i = v_i\,\mathbf{e}^i \tag{15.23}$$

$$\mathbf{e}_k = g_{ki}\,\mathbf{e}^i \qquad \mathbf{e}^k = g^{ki}\,\mathbf{e}_i \tag{15.24}$$

$$g^{ki}\,g_{i\ell} = \delta^k_\ell \tag{15.25}$$

$$v_k = g_{ki}\,v^i \qquad v^k = g^{ki}\,v_i \tag{15.26}$$

The economy and power of the new versions is apparent.

A complement to the running sum convention is the parity check device which requires that running indices must always occur in pairs, and that one index of each pair must be a superscript, the other a subscript. This requirement follows from the necessity that the transformation property of vectors in the space shall be preserved. A formula or other mathematical expression can then be kept homogeneous as is desired.

For example, in the transformation equation

$$v_k = g_{ki}\, v^i$$

the index i is summed over all values from 1 to n, and in the end is "summed out" because no residual i is found on the left-hand side of the equation. In checking an equation for homogeneity, a scanner can be allowed to run on all upper indices, and if there is a corresponding lower index of the same name for each running index, then the formulation may be correct.

A third element of the summation convention is the change of index from upper to lower (and vice-versa) as the quantity bearing the index crosses from numerator to denominator (and vice-versa) in a fraction. In other words, a superscript in the denominator is equivalent to a subscript in the numerator and a superscript in the numerator is equivalent to a subscript in the denominator. This element is a frequent cause of confusion to those who encounter it for the first time.

Various devices have been used to lessen the confusion, including mnemonics such as the old "co go below and the bars also" from the days when overbarred quantities were the "new" or transformed coordinates in differential geometries.

15.2.5 SCALAR PRODUCTS

By utilizing the notational conventions above, the product of two vectors which results in a scalar outcome can be shown as follows.

$$\mathbf{u} = u^k\, \mathbf{e}_k \qquad \text{and} \qquad \mathbf{v} = v_\ell\, \mathbf{e}^\ell \tag{15.27}$$

The *inner* or *scalar* or *dot* product of these two vectors is then

$$\begin{aligned} \mathbf{u} \cdot \mathbf{v} &= u^k\, \mathbf{e}_k \cdot v_\ell\, \mathbf{e}^\ell = u^k v_\ell\, \mathbf{e}_k \cdot \mathbf{e}^\ell = u^k v_\ell\, \delta^\ell_k \\ &= u^k\, v_k = u^k\, g_{ki}\, v^i = g^{ki} u_i\, v_k \end{aligned} \tag{15.28}$$

One can employ any of these forms, as appropriate to the circumstances of the moment. If the space has a metric, and the metric tensor components g^{ki}, $g^i_{\ j}$, and g_{ki} are known, the last two expressions will normally be used, but if there is no metric, the linear form $u^k\, v_k$ is the necessary version. Since it is a scalar, the result is an invariant which has the same value irrespective of a change of axes or other transformation.

The quantity associated with a vector is its *magnitude*, or its size or length. The usual name is the *norm* of the vector. It is denoted $|\,\mathbf{v}\,|$ or $||\,\mathbf{v}\,||$ or just v. The norm has these properties:

1. $|\,\mathbf{v}\,| \geq 0$ and $|\,\mathbf{v}\,| = 0$ if and only if $\mathbf{v} = \mathbf{0}$,

2. $|\alpha\mathbf{v}| = |\alpha||\mathbf{v}|$ in which α is a scalar number,
3. $|\mathbf{v}^1 + \mathbf{v}^2| \leq |\mathbf{v}^1| + |\mathbf{v}^2|$.

When the norm has been furnished in some suitable manner, the vector space is said to be *normed*. Among the various ways this can be accomplished is a common one. It utilizes the scalar product of vectors of the space, such as $\mathbf{v}^1$ and $\mathbf{v}^2$, whose scalar product is a real number with the following properties:

1. $\mathbf{v}^1 \cdot \mathbf{v}^2 = \mathbf{v}^2 \cdot \mathbf{v}^1$,
2. $\mathbf{v} \cdot (\mathbf{v}^1 + \mathbf{v}^2) = (\mathbf{v} \cdot \mathbf{v}^1) + (\mathbf{v} \cdot \mathbf{v}^2)$,
3. $k\,(\mathbf{v}^1 \cdot \mathbf{v}^2) = (k\,\mathbf{v}^1) \cdot \mathbf{v}^2 = \mathbf{v}^1 \cdot (k\mathbf{v}^2)$ where k is a real number,
4. $\mathbf{v} \cdot \mathbf{v} \geq 0$ and $\mathbf{v} \cdot \mathbf{v} = 0$ if and only if $\mathbf{v} = \mathbf{0}$.

The definition of the scalar product is

$$\mathbf{v}^1 \cdot \mathbf{v}^2 = |\mathbf{v}^1||\mathbf{v}^2| \cos(\mathbf{v}^1, \mathbf{v}^2) \tag{15.29}$$

in which the notation $\cos(\mathbf{v}^1, \mathbf{v}^2)$ identifies the cosine of the angle between the lines of action of $\mathbf{v}^1$ and $\mathbf{v}^2$. The definition obviously satisfies the conditions of a norm defined above.

15.2.6 VECTOR PRODUCTS

The product of two vectors which results in a vector outcome, called the *vector-* or *outer-* or *cross- product*, is defined by the following properties:

1. $\mathbf{v}^1 \cdot (\mathbf{v}^1 \times \mathbf{v}^2) = \mathbf{v}^2 \cdot (\mathbf{v}^1 \times \mathbf{v}^2) = 0$. This means that the answer-product vector $\mathbf{v}^1 \times \mathbf{v}^2$ is at right angles to both of the vectors $\mathbf{v}^1$ and $\mathbf{v}^2$ from which it was generated.
2. The product $\mathbf{v}^1 \times \mathbf{v}^2$ has a sense which is given by the so-called *right-hand screw rule*. In this convention, the product-vector has a direction that is along the third finger of the right hand when the thumb is aligned with vector $\mathbf{v}^1$ and the index finger is along vector $\mathbf{v}^2$. Figure 15.2(a) shows this.

   Alternatively, the so-called *corkscrew rule* can be invoked. According to this device, the sense of the product vector is along the line of advance of a corkscrew when it is rotated in a right-handed sense. See Figure 15.2(b).
3. $|\mathbf{v}^1 \times \mathbf{v}^2| = |\mathbf{v}^1||\mathbf{v}^2| \sin\phi$. Again, ϕ is the angle between the vectors $\mathbf{v}^1$ and $\mathbf{v}^2$, as seen in Figure 15.2.

As defined above, the cross product has the following properties:

1. $\mathbf{v}^1 \times \mathbf{v}^2 = -\mathbf{v}^2 \times \mathbf{v}^1$. That is, it does not commute,
2. $(k\mathbf{v}^1) \times \mathbf{v}^2 = k\,(\mathbf{v}^1 \times \mathbf{v}^2)$ in which k is a scalar,
3. $\mathbf{v} \times (\mathbf{v}^1 + \mathbf{v}^2) = \mathbf{v} \times \mathbf{v}^1 + \mathbf{v} \times \mathbf{v}^2$.

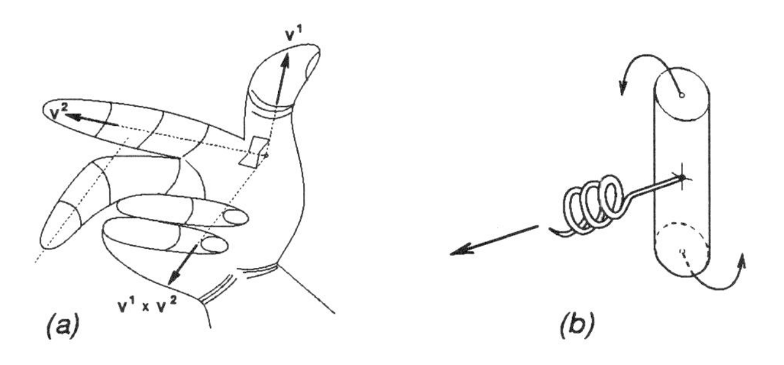

Figure 15.2: (a) Right-Hand Rule. (b) The Corkscrew Rule.

The base vectors can be subjected to this definition to yield

$$\mathbf{e}_i \times \mathbf{e}_j = \epsilon_{ijk}\,\mathbf{e}_k \tag{15.30}$$

wherein ϵ_{ijk} is the *Levi-Civita density* or *permutation symbol*

$$\epsilon_{ijk} = +1 \qquad \text{if } i,j,k \text{ are of cyclic order; e.g., 1, 2, 3, } \ldots, \tag{15.31}$$

$$= \ 0 \qquad \text{if any two of } i,j,k \text{ are the same}, \tag{15.32}$$

$$= -1 \qquad \text{if } i,j,k \text{ are of acyclic order; e.g., 3, 2, 1, } \ldots. \tag{15.33}$$

This result allows the vector product $\mathbf{c}$ of two vectors $\mathbf{a}$ and $\mathbf{b}$ to be written as

$$\mathbf{c} = \mathbf{a} \times \mathbf{b} = a^i\,\mathbf{e}_i \times b^j\mathbf{e}_j = a^i\,b^j\,\epsilon_{ijk}\,\mathbf{e}_k \tag{15.34}$$

Obviously, the components of $\mathbf{c}$ are

$$c_k = \epsilon_{ijk}\,a^i\,b^j \tag{15.35}$$

15.2.7 DYADIC PRODUCTS

It was suggested above that a further product between vectors is possible. This product, often called the *dyadic-* or *open-* or *tensor-product* may have been an unexpected outcome in the beginning, but is soon seen to be an appropriate result of the operation of one vector upon another in the sense of multiplication.

Let the vectors of the space be $\mathbf{a}$ and $\mathbf{b}$ and denote the intended product as

$$\mathbf{c} = \mathbf{a} \otimes \mathbf{b} \tag{15.36}$$

When the vectors are written as

$$\mathbf{a} = a^i\,\mathbf{e}_i \qquad \mathbf{b} = b^j\,\mathbf{e}_j \qquad \mathbf{c} = c^k\,\mathbf{e}_k \tag{15.37}$$

the open product is

$$\mathbf{c} = \mathbf{a} \otimes \mathbf{b} = (a^i\, \mathbf{e}_i) \otimes (c^j\, \mathbf{e}_j) = a^i\, c^j\, \mathbf{e}_i \otimes \mathbf{e}_j \tag{15.38}$$

This leaves the meaning of $\mathbf{e}_i \otimes \mathbf{e}_j$ "open," which is where the name *open-product* comes from. This new synbol $\otimes$ is used to indicate that it is different than either the scalar or the vector product of the two vectors. It also allows the meaning of the open product to be assigned at the end of the operation. Customarily, the new symbol is

$$\mathbf{e}_i \otimes \mathbf{e}_j = \mathbf{g}_{ij} \qquad i, j = 1, 2, \ldots, \mathrm{n};\ 1, 2, \ldots, \mathrm{m} \tag{15.39}$$

It is seen immediately that the new quantity is of higher order than its constituent vectors, because two indices are required to give it meaning. These $\mathbf{g}_{ij}$ are in fact the basis of a *tensor-valued* space, of which the open product is a member. The space has a dimension mn, and the product itself is defined by an array of $n \times m$ numbers. Ordinarily, this array is represented in a matrix form which displays the components of the higher-ordered entity called a *tensor*.

That is,

$$\mathbf{c} = \mathbf{a} \otimes \mathbf{b} = a^i\, b^j\, \mathbf{g}_{ij} = a_i\, b^j\, \mathbf{g}^i_j = a^i\, b_j\, \mathbf{g}^j_i = a_i\, b_j\, \mathbf{g}^{ij} \tag{15.40}$$

or

$$\mathbf{c} = \begin{bmatrix} a^1 b_1 & a^1 b_2 & \cdots & a^1 b_m \\ a^2 c_1 & a^2 b_2 & \cdots & a^2 b_m \\ \cdot & \cdot & \cdots & \cdot \\ \cdot & \cdot & a^i b_j & \cdot \\ \cdot & \cdot & \cdots & \cdot \\ a^n b_1 & a^n b_2 & \cdots & a^n b_m \end{bmatrix} = \begin{bmatrix} c^1_1 & c^1_2 & \cdots & c^1_m \\ c^2_1 & c^2_2 & \cdots & c^2_m \\ \cdot & \cdot & \cdots & \cdot \\ \cdot & \cdot & c^i_j & \cdot \\ \cdot & \cdot & \cdots & \cdot \\ c^n_1 & c^n_2 & \cdots & c^n_m \end{bmatrix} \tag{15.41}$$

When written in matrix form, a tensor displays its elements as the array, without the need to include the $\mathbf{g}_{ij}$ associated with each ij element. They are, in fact, "understood" to be present.

The higher order entity called *tensor* was identified first, apparently, by one W. Voigt, a Swiss crystallographer who coined the word *tenseur* in reference to the collection of tensions or stresses in an anisotropic solid body.[1]

15.2.8 TENSOR PROPERTIES

The tensor or dyadic or open product of vectors has the following properties:

$$\mathbf{v} \otimes (\mathbf{v}^1 + \mathbf{v}^2) = \mathbf{v} \otimes \mathbf{v}^1 + \mathbf{v} \otimes \mathbf{v}^2 \tag{15.42}$$

$$(\mathbf{v}^1 + \mathbf{v}^2) \otimes \mathbf{v} = \mathbf{v}^1 \otimes \mathbf{v} + \mathbf{v}^2 \otimes \mathbf{v} \tag{15.43}$$

$$k\mathbf{v}^1 \otimes \mathbf{v}^2 = \mathbf{v}^1 \otimes k\mathbf{v}^2 = k(\mathbf{v}^1 \otimes \mathbf{v}^2) \tag{15.44}$$

Athough it had been unexpected, the discovery of the tensor at once presents the possibility that it might operate upon a vector of the space in question in the

[1] Léon Brillouin, *Les Tenseurs en Mécanique et en Elasticité*, Mason et Cie, Editeurs, Paris, 1938, Préface and p. 212.

sense of multiplication and produce a new vector as a result. Both vectors would be elements of appropriate subspaces of the tensor space. Call the tensor $\mathbf{T}$, and the vectors $\mathbf{u}$ and $\mathbf{v}$, so that

$$\mathbf{v} = \mathbf{T}\,\mathbf{u} \tag{15.45}$$

is the postulated relationship. This operation is called a *linear transformation*, a *linear operator*, or, simply, a *tensor*. It has the following properties:

$$\mathbf{T}(\mathbf{u} + \mathbf{v}) = \mathbf{T}\mathbf{u} + \mathbf{T}\mathbf{v} \tag{15.46}$$

$$\mathbf{T}(k\,\mathbf{v}) = k\,\mathbf{T}\mathbf{v} \tag{15.47}$$

These properties allow the definition of the *inverse* of the transformation $\mathbf{T}^{-1}$ as follows:

$$\mathbf{T}\mathbf{T}^{-1} = \mathbf{T}^{-1}\mathbf{T} = \mathbf{1} = \mathbf{I} \tag{15.48}$$

wherein $\mathbf{1} \equiv \mathbf{I}$ is the *identity transformation* or *unit linear transformation* or *substitution tensor*. It has the special property

$$\mathbf{I}\mathbf{v} = \mathbf{v} \tag{15.49}$$

which maps a vector upon itself, or, in other words, it serves as the unit tensor.

The transformation of an Euclidean space onto itself is accomplished by invoking the *transpose* $\mathbf{T}^t$ of the operator $\mathbf{T}$, as defined by

$$\mathbf{T}\mathbf{u} \cdot \mathbf{v} = \mathbf{u} \cdot \mathbf{T}^t\mathbf{v} \tag{15.50}$$

where $\mathbf{u}$ and $\mathbf{v}$ are all pairs of vectors of the space. Then there is this relationship for the transpose

$$(\mathbf{T}\mathbf{S})^t = \mathbf{S}^t\mathbf{T}^t \tag{15.51}$$

and further

$$(\mathbf{T}^{-1})^t = (\mathbf{T}^t)^{-1} = \mathbf{T}^{-t} \tag{15.52}$$

If there is an orthogonal system, the linear transformation has the following special relationship:

$$\mathbf{T}^t = \mathbf{T}^{-1} \tag{15.53}$$

and consequently

$$\mathbf{T}\mathbf{T}^t = \mathbf{I} \tag{15.54}$$

This latter result is especially valuable in working with vector spaces for which it exists, as it provides for the equivalent of division among vectors of such spaces.

15.2.9 GENERAL TRANSFORMATIONS

Consider the general linear transformation between vectors $\mathbf{a}$ and $\mathbf{c}$, elements of subspaces of the tensor space of which $\mathbf{T}$ is an element. The operation is

$$\mathbf{a} = \mathbf{T}\mathbf{c} \tag{15.55}$$

with

$$\mathbf{a} = a^i\,\mathbf{e}_i \qquad \mathbf{c} = c^j\,\mathbf{e}_j \tag{15.56}$$

so that

$$a^i \, \mathbf{e}_i = \mathbf{T} \, c^j \, \mathbf{e}_j = T^i_{\,j} \, c^j \, \mathbf{e}_i \tag{15.57}$$

from which

$$a^i = T^i_{\,j} \, c^j \tag{15.58}$$

This very important relationship furnishes the fundamental test of tensor character, and also gives the mechanism for converting the components of the vectors in question. Note that each component of vector **a** is produced as a running sum over the products of the tensor components and those of vector **c**.

When this law of transformation has been made available, the next step is to consider the transformation which converts a vector of a space erected upon one basis into the same vector drawn upon a different basis. If the space is Cartesian, the transformation may be said to be orthogonal. Please see Figure 15.3.

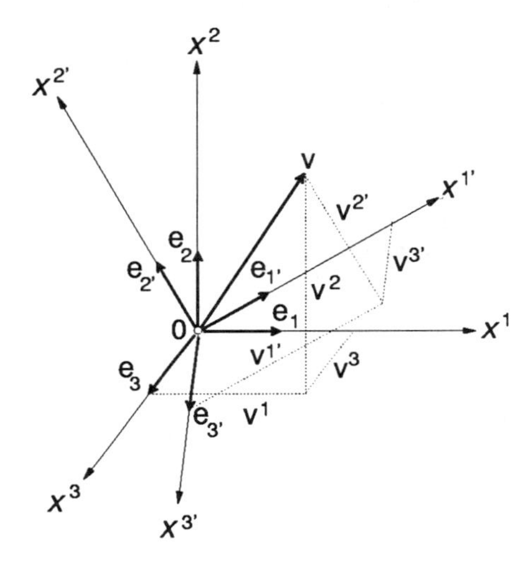

Figure 15.3: Rotated Coordinate System

Let the bases be $\mathbf{e}_i$ and $\mathbf{e}_{i'}$, both of which are of the same space. Then the bases transform according to

$$\mathbf{e}_{i'} = \mathbf{A}\mathbf{e}_i \qquad i = 1, 2, 3, \ldots, n \qquad i' = 1, 2, 3, \ldots, n \tag{15.59}$$

in which **A** is the transformation operator. Obviously, it is a "rotation" of the coordinate frame from the one formed to the basis $\mathbf{e}_i$ to the other one $\mathbf{e}_{i'}$. Therefore,

$$A^j_{\,i} = \delta^j_{\,i} \tag{15.60}$$

Next write the vector **v** in both of its representations as

$$\mathbf{v} = v^i \, \mathbf{e}_i = v^{i'} \, \mathbf{e}_{i'} \tag{15.61}$$

and enter the bases transformation as

$$\mathbf{e}_{i'} = \mathbf{A} \, \mathbf{e}_i = A^j_{\,i'} \, \mathbf{e}_j \tag{15.62}$$

then inject this into the previous equation to get

$$\mathbf{v} = v^i\,\mathbf{e}_i = v^{i'}\,A^j_{\ i'}\,\mathbf{e}_j = v^{i'}\,A^j_{\ i'}\,A^i_{\ j}\,\mathbf{e}_i$$
$$= v^{i'}\,A^j_{\ i'}\,\delta^i_{\ j}\,\mathbf{e}_i = v^{i'}\,A^i_{\ i'}\,\mathbf{e}_i = A^i_{\ j}\,v^j\,\mathbf{e}_i \tag{15.63}$$

Upon comparing the first and last terms containing $\mathbf{e}_i$, it is seen that

$$v^i = A^i_{\ j}\,v^j \tag{15.64}$$

In a similar manner, using the relationship $\mathbf{A}\mathbf{A}^{-1} = \mathbf{I}$,

$$v^{i'} = (A^{-1})^{i'}_{\ j}\,v^j \tag{15.65}$$

with $(A^{-1})^{i'}\,j$ as the element of $\mathbf{A}^{-1}$.

These relationships provide the mechanism for converting the components of a vector drawn to one basis into the components of the same vector with some different basis and the inverse conversions and are thus most valuable in a practical sense, and, indeed, also serve as the fundamental definition of this type of transformation.

The next level of transformation from one basis to another is that of the tensor components. Take

$$\mathbf{T}\,\mathbf{e}_i = \mathbf{T}^j_{\ i}\,\mathbf{e}_j \qquad \mathbf{T}\,\mathbf{e}_{i'} = T^{j'}_{\ i'}\,\mathbf{e}_{j'} \tag{15.66}$$

and with $\mathbf{e}_{i'} = \mathbf{A}\,\mathbf{e}_i = A^j_{\ i'}\,\mathbf{e}_j$,

$$\mathbf{T}\,\mathbf{e}_{i'} = \mathbf{T}\,A^j_{\ i'}\,\mathbf{e}_j = A^j_{\ i'}\mathbf{T}\,\mathbf{e}_j = A^j_{\ i'}\,T^k_{\ j}\,\mathbf{e}_k = T^{j'}_{\ i'}\,\mathbf{e}_{j'} = T^{j'}_{\ i'}\,A^k_{\ j'}\,\mathbf{e}_k \tag{15.67}$$

Upon comparing the coefficients of $\mathbf{e}_k$ in this equation, it can be seen that

$$T^k_{\ j}\,A^j_{\ i'} = T^{j'}_{\ i'}\,A^k_{\ j'} \tag{15.68}$$

and when it is recalled that $\mathbf{A}\mathbf{A}^{-1} = \mathbf{I}$, which is

$$A^i_{\ j}\,(A^{-1})^j_{\ k} = \delta^i_{\ k} \tag{15.69}$$

it is further evident that upon multiplying both sides of the first equation by $(A^{-1})^{i'}_{\ m}$, there is

$$T^k_{\ j}\,A^j_{\ i'} = T^{j'}_{\ i'}\,A^k_{\ j'} = A^k_{\ j'}\,T^{j'}_{\ i'} \tag{15.70}$$

$$T^k_{\ j}\,A^j_{\ i'}\left[(A^{-1})^{i'}_{\ m}\right] = A^k_{\ j'}\,T^{j'}_{\ i'}\left[(A^{-1})^{i'}_{\ m}\right] \tag{15.71}$$

$$T^k_{\ j}\left[A^j_{\ i'}\,(A^{-1})^{i'}_{\ m}\right] = T^k_{\ j}\,\delta^j_{\ m} = T^k_{\ m} = A^k_{\ j'}\,T^{j'}_{\ \ell'}\,(A^{-1})^{\ell'}_{\ m} \tag{15.72}$$

This result contains running sums over dummy indices. For the sake of convenience, let the k be changed to i,

$$T^i_{\ m} = A^i_{\ j'}\,T^{j'}_{\ \ell'}\,(A^{-1})^{\ell'}_{\ m} \tag{15.73}$$

and then, for the same reason, change j' to k'

$$T^i_{\ m} = A^i_{\ k'}\,T^{k'}_{\ \ell'}\,(A^{-1})^{\ell'}_{\ m} \tag{15.74}$$

and, lastly, change m to j

$$T^i_j = A^i_{k'}\, T^{k'}_{\ell'}\, (A^{-1})^{\ell'}_j \tag{15.75}$$

This most important result defines the manner of converting the components of a tensor drawn to one basis into its components with respect to another basis, all the while serving as the fundamental definition of a proper transformation of the tensor. In fact, it is frequently the test or litmus of proper tensor character. That is, if the components transform in accordance with this formula, the quantity being transformed is a proper tensor.

The transformation is seen to depend directly upon the geometry of the transformation through its dependence upon the As and (A^{-1})s, or rotation matrix components.

The final result is also often called a *similarity transformation*, because the original and final versions of the tensor are similar to each other.

15.2.10 INVARIANTS

Among the more important properties associated with the family of vectors and tensors is that of invariance of certain quantities. That is, the quantities in question are not dependent upon the coordinate systems employed to represent the vectors and tensors. To investigate these properties, let the spaces in question be such that an orthogonal transformation is possible.

Let the transformation be named $\mathbf{R}$, with the presumption that it is a rotation of the axis set from one orientation into another. From the definitions previously given

$$\mathbf{R}^t = \mathbf{R}^{-1} \qquad \text{and} \qquad \mathbf{R}\mathbf{R}^t = \mathbf{I} \tag{15.76}$$

for this transformation.

Consider these invariants:

1. Take two vectors $\mathbf{v}^1$ and $\mathbf{v}^2$ which are transformed into $\overset{*}{\mathbf{v}}{}^1$ and $\overset{*}{\mathbf{v}}{}^2$ by $\mathbf{R}$ in the following manner:

$$\overset{*}{\mathbf{v}}{}^1 = \mathbf{R}\mathbf{v}^1 \qquad \overset{*}{\mathbf{v}}{}^2 = \mathbf{R}\mathbf{v}^2 \tag{15.77}$$

   The scalar product is formed as

$$\overset{*}{\mathbf{v}}{}^1 \cdot \overset{*}{\mathbf{v}}{}^2 = \mathbf{R}\mathbf{v}^1 \cdot \mathbf{R}\mathbf{v}^2 = \mathbf{v}^1 \cdot \mathbf{R}^t(\mathbf{R}\mathbf{v}^2) = \mathbf{v}^1 \cdot (\mathbf{R}^t\mathbf{R})\mathbf{v}^2 = \mathbf{v}^1 \cdot \mathbf{I}\mathbf{v}^2 = \mathbf{v}^1\mathbf{v}^2 \tag{15.78}$$

   and it is seen rightaway that it has the same form in both original and transformed conditions. That is, it is an invariant under orthogonal transformations.

   A special case of this is the magnitude of a vector subjected to an orthogonal transformation. Then

$$|\mathbf{v}| = |\mathbf{R}\mathbf{v}| \tag{15.79}$$

   Thus the magnitude of a vector is invariant when transformed by an orthogonal transformation. This was to have been expected, of course, since the magnitude of a vector is a scalar quantity.

2. Take, as a further example, a tensor $\mathbf{T}$ whose elements are $T^i_{\ j}$. There is the quantity called the *trace*

$$tr\,\mathbf{T} = T^i_{\ i} \tag{15.80}$$

When the basis is changed from one to another one, it is to be expected that the form will remain $tr\mathbf{T} = T^i_{\ i}$. The general transformation law for tensors then gives

$$T^i_{\ i} = A^i_{\ k'}\, T^{k'}_{\ \ell'}\, (A^{-1})^{\ell'}_{\ i} = A^i_{\ k'}\, (A^{-1})^{\ell'}_{\ i} \cdot T^{k'}_{\ \ell'} = \delta^{\ell'}_{\ k'}\, T^{k'}_{\ \ell'} = T^{\ell'}_{\ \ell'} \tag{15.81}$$

The trace of a tensor has therefore been shown to be an invariant, as expected.

This idea of magnitude joined to members of the vector family can be extended to tensors as

$$|\mathbf{T}| = \ [tr(\mathbf{T}\mathbf{T}^t)]^{\frac{1}{2}} \tag{15.82}$$

3. The determinant formed from the matrix of a tensor is also an invariant. To show this take

$$det\,\mathbf{T} = det\,T^i_{\ j} = det\,A^i_{\ k'}\, T^{k'}_{\ \ell'}\, (A^{-1})^{\ell'}_{\ j} = det\,\mathbf{A}\, det\,\mathbf{T}'\, det\,\mathbf{A}^{-1}$$

$$= det\,\mathbf{A}\,\mathbf{A}^{-1}\, det\,\mathbf{T}' = det\,\mathbf{T}' \tag{15.83}$$

15.2.11 EIGENVALUES

There is a higher form of invariant associated with tensors—one which might not be anticipated before the fact. This special property is of the nature of stationary values which remain invariant as the tensor is transformed. These are called *eigenvalues*, a half-German, half-English noun that is often rendered *proper-values*, or sometimes as *singular values*.

The existence and attributes of eigenvalues can be pursued by postulating a thesis in the form of a question as follows:

> *Do there exist particular directions* **e** *for which a transformation* **T** *operating upon* **e** *produce new vectors which are scalar multiples of these original* **e***s?*

The question can be formulated mathematically as

$$\mathbf{T}\,\mathbf{e} = \lambda\,\mathbf{e} \tag{15.84}$$

in which the parameter λ is a real scalar number that is at this point unknown. Rearrange the question to

$$(\mathbf{T} - \lambda\,\mathbf{I})\,\mathbf{e} = \mathbf{0} \tag{15.85}$$

in vector notation, or

$$(T^i_{\ j} - \lambda\,\delta^i_{\ j}) = 0 \tag{15.86}$$

in its scalar version.

In the latter description, the question is now seen to be expressed as a set of simultaneous linear equations with the unknown **e**s as the solution. But λ, the parameter dominating the set, is also unknown. The customary procedure at this point is to change emphasis from the unknown **e**s and focus upon the unknown λ. Note, first, that a solution (other than the trivial one) of the set of equations exists if and only if

$$det\,(T^i_{\;j} - \lambda\,\delta^i_{\;j}) = 0 \tag{15.87}$$

Of course, since the determinant of the matrix of a tensor is invariant, as seen in the last section, there is

$$det\,(\mathbf{T} - \lambda\,\mathbf{I}) = 0 \tag{15.88}$$

To facilitate the further expansion of this determinant, assume that the discussion pertains to a 3-dimensional vector space. In this case,

$$det\,(\mathbf{T} - \lambda\,\mathbf{I}) = f(\lambda) = -\,\lambda^3 + I_1\,\lambda^2 - \,I_2\,\lambda + I_3 = 0 \tag{15.89}$$

in which there are three invariants I_1, I_2, and I_3. Specifically,

$$I_1 = tr\,\mathbf{T} = T^i_{\;i} \tag{15.90}$$

$$I_2 = \tfrac{1}{2}\,(tr^2\mathbf{T} - tr\,\mathbf{T}^2) = \tfrac{1}{2}\,(T^i_{\;i}\,T^j_{\;j} - \,T^i_{\;j}\,T^j_{\;i}) \tag{15.91}$$

$$I_3 = det\,\mathbf{T} = \epsilon_{ijk}\,T^i_{\;1}\,T^j_{\;2}\,T^k_{\;3} \tag{15.92}$$

Of course, since they are invariants, the Is can be computed in whatever is the more convenient reference frame.

Now the equation

$$f(\lambda) = -\,\lambda^3 + I_1\,\lambda^2 - \,I_2\,\lambda + I_3 = 0 \tag{15.93}$$

because it is a cubic equation, has three roots. Either one of them or all three are real numbers, and for a real symmetrical tensor $\mathbf{T}$, all three are real. Therefore, there are three ***eigenvalues*** of $\mathbf{T}$. By custom, these are designated in accordance with the rank ordering $\lambda_1 > \lambda_2 > \lambda_3$. When these three numbers—the ***principal*** or ***proper*** values—have been specified, the stationary transformation of the tensor exists. Also, since the invariants can be computed in whatever frame of reference is convenient, the principal axis frame can be chosen so that

$$T^1_{\;1} = \lambda_1 \qquad T^2_{\;2} = \lambda_2 \qquad T^3_{\;3} = \lambda_3 \qquad T^i_{\;j} = 0 \quad i \neq j \tag{15.94}$$

so it can be seen immediately that this leads to the simplest expressions for the invariants:

$$I_1 = \lambda_1 + \lambda_2 + \lambda_3 \tag{15.95}$$

$$I_2 = \lambda_1\,\lambda_2 + \lambda_1\,\lambda_3 + \lambda_2\,\lambda_3 \tag{15.96}$$

$$I_3 = \lambda_1\,\lambda_2\,\lambda_3 \tag{15.97}$$

When the secular equation $f(\lambda)$ has been solved for the three roots λ_1, λ_2, and λ_3, these roots may be entered into the eigenvalue equation individually to obtain the result

$$(\mathbf{T} - \lambda_1\,\mathbf{I})\,\mathbf{e}_1 = \mathbf{0} \qquad (\mathbf{T} - \lambda_2\,\mathbf{I})\,\mathbf{e}_2 = \mathbf{0} \qquad (\mathbf{T} - \lambda_3\,\mathbf{I})\,\mathbf{e}_3 = \mathbf{0} \tag{15.98}$$

These vector equations each subsume three scalar equations for the determination of the direction cosines of the $\mathbf{e}_1$, $\mathbf{e}_2$, and $\mathbf{e}_3$ axes with respect to the $\mathbf{e}_{1'}$, $\mathbf{e}_{2'}$, $\mathbf{e}_{3'}$ axes of the transformation in question. Thus

$$\mathbf{e}_1 = \ell^{j'}_{\ 1}\, \mathbf{e}_{j'} \qquad \mathbf{e}_2 = \ell^{j'}_{\ 2}\, \mathbf{e}_{j'} \qquad \mathbf{e}_3 = \ell^{j'}_{\ 3}\, \mathbf{e}_{j'} \tag{15.99}$$

Each of these equations contains a running sum over the dummy index j'.

But the equations are not independent, because in each case they have already been used in the solution for the eigenvalue. However, at least in the case where the geometry of the space is known, that geometry can be adjoined to assist in their solution. For example, if the space is Cartesian, there is the well-known relationship betwen the direction cosines of each axis (usually called the *normalization condition*) given by

$$\ell^k_{\ 1}\, \ell^k_{\ 1} = 1 \qquad \ell^k_{\ 2}\, \ell^k_{\ 2} = 1 \qquad \ell^k_{\ 3}\, \ell^k_{\ 3} = 1 \tag{15.100}$$

This furnishes the additional condition needed to replace the one used up in finding λ. Should the geometry of the space not be known, there is no such supplemental equation available, therefore only the ratio of the direction cosines can be found.

The augmented sets of equations can be displayed as

$$\left(T^1_{\ 1} - \lambda_1\right) \ell^1_{\ 1} + T^1_{\ 2}\, \ell^2_{\ 1} + T^1_{\ 3}\, \ell^3_{\ 1} = 0 \tag{15.101}$$

$$T^2_{\ 1}\, \ell^1_{\ 1} + \left(T^2_{\ 2} - \lambda_1\right) \ell^2_{\ 1} + T^2_{\ 3}\, \ell^3_{\ 1} = 0 \tag{15.102}$$

$$T^3_{\ 1}\, \ell^1_{\ 1} + T^3_{\ 2}\, \ell^2_{\ 1} + \left(T^3_{\ 3} - \lambda_1\right) \ell^3_{\ 1} = 0 \tag{15.103}$$

$$\left(\ell^1_{\ 1}\right)^2 + \left(\ell^2_{\ 1}\right)^2 + \left(\ell^3_{\ 1}\right)^2 = 1 \tag{15.104}$$

$$\left(T^1_{\ 1} - \lambda_2\right) \ell^1_{\ 2} + T^1_{\ 2}\, \ell^2_{\ 2} + T^1_{\ 3}\, \ell^3_{\ 2} = 0 \tag{15.105}$$

$$T^2_{\ 1}\, \ell^1_{\ 2} + \left(T^2_{\ 2} - \lambda_2\right) \ell^2_{\ 2} + T^2_{\ 3}\, \ell^3_{\ 2} = 0 \tag{15.106}$$

$$T^3_{\ 1}\, \ell^1_{\ 2} + T^3_{\ 2}\, \ell^2_{\ 2} + \left(T^3_{\ 3} - \lambda_2\right) \ell^3_{\ 2} = 0\,, \tag{15.107}$$

$$\left(\ell^1_{\ 2}\right)^2 + \left(\ell^2_{\ 2}\right)^2 + \left(\ell^3_{\ 2}\right)^2 = 1 \tag{15.108}$$

$$\left(T^1_{\ 1} - \lambda_3\right) \ell^1_{\ 3} + T^1_{\ 2}\, \ell^2_{\ 3} + T^1_{\ 3}\, \ell^3_{\ 3} = 0 \tag{15.109}$$

$$T^2_{\ 1}\, \ell^1_{\ 3} + \left(T^2_{\ 2} - \lambda_3\right) \ell^2_{\ 3} + T^2_{\ 3}\, \ell^3_{\ 3} = 0 \tag{15.110}$$

$$T^3_{\ 1}\, \ell^1_{\ 3} + T^3_{\ 2}\, \ell^2_{\ 3} + \left(T^3_{\ 3} - \lambda_3\right) \ell^3_{\ 3} = 0 \tag{15.111}$$

$$\left(\ell^1_{\ 3}\right)^2 + \left(\ell^2_{\ 3}\right)^2 + \left(\ell^3_{\ 3}\right)^2 = 1 \tag{15.112}$$

any two of the first three of which, plus the fourth in each set, constitute a set of three independent equations for determining the three ℓs. These ℓs are then the components of the ***unit eigenvectors*** which correspond, respectively, to the eigenvalues

$$\lambda_1 > \lambda_2 > \lambda_3 \tag{15.113}$$

In other words, there are three base (unit) vectors $\mathbf{e}_1$, $\mathbf{e}_2$, $\mathbf{e}_3$, which describe a new coordinate system called the ***principal axes*** associated with the transformation of the tensor.

Moreover, it is possible to demonstrate that these eigenvalues are an orthonormal system in which each axis is at right angles to the other two, so that there is a Cartesian coordinate system formed by them. Return to the basic equation of the eigenvalue question

$$(\mathbf{T} - \lambda_k \mathbf{I})\,\mathbf{e}_k \;=\; \mathbf{0}, \qquad k \;=\; 1,\,2,\,3 \tag{15.114}$$

and let it be written in componental form

$$T^i_{\;j}\,\ell^j_{\;k} - \lambda_k\,\delta^i_{\;j}\,\ell^j_{\;k} = 0, \qquad i,j \;=\; 1,\,2,\,3 \tag{15.115}$$

Apply this first to $k = 1$

$$T^i_{\;j}\,\ell^j_{\;1} - \lambda_1\,\delta^i_{\;j}\,\ell^j_{\;1} \;=\; 0 \tag{15.116}$$

and then to $k = 2$

$$T^i_{\;j}\,\ell^j_{\;2} - \lambda_2\,\delta^i_{\;j}\,\ell^j_{\;2} \;=\; 0 \tag{15.117}$$

Next, multiply the first equation by $\ell^j_{\;2}$ and the second by $\ell^j_{\;1}$ to get

$$T^i_{\;j}\,\ell^j_{\;1}\,\ell^j_{\;2} - \lambda_1\,\delta^i_{\;j}\,\ell^j_{\;1}\,\ell^j_{\;2} \;=\; 0 \tag{15.118}$$

$$T^i_{\;j}\,\ell^j_{\;1}\,\ell^j_{\;2} - \lambda_2\,\delta^i_{\;j}\,\ell^j_{\;1}\,\ell^j_{\;2} \;=\; 0 \tag{15.119}$$

Now subtract these two equations and obtain

$$T^i_{\;j}\,\left(\ell^j_{\;1}\,\ell^j_{\;2} - \ell^j_{\;1}\,\ell^j_{\;2}\right) - (\lambda_1 - \lambda_2)\,\ell^j_{\;1}\,\ell^j_{\;2}\,\delta^i_{\;j} = 0 \tag{15.120}$$

On the face of it, the first term is equal to 0, and this leaves

$$-\,(\lambda_1 - \lambda_2)\,\ell^j_{\;1}\,\ell^j_{\;2}\,\delta^i_{\;j} = 0 \tag{15.121}$$

However, λ_1 and λ_2 are, in general, distinct roots (that is, they are not equal to each other), $\lambda_1 - \lambda_2 \neq 0$, and this requires that

$$\ell^j_{\;1}\,\ell^j_{\;2} = \ell^i_{\;1}\,\ell^i_{\;2} = 0 \tag{15.122}$$

But this is just the requirement which governs direction cosines in an orthogonal system. When this analysis is repeated for the other two axis sets, it shows that

$$\ell^i_{\;1}\,\ell^i_{\;3} = 0\,, \qquad \ell^i_{\;2}\,\ell^i_{\;3} = 0 \tag{15.123}$$

That is to say, the eigenvectors $\mathbf{e}_1$, $\mathbf{e}_2$, $\mathbf{e}_3$ constitute a mutually orthogonal right-handed basis for a Cartesian axis set, which is called the ***principal axis system***.

Because each eigenvector is different in value from the other two, this principal axis system forms the semi-axes of a second-degree surface which could be described as an ***ellipsoidal eigenstate***. This is illustrated in Figure 15.4.

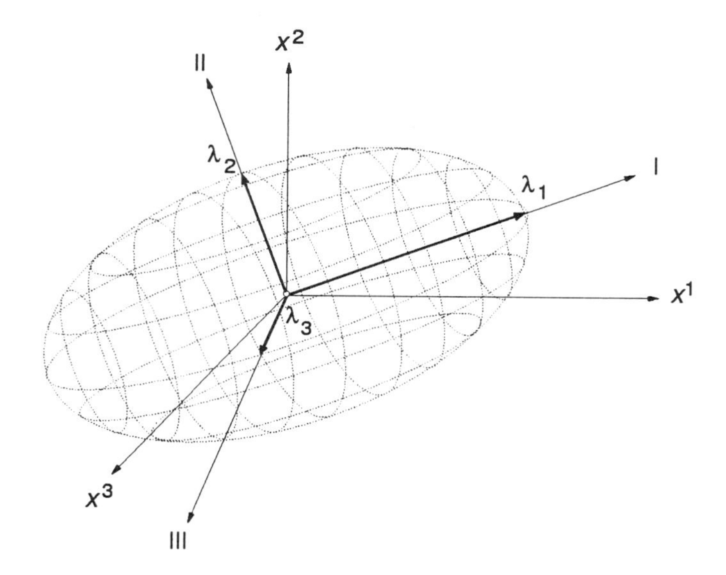

Figure 15.4: Eigenstate Ellipsoid

Naturally, if λ_k are not all distinct, this proof of orthogonality fails, and it becomes necessary to begin the analysis again. This time, assume that λ_1 and λ_2 are equal, so that it is λ_3 which is distinct. The result then is

$$\ell^j_{\ 1}\, \ell^j_{\ 3} \;=\; 0\,, \qquad \ell^j_{\ 2}\, \ell^j_{\ 3} \;=\; 0 \tag{15.124}$$

so that there remain two unit eigenvectors which are at right angles to each other. In this case it is $\mathbf{e}_1$ and $\mathbf{e}_3$ which are orthogonal, and so are $\mathbf{e}_2$ and $\mathbf{e}_3$, but $\mathbf{e}_1$ and $\mathbf{e}_2$ need not be.

Next, suppose that $\mathbf{e}_1$ and $\mathbf{e}_3$ are perpendicular to each other, and name them $\mathbf{e}_{1'}$, $\mathbf{e}_{3'}$, bases of a new (transformed) coordinate system. Let $\mathbf{e}_{2'}$ to be the third unit vector of this reference triad. Evidently, $\mathbf{e}_{2'} \;=\; \mathbf{e}_{3'} \;\times\; \mathbf{e}_{1'}$, so an orthogonal triad is still possible. Using these in the basic equation,

$$\mathbf{T}\;\mathbf{e}_1 \equiv \mathbf{T}\;\mathbf{e}_{1'} = \lambda_1\;\mathbf{I}\;\mathbf{e}_{1'} \;\equiv\; \lambda_1\;\mathbf{e}_{1'} = \lambda_2\;\mathbf{e}_{1'} \tag{15.125}$$

$$(\text{since}\;\; \lambda_1 = \lambda_2) \tag{15.126}$$

Let there be formed a linear combination of these eigenvectors by multiplying the first by some arbitrary scalar constant c_1 and the second by another arbitrary scalar constant c_2 to obtain

$$c_1 \mathbf{T}\;\mathbf{e}_{1'} = c_1\;\lambda_2\;\mathbf{e}_{1'} \tag{15.127}$$

$$c_2 \mathbf{T}\;\mathbf{e}_{2'} = c_2\;\lambda_2\;\mathbf{e}_{2'} \tag{15.128}$$

When these two are added

$$\mathbf{T}\,(c_1\;\mathbf{e}_{1'} + c_2\;\mathbf{e}_{2'}) = \lambda_2\,(c_1\;\mathbf{e}_{1'} + c_2\;\mathbf{e}_{2'}) \tag{15.129}$$

and with $\mathbf{e} = c_1\ \mathbf{e}_{1'} + c_2\ \mathbf{e}_{2'}$, this becomes

$$\mathbf{T}\ \mathbf{e} = \lambda_2\ \mathbf{e} \tag{15.130}$$

This shows that $\mathbf{e}$ is an eigenvector. But since c_1 and c_2 are arbitrary, $\mathbf{e}$ can be any vector in the plane formed by $\mathbf{e}_{1'}$ and $\mathbf{e}_{2'}$. Hence, all vectors in this plane are eigenvectors, and any two of them may be chosen along with $\mathbf{e}_{3'}$ to create the orthogonal triad which is designated as the principal axis system. Sometimes this condition will be named a ***cylindrical eigenstate***. Please refer to Figure 15.5.

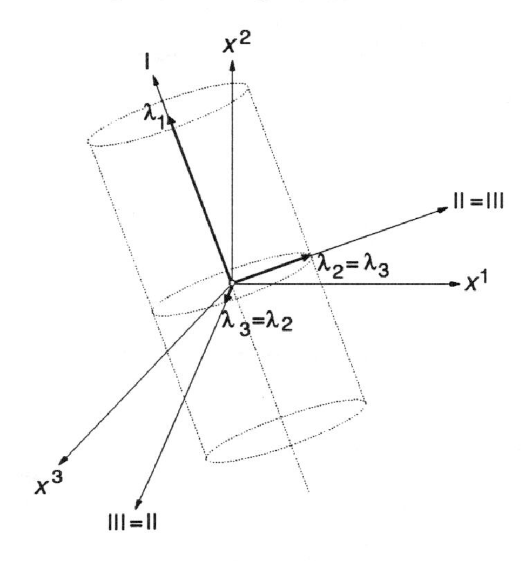

Figure 15.5: Cylindrical Eigenstate

For the case in which there are no distinct roots, (λ_1, λ_2, λ_3 are all equal to λ), a similar proof to that employed in the prior case can be used to show that all directions in space are eigenvectors. This means that every unit vector is an eigenvector, and each triad of reference is a principal axis system. Such a system might be referred to as a ***spherical eigenstate***. Please refer to Figure 15.6.

Note that when presented in the form of its principal axis configuration, the Cartesian tensor matrix is ***diagonalized***; that is, only elements on the principal diagonal have nonzero values. Therefore

$$\mathbf{T} = \begin{bmatrix} T^1_1 & T^1_2 & T^1_3 \\ T^2_1 & T^2_2 & T^2_3 \\ T^3_1 & T^3_2 & T^3_3 \end{bmatrix} = \begin{bmatrix} T_I & 0 & 0 \\ 0 & T_{II} & 0 \\ 0 & 0 & T_{III} \end{bmatrix} = \begin{bmatrix} \lambda_1 & 0 & 0 \\ 0 & \lambda_2 & 0 \\ 0 & 0 & \lambda_3 \end{bmatrix} \tag{15.131}$$

are all equivalent ways of expressing the same tensor quantity.

It must be emphasized, however, that the distinction between a tensor and the various forms of its representation must be scrupulously preserved, and that the tensor possesses its fundamental properties whether or not any particular one of these modes of representation is employed.

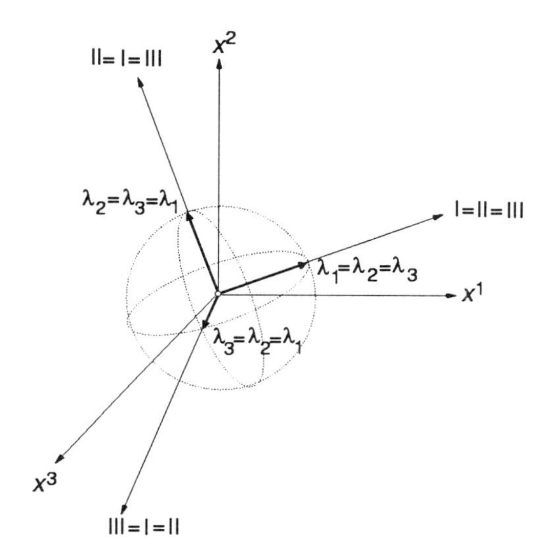

Figure 15.6: Spherical Eigenstate

A method of computing the eigenvalues and eigenvectors of a matrix tensor is contained in Appendix K.

15.2.12 VECTOR CALCULUS

The dynamics of state space systems is formulated in terms of vector and tensor quantities that are subjected to differentiating and integrating operators. These operators are logical extensions of those employed in the calculus of real scalar-valued variables. But it is obvious that this vector and tensor calculus is more general than that involving only scalars. Thus a more general start must be made in its development.

As a first step, note that for a vector function of a scalar variable the vector-valued *derivative* is, by analogy with that of the scalar derivative, defined as

$$\frac{d}{ds}\mathbf{v}(s) = \lim_{\Delta s \to 0} \frac{\mathbf{v}(s+\Delta s) - \mathbf{v}(s)}{\Delta s} \tag{15.132}$$

In the case of both scalar and vector products, differentiation of vector and tensor quantities obeys the common rules for products as

$$\frac{d}{ds}(\mathbf{u}\cdot\mathbf{v}) = \frac{d\mathbf{u}}{ds}\cdot\mathbf{v} + \mathbf{u}\cdot\frac{d\mathbf{v}}{ds} \tag{15.133}$$

and

$$\frac{d}{ds}(\mathbf{u}\times\mathbf{v}) = \frac{d\mathbf{u}}{ds}\times\mathbf{v} + \mathbf{u}\times\frac{d\mathbf{v}}{ds} \tag{15.134}$$

where s is the scalar-valued variable. When the question is that of a vector-valued function of a vector-valued argument, there is available the *chain rule*. Using

$$\mathbf{w}(\mathbf{v}) = \mathbf{u}[f(\mathbf{v})] \tag{15.135}$$

the result obtained is

$$\frac{\partial \mathbf{w}}{\partial \mathbf{v}} = \frac{\partial \mathbf{u}}{\partial f}\frac{\partial f}{\partial \mathbf{v}} \tag{15.136}$$

Base vectors $\mathbf{e}_i$ are functions of the field position as expressed by field coordinates x^i, and these are in turn components of the ***position vector*** $\mathbf{r}$ from the coordinate origin to the field point. The position vector is

$$\mathbf{r} = x^1\mathbf{e}_1 + x^2\mathbf{e}_2 + x^3\mathbf{e}_3 + \cdots + x^n\mathbf{e}_n \equiv x^i\mathbf{e}_i \tag{15.137}$$

so it is seen that

$$\frac{\partial \mathbf{r}}{\partial x^i} = \mathbf{e}_i \tag{15.138}$$

which may be taken as a definition of the base vector $\mathbf{e}_i$, or, alternatively, of the ***partial derivative***.

Inertial derivatives of the position vector are formed as

$$\frac{d\,\mathbf{r}}{dt} = \begin{bmatrix} \dfrac{dx^1}{dt} \\ \dfrac{dx^2}{dt} \\ \cdot \\ \cdot \\ \dfrac{dx^n}{dt} \end{bmatrix} \tag{15.139}$$

Inertial derivatives of higher-order entities such as tensors can be displayed by using their matrix representations

$$\mathbf{T}(t) = [T^i_j(t)] \tag{15.140}$$

and then

$$\frac{d\mathbf{T}(t)}{dt} \doteq \left[\frac{dT^i_j}{dt}\right] \tag{15.141}$$

Some additional results are:

$$\frac{d\,(\mathbf{Tr})}{dt} = \frac{d\mathbf{T}}{dt}\mathbf{r} + \mathbf{T}\frac{d\mathbf{r}}{dt} \tag{15.142}$$

$$\frac{d\,(\mathbf{T}^1\mathbf{T}^2)}{dt} = \frac{d\mathbf{T}^1}{dt}\mathbf{T}^2 + \mathbf{T}^1\frac{d\mathbf{T}^2}{dt} \tag{15.143}$$

$$\frac{d\mathbf{T}^{-1}}{dt} = -\mathbf{T}^{-1}\frac{d\mathbf{T}}{dt}\mathbf{T}^{-1} \tag{15.144}$$

Likewise, the integration of a vector—say, a position vector—is given by

$$\int^t \mathbf{r}(\tau)\,d\tau \doteq \begin{bmatrix} \int^t x^1(\tau)\,d\tau \\ \int^t x^2(\tau)\,d\tau \\ \cdot \\ \cdot \\ \int^t x^n(\tau)\,d\tau \end{bmatrix} \tag{15.145}$$

Integration with respect to tensor-valued matrix quantities proceeds according to

$$\int^t \mathbf{T}(\tau)\,d\tau \doteq \left[\int^t T^i_j(\tau)\,d\tau\right] \tag{15.146}$$

With the exception that, in general, matrices do not commute, most operations of scalar calculus can be appropriated to the levels of vectors and tensors insofar as they are represented by suitable matrices.

15.3 LAPLACE TRANSFORM

One other higher order operation which is ubiquitous in state space systems work is that of the Laplace transform. For a scalar-valued quantity, this is defined by the integral

$$\mathcal{L}[f(t)] = \int_0^\infty f(t)\,e^{-st}\,dt = F(s) \tag{15.147}$$

The value of and purpose in making such a transformation is to remove the function being transformed ($f(t)$ in the example above) from a realm in which operations upon it are difficult to a transformed region where, hopefully, these operations may be made simpler. Of course, to be useful in this manner there needs to be a dictionary of inverse transformations to bring the quantity after operations upon it have been performed back to the original domain, where it can then be interpreted.

Thus, the inverse transform is

$$\mathcal{L}^{-1}[F(s)] = f(t) = \frac{1}{2\pi i}\int_{\gamma-i\infty}^{\gamma+i\infty} F(s)\,e^{st}\,ds \tag{15.148}$$

The latter is a contour integral derived from Cauchy's fundamental theorem of the complex variable.[2] The pair $\mathcal{L}[f(t)]$ and $\mathcal{L}^{-1}[F(s)]$ are therefore seen to be operators which convert (map) a function $f(t)$ of the t-domain into its image $F(s)$ in the s-domain and the inverse map of $F(s)$ in the s-domain into its image $f(t)$ in the t-domain.

Although the present status of this transform theory represents the distillation of nearly two centuries of improvements, the original work on it was done by Pierre Simon, Marquis de Laplace. Major contributions by way of application of the method to problems of technology were made by Oliver Heaviside in 1894. He called his methods *Operational Calculus*, but these were driven by pragmatic concerns not shared by academic mathematicians and embroiled him in long-running disputes with them as to the validity of the methods. In the end, the merits of the Heaviside calculus won out over the objections to it, but the subject is now subsumed in the *Laplace Transform Theory*. It seems fair to say that this topic has driven both mathematical theory and technological applications to vastly more productive regions.

---

[2] Sneddon, I. N., *The Use of Integral Transforms*, Chapter 3, McGraw-Hill Book Company, New York, 1972, xii+539 pp.

When applied to matrix tensor quantities, the Laplace transform is

$$\mathcal{L}[\mathbf{T}(t)] = \int_0^\infty \mathbf{T}(t)\, e^{-st}\, dt = \left[\int_0^\infty T^i_j(t)\, e^{-st}\, dt\right] = [T^i_j(s)] = \mathbf{T}(s) \qquad (15.149)$$

Suppose that there is a matrix tensor $\mathbf{T}^1$ which is a constant with respect to time. The Laplace transform of the product of $\mathbf{T}^1$ and a second matrix tensor $\mathbf{T}^2$ that is time-varying can be obtained as follows:

$$\mathcal{L}[\mathbf{T}^1\, \mathbf{T}^2] = \mathbf{T}^1\, \mathcal{L}[\mathbf{T}^2] \qquad (15.150)$$

The frequent occurrence of motional quantities such as velocity and acceleration can also be subjected to Laplace transformation. For instance,

$$\mathcal{L}[\mathbf{v}] = \mathcal{L}\left[\frac{d\mathbf{r}}{dt}\right] = \begin{bmatrix} \mathcal{L}\left[\dfrac{dx^1}{dt}\right] \\ \mathcal{L}\left[\dfrac{dx^2}{dt}\right] \\ \cdot \\ \cdot \\ \cdot \\ \mathcal{L}\left[\dfrac{dx^n}{dt}\right] \end{bmatrix} = \begin{bmatrix} s\,x^1(s) - x^1(0) \\ s\,x^2(s) - x^2(0) \\ \cdot \\ \cdot \\ \cdot \\ s\,x^n(s) - x^n(0) \end{bmatrix} = s\,\mathbf{r}(s) - \mathbf{r}(0) \qquad (15.151)$$

In this formula, $\mathbf{r}(0)$ is the ***initial value*** of $\mathbf{r}$ at the moment the stop-watch begins running; i.e., at time zero.

It is clear that the Laplace transform converts the time-varying velocity into a purely algebraic non-time-varying transformed variable s, so that manipulations of it can be carried out as algebra, rather than as calculus. This might often be a very large advantage.

Similarly, the acceleration vector is transformed by

$$\mathcal{L}[\mathbf{a}] = \mathcal{L}\left[\frac{d\mathbf{v}}{dt}\right] = s^2\,\mathbf{r}(s) - s\,\mathbf{r}(0) - \frac{d\mathbf{r}}{dt}(0) \qquad (15.152)$$

where $\frac{d\mathbf{r}}{dt}(0)$ is the initial velocity $\mathbf{v}(0) = \dot{\mathbf{r}}(0)$.

The process can be continued repetitively to obtain the higher derivatives

$$\mathcal{L}[\mathbf{r}^{(n)}(t)] = s^n\,\mathbf{r}(s) - s^{n-1}\frac{d\mathbf{r}}{dt}(0) - s^{n-2}\frac{d^2\mathbf{r}}{dt^2}(0) - \cdots - \frac{d^{n-1}\mathbf{r}}{dt^{n-1}}(0) \qquad (15.153)$$

Obviously, derivatives of all orders are converted by the transformation into algebraic polynomials in the transformed variable s. The import of this finding is that systems of linear differential equations are converted into systems of algebraic equations which are simultaneous in the transformed variable.

The Laplace transforms of some elementary functions are:

- $\mathcal{L}[1] = \int_0^\infty e^{-st}\,dt = [-s^{-1}\,e^{-st}]_0^\infty = 1/s$,
- $\mathcal{L}[t] = \int_0^\infty t\,e^{-st}\,dt = [-s^{-1}\,e^{-st}]_0^\infty + 1/s\int_0^\infty e^{-st}\,dt = 1/s^2$,
- $\mathcal{L}[t^n] = \int_0^\infty t^n\,e^{-st}\,dt = 1/s^{n+1}\,\Gamma(n+1) = n!/s^{n+1}$ if n is an integer ,
- $\mathcal{L}[sin\,\omega t] = \omega/(s^2+\omega^2)$,
- $\mathcal{L}[cos\,\omega t] = s/(s^2+\omega^2)$,
- $\mathcal{L}[e^{at}] = 1/(s-a)$.

A few elementary inverse transforms are:

- $\mathcal{L}^{-1}[\beta/(s^2-\beta^2)] = Sinh\,\beta t$,
- $\mathcal{L}^{-1}[s/(s^2-\beta^2)] = Cosh\,\beta t$,
- $\mathcal{L}^{-1}[s/(s^4+4a^4)] = 1/2a^2\,sin\,at\,Sinh\,at$,
- $\mathcal{L}^{-1}[s^3/(s^4+4a^4)] = cos\,at\,Cosh\,at$,
- $\mathcal{L}^{-1}[\ln\frac{s+a}{s+b}] = \frac{1}{t}(e^{-bt}-e^{-at})$,
- $\mathcal{L}^{-1}[\ln\sqrt{\frac{s^s+b^2}{s^s-a^2}}] = \frac{1}{t}(Cosh\,at - cos\,bt)$.

In fact, lengthy tables of transforms for a variety of algebraic, trigonometric, orthogonal, and transcendental functions are available in the literature.[3]

15.4 STATE SPACE

The state of a system has a formal definition which follows that originally given by Kalman:[4]

> The *state* of a system is the least set of quantities (*state variables*) that furnish adequate information about the system to permit the computation of its future dynamic.

The specification of the state is most appropriately given in terms of the *state vector* $\mathbf{x}$. This vector is composed of the state variables as

$$\mathbf{x}(t) = [x^1(t)\;\; x^2(t)\;\; x^3(t)\;\; \cdots\;\; x^n(t)] \tag{15.154}$$

The state is also a function of the *initial condition*, or *initial state*,

$$\mathbf{x}(t_0) = [x^1(t_o)\;\; x^2(t_0)\;\; x^3(t_0)\;\; \cdots\;\; x^n(t_0)] \tag{15.155}$$

---

[3]Doetsch, G., *Theorie und Anwendung der Laplace-Transformation*, Springer-Verlag, Berlin, 1937.

[4]Kalman, R. E., "On the General Theory of Control Systems," *Proceedings of the First International Congress on Automatic Control*, Butterworth & Company (publishers), Ltd., London, Volume 1, 1961, pp. 481-493.

The system's dynamic is then expressed by the ***state transition equation***

$$\mathbf{x}(t) = \mathcal{F}[\mathbf{x}(t_0); \mathbf{u}(t)] \tag{15.156}$$

and the ***input-output-state relation***

$$\mathbf{y}(t) = \mathcal{G}[\mathbf{x}(t_0); \mathbf{u}(t)] \tag{15.157}$$

wherein $\mathbf{x}(t_0)$ is the initial state, $\mathbf{x}(t)$ is the state at time t, $\mathbf{u}(t)$ is the input vector, and $\mathbf{y}(t)$ is the output vector. $\mathcal{F}$ and $\mathcal{G}$ are some specific functions.

Implied in these definitions is the requirement that the system state at time t is ***uniquely*** determined by the prior state at t_0 and the input $\mathbf{u}(t)$; that is, $\mathcal{F}$ and $\mathcal{G}$ are single-valued functions of their arguments.

In the remainder of this chapter, the systems which are treated are assumed to be those for which the state equations are differential equations with the form

$$\dot{\mathbf{x}}(t) = \mathbf{f}[\mathbf{x}(t), \mathbf{u}(t), t] \tag{15.158}$$

$$\mathbf{y}(t) = \mathbf{g}[\mathbf{x}(t), \mathbf{u}(t), t] \tag{15.159}$$

Example Let the system be represented by the differential equation

$$m\,\ddot{y} + c\,\dot{y} + k\,y = f\,cos\,\omega t \tag{15.160}$$

in which m is a mass, c is a damping constant, k is a spring modulus, and y is the displacement of the mass from its equilibrium position. There is a driving force, $f\,cos\,\omega t$ which acts upon the mass. The system is illustrated in Figure 15.7.

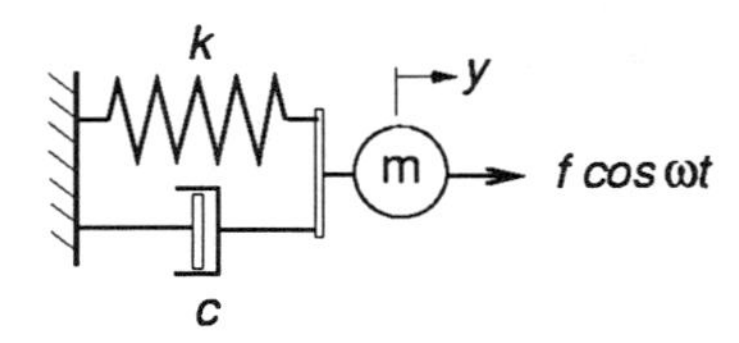

Figure 15.7: Spring, Mass, and Dashpot System

Choose the state variables to be $x^1 = y$, and $x^2 = \dot{y}$, and there are then

$$\dot{x}^1 = \dot{y} = x^2 \tag{15.161}$$

$$\dot{x}^2 = \frac{1}{m}(f\,cos\,\omega t - c\,\dot{x}^1 - k\,x^1) = \frac{1}{m}(f\,cos\,\omega t - c\,x^2 - k\,x^1) \tag{15.162}$$

Then, with the state vector $\mathbf{x} = [x^1 \;\; x^2]$, and input $\mathbf{u} = [0 \;\; f\,cos\,\omega t]$, there is

$$\begin{bmatrix} \dot{x}^1 \\ \dot{x}^2 \end{bmatrix} = \begin{bmatrix} x^2 \\ -c\,x^2 - m\,x^1 + f\,cos\,\omega t \end{bmatrix} \tag{15.163}$$

or, in functional form

$$\dot{\mathbf{x}} = \mathbf{f}(\mathbf{x}, \mathbf{u}) \tag{15.164}$$

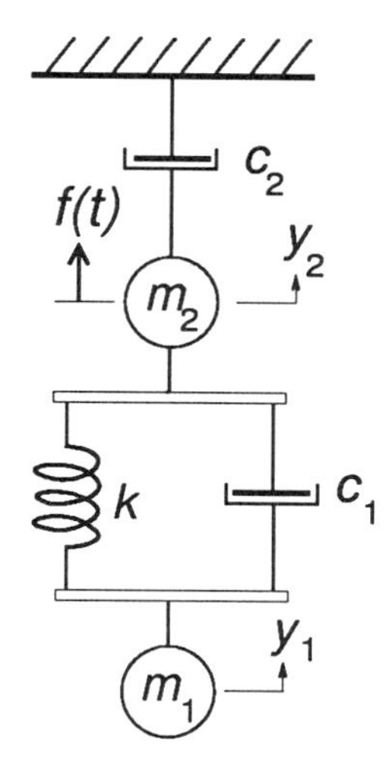

Figure 15.8: Spring, Masses, and Dashpots System

State space methods are applicable to a very large number of system types. In the example above, the system was represented by a differential equation. Some other common types are called *discrete-time* or *sample-data*, and there are also *difference equations* instead of differential equations. Each of these has many applications, and merits an entire text in its own right. Moreover, the system can be designated as *discrete-element* or *continuous-element*, to signify the composition of the physical or material distribution within the system.

The systems which will be the subject of the following development will be, largely, of the type known as *linear differential systems*, meaning that they are describable by sets of linear differential equations with the dynamical elements continuous. For such systems the state equations can be set into canonical form, again, after Kalman,[5]

$$\dot{\mathbf{x}}(t) = \mathbf{A}(t)\,\mathbf{x}(t) + \mathbf{B}(t)\,\mathbf{u}(t) \tag{15.165}$$

$$\mathbf{y}(t) = \mathbf{C}(t)\,\mathbf{x}(t) + \mathbf{D}(t)\,\mathbf{u}(t) \tag{15.166}$$

In these equations, $\mathbf{A}(t)$, $\mathbf{B}(t)$, $\mathbf{C}(t)$, and $\mathbf{D}(t)$ are matrix operators of appropriate ranks.

Example To illustrate the application of the canonical form modeling, consider the mechanical system represented in Figure 15.8.

There are two masses, m_1 and m_2, joined to each other by a spring which has a modulus k N/m. The masses also experience damping as they are moving relative to each other. This dissipation of energy can be represented by a dashpot (or shock absorber) which has a damping constant c_1. Mass m_2 also produces a damping loss between itself and

[5]Kalman, R. E., "Mathematical Description of Linear Dynamic Systems," *SIAM Journal of Control*, Volume 1, 1963, pp. 152-192.

the foundation (ground) with constant c_2. Mass m_2 is subjected to a force $f(t)$.

The displacements of the two masses (with respect to ground) are denoted $y_1(t)$ and $y_2(t)$, respectively.

Upon invoking the Newtonian laws of motion, the equations of motion for both masses are:

$$m_1 \frac{dv_1(t)}{dt} = k[y_2(t) - y_1(t)] + c_1[v_2(t) - v_1(t)] - m_1 g \qquad (15.167)$$

$$m_2 \frac{dv_2(t)}{dt} = f(t) - k[y_2(t) - y_1(t)] - c_2 v_2(t) - c_1[v_2(t) - v_1(t)] - m_2 g \qquad (15.168)$$

There are usually a number of different ways in which the state variables can be selected for a given system. The ease and sometimes even the power of solving the system equations may be dependent upon a judicious choice of these state variables. A good policy to follow in the case of mechanical systems is to choose the displacements and velocities of each of the masses in the system as state variables. In the case under study, there are two masses, therefore two state variables based upon their displacements, and two more for the velocities. Thus in this problem there are four state variables.

Note that the relative velocity between the masses is available as

$$\dot{y}^2(t) - \dot{y}^1(t) \qquad (15.169)$$

and there is also the displacement of mass m_1. Therefore let the state variables be

$$\dot{x}^1(t) = \dot{v}^1(t) = \frac{k}{m_1} x^3(t) + \frac{c_1}{m_1}[x^2(t) - x^1(t)] - g \qquad (15.170)$$

$$\dot{x}^2(t) = \dot{v}^2(t) = \frac{1}{m_2}\{f(t) - k\, x^2(t) - c_2\, x^2(t) - c_1[x^2(t) - x^1(t)]\} - g \qquad (15.171)$$

$$\dot{x}^3(t) = \dot{y}^2(t) - \dot{y}^1(t) = x^2(t) - x^1(t) \qquad (15.172)$$

$$\dot{x}^4(t) = x^1(t) \qquad (15.173)$$

Now it can be seen that the matrices **A**, **B**, **C**, and **D** of the canonical form for this system are

$$\mathbf{A} = \begin{bmatrix} -\frac{c_1}{m_1} & \frac{c_2}{m_1} & \frac{k}{m_1} & 0 \\ \frac{c_1}{m_2} & -\frac{c_1+c_2}{m_2} & -\frac{k}{m_2} & 0 \\ -1 & 1 & 0 & 0 \\ -1 & 1 & 0 & 0 \end{bmatrix} \qquad \mathbf{B} = \begin{bmatrix} 0 \\ \frac{1}{m_2} \\ 0 \\ 0 \end{bmatrix} \tag{15.174}$$

$$\mathbf{C} = [\,0\,0\,0\,1\,] \qquad \mathbf{D} = [\,0\,0\,0\,0\,] \tag{15.175}$$

The state vector and its derivative are

$$\mathbf{x} = \begin{bmatrix} x^1 \\ x^2 \\ x^3 \\ x^4 \end{bmatrix} \qquad \dot{\mathbf{x}} = \begin{bmatrix} \dot{x}^1 \\ \dot{x}^2 \\ \dot{x}^3 \\ \dot{x}^4 \end{bmatrix} \tag{15.176}$$

and the input and output vectors are

$$\mathbf{u} = [0 \;\; f(t) \;\; 0 \;\; 0], \qquad \mathbf{y} = [0 \;\; x^2 \;\; 0 \;\; 0] \tag{15.177}$$

The state equations are seen to be

$$\begin{bmatrix} \dot{x}^1 \\ \dot{x}^2 \\ \dot{x}^3 \\ \dot{x}^4 \end{bmatrix} = \begin{bmatrix} -\frac{c_1}{m_1} & \frac{c_2}{m_1} & \frac{k}{m_1} & 0 \\ \frac{c_1}{m_2} & -\frac{c_1+c_2}{m_2} & -\frac{k}{m_2} & 0 \\ -1 & 1 & 0 & 0 \\ -1 & 1 & 0 & 0 \end{bmatrix} \begin{bmatrix} x^1 \\ x^2 \\ x^3 \\ x^4 \end{bmatrix} + \begin{bmatrix} 0 \\ \frac{1}{m_2} \\ 0 \\ 0 \end{bmatrix} [0 \; f(t) \; 0 \; 0]$$

$$= \begin{bmatrix} -\frac{c_1}{m_1}x^1 + \frac{c_2}{m_1}x^2 + \frac{k}{m_1}x^3 + 0\,x^4 \\ \frac{c_1}{m_2}x^1 - \frac{c_1+r_2}{m_2}x^2 - \frac{k}{m_2}x^3 + \frac{1}{m_2}f(t) \\ -x^1 + x^2 \\ -x^1 + x^2 \end{bmatrix} \tag{15.178}$$

and

$$\begin{bmatrix} 0 \\ x^2 \\ 0 \\ 0 \end{bmatrix} = [0\,0\,0\,1] \begin{bmatrix} x^1 \\ x^2 \\ x^3 \\ x^4 \end{bmatrix} + [0\,0\,0\,0] \begin{bmatrix} 0 \\ f(t) \\ 0 \\ 0 \end{bmatrix} \tag{15.179}$$

These therefore confirm the canonical form of the state equations for this system.

Example Consider another representative state space modeling of a system; this time it is an electrical network. There are two loops, as seen in Figure 15.9. The first circuit consists of a signal generator which produces a voltage $e(t)$ that is applied to a capacitor in series with a resistor, R_1. In shunt with this resistor is a second circuit comprised of an inductance L and a second resistor, R_2. The currents in the two loops are designated i_1 and i_2.

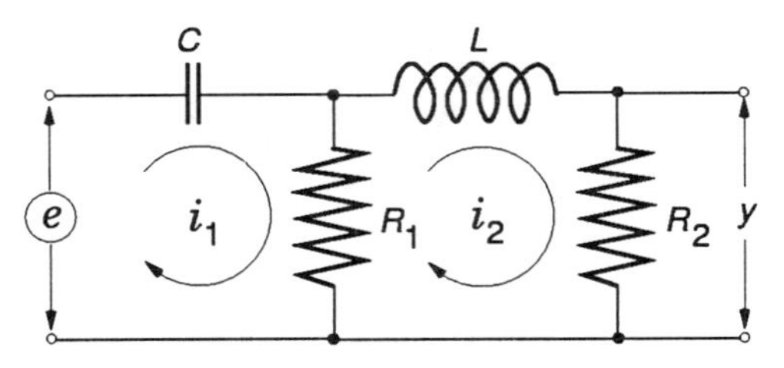

Figure 15.9: System Composed of Capacitor, Inductance, Resistors

From the elementary theory of electrical elements, the voltage across the inductor and the current through it are related by

$$v_L = L\frac{di_L}{dt} \tag{15.180}$$

and the voltage and current relationship for the capacitor is given by

$$i_C = C\frac{dv_C}{dt} \tag{15.181}$$

Because the signal source produces a voltage that is continuously varying in time, the circuits comprising the system are dynamic. However, the circuit elements themselves are not time-varying in this example. The first task is to designate a state vector for the system. It is convenient to take the two loop currents $x^1 = i_1$ and $x^2 = i_2$ as the two components of the state vector in this example. Next, write Kirchoff's laws as applied to each of the loops as follows:

$$e(t) = \frac{1}{C}\int i_1\,dt + R_1(i_1 - i_2) \tag{15.182}$$

$$0 = R_1(i_2 - i_1) + L\frac{d\,i_2}{dt} + R_2\,i_2 \tag{15.183}$$

Then differentiate the first equation and afterwards substitute $x^1 = i_1$, $x^2 = i_2$, (and also $\dot{x}^1 = d\,i_1/dt$, $\dot{x}^2 = d\,i_2/dt$), and obtain the result

$$\begin{bmatrix}\dot{x}^1\\ \dot{x}^2\end{bmatrix} = \begin{bmatrix}\dfrac{R_1}{L} - \dfrac{1}{R_1C} & -\dfrac{R_1+R_2}{L}\\ \dfrac{R_1}{L} & -\dfrac{R_1+R_2}{L}\end{bmatrix}\begin{bmatrix}x^1\\ x^2\end{bmatrix} + \begin{bmatrix}\dfrac{\dot{e}}{R_1}\\ 0\end{bmatrix} \tag{15.184}$$

or

$$\begin{bmatrix} \dot{x}^1 \\ \dot{x}^2 \end{bmatrix} = \begin{bmatrix} \dfrac{R_1}{L} - \dfrac{1}{R_1 C} & -\dfrac{R_1 + R_2}{L} \\ \dfrac{R_1}{L} & -\dfrac{R_1 + R_2}{L} \end{bmatrix} \begin{bmatrix} x^1 \\ x^2 \end{bmatrix} + \begin{bmatrix} \dfrac{\dot{e}}{R_1} \\ 0 \end{bmatrix} \begin{bmatrix} 1 & 0 \\ 0 & 1 \end{bmatrix} \tag{15.185}$$

If the output from this system is taken to be the voltage y across the terminals at R_2, the result is

$$y = i_2 R_2 = R_2 x^2 \tag{15.186}$$

or, in matrix form,

$$\mathbf{y} = [\,0 \; R_2\,] \begin{bmatrix} x^1 \\ x^2 \end{bmatrix} = \mathbf{C}\,\mathbf{x} \tag{15.187}$$

The linear differential state equations for this system are seen to be

$$\dot{\mathbf{x}} = \mathbf{A}(t)\,\mathbf{x} + \mathbf{B}(t)\,\mathbf{u} \tag{15.188}$$

$$\mathbf{y} = \mathbf{C}(t)\,\mathbf{x} + \mathbf{D}(t)\,\mathbf{u} \tag{15.189}$$

in which the input vector is the signal source

$$\mathbf{u} = \begin{bmatrix} \dfrac{\dot{e}}{R_1} \\ 0 \end{bmatrix} \tag{15.190}$$

The matrices **A**, **B**, **C**, and **D** are given by

$$\mathbf{A} = \begin{bmatrix} \dfrac{R_1}{L} - \dfrac{1}{R_1 C} & -\dfrac{R_1 + R_2}{L} \\ \dfrac{R_1}{L} & -\dfrac{R_1 + R_2}{L} \end{bmatrix} \qquad \mathbf{B} = \begin{bmatrix} 1 & 0 \\ 0 & 1 \end{bmatrix} \tag{15.191}$$

$$\mathbf{C} = [\,0 \quad R_2\,] \qquad \mathbf{D} = \mathbf{0} \tag{15.192}$$

The canonical form of the state equations is thereby shown explicitly.

15.5 SINGULARITY FUNCTIONS

Certain types of excitation prove to be of special virtue in analyzing the response of systems when modeled in state space form. These are represented by functions of specified nature, and are the now-canonical forms which arose, as did much of what is known in science, in a piecemeal way from various specialties and in differing times.

The family of such functions is often referred to as ***singularity functions***, although it has sometimes been fashionable to call them ***pathological functions***. The latter reference, particularly, has some history in connection with the functions originally employed by Oliver Heaviside in 1894 in his studies of the transient response of electrical circuits. Heaviside proposed the elements of an "experimental mathematics," which worked very well to predict the desired responses, but were lacking in mathematical rigor and sophistication.

Predictably, Heaviside's *Operational Mathematics* was found wanting in the learned mathematical circles of his time, and even for many years later. He dissipated his energies in warring with "the stuffed shirts at Cambridge and Oxford," and was even more villified by the continentals. In the end, Heaviside's calculus was incorporated into what is now known as the theory of the *Laplace Transform*. The similar heuristic proposals advanced by J. B. J. Fourier escaped a similar opprobrium, and emerged as *Fourier Transforms*.

One of the more important of the pathological functions was utilized by P. A. M. Dirac, the English physicist, and has often been called the *Dirac Delta Function*.[6] To some, it seemed the most pathological of all the family of such functions, and, in fact, it challenged the mathematical definition of a function at a most basic level.

The contributions of Fourier, Heaviside, and Dirac, taken together, served to reform the mathematical definition of a function, and resulted in this new class now called either singularity functions or possibly *generalized functions*.[7]

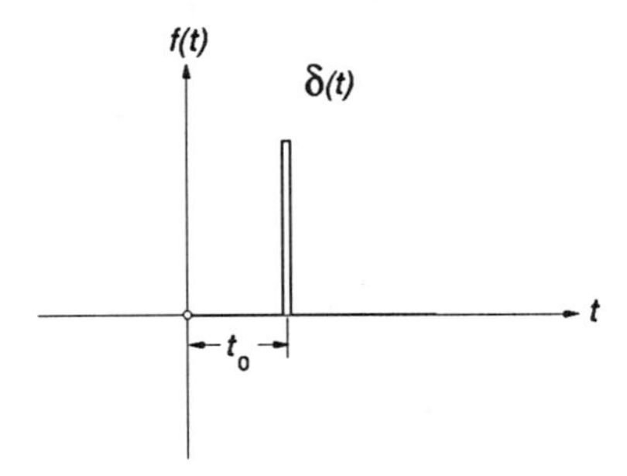

Figure 15.10: The Dirac Delta Function

The Dirac Delta Function, or *unit impulse function*, is a weird animal such that it is everywhere (in time) equal to zero except at the time t_0, where it is of undetermined but presumed-to-be-very-large magnitude. See Figure 15.10.

$$\delta(t - t_0) = \text{undefined} \qquad t = t_0 \tag{15.193}$$

$$= 0 \qquad t \neq t_0 \tag{15.194}$$

The further stipulation is that

$$\int_{-\infty}^{+\infty} \delta(t - t_0)\, dt = 1 \tag{15.195}$$

[6]Dirac, P. A. M., *The Principles of Quantum Mechanics*, Clarendon Press, Oxford, 1947, p. 58.

[7]Sneddon, I. N., *The Use of Integral Transforms*, McGraw-Hill Book Company, New York, 1972, p. 9.

In other words, the area under the waveform of this function is normalized to unity, hence the name ***unit*** impulse function. This concept arises from the definition of impulse in mechanics, where

$$\int_0^t F(t)\,dt = I \tag{15.196}$$

is the impulse of the force $F(t)$ acting through the time increment dt, and I is that impulse. It is thus characteristic of those forces in mechanics which act for a short duration of time with a large magnitude of force. The generalization to systems of other than mechanical nature is a matter of rhetoric only.

Mathematicians approach the definition of the function in a limiting context. Take, for example, the function in the form

$$\delta_\epsilon(t) = 0 \qquad t \leq 0 \tag{15.197}$$

$$= 1/\epsilon \qquad 0 \leq t \leq \epsilon \tag{15.198}$$

$$= 0 \qquad t \geq \epsilon \tag{15.199}$$

For a function $f(t)$ continuous over an interval containing the sub-interval $(0, \epsilon)$, there is then

$$\int_{-\infty}^{+\infty} \delta_\epsilon(t)\, f(t)\,dt = \frac{1}{\epsilon}\int_0^\epsilon f(t)\,dt \tag{15.200}$$

This gives (using the mean value theorem of the calculus)

$$\int_{-\infty}^{+\infty} \delta_\epsilon(t)\, f(t)\,dt = f(k\epsilon)\,, \qquad 0 \leq k \leq 1 \tag{15.201}$$

and in the limit as $\epsilon \to 0$,

$$\int_{-\infty}^{+\infty} \delta(t)\, f(t)\,dt = f(0) \tag{15.202}$$

for functions that are continuous in the neighborhood of the origin.

This last result has the obvious property that it selects the particular value of $f(t)$ at the place where $t = 0$; namely, $f(0)$. In recognition of this selective attribute, it was named the ***sifting function*** by Balthasar van der Pol.[8]

Now a quick change of variable from t to τ where $\tau = t + t_0$ gives

$$f(t_0) = \int_{-\infty}^{+\infty} \delta(\tau - t_0)\, f(\tau)\,d\tau \tag{15.203}$$

The Delta function is also seen to be the singularity function obtained from another such function by taking the "derivative" of it, whatever that may mean. Consider the integral

$$\int_{-\infty}^{+\infty} F'(t)\, f(t)\,dt\,. \tag{15.204}$$

[8]Van der Pol, B., and Bremmer, H., ***Operational Calculus Based on the Two-Sided Laplace Integral***, Cambridge University Press, New York, 1950.

It can be integrated by parts as

$$\int_{-\infty}^{+\infty} F'(t)\, f(t)\, dt = [F(t)\, f(t)]_{-\infty}^{+\infty} - \int_{-\infty}^{+\infty} F(t)\, f'(t)\, dt \tag{15.205}$$

Now introduce the ***unit step function*** shown in Figure 15.11;

$$\mathbf{1}(t-t_0) = 0 \quad t < t_0 \tag{15.206}$$

$$= 1 \quad t > t_0 \tag{15.207}$$

This was the function originally employed by Heaviside. Let the unit step function play the role of $F'(t)$ in the previous integral.

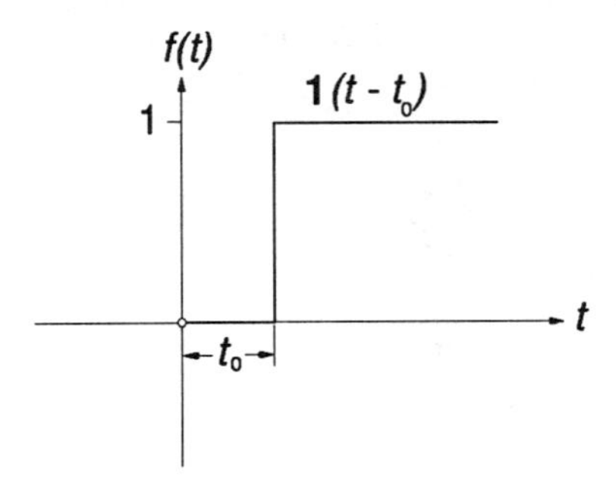

Figure 15.11: The Unit Step Function

Then it is seen that

$$\int_{-\infty}^{+\infty} \mathbf{1}'(t-t_0)\, f(t)\, dt = -\int_{0}^{+\infty} f'(t)\, dt = f(t_0) = \int_{-\infty}^{+\infty} \delta(t-t_0)\, f(t)\, dt \tag{15.208}$$

since the substitution of 1 and 0 for $\mathbf{1}(t-t_0)$ in the integral by parts eliminates the first evaluation by parts, and reduces the lower limit of the second parts integral to 0. (This assumes, of course, that $f(t) \to 0$ as $t \to \infty$.) Therefore

$$\delta(t-t_0) = \mathbf{1}'(t-t_0) \tag{15.209}$$

or

$$\mathbf{1}(t-t_0) = \int_{-\infty}^{t} \delta(t-t_0)\, dt \tag{15.210}$$

Higher derivatives of the unit step function are also sometimes employed. The usual nomenclature is "doublet," "triplet," etc., for the second, third, and higher ones. In general, a singularity function can be represented in terms of its next lower order function by the recurrence formula

$$\mathsf{f}_{n+1}(t-\tau) = \int_{-\infty}^{t} \mathsf{f}_n(\theta-\tau)\, d\theta \qquad n = 0, 1, 2, \ldots \tag{15.211}$$

Those readers who may be concerned about the incorporation of these elementary pathological functions into standard methods of the calculus, as illustrated above, may wish to refer to more rigorous treatments.[9]

[9]Sneddon, I. N., op cit, pp. 489 et seq.

15.6 CONVOLUTION INTEGRAL

A large proportion of systems can be described as linear. Likewise, a significant fraction—possibly a preponderance—of system inputs and outputs can be obtained as the result of adding several or many specific inputs. Linear systems with such inputs and outputs are often said to obey the principle of superposition. That is, one element can be superposed upon another in the sense of addition.

In the world of differential calculus, elemental quantities can be assumed to behave in accordance with a linear superposition. To investigate the superposition of elementary impulses, consider the following depiction, Figure 15.12.

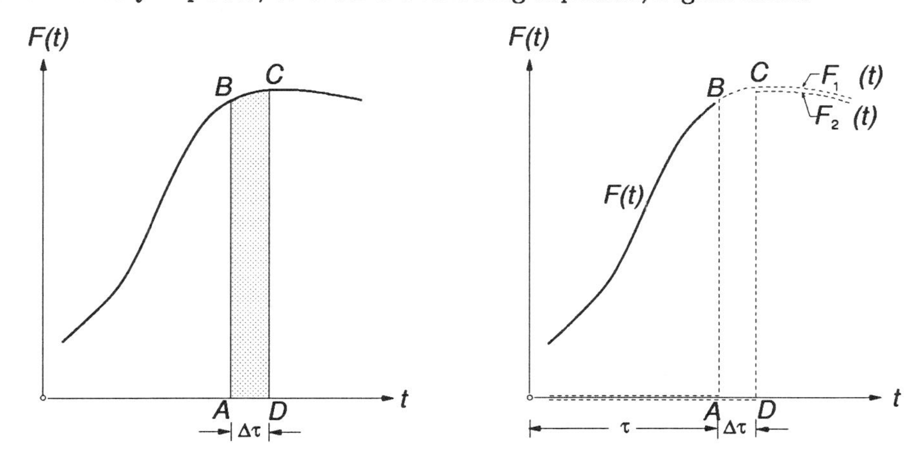

Figure 15.12: Superposition of Elemental Impulses

Let the impulse $ABCD$ be obtained as the subtraction of function F_2 from F_1, and represent the two functions in terms of the common function $F(t)$ by employing unit stepfunctions

$$F_1(t) = F(t)\,\mathbf{1}(t-\tau) \tag{15.212}$$

$$F_2(t) = F(t)\,\mathbf{1}[t-(\tau+\Delta\tau)] \tag{15.213}$$

These follow from the definition of $\mathbf{1}(t-t_0)$ since

$$F_1(t) = 0 \qquad t<\tau \qquad t-\tau<0 \tag{15.214}$$

$$= F(t) \qquad t>\tau \qquad t-\tau>0 \tag{15.215}$$

$$F_2(t) = 0 \qquad t<\tau+\Delta\tau \tag{15.216}$$

$$= F(t) \qquad t>\tau+\delta\tau \tag{15.217}$$

Therefore

$$ABCD = F_1(t) - F_2(t) \tag{15.218}$$

$$= F(t)\{\mathbf{1}(t-\tau) - \mathbf{1}[t-(\tau+\Delta\tau)]\} \tag{15.219}$$

$$= F(t)\{\mathbf{1}(t-\tau) - \mathbf{1}[(t-\tau)-\Delta\tau]\} \tag{15.220}$$

$$= F(t)\,\Delta\tau\frac{\mathbf{1}(t-\tau) - \mathbf{1}[(t-\tau)-\Delta\tau]}{\Delta\tau} \tag{15.221}$$

Upon passing to the limit as $\Delta\tau \to 0$,

$$\lim_{\Delta\tau\to 0} \frac{\mathbf{1}(t-\tau) - \mathbf{1}[(t-\tau)-\Delta\tau]}{\Delta\tau} = \frac{d}{d\tau}\mathbf{1}(t-\tau) \tag{15.222}$$

It will be recalled, from the previous section, that

$$\frac{d}{d\tau}\mathbf{1}(t-\tau) = \delta(t-\tau) \tag{15.223}$$

hence

$$\lim_{\Delta\tau\to 0}(ABCD) = F(t)\,\delta(t-\tau)\,d\tau \tag{15.224}$$

Then the total impulse $I(t)$ from the force $F(t)$ is given by

$$I(t) = \int_0^t F(\tau)\,\delta(t-\tau)\,d\tau \tag{15.225}$$

In a general form, such a functional is known as the *convolution integral.*

$$f(t) = f_1 * f_2 = \int_a^b f_1(\tau)\,f_2(t-\tau)\,d\tau \tag{15.226}$$

It is also known as the Boltzman superposition theorem. What it says in words is that the total effect is obtained as the linear summation of the many lessor componental effects.

15.7 SYSTEM DESCRIPTORS

It is convenient to erect a nomenclature for systems which allows for a facile reference to their properties, with important conditions understood and incorporated. Therefore, let the representation be

$$\mathbf{y}(t) = \mathbf{S}[\mathbf{x}(t_0); \mathbf{u}(t)] \tag{15.227}$$

or, for further simplicity,

$$\mathbf{y} = \mathbf{S}(\mathbf{x}_0; \mathbf{u}) \tag{15.228}$$

As before, $\mathbf{y}$ is the output vector, $\mathbf{u}$ is the input, and $\mathbf{x}(t_0)$ or $\mathbf{x}_0$ is the initial state of the system. $\mathbf{S}$ is the symbol for the system itself.

There are some special states which should be described before proceeding to more complex circumstances.

- The *zero state* of a system occurs when the initial state is zero and the input is also zero. The output will then be zero as well. Further, the output can be zero with no input for an initial state which is not zero. The zero state is therefore defined:[10]

[10] Zadeh, L. A., and Desoer, C. A., *Linear System Theory*, McGraw-Hill Book Company, New York, 1963, p. 108.

The zero state is an initial state such that the response of the system to zero input is zero.

$$\mathbf{\Psi} \text{ is a zero state if } \mathbf{0} = \mathbf{S}(\mathbf{\Psi}; \mathbf{0}) \tag{15.229}$$

- The *equilibrium state* is that for which $\mathbf{x}_e(t)$ is the state corresponding to $\mathbf{x}(t_0)$ when $\mathbf{u} = \mathbf{0}$.

$$\mathbf{x}_e = \mathbf{S}[\mathbf{x}_e; \mathbf{0}] \qquad t > t_0 \tag{15.230}$$

- In considering the ultimate state of a system, it is sometimes a function of the initial state when there is no input. When it is not such a function, the end state is called the *ground state*. Thus

$$\mathbf{x}_g = \lim_{t \to \infty} \mathbf{x}(t; \mathbf{x}_0, \mathbf{0}) \tag{15.231}$$

  is that ground state, where it is understood that this limit exists and is independent of $\mathbf{x}(t_0)$.

Some specific system properties which will be of importance in the work to follow are:

- The *time invariant* or *fixed* system is one for which

$$\mathbf{y}(t) = \mathbf{y}(t + T) \tag{15.232}$$

  for all times T, for all $\mathbf{u}(t_0, t)$, $\mathbf{y}_0$, and t. If this condition is not met, the system is said to be *time varying*.

- A *homogeneous* system is one for which

$$\mathbf{S}(\mathbf{\Psi}; k\mathbf{u}) = k\,\mathbf{S}(\mathbf{\Psi}; \mathbf{u}) \tag{15.233}$$

  for all values of k and $\mathbf{u}$, where k is a scalar number.

- An *additive* system fulfils the requirement

$$\mathbf{S}(\mathbf{\Psi}; \mathbf{u}_1 + \mathbf{u}_2) = \mathbf{S}(\mathbf{\Psi}; \mathbf{u}_1) + \mathbf{S}(\mathbf{\Psi}; \mathbf{u}_2) \tag{15.234}$$

  This is another version of superposition.

- The *zero-state linear* system is both homogeneous and additive. It is

$$\mathbf{S}[\mathbf{\Psi}; k\,(\mathbf{u}_1 + \mathbf{u}_2)] = k\,\mathbf{S}(\mathbf{\Psi}; \mathbf{u}_1) + k\,\mathbf{S}(\mathbf{\Psi}; \mathbf{u}_2) \tag{15.235}$$

  where k is a scalar number.

- The *zero-input linear* system is also both homogeneous and additive. For homogeneity,

$$\mathbf{S}(k\,\mathbf{x}_0; \mathbf{0}) = k\,\mathbf{S}(\mathbf{x}_0; \mathbf{0}) \tag{15.236}$$

  and for the additive property

$$\mathbf{S}(\mathbf{x}_1 + \mathbf{x}_2; \mathbf{0}) = \mathbf{S}(\mathbf{x}_1; \mathbf{0}) + \mathbf{S}(\mathbf{x}_2; \mathbf{0}) \tag{15.237}$$

  That system which satisfies both simultaneously is the zero-input system.

- The *linear* system is that one which is both zero-state linear and zero-input linear for any initial state $\mathbf{x}_0$. That is, if

$$\mathbf{S}[\mathbf{\Psi}; k\,(\mathbf{u}_1 + \mathbf{u}_2)] = k\,\mathbf{S}(\mathbf{x}_0; \mathbf{u}_1) + k\,\mathbf{S}(\mathbf{x}_0; \mathbf{u}_2) \tag{15.238}$$

and

$$\mathbf{S}(\mathbf{x}_1 + \mathbf{x}_2; \mathbf{0}) = \mathbf{S}(\mathbf{x}_1; \mathbf{0}) + \mathbf{S}(\mathbf{x}_2; \mathbf{0}) \tag{15.239}$$

15.8 IMPULSE RESPONSE

Using the equipment developed in the two previous sections, the zero-state response of a system can be represented as

$$\begin{aligned}\mathbf{y}(t) = \mathbf{S}[\mathbf{u}(t)] &= \mathbf{S}\left\{\int_{-\infty}^{+\infty} \mathbf{u}(\tau)\,\delta(t-\tau)\,d\tau\right\} \\ &= \int_{-\infty}^{+\infty} \mathbf{u}(\tau)\,\mathbf{S}[\delta(t-\tau)]\,d\tau \\ &= \int_{-\infty}^{+\infty} \mathbf{u}(\tau)\,\mathbf{h}(t,\tau)\,d\tau \end{aligned} \tag{15.240}$$

in which $\mathbf{h}(t,\tau)$ is the zero-state response of the system (seen at time t) to a unit impulse which occurs at time τ.

This response is named the *impulse response function*.

Similarly, the prior developments can be invoked to obtain the system response to more general excitations. Note first that the unit step function can be employed to represent the general time-varying function

$$f(t) = f(t_0) + \int_{t_0}^{t} \dot{f}(\tau)\,d\tau \qquad t \geq t_0$$

as

$$f(t) = f(t_0)\,\mathbf{1}(t-t_0) + \int_{t_0}^{\infty} \dot{f}(\tau)\,\mathbf{1}(t-\tau)\,d\tau \qquad t \geq t_0 \tag{15.241}$$

Herein $f(t)$ is furnished as a linear superposition of unit step functions.

Likewise, a system's response to a general input can be composed as

$$\mathbf{y}(t) = \mathbf{S}[\mathbf{u}(t)] = \mathbf{u}(t_0)\,\mathbf{S}[\mathbf{1}(t-t_0)] + \int_{t_0}^{\infty} \dot{\mathbf{u}}(\tau)\,\mathbf{S}[\mathbf{1}(t-\tau)]\,d\tau \tag{15.242}$$

Then if $\mathbf{g}(t,\tau)$ is the zero-state response, seen at time t to a unit step which begins at time τ,

$$\mathbf{y}(t) = \mathbf{u}(t_0)\,\mathbf{g}(t,t_0) + \int_{t_0}^{\infty} \dot{\mathbf{u}}(\tau)\,\mathbf{g}(t,\tau)\,d\tau \tag{15.243}$$

This is called the *unit step response function*.

There is a relationship between the two response functions mentioned above. It is seen that

$$\mathbf{g}(t,\tau) = \mathbf{S}[\mathbf{1}(t-\tau)] = \int_{-\infty}^{+\infty} \mathbf{1}(\theta-\tau)\,\mathbf{h}(t,\theta)\,d\theta = \int_{\tau}^{\infty} \mathbf{h}(t,\theta)\,d\theta \tag{15.244}$$

so the step response is the integral of the impulse response (with respect to the variable θ.) Also, upon taking the partial derviative with respect to τ,

$$\frac{\partial \mathbf{g}(t,\tau)}{\partial \tau} = \frac{\partial}{\partial \tau}\left[-\int_{\infty}^{\tau} \mathbf{h}(t,\theta)\, d\theta\right] = -\mathbf{h}(t,\tau) \tag{15.245}$$

so that the impulse response is the negative of the step response derived with respect to θ.

Example Consider the linear system for which the unit step response is as shown in Figure 15.13.

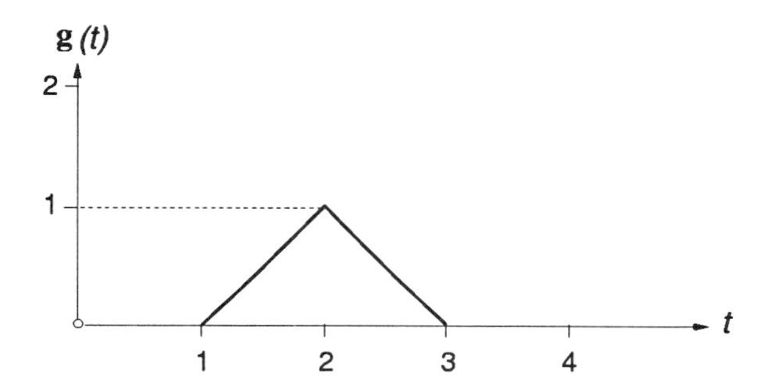

Figure 15.13: System Unit Step Response

Now let the input vector to this system be

$$\mathbf{u}(t) = t\,\mathbf{1}(t-1) - (t-2)\,\mathbf{1}(t-2) \tag{15.246}$$

To show the power of the unit step function as an operator, let this input vector be reconstructed as

$$\begin{aligned}\mathbf{u}(t) &= (t-1)\,\mathbf{1}(t-1) + \mathbf{1}(t-1) - (t-2)\,\mathbf{1}(t-2)\\ &= \mathsf{f}_2(t-1) + \mathsf{f}_1(t-1) - \mathsf{f}_2(t-2)\end{aligned} \tag{15.247}$$

This representation involves subtracting and then adding (or superposing) a term in $\mathbf{1}(t-1)$.

Similarly, the unit step response function can be described analytically (using the picture in Figure 15.13) as

$$\mathbf{g}(t) = \mathsf{f}_2(t-1) - 2\,\mathsf{f}_2(t-2) + \mathsf{f}_2(t-3) \tag{15.248}$$

But $\mathbf{g}(t) = \mathbf{S}[\mathsf{f}_1(t)]$, and

$$\mathbf{S}[\mathsf{f}_2(t)] = \int_{-\infty}^{t} \mathbf{g}(\tau)\, d\tau = \mathsf{f}_3(t-1) - 2\,\mathsf{f}_3(t-2) + \mathsf{f}_3(t-3) \tag{15.249}$$

in which $\mathsf{f}_3(t) = \frac{1}{2}t^2\,\mathbf{1}(t)$.

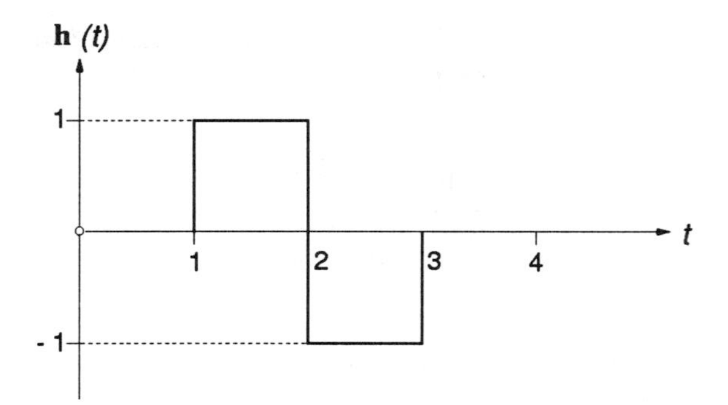

Figure 15.14: System Impulse Response

For a fixed linear system

$$\mathbf{y}(t) = \mathbf{S}[\mathbf{u}(t)] = \mathbf{S}[\mathsf{f}_2(t-1)] + \mathbf{S}[\mathsf{f}_1(t-1)] - \mathbf{S}[\mathsf{f}_2(t-2)] \quad (15.250)$$

with $\mathbf{g}(t)$ as previously represented.

The step function which is associated with this system's response is

$$\mathbf{h}(t) = \frac{d\,\mathbf{g}(t)}{dt} = \mathsf{f}_1(t-1) - 2\,\mathsf{f}_1(t-2) + \mathsf{f}_1(t-3)$$
$$= \mathbf{1}(t-1) - 2\,\mathbf{1}(t-2) + \mathbf{1}(t-3) \quad (15.251)$$

and has the graph shown in Figure 15.14.

The example displays the efficacy of the singularity functions.

Example A simple system is composed of a spring and a dashpot in parallel, as seen in Figure 15.15. It is subjected to a force f. It is desired to compute the zero-state response and the impulse response.

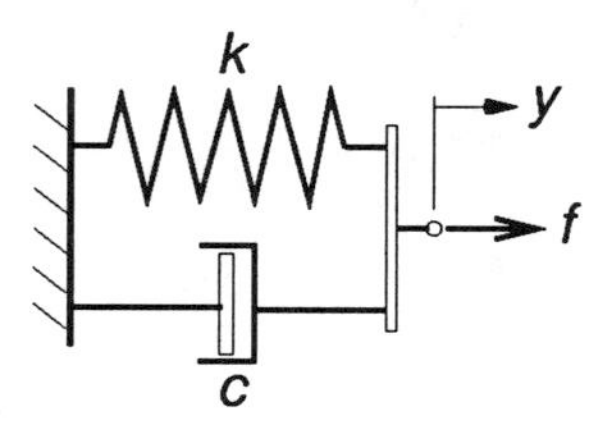

Figure 15.15: Spring-Dashpot System

The system is governed by the equation of motion

$$c\,\frac{dy}{dt} + k\,y = f \quad (15.252)$$

in which c is a damping factor, dy/dt is the velocity of the piston in the dashpot, and k is the spring modulus. Since this is a linear, first-order differential equation with constant coefficients, it can be solved by employing an integrating factor

$$w(t,t_0) = exp\left[\int_{t_0}^{t} \frac{d\tau}{c}\right] \qquad (15.253)$$

Form the product

$$w(t,t_0)\,y(t) = exp\left[\int_{t_0}^{t} \frac{d\tau}{c}\right]\,y(t) \qquad (15.254)$$

so that

$$\begin{aligned}
\frac{d}{dt}[w(t,t_0)\,y(t)] &= w(t,t_0)\,\frac{dy}{dt} + y(t)\,\frac{dw(t,t_0)}{dt} \\
&= w(t,t_0)\,\frac{f-y}{c} + y(t)\,\left[\frac{d}{dt}\int_{t_0}^{t}\frac{d\tau}{c}\right]\,exp\left[\int_{t_0}^{t}\frac{d\tau}{c}\right] \\
&= w(t,t_0)\,\frac{f-y}{c} + y(t)\,w(t,t_0)\,\frac{1}{c} \\
&= w(t,t_0)\,\frac{f-y}{c} + w(t,t_0)\,\frac{y(t)}{c} \\
&= w(t,y_0)\frac{f}{c} \qquad (15.255)
\end{aligned}$$

When this is integrated, the result is

$$w(t,t_0)\,y(t) = \int_{t_0}^{t} \frac{f(\theta)}{c}\,w(\theta,t_0)\,d\theta$$

and then

$$y(t) = \frac{1}{w(t,t_0)}\int_{t_0}^{t} \frac{w(\theta,t_0)}{c}\,f(\theta)\,d\theta \qquad (15.256)$$

yields the zero-state response.

The impulse response is obtained by putting $f = \delta(t)$. The sifting function is invoked to get

$$\begin{aligned}
\mathbf{1}(t-t_0) &= \frac{1}{w(t,t_0)}\left[\frac{w(\theta,y_0)}{c}\right] && t>\theta \\
&= \frac{1}{c}\,exp\left[-\int_{\theta}^{t}\frac{d\tau}{c}\right] && t>\theta \\
&= 0 && t<\theta \qquad (15.257)
\end{aligned}$$

For a constant damping coefficient c,

$$\mathbf{1}(t,\theta) = \mathbf{1}(t-\theta) = \frac{1}{c}\,exp\left(\frac{\theta-t}{c}\right) \qquad t>\theta \qquad (15.258)$$

$$= 0 \qquad t<\theta \qquad (15.259)$$

15.9 GENERAL RESPONSE

In the previous sections the emphasis has been on the system's response to excitation by impulse functions and unit step functions. These have been seen to be indicative of the character of the response for a given system, and therefore of great importance to the system analyst. However, the analyst is concerned with the response to some particular inputs which are possibly rather complex, and almost certainly not as pure as the singularity functions. Therefore, it is time to turn attention to the general conditions.

As before, the state equations will be taken in the form

$$\dot{\mathbf{x}}(t) = \mathbf{A}(t)\,\mathbf{x}(t) + \mathbf{B}(t)\,\mathbf{u}(t) \tag{15.260}$$

$$\mathbf{y}(t) = \mathbf{C}(t)\,\mathbf{x}(t) + \mathbf{D}(t)\,\mathbf{u}(t) \tag{15.261}$$

with the understanding that in the further analysis the initial state $\mathbf{x}(t_0) = \mathbf{x}_0$ is arbitrary, and so is the input $\mathbf{u}(t)$.

15.9.1 FREE RESPONSE

To start with, look at the free system, for which $\mathbf{u} = \mathbf{0}$. The relevant equation to be satisfied is

$$\dot{\mathbf{x}}(t) = \mathbf{A}(t)\,\mathbf{x}(t) \tag{15.262}$$

The object is to find an equation for its motion

$$\mathbf{x}(t) = \mathbf{x}(t; \mathbf{x}_0) \tag{15.263}$$

Call this function $\mathbf{x}(t; \mathbf{x}_0) = \mathbf{\Phi}(t, t_0)$, which is then a solution of

$$\dot{\mathbf{\Phi}}(t, t_0) = \mathbf{A}(t)\,\mathbf{\Phi}(t, t_0) \tag{15.264}$$

There have been a number of different names for the $\mathbf{\Phi}(t, t_0)$ function, but the one in current use is ***state transition matrix***. Mathematically speaking, it is a transformation which is a mapping of an initial state $\mathbf{x}_0$ into $\mathbf{x}(t)$, the state at time t. It is also known as a ***linear operator***, and it has the property $\mathbf{\Phi}(t_0, t_0) = \mathbf{I}$.

The solution space for the state equation is an n-dimensional vector space. Assume that there is a basis $\mathbf{e}_1$, $\mathbf{e}_2$, ..., $\mathbf{e}_n$ for this space so that any vector of it can be represented as

$$\mathbf{x}_0 = \sum_{i=1}^{n} x^i\,\mathbf{e}_i \equiv x^i\,\mathbf{e}_i \tag{15.265}$$

Because $\mathbf{\Phi}(t, t_0)$ is a linear operator, when allowed to operate upon any vector of the space the result is

$$\mathbf{\Phi}(t, t_0) = \sum_{i=1}^{n} x^i\,\mathbf{\Phi}(t, t_0)\,\mathbf{e}_i \tag{15.266}$$

hence

$$\mathbf{x}(t) = \sum_{i=1}^{n} x^i\,\mathbf{x}(t, \mathbf{e}_i) \equiv x^i\,\mathbf{x}(t, \mathbf{e}_i) \tag{15.267}$$

since

$$\dot{\mathbf{x}}(t) = \mathbf{A}(t)\,\mathbf{x}(t) \tag{15.268}$$

This shows that any general free motion of the system is amenable to representation as a linear superposition of the n zero-input responses $\mathbf{x}(t;\mathbf{e}_i)$.

15.9.2 FORCED RESPONSE

When the free response of the system is in hand, it is appropriate to turn next to the forced motion; i.e., when there is an input $\mathbf{u}(t)$. The method known as variation of parameters is available to pursue this objective. The method is sometimes attributed to Lagrange.

The process assumes that a solution for the forced motion can be generated in such a form that it resembles that for the free motion, only that there is a parameter in the more general solution which reduces to that of the free motion parameter when the input vanishes. That is, there is a continuous variation of the parameter over a domain which includes both free and forced conditions. The free solution is known to be $\mathbf{x}(t) = \mathbf{\Phi}(t;t_0)\,\mathbf{x}_0$, so take the general solution in the form

$$\mathbf{x}(t) = \mathbf{\Phi}(t,t_0)\,\mathbf{f}(t) \tag{15.269}$$

and investigate the nature of $\mathbf{f}$ which makes it so.

First, differentiate the general state equation to obtain

$$\begin{aligned}\dot{\mathbf{x}}(t) &= \dot{\mathbf{\Phi}}(t,t_0)\,\mathbf{f}(t) + \mathbf{\Phi}(t,t_0)\,\dot{\mathbf{f}}(t)\\ &= \mathbf{A}(t)\,\mathbf{\Phi}(t,t_0)\,\mathbf{f}(t) + \mathbf{\Phi}(t,t_0)\,\dot{\mathbf{f}}(t)\\ &= \mathbf{A}(t)\,\mathbf{x}(t) + \mathbf{\Phi}(t,t_0)\,\dot{\mathbf{f}}(t)\end{aligned} \tag{15.270}$$

Next, let this result be put in the place of $\dot{\mathbf{x}}$ on the left-hand side of the state equation to yield

$$\mathbf{A}(t,t_0)\,\mathbf{x}(t) + \mathbf{\Phi}(t,t_0)\,\dot{\mathbf{f}}(t) = \mathbf{A}(t,t_0)\,\mathbf{x}(t) + \mathbf{B}(t)\,\mathbf{u}(t) \tag{15.271}$$

or

$$\mathbf{\Phi}(t,t_0)\,\dot{\mathbf{f}}(t) = \mathbf{B}(t)\,\mathbf{u}(t) \tag{15.272}$$

Therefore, operating from the left on both sides of the equation with $\mathbf{\Phi}^{-1}(t,t_0)$,

$$\dot{\mathbf{f}}(t) = \mathbf{\Phi}^{-1}(t,t_0)\,\mathbf{B}(t)\,\mathbf{u}(t) = \mathbf{\Phi}(t_0,t)\,\mathbf{B}(t)\,\mathbf{u}(t) \tag{15.273}$$

because $\mathbf{\Phi}^{-1}(t,t_0) = \mathbf{\Phi}(t_0,t)$. Now integrate the expression for $\dot{\mathbf{f}}(t)$ and find

$$\mathbf{f}(t) = \mathbf{f}(t_0) + \int_{t_0}^{t} \mathbf{\Phi}(t_0,\tau)\,\mathbf{B}(\tau)\,\mathbf{u}(\tau)\,d\tau \tag{15.274}$$

When this is inserted into the original equation above, there is

$$\mathbf{x}(t) = \mathbf{\Phi}(t,t_0)\,\mathbf{f}(t_0) + \int_{t_0}^{t} \mathbf{\Phi}(t,t_0)\,\mathbf{\Phi}(t_0,\tau)\,\mathbf{B}(\tau)\,\mathbf{u}(\tau)\,d\tau \tag{15.275}$$

which is, because $\mathbf{\Phi}(t, t_0)\,\mathbf{\Phi}(\tau, t_0) = \mathbf{\Phi}(t, \tau)$,

$$\mathbf{x}(t) = \mathbf{\Phi}(t, t_0)\,\mathbf{x}_0 + \int_{t_0}^{t} \mathbf{\Phi}(t, \tau)\,\mathbf{B}(\tau)\,\mathbf{u}(\tau)\,d\tau \tag{15.276}$$

Once this is available, the output can be obtained by returning to the second state equation and inserting $\mathbf{x}(t)$ in it as follows:

$$\mathbf{y}(t) = \mathbf{C}(t)\,\mathbf{\Phi}(t, t_0)\,\mathbf{x}_0 + \int_{t_0}^{t} \mathbf{C}(t)\,\mathbf{\Phi}(t, \tau)\,\mathbf{B}(\tau)\,\mathbf{u}(\tau)\,d\tau + \mathbf{D}(t)\,\mathbf{u}(t) \tag{15.277}$$

Note that this is the sum of the zero-input and the zero-state responses.

The second term—the zero-state response—is

$$\mathbf{y}(t) = \int_{-\infty}^{t} \mathbf{C}(t)\,\mathbf{\Phi}(t, \tau)\,\mathbf{B}(\tau)\,\mathbf{u}(\tau)\,d\tau + \mathbf{D}(t)\,\mathbf{u}(t) \tag{15.278}$$

because $t_0 = -\infty$ corresponds to $\mathbf{x}_0 = \mathbf{0}$, and when a unit step function is introduced in the kernel of the integral and with the representation

$$\mathbf{u}(t) = \int_{-\infty}^{+\infty} \mathbf{u}(\tau)\,\delta(t - \tau)\,d\tau \tag{15.279}$$

the output vector becomes

$$\mathbf{y}(t) = \int_{-\infty}^{+\infty} [\mathbf{C}(t)\,\mathbf{\Phi}(t, \tau)\,\mathbf{B}(\tau)\,\mathbf{1}(t - \tau) + \mathbf{D}(t)\,\delta(t - \tau)]\,\mathbf{u}(\tau)\,d\tau \tag{15.280}$$

In this form it appears as the convolution integral for the zero-state response in which the impulse response is

$$\mathbf{F}(t, \tau) = \mathbf{C}(t)\,\mathbf{\Phi}(t, \tau)\,\mathbf{B}(\tau)\,\mathbf{1}(t - \tau) + \mathbf{D}(t)\,\delta(t - \tau) \tag{15.281}$$

By this means $\mathbf{F}(t, \tau)$ is constructed directly from the constitutive matrices of the state equations.

For the special case of fixed systems (i.e., $\mathbf{A}$, $\mathbf{B}$, $\mathbf{C}$, and $\mathbf{D}$ are not functions of time),

$$\mathbf{F}(t - \tau) = \mathbf{C}\,\mathbf{\Phi}(t - \tau)\,\mathbf{B}\,\mathbf{1}(t - \tau) + \mathbf{D}\,\delta(t - \tau) \tag{15.282}$$

15.10 FIXED SYSTEM RESPONSE

In the developments outlined in the previous section, it is evident that in order to produce the transition matrix the free system equations must be integrated. There is no assurance that this can be done in closed form in the general case. However, for the *fixed* system, with constant matrices, there are processes which allow for the determination of the transition matrix.

Ultimately, the labor of rendering solutions for the transition matrix may be done by computers, and it is customary to put the machinery in a form most suitable for this purpose.

What is required is the solution of the equation

$$\dot{\mathbf{\Phi}}(t) = \mathbf{A}\,\mathbf{\Phi}(t)\,, \qquad \mathbf{\Phi}(\mathbf{0}) = \mathbf{I} \tag{15.283}$$

If the equation were a simple ordinary differential equation in one scalar variable

$$\dot{x}(t) = a\,x(t)\,, \qquad x(0) = 1 \tag{15.284}$$

almost everyone would recognize the answer right away. It is $x(t) = e^{at}$. It can be shown that a suitable generalization of the exponential function serves to solve the matrix equation.

For the ordinary exponential function, there is

$$e^{at} = 1 + a\,t + \frac{a^2\,t^2}{2!} + \cdots + \frac{a^i\,t^i}{i!} + \cdots = \sum_{i=1}^{\infty} \frac{a^i\,t^i}{i!} \tag{15.285}$$

The analogous development for the so-called *matrix exponential* is

$$e^{\mathbf{A}t} \doteq \mathbf{I} + \mathbf{A}\,t + \frac{\mathbf{A}^2\,t^2}{2!} + \cdots + \frac{\mathbf{A}^i\,t^i}{i!} + \cdots = \sum_{i=1}^{\infty} \frac{\mathbf{A}^i\,t^i}{i!} \tag{15.286}$$

This series has been shown to converge absolutely (for finite t), and is taken to be the definition of the matrix exponential.

Some properties of the matrix exponential are:

$$e^{\mathbf{A}(t+\tau)} = e^{\mathbf{A}t}\,e^{\mathbf{A}\tau} \qquad e^{(\mathbf{A}+\mathbf{B})t} = e^{\mathbf{A}t}\,e^{\mathbf{B}t} \quad \text{if} \quad \mathbf{AB} = \mathbf{BA} \tag{15.287}$$

$$e^{\mathbf{A}(t-t)} = e^{\mathbf{A}t}\,e^{-\mathbf{A}t} \qquad e^{\mathbf{A}t}\,e^{-\mathbf{A}t} = \mathbf{I} \tag{15.288}$$

To show that the matrix exponential satisfies the required state equation, let the series expansion be differentiated term-by-term to obtain

$$\begin{aligned} \frac{d}{dt}e^{\mathbf{A}t} &= \mathbf{A} + \mathbf{A}^2 t + \frac{\mathbf{A}^3 t^3}{2!} + \cdots + \frac{\mathbf{A}^i t^{i-1}}{(i-1)!} + \cdots \\ &= \mathbf{A}\,[\,\mathbf{I} + \mathbf{A}t + \frac{\mathbf{A}^2 t^2}{2!} + \cdots + \frac{\mathbf{A}^{i-1} t^{i-1}}{(i-1)!} + \cdots] \\ &= \mathbf{A}\,e^{\mathbf{A}t} \end{aligned} \tag{15.289}$$

15.10.1 FREE FIXED SYSTEM RESPONSE

What has been shown is that

$$\dot{\mathbf{x}} = \mathbf{A}\mathbf{x}\,, \qquad \mathbf{x}(t_0) = \mathbf{x}_0 \tag{15.290}$$

is solved by

$$\mathbf{x}(t) = \mathbf{x}_0\,exp[\mathbf{A}(t - t_0)] \tag{15.291}$$

Using the infinite series form above is useful in demonstrating the properties of the matrix exponential, but is not so useful in actually performing the computation of the transition matrix as it hides the closed form. Should a computer be dedicated to the task, it requires only a suitable algorithm to accomplish the work. For an analytical version of the answer, however, some other tack must be taken.

EIGENVALUE EXPANSION

The usual approach to discovering solutions of the system $\dot{\mathbf{x}} = \mathbf{A}\mathbf{x}$ which satisfy $\mathbf{x}(t_0) = \mathbf{x}_0$ is to substitute a presumed answer of the form $\mathbf{x}(t) = e^{\lambda t}\,\mathbf{e}$, in which $\mathbf{e}$ is some constant vector. This yields

$$\dot{\mathbf{x}}(t) = \lambda e^{\lambda t}\mathbf{e} = \mathbf{A}e^{\lambda t}\mathbf{e} = e^{\lambda t}\mathbf{A}\mathbf{e} \qquad (15.292)$$

Dividing the first and last terms by $e^{\lambda t}$, which is allowed since $e^{\lambda t} \neq 0$, for finite times t, it is seen that

$$\lambda\mathbf{e} = \mathbf{A}\mathbf{e} \qquad (15.293)$$

or, what is the same thing,

$$(\lambda\mathbf{I} - \mathbf{A})\mathbf{e} = \mathbf{0} \qquad (15.294)$$

Solutions of this latter equation, other than the trivial $\mathbf{e} = \mathbf{0}$, yield the eigenvalues, as it is an eigenvalue equation. As was shown earlier in this chapter, there is the characteristic equation

$$f(\lambda) = det[\lambda\mathbf{I} - \mathbf{A}] = \lambda^n + a_{n-1}\lambda^{n-1} + a_{n-2}\lambda^{n-2} + \cdots + a_1\lambda + a_0 \qquad (15.295)$$

The n roots of this equation are the eigenvalues, and the $\mathbf{e}_i$ are the eigenvectors, where $i = 1, 2, \ldots, n$. In other words, there is one set of eigenvector components for each eigenvalue.

Example Suppose there is a transition matrix

$$\mathbf{A} = \begin{bmatrix} 0 & 1 \\ -3 & -4 \end{bmatrix} \qquad (15.296)$$

such that

$$det[\lambda\mathbf{1} - \mathbf{A}] = det\begin{bmatrix} \lambda & -1 \\ 3 & \lambda + 4 \end{bmatrix} = \lambda(\lambda+4) - (-1)(3) = 0 \qquad (15.297)$$

This yields, when expanded,

$$\lambda^2 + 4\lambda + 3 = (\lambda+3)(\lambda+1) = 0 \qquad (15.298)$$

The roots of this equation (the eigenvalues) are $\lambda_1 = -1$, and $\lambda_2 = -3$.

For the first eigenvalue, $\lambda_1 = -1$, the eigenvector components are the solution of the equations

$$\begin{bmatrix} -1 & -1 \\ 3 & 3 \end{bmatrix}\begin{bmatrix} e_1 \\ e_2 \end{bmatrix} = \begin{bmatrix} 0 \\ 0 \end{bmatrix} \qquad (15.299)$$

That is, $e_1 = -e_2$. The eigenvector is therefore

$$\mathbf{e}_1 = k_1\begin{bmatrix} 1 \\ -1 \end{bmatrix} \qquad (15.300)$$

Herein, k_1 is some indefinite constant affiliated with this eigenvalue.

Similarly, for $\lambda = -3$, the eigenvector components come from

$$\begin{bmatrix} -3 & -1 \\ 3 & -1 \end{bmatrix} \begin{bmatrix} e_1 \\ e_2 \end{bmatrix} = \begin{bmatrix} 0 \\ 0 \end{bmatrix} \tag{15.301}$$

which has the solution $3e_1 = -e_2$. In this case the eigenvector is

$$\mathbf{e}_2 = k_2 \begin{bmatrix} 1 \\ 3 \end{bmatrix} \tag{15.302}$$

wherein k_2 is some indefinite constant associated with this eigenvalue.

Consider the more general case where there are n distinct roots of the characteristic equation, therefore n eigenvalues. Using the principle of superposition, the state vector can be constructed as

$$\mathbf{x}(t) = k_1 e^{\lambda_1 t}\mathbf{e}_1 + k_2 e^{\lambda_2 t}\mathbf{e}_2 + \cdots + k_n e^{\lambda_n t}\mathbf{e}_n \tag{15.303}$$

and this can be shown explicitly to solve the state equation

$$\begin{aligned} \dot{\mathbf{x}}(t) &= k_1\lambda_1 e^{\lambda_1 t}\mathbf{e}_1 + k_2\lambda_2 e^{\lambda_2 t}\mathbf{e}_2 + \cdots + k_n\lambda_n e^{\lambda_n t}\mathbf{e}_n \\ &= k_1 e^{\lambda_1 t}\mathbf{A}\mathbf{e}_1 + k_2 e^{\lambda_2 t}\mathbf{A}\mathbf{e}_2 + \cdots + k_n e^{\lambda_n t}\mathbf{A}\mathbf{e}_n \\ &= \mathbf{A}\mathbf{x}(t) \end{aligned}$$

It remains to show that the solution satisfies the initial condition $\mathbf{x}(0) = \mathbf{x}_0$. This is accomplished by a suitable choice of the k_i. Thus,

$$\begin{aligned} \mathbf{x}(0) &= k_1\mathbf{e}_1 + k_2\mathbf{e}_2 + \cdots + k_n\mathbf{e}_n \\ &= [\mathbf{e}_1\ \mathbf{e}_2\ \cdots\ \mathbf{e}_n] \begin{bmatrix} k_1 \\ k_2 \\ \vdots \\ k_n \end{bmatrix} = \mathbf{T}\mathbf{k} = \mathbf{x}_0 \end{aligned} \tag{15.304}$$

$\mathbf{T}$ is an $n \times n$ matrix whose elements are constants.

In the theory of matrices it is shown that the eigenvectors belonging to distinct eigenvalues are linearly independent, and also that an $n \times n$ matrix having linearly independent columns is nonsingular. That is, it has an inverse. Therefore, since there is a $\mathbf{T}^{-1}$, the latter equation, Equation 15.304 above, can be operated upon from the left with $\mathbf{T}^{-1}$ to have

$$\mathbf{T}^{-1}\mathbf{T}\mathbf{k} = \mathbf{1}\mathbf{k} = \mathbf{k} = \mathbf{T}^{-1}\mathbf{x}_0 \tag{15.305}$$

This, in turn, can be entered into the state vector equation as

$$\begin{aligned} \mathbf{x}(t) &= k_1 e^{\lambda_1 t}\mathbf{e}_1 + k_2 e^{\lambda_2 t}\mathbf{e}_2 + \cdots + k_n e^{\lambda_n t}\mathbf{e}_n \\ &= [e^{\lambda_1 t}\mathbf{e}_1\ e^{\lambda_2 t}\mathbf{e}_2\ \cdots\ e^{\lambda_n t}\mathbf{e}_n]\mathbf{k} \\ &= [\mathbf{e}_1\ \mathbf{e}_2\ \cdots\ \mathbf{e}_n] \begin{bmatrix} e^{\lambda_1 t} & 0 & \cdots & 0 \\ 0 & e^{\lambda_2 t} & \cdots & 0 \\ \vdots & \vdots & e^{\lambda_i t} & \vdots \\ 0 & 0 & \cdots & e^{\lambda_n t} \end{bmatrix} \mathbf{k} \end{aligned} \tag{15.306}$$

When $\mathbf{k} = \mathbf{T}^{-1}\mathbf{x}_0$ is inserted, this latest becomes

$$\mathbf{x}(t) = \mathbf{T}\begin{bmatrix} e^{\lambda_1 t} & 0 & \cdots & 0 \\ 0 & e^{\lambda_2 t} & \cdots & 0 \\ \vdots & \vdots & e^{\lambda_i t} & \vdots \\ 0 & 0 & \cdots & e^{\lambda_n t} \end{bmatrix}\mathbf{T}^{-1}\mathbf{x}_0 = \mathbf{T}e^{\mathbf{\Lambda}t}\mathbf{T}^{-1}\mathbf{x}_0 \tag{15.307}$$

in which

$$e^{\mathbf{\Lambda}t} = \begin{bmatrix} e^{\lambda_1 t} & 0 & \cdots & 0 \\ 0 & e^{\lambda_2 t} & \cdots & 0 \\ \vdots & \vdots & e^{\lambda_i t} & \vdots \\ 0 & 0 & \cdots & e^{\lambda_n t} \end{bmatrix} \tag{15.308}$$

with

$$\mathbf{\Lambda} = \begin{bmatrix} \lambda_1 & 0 & \cdots & 0 \\ 0 & \lambda_2 & \cdots & 0 \\ \vdots & \vdots & \lambda_i & \vdots \\ 0 & 0 & \cdots & \lambda_n \end{bmatrix} \tag{15.309}$$

Now compare $\mathbf{x}(t) = e^{\mathbf{A}(t-t_0)}\mathbf{x}_0$ and $\mathbf{x}(t) = \mathbf{T}e^{\mathbf{\Lambda}t}\mathbf{T}^{-1}\mathbf{x}_0$, to see that

$$\mathbf{\Phi}(t) = e^{\mathbf{A}t} = \mathbf{T}e^{\mathbf{\Lambda}t}\mathbf{T}^{-1} \tag{15.310}$$

This result gives the transition matrix as the matrix exponential $e^{\mathbf{A}t}$ in terms of its representation $e^{\mathbf{\Lambda}t}$ through a similarity transformation involving the matrix $\mathbf{T}$ and its inverse $\mathbf{T}^{-1}$ in which $\mathbf{T}$ is a matrix whose columns are the eigenvectors of $\mathbf{A}$.

15.10.2 FORCED FIXED SYSTEM RESPONSE

The forced motion of a fixed or time-invariant system

$$\dot{\mathbf{x}} = \mathbf{A}\mathbf{x} + \mathbf{B}\mathbf{u} \tag{15.311}$$

is given by

$$\mathbf{x}(t) = \mathbf{\Phi}(t, t_0)\,\mathbf{x}_0 + \int_{t_0}^{t} \mathbf{\Phi}(t, \tau)\,\mathbf{B}(\tau)\,\mathbf{u}(\tau)\,d\tau \tag{15.312}$$

In this equation, the transition matrix is a matrix exponential

$$\mathbf{\Phi}(t, t_0) = e^{\mathbf{A}(t-t_0)} \tag{15.313}$$

Since the system is fixed, $\mathbf{B}$ is a constant matrix, not a function of t (or τ). Further, all of the methods outlined above for treating the matrix exponential are readily available for the forced case as well.

15.10.3 TRANSFER FUNCTION RESPONSE

Let there be a system decribed by the linear differential equation

$$a_n \frac{d^n y}{dt^n} + a_{n-1} \frac{d^{n-1} y}{dt^{n-1}} + \cdots + a_0 \, y = u \tag{15.314}$$

so that by choosing as state variables

$$x^1 = y \tag{15.315}$$
$$x^2 = \frac{dy}{dt} \tag{15.316}$$
$$\cdots = \cdots \tag{15.317}$$
$$x^n = \frac{d^{n-1} y}{dt^{n-1}} \tag{15.318}$$

and differentiating

$$\dot{x}^1 = \frac{dy}{dt} = x^2 \tag{15.319}$$
$$\dot{x}^2 = \frac{d^2 y}{dt^2} = x^3 \tag{15.320}$$
$$\cdots = \cdots \tag{15.321}$$
$$\dot{x}^n = \frac{d^n y}{dt^n} = -\frac{1}{a_n}(a_0 \, x^1 + \cdots + a_{n-1} \, x^n) + \frac{u}{a_n} \tag{15.322}$$

This can be put into the canonical matrix format as

$$\dot{\mathbf{x}} = \mathbf{A}(t) \, \mathbf{x} + \mathbf{B}(t) \, \mathbf{u} \tag{15.323}$$
$$\mathbf{y} = \mathbf{C}(t) \, \mathbf{x} \tag{15.324}$$

wherein

$$\mathbf{A} = \begin{bmatrix} 0 & 1 & 0 & \cdots & 0 \\ 0 & 0 & 1 & \cdots & 0 \\ \cdots & \cdots & \cdots & \cdots & \cdots \\ -\dfrac{a_o}{a_n} & \cdots & \cdots & \cdots & -\dfrac{a_{n-1}}{a_n} \end{bmatrix} \tag{15.325}$$

$$\mathbf{B} = \begin{bmatrix} 0 \\ 0 \\ \vdots \\ \dfrac{1}{a_n} \end{bmatrix} \qquad \mathbf{C} = [1 \; 0 \; \cdots \; 0]. \tag{15.326}$$

There is no matrix $\mathbf{D}$ in this case.

Therefore, the motion of $\mathbf{x}$ is supplied by

$$\mathbf{x}(t) = \mathbf{\Phi}(t, t_0) \, \mathbf{x}(t_0) + \int_{t_0}^{t} \mathbf{\Phi}(t, \tau) \, \mathbf{B}(\tau) \, \mathbf{u}(\tau) \, d\tau \tag{15.327}$$

and the response function (output) is

$$\mathbf{y}(t) = \mathbf{C}(t)\,\mathbf{x}(t) = \mathbf{C}(t)\,\mathbf{\Phi}(t,t_0)\,\mathbf{x}(t_0) + \int_{t_0}^{t} \mathbf{C}(t)\,\mathbf{\Phi}(t,\tau)\,\mathbf{B}(\tau)\,\mathbf{u}(\tau)\,d\tau \quad (15.328)$$

This is recognized as the impulse response function

$$\mathbf{y}(t) = \int_{-\infty}^{+\infty} \mathbf{u}(\tau)\,\mathbf{h}(t,\tau)\,d\tau \quad (15.329)$$

where the function

$$\mathbf{h}(t,\tau) = \mathbf{C}(t)\,\mathbf{\Phi}(t,\tau)\,\mathbf{B}(\tau) \quad (15.330)$$

Moreover, the zero-input response is

$$\mathbf{C}(t)\,\mathbf{\Phi}(t,t_0)\,\mathbf{x}(t_0) = \phi_{1i}\,x^i(t_0) \quad (15.331)$$

Accordingly, the general solution of the motion of the system in its canonical form

$$\dot{\mathbf{x}} = \mathbf{A}(t)\,\mathbf{x} + \mathbf{B}(t)\,\mathbf{u} \quad (15.332)$$

is presented as

$$\mathbf{x}(t) = \mathbf{\Phi}(t,t_0)\,\mathbf{x}(t_0) + \int_{t_0}^{t} \mathbf{\Phi}(t,\tau)\,\mathbf{B}(\tau)\,\mathbf{u}(\tau)\,d\tau \quad (15.333)$$

A customary notation is

$$\mathbf{x}(t) = \mathbf{x}_0(t) + \int_{t_0}^{t} \mathbf{H}_x(t,\tau)\,\mathbf{u}(\tau)\,d\tau \quad (15.334)$$

in which

$$\mathbf{x}_0(t) = \mathbf{\Phi}(t,t_0)\,\mathbf{x}(t_0) \quad (15.335)$$

and

$$\mathbf{H}_x(t,\tau) = \mathbf{\Phi}(t,\tau)\,\mathbf{B}(\tau) \quad (15.336)$$

In the same way,

$$\mathbf{y}(t) = \mathbf{y}_0(t) + \int_{t_0}^{t} \mathbf{H}_y(t,\tau)\,\mathbf{u}(\tau)\,d\tau + \mathbf{D}(t)\,\mathbf{u}(t) \quad (15.337)$$

with

$$\mathbf{y}_0(t) = \mathbf{C}(t)\,\mathbf{x}_0(t) \quad (15.338)$$

and

$$\mathbf{H}_y(t,\tau) = \mathbf{C}(t)\,\mathbf{\Phi}(t,\tau)\,\mathbf{B}(\tau) \quad (15.339)$$

The function $\mathbf{H}_y(t,\tau) = \mathbf{C}(t)\,\mathbf{\Phi}(t,\tau)\,\mathbf{B}(\tau)$ is the impulse-response function of a system for which $\mathbf{D}(t) = 0$. That is, a system with no nondynamic element.

It is at this juncture that the formulation yields dividends: The *transfer-function* is, by definition, created from the impulse-response function by performing upon it the Laplace transform. Thus, for a fixed system, $(\mathbf{C} \neq f(t))$, $(\mathbf{B} \neq f(t))$,

$$\mathcal{L}[\mathbf{H}_y(t)] = \mathbf{H}(s) = \mathbf{C}\,\mathcal{L}[\mathbf{\Phi}(t)]\,\mathbf{B} = \mathbf{C}\,\mathbf{\Phi}(s)\,\mathbf{B} \quad (15.340)$$

15.10.4 FREQUENCY-DOMAIN ANALYSIS

Using the transfer function allows the analysis of a fixed system to be accomplished by application of the Laplace transform directly. Consider, for example, a system with an oscillatory input function $\mathbf{u}(t)$ which has a period T. (It is necessary to add the caveat that this input satisfies the Dirichlet conditions: to wit;

- $\mathbf{u}(t)$ has a finite number of maxima and minima,
- $\mathbf{u}(t)$ has a finite number of discontinuities, and
- $\int_{-\frac{T}{2}}^{+\frac{T}{2}} |\mathbf{u}(t)| \, dt \leq \infty$.

The input function can be represented as a Fourier series

$$\mathbf{u}(t) = \sum_{n=-\infty}^{n=+\infty} \mathbf{U}_n \, e^{in\omega_o t} \tag{15.341}$$

with

$$\omega_o = \frac{2\pi}{T} \tag{15.342}$$

In this representation, the functions $\mathbf{U}_n$ are the complex amplitudes

$$\mathbf{U}_n = \frac{1}{T} \int_{-\frac{T}{2}}^{+\frac{T}{2}} \mathbf{u}(t) \, e^{-in\omega_o t} \, dt \tag{15.343}$$

So if this form of input is injected into the system, the output can be seen to be

$$\mathbf{y}(t) = S[\mathbf{u}(t)] = \sum_{n=-\infty}^{n=+\infty} \mathbf{U}_n \, S[e^{in\omega_o t}] = \sum_{n=-\infty}^{+\infty} \mathbf{U}_n \mathbf{H}(in\omega_o) \, e^{in\omega_o t} \tag{15.344}$$

and the complex amplitudes of the output function are

$$\mathbf{Y}_n = \mathbf{H}(in\omega_0) \, \mathbf{U}_n \tag{15.345}$$

That is, the n-th output amplitude $\mathbf{Y}_n$ is just the n-th input amplitude $\mathbf{U}_n$ multiplied by the transfer function $\mathbf{H}(s)$ with s replaced by the frequency $s = in\omega_0$.

If the system input $\mathbf{u}(t)$ is transient (i.e., not periodic), the Laplace transform for the amplitude is

$$\mathbf{u}(s) = \int_0^{\infty} \mathbf{u}(t) \, e^{-st} \, dt \tag{15.346}$$

and then the inverse transform is given by

$$\mathbf{u}(t) = \mathcal{L}^{-1}[\mathbf{u}(s)] = \frac{1}{2\pi i} \int_{\gamma - i\infty}^{\gamma + i\infty} \mathbf{u}(s) \, e^{st} \, ds \tag{15.347}$$

Note that in this case, $\mathbf{u}(t)$ must fulfill the Dirichlet conditions on the range $[0, \infty]$ with the additional requirement that

$$\int_0^{\infty} |\, \mathbf{u}(t) \,| \; e^{-\rho t} \, dt < \infty \, , \qquad \gamma < \rho \, , \qquad \rho = \mathcal{R}e \, s \tag{15.348}$$

One can move directly to the state equations for a fixed linear differential system and make the Laplace transforms at that point. Thus, for

$$\dot{\mathbf{x}}(t) = \mathbf{A}\,\mathbf{x}(t) + \mathbf{B}\,\mathbf{u}(t) \tag{15.349}$$
$$\mathbf{y}(t) = \mathbf{C}\,\mathbf{x}(t) + \mathbf{D}\,\mathbf{u}(t) \tag{15.350}$$

the transform is applied to both sides of both equations to yield

$$\mathcal{L}[\dot{\mathbf{x}}(t)] = \mathbf{A}\,\mathcal{L}[\mathbf{x}(t)] + \mathbf{B}\,\mathcal{L}[\mathbf{u}(t)] \tag{15.351}$$
$$\mathcal{L}[\mathbf{y}(t)] = \mathbf{C}\,\mathcal{L}[\mathbf{x}(t)] + \mathbf{D}\,\mathcal{L}[\mathbf{u}(t)] \tag{15.352}$$

Recall that

$$\mathcal{L}[\dot{\mathbf{x}}(t)] = s\,\mathbf{x}(s) - \mathbf{x}(0) \tag{15.353}$$

therefore the state equations when transformed are

$$s\,\mathbf{x}(s) - \mathbf{x}(0) = \mathbf{A}\,\mathbf{x}(s) + \mathbf{B}\,\mathbf{u}(s) \tag{15.354}$$
$$\mathbf{y}(s) = \mathbf{C}\,\mathbf{x}(s) + \mathbf{D}\,\mathbf{u}(s) \tag{15.355}$$

These are now *algebraic* equations in the transformed variable s, and can be solved as such to have

$$\mathbf{x}(s) = \left[\frac{1}{s\,\mathbf{I} - \mathbf{A}}\right]\,\mathbf{x}(0) + \left[\frac{1}{s\,\mathbf{I} - \mathbf{A}}\right]\,\mathbf{B}\,\mathbf{u}(s) \tag{15.356}$$
$$\mathbf{y}(s) = \mathbf{C}\,\left[\frac{1}{s\,\mathbf{I} - \mathbf{A}}\right]\,\mathbf{x}(0) + \left[\mathbf{C}\left(\frac{1}{s\,\mathbf{I} - \mathbf{A}}\right)\mathbf{B} + \mathbf{D}\right]\,\mathbf{u}(s) \tag{15.357}$$

It was previously shown that a direct integration of the state equations produced the results

$$\mathbf{x}(t) = \mathbf{\Phi}(t - t_0)\,\mathbf{x}(t_0) + \int_{t_0}^{t} \mathbf{\Phi}(t - \tau)\,\mathbf{B}\,\mathbf{u}(\tau)\,d\tau \tag{15.358}$$

and

$$\mathbf{y}(t) = \mathbf{C}\,\mathbf{\Phi}(t - t_0)\,\mathbf{x}(t_0) + \int_{t_0}^{t} \mathbf{C}\,\mathbf{\Phi}(t - \tau)\,\mathbf{B}\,\mathbf{u}(\tau)\,d\tau + \mathbf{D}\,\mathbf{u}(t) \tag{15.359}$$

Let the stop-watch be set running from time $t = 0$, (i.e., $t_0 = 0$), and construct the Laplace transforms of both equations to arrive at

$$\begin{aligned}\mathbf{x}(s) &= \mathcal{L}[\mathbf{\Phi}(t)]\,\mathbf{x}(0) + \mathcal{L}\left[\int_0^t \mathbf{\Phi}(t - \tau)\,\mathbf{B}\,\mathbf{u}(\tau)\,d\tau\right] \\ &= \mathcal{L}[\mathbf{\Phi}(t)]\,\mathbf{x}(0) + \mathcal{L}[\mathbf{\Phi}(t)]\,\mathbf{B}\,\mathbf{u}(s)\end{aligned} \tag{15.360}$$

and

$$\mathbf{y}(s) = \mathbf{C}\,\mathcal{L}[\mathbf{\Phi}(\tau)]\,\mathbf{x}(0) + [\mathbf{C}\,\mathcal{L}[\mathbf{\Phi}(t)]\,\mathbf{B} + \mathbf{D}]\,\mathbf{u}(s) \tag{15.361}$$

Now compare these latter two equations with the previous two obtained by transforming the state equations to find that

$$\mathcal{L}[\mathbf{\Phi}(t)] = \mathbf{\Phi}(s) = \frac{1}{s\,\mathbf{I} - \mathbf{A}} \tag{15.362}$$

Transforming the state equations directly reveals that $(s\,\mathbf{I}-\mathbf{A})^{-1}\,\mathbf{x}(0)$ is the zero-input response and $(s\,\mathbf{I}-\mathbf{A})^{-1}\,\mathbf{B}\,\mathbf{u}(s)$ is the zero-state response. The transformed transfer function is $\mathbf{H}_x(s) = (s\,\mathbf{I}-\mathbf{A})^{-1}\,\mathbf{B}$.

Example Given the simple system

$$\dot{x}^1 = 2\,x^2 = 0\,x^1 + x^2 \tag{15.363}$$

$$\dot{x}^2 = -2\,x^2 + 2\,u = 0\,x^1 - 2\,x^2 + 2\,u \tag{15.364}$$

$$y = 2\,x^1 = 2\,x^1 + 0\,x^2 \tag{15.365}$$

the constitutive matrices are identified as

$$\mathbf{A} = \begin{bmatrix} 0 & 2 \\ 0 & -2 \end{bmatrix} \quad \mathbf{B} = \begin{bmatrix} 0 \\ 2 \end{bmatrix} \quad \mathbf{C} = [2\;0] \quad \mathbf{D} = [0\;0] \tag{15.366}$$

Then

$$\begin{aligned}
\mathbf{\Phi}(s) &= \frac{1}{s\,\mathbf{I}-\mathbf{A}} = \frac{1}{s\begin{bmatrix} 1 & 0 \\ 0 & 1 \end{bmatrix} - \begin{bmatrix} 0 & 2 \\ 0 & -2 \end{bmatrix}} = \frac{1}{\begin{bmatrix} s & -2 \\ 0 & s+2 \end{bmatrix}} \\
&= \frac{1}{s(s+2)\begin{bmatrix} \dfrac{1}{s+2} & \dfrac{-2}{s(s+2)} \\ 0 & \dfrac{1}{s} \end{bmatrix}} \\
&= \frac{1}{s(s+2)\,\mathbf{M}}
\end{aligned}$$

in which $\mathbf{M}$ is the matrix

$$\mathbf{M} = \begin{bmatrix} \dfrac{1}{s+2} & \dfrac{-2}{s(s+2)} \\ 0 & \dfrac{1}{s} \end{bmatrix}$$

The inverse of $\mathbf{M}$ is found from $\mathbf{M}\mathbf{M}^{-1} = \mathbf{1}$, so that

$$\mathbf{M}^{-1} = \begin{bmatrix} (s+2) & 2 \\ 0 & s \end{bmatrix}$$

whence

$$\begin{aligned}
\mathbf{\Phi}(s) &= \frac{\mathbf{M}^{-1}}{s(s+2)} = \frac{1}{s(s+2)}\begin{bmatrix} (s+2) & 2 \\ 0 & s \end{bmatrix} \\
&= \begin{bmatrix} \dfrac{s+2}{s(s+2)} & \dfrac{2}{s(s+2)} \\ 0 & \dfrac{s}{s(s+2)} \end{bmatrix} \\
&= \begin{bmatrix} \dfrac{1}{s} & \dfrac{1}{s(s+1)} \\ 0 & \dfrac{1}{s+1} \end{bmatrix}
\end{aligned} \tag{15.367}$$

From this the inverse transform yields

$$\mathbf{\Phi}(t) = e^{\mathbf{A}t} = \mathcal{L}^{-1}[\mathbf{\Phi}(s)] = \mathcal{L}^{-1}\begin{bmatrix} \frac{1}{s} & \frac{2}{s(s+2)} \\ 0 & \frac{1}{(s+2)} \end{bmatrix}$$

$$= \begin{bmatrix} \mathcal{L}^{-1}\left[\frac{1}{s}\right] & \mathcal{L}^{-1}\left[\frac{2}{s(s+2)}\right] \\ \mathcal{L}^{-1}\left[0\right] & \mathcal{L}^{-1}\left[\frac{1}{s+2}\right] \end{bmatrix}$$

$$= \begin{bmatrix} 1 & 1-e^{-2t} \\ 0 & e^{-2t} \end{bmatrix} \tag{15.368}$$

and

$$\mathbf{H}(s) = \mathbf{C}\frac{1}{s\mathbf{I}-\mathbf{A}}\mathbf{B} = [1\ 0]\begin{bmatrix} \frac{1}{s} & \frac{1}{s(s+2)} \\ 0 & \frac{1}{s+2} \end{bmatrix}\begin{bmatrix} 0 \\ 1 \end{bmatrix}$$

$$= \frac{1}{s(s+2)} \tag{15.369}$$

$$h(t) = 1 - e^{-2t} \text{ when } t \geq 0 \tag{15.370}$$

Example An alternate method for obtaining the transfer function is exhibited in the following example. Say the system equations are

$$\begin{bmatrix} \dot{x}^1 \\ \dot{x}^2 \end{bmatrix} = \begin{bmatrix} a_1^1 & a_2^2 \\ a_1^2 & a_2^2 \end{bmatrix}\begin{bmatrix} x^1 \\ x^2 \end{bmatrix} + \begin{bmatrix} u^1 \\ u^2 \end{bmatrix} \tag{15.371}$$

$$\begin{bmatrix} y^1 \\ y^2 \end{bmatrix} = \begin{bmatrix} c_1^1 & c_2^1 \\ 0 & c_2^2 \end{bmatrix}\begin{bmatrix} x^1 \\ x^2 \end{bmatrix} \tag{15.372}$$

The next step in the process is to determine the inverse of the matrix

$$s\mathbf{I} - \mathbf{A} = s\begin{bmatrix} 1 & 0 \\ 0 & 1 \end{bmatrix} - \begin{bmatrix} a_1^1 & a_2^1 \\ a_1^2 & a_2^2 \end{bmatrix} = \begin{bmatrix} s & 0 \\ 0 & s \end{bmatrix} - \begin{bmatrix} a_1^1 & a_2^1 \\ a_1^2 & a_2^2 \end{bmatrix}$$

$$= \begin{bmatrix} s - a_1^1 & -a_2^1 \\ -a_1^2 & s - a_2^2 \end{bmatrix} \tag{15.373}$$

Using the method outlined in Appendix E, the inverse can be obtained as

$$\mathbf{H}_x(s) = (s\mathbf{I} - \mathbf{A})^{-1} = \frac{(s\mathbf{I} - \mathbf{A})^*}{|\, s\mathbf{I} - \mathbf{A}\,|} \tag{15.374}$$

where $(s\mathbf{I} - \mathbf{A})^*$ is the adjoint matrix and $|\, s\mathbf{I} - \mathbf{A}\,|$ is the determinant of the matrix.

The determinant is obtained as

$$|\, s\mathbf{I} - \mathbf{A}\,| = (s - a_1^1)(s - a_2^2) = s^2 - (a_1^1 + a_2^2)\, s + a_1^1 a_2^2 - a_1^2 a_2^1 \tag{15.375}$$

and this is also the characteristic equation for the **A** matrix.

The adjoint matrix is

$$(s\mathbf{I} - \mathbf{A})^* = \begin{bmatrix} s - a_2^2 & a_2^1 \\ a_1^2 & s - a_1^1 \end{bmatrix} \tag{15.376}$$

From these two results the transfer-function matrix elements can be built individually as

$$\frac{x^1(s)}{u^1(s)} = h_1^1(s) = \frac{s - a_2^2}{|s\mathbf{I} - \mathbf{A}|} = \frac{s - a_2^2}{(s - a_1^1)(s - a_2^2)} \tag{15.377}$$

$$\frac{x^1(s)}{u^2(s)} = h_2^1(s) = \frac{a_2^1}{|s\mathbf{I} - \mathbf{A}|} = \frac{a_2^1}{(s - a_1^1)(s - a_2^2)} \tag{15.378}$$

$$\frac{x^2(s)}{u^1(s)} = h_1^2(s) = \frac{a_1^2}{|s\mathbf{I} - \mathbf{A}|} = \frac{a_1^2}{(s - a_1^1)(s - a_2^2)} \tag{15.379}$$

$$\frac{x^2(s)}{u^2(s)} = h_2^2(s) = \frac{s - a_1^1}{|s\mathbf{I} - \mathbf{A}|} = \frac{s - a_1^1}{(s - a_1^1)(s - a_2^2)} \tag{15.380}$$

15.11 A BEAM VIBRATION

Consider the motion of a simply-supported prismatic beam shown in Figure 15.16. It has a mass density ρ kg/m^3, a cross-sectional area A m^2, and a flexural rigidity EI N·m^2. The transverse deflection is denoted $y(x)$ at the location x. Its length is ℓ m.

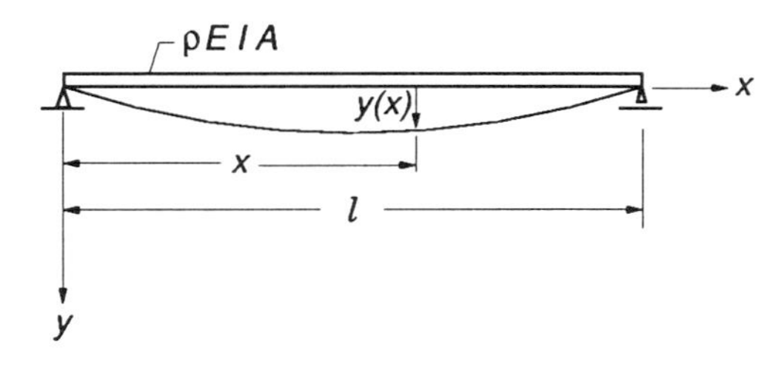

Figure 15.16: Vibrating Beam

The deflection of the beam can be represented by the series

$$y = \sum_{n=1}^{\infty} a_n \, u_n \, sin \frac{n\pi x}{\ell} \tag{15.381}$$

where the u_n are functions of time only, and are to be determined. Each mode has an amplitude coefficient, a_n. There are thus infinitely many modes of vibration each of the shape of a sine wave, so that the end conditions are met; that is, zero deflection at $x = 0$ and $x = \ell$.

This separation of the modes, or modal decomposition, as it is called, amounts to a semi-inverse technique in which a portion of the answer to the problem is assumed, while the remainder is to be solved for. It is also clearly a method of separating functions of the variables x and t in advance. In the language of mathematics, there is ***a two-point boundary-value problem*** in this case.

15.11.1 BEAM ENERGIES

During the vibration of the beam, the energies are:

Kinetic energy

$$T = \tfrac{1}{2}\,\rho A \int_0^\ell \left(\frac{\partial y}{\partial t}\right)^2 dx \tag{15.382}$$

and this can be integrated directly by inserting the formula for y given above in Equation (15.381) to obtain

$$T = \frac{\rho A \ell}{4} \sum_{n=1}^{\infty} a_n^2\,\dot{u}_n^2 \tag{15.383}$$

The usual notation $\dot{u}_n$ is employed for the time derivative of u_n.

Potential energy of bending

$$V = \tfrac{1}{2} EI \int_0^\ell \left(\frac{\partial^2 y}{\partial x^2}\right)^2 dx = \frac{\pi^4 EI}{4\ell^3} \sum_{n=1}^{\infty} n^4\,a_n^2\,u_n^2 \tag{15.384}$$

The Lagrangean function

$$L = T - V$$

can now be formed as

$$L = \frac{\rho A\ell}{4} \sum_{n=1}^{\infty} a_n^2\,\dot{u}_n^2 - \frac{\pi^4 EI}{4\ell^3} \sum_{n=1}^{\infty} n^4\,a_n^2 u_n^2 \tag{15.385}$$

15.11.2 EQUATIONS OF MOTION

The Lagrangean equations of motion for the vibration of this system are[11]

$$\frac{d}{dt}\left(\frac{\partial L}{\partial \dot{u}_n}\right) - \frac{\partial L}{\partial u_n} = Q_n \qquad n = 1, 2, 3, \ldots \tag{15.386}$$

wherein the u_n play the role of generalized coordinates, and Q_n is the generalized force corresponding to the u_n generalized coordinate.

After Equation (15.385) is introduced into Equation (15.386), the result is

$$\ddot{u}_n + \alpha\, n^4\, u_n = \frac{2}{\rho A \ell\, a_n^2} Q_n \tag{15.387}$$

[11]McDonald, P. H., *Continuum Mechanics*, PWS Publishing Company, Boston, 1996, p. 14.

In this equation,

$$\alpha = \frac{\pi^4 EI}{\rho A \ell^4} \tag{15.388}$$

Equations (15.387) are linear, second-order, ordinary differential equations with constant coefficients. They can be solved in the usual way in terms of the complementary part and the particular integral as

$$u_n = (u_n)_c + (u_n)_p \tag{15.389}$$

where the complementary part satisfies

$$(\ddot{u}_n)_c + \alpha\, n^4\, (u_n)_c = 0$$

and is

$$(u_n)_c = cos(\omega_n t + \lambda_n)$$

The eigenvalues ($\omega_n^2 = \alpha\, n^4$) are seen immediately to be given by

$$\omega_n^2 = \frac{n^4 \pi^4 EI}{\rho A \ell^4} \tag{15.390}$$

in which the natural frequencies are the ω_ns corresponding to each mode.

The particular integral depends upon the nature of the driving force acting upon the beam, and is therefore seen to be associated with the forced motion, while the $cos(\omega_n t + \lambda)$ part is the contribution of the free vibration. λ_n is a phase shift appropriate to the n-th mode.

Both a_n, the amplitude coefficient, and λ_n, the phase angle, are integration constants to be determined from the initial conditions. Two constants are necessary for second-order equations, and one (the a_n) is established by the initial deflection curve of the beam, while the other (λ_n) derives from the initial velocities of the modes.

To complete the solution to the motion problem, the initial conditions and the nature of the driving force must be adjoined to the development. Suppose, for example, that the beam is driven by a force $P\,cos\,\Omega t$ located at the center of the beam ($x = \ell/2$). Let the motion begin from rest at time zero at the deflected position produced by the concentrated load P.

15.11.3 THE FREE VIBRATION

To find the integration constants from the initial conditions, note the data as displayed in Figure 15.17. The constant force P deflects the beam symmetrically at rest. The zero state of the beam is defined by the following initial conditions:

$$t = 0 \quad y = y_0(x) = \sum_{n=1}^{\infty} a_n\, sin \frac{n\pi x}{\ell} \tag{15.391}$$

$$\frac{dy}{dt} = \left(\frac{dy}{dy}\right)_0 = 0 \quad \text{for all values of } x \tag{15.392}$$

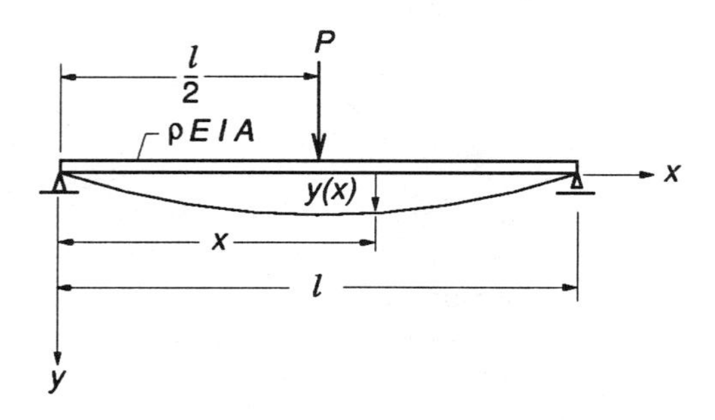

Figure 15.17: Initial Conditions

The a_n are those which exist at the time $t = 0$, and are determined by an examination of the elastic potential energy of the beam and the work of the force P at the initial instant as follows, using the principle of virtual work.

$$V(0) = \frac{\pi^4 EI}{4\ell^3} \sum_{n=1}^{\infty} n^4\, a_n^2 \tag{15.393}$$

since $u_n(0) = 1$. For the virtual displacement δa_n there are:

1. The work of the force P

$$\begin{aligned} W(0) &= P\, sin \frac{n\pi\ell/2}{\ell}\, \delta a_n \\ &= P\, sin \frac{n\pi}{2}\, \delta a_n \\ &= (-1)^{\frac{n-1}{2}}\, P\, \delta a_n \qquad \text{if } n \text{ is odd} \qquad (15.394) \\ &= 0 \qquad \text{if } n \text{ is even} \qquad (15.395) \end{aligned}$$

2. The increase of strain energy

$$dV(0) = \frac{n^4 \pi^4 EI}{2\ell^3}\, a_n\, \delta a_n \tag{15.396}$$

These must be equal, hence

$$\begin{aligned} a_n &= (-1)^{\frac{n-1}{2}} \frac{2P\ell^3}{n^4 \pi^4 EI} \qquad \text{if } n \text{ is odd} \qquad (15.397) \\ &= 0 \qquad \text{if } n \text{ is even} \qquad (15.398) \end{aligned}$$

The phase shifts λ_n are required to be all zero, as this is the only manner in which the condition in Equation (15.392) can be satisfied.

The complete equation for the deflection of the beam as it undergoes its free vibration can now be written

$$y(x,t) = \left(\frac{2P\ell^3}{\pi^4 EI}\right) \sum_{n=1,3,5,\ldots}^{\infty} (-1)^{\frac{n-1}{2}} \frac{1}{n^4}\, cos(\omega_n t)\, sin \frac{n\pi x}{\ell} \tag{15.399}$$

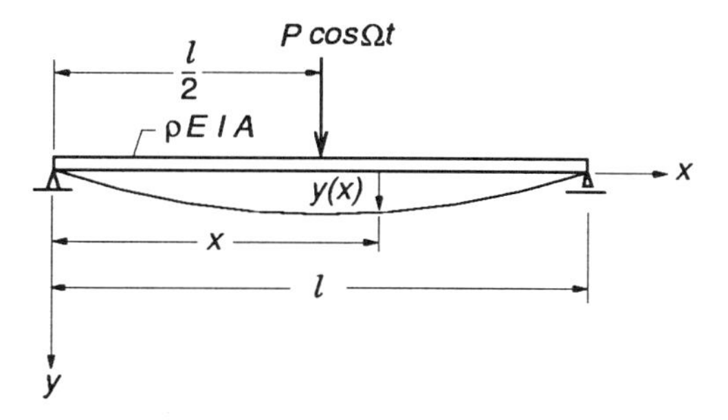

Figure 15.18: Driving Force

15.11.4 THE FORCED VIBRATION

Next, turn to the investigation of the particular integral $(u_n)_p$, which is identified as the forced vibration. The figure appropriate to this case is Figure 15.18. The generalized force Q_n is, by definition, that quantity which, when it multiplies δu_n, equals the virtual work done by the actual force $P\cos\Omega t$. Therefore,

$$Q_n\,\delta u_n = P\cos\Omega t\,\delta y = P\cos\Omega t\,a_n \sin\frac{n\pi\ell/2}{\ell}\,\delta u_n$$

or

$$\begin{aligned} Q_n &= a_n\,P\cos\Omega t\sin\frac{n\pi}{2} \\ &= a_n\,(-1)^{\frac{n-1}{2}}\,P\cos\Omega t \qquad \text{if } n \text{ is odd} \\ &= 0 \qquad\qquad\qquad\qquad\quad \text{if } n \text{ is even} \end{aligned}$$

With this result, the particular (complementary) solution to the equation of motion can be formulated, as is customary, by taking an assumed form

$$(u_n)_p = \mathcal{A}_n\,(-1)^{\frac{n-1}{2}}\,P\cos\Omega t \tag{15.400}$$

where $\mathcal{A}_n$ is an amplitude coefficient to be determined, and substitute this into Equation (15.387) to obtain

$$-\Omega^2\mathcal{A}_n(-1)^{\frac{n-1}{2}}P\cos\Omega t + \alpha n^4\mathcal{A}_n(-1)^{\frac{n-1}{2}}P\cos\Omega t = \frac{2}{\rho A\ell a_n^2}a_n(-1)^{\frac{n-1}{2}}P\cos\Omega t$$

The $\mathcal{A}_n$ which solves this is

$$\mathcal{A}_n = \frac{2}{\rho\,A\,\ell\,a_n\,(\alpha\,n^4 - \Omega^2)} \tag{15.401}$$

Using the result given in Equation (15.390), this can be re-written as

$$\mathcal{A}_n = \frac{2}{\rho\,A\,\ell\,a_n\,(\omega_n^2 - \Omega^2)} \tag{15.402}$$

and, further, as

$$\mathcal{A}_n = \left(\frac{2\,\ell^3}{n^4\,\pi^4\,E\,I\,a_n} \right) \left(\frac{1}{1 - \frac{\Omega^2}{\omega_n^2}} \right) \tag{15.403}$$

Obviously, the dimensionality of $\mathcal{A}_n$ is carried within the first bracketed part of this equation, and the second portion is that which is called the *magnification factor*, because it is a multiplier which amplifies the "statical" deflection of the beam under its "statically equivalent" load P, to account for the dynamic effect of it. As is quickly apparent, this factor becomes very great as the frequency of the applied load approaches that of the natural frequency, a phenomenon called *resonance*. Of course, in the case of real beams, the deflection never reaches the indefinitely large scale predicted by the magnification factor, since no provision has been made in this model for friction forces in the real beam. It is therefore an idealization, although it does correctly indicate that there will be large motions in and near resonance. There are as many resonances as there are modes, of course, each one with its own resonant frequency.

The full equation of motion for the beam vibration is then

$$\begin{aligned} y(x,t) &= \sum_{n=1,3,5,\ldots}^{\infty} a_n\,u_n\,sin\frac{n\pi x}{\ell} \\ &= \sum_{n=1,3,5,\ldots}^{\infty} a_n\left[(u_n)_c + (u_n)_p\right] sin\frac{n\pi x}{\ell} \\ &= \sum_{n=1,3,5,\ldots}^{\infty} a_n\,cos(\omega_n t + \lambda_n)\,sin\frac{n\pi x}{\ell} \\ &+ \sum_{n=1,3,5,\ldots}^{\infty} a_n\,\frac{2\,P\,\ell^3}{n^4\,\pi^4\,E\,I\,a_n}\left(\frac{1}{1-\frac{\Omega^2}{\omega_n^2}}\right) cos\,\Omega t\,sin\frac{n\pi x}{\ell} \end{aligned}$$

The forced vibration—the steady-state portion of the motion—is

$$y(x,t) = \sum_{n=1,3,5,\ldots}^{\infty} (-1)^{\frac{n-1}{2}}\,\frac{2\,P\,\ell^3}{n^4\,\pi^4\,E\,I}\left(\frac{1}{1-\frac{\Omega^2}{\omega_n^2}}\right) P\,cos\,\Omega\,t\,sin\frac{n\pi x}{\ell} \tag{15.404}$$

15.12 A BEAM VIBRATION—STATE SPACE STYLE

The treatment outlined in the previous section may be called a conventional or classical approach to the analysis of a beam vibration. It has the merit that the physical elements of the analysis are kept in the forefront of developments. Once these elements have been absorbed, however, the model can be revisited as it is formulated in terms of a state space representation. For this purpose the starting point may be taken to be that of the so-called *Timoshenko beam equation*[12,13]

$$EI\frac{\partial^4 y(x,t)}{\partial x^4} - \rho I\left(1+\frac{\alpha E}{G}\right)\frac{\partial^4 y(x,t)}{\partial x^2 \partial t^2} + \frac{\alpha \rho^2 I}{G}\frac{\partial^4 y(x,t)}{\partial t^4} + \rho A\frac{\partial^2 y(x,t)}{\partial t^2} = 0$$

in which G is the shear modulus and α is a dimensionless shape factor. The solution of this equation was investigated by R. A. Anderson,[14] who may be largely responsible for the nomenclature *Timoshenko!beam*.

This equation, first propounded by the famous Professor Timoshenko, includes the effects of rotary motion and shearing deformations, both of which are sometimes considered to be secondary effects. When these effects are neglected, and provision is made for the addition of applied force distributed along the length of the beam, the result is

$$EI\frac{\partial^4 y(x,t)}{\partial x^4} + \rho A\frac{\partial^2 y(x,t)}{\partial t^2} = w(x,t) \tag{15.405}$$

The physical quantities involved have the same meaning as previously, except that $w(x,t)$ is now a ***distributed*** loading applied to the beam as a function of both its length and time. The equation is still generally referred to as a *Timoshenko beam equation*.

With the fore-knowledge gained in the previous section, let the deflection of the beam be represented as the sum of an infinite number of modes

$$y(x,t) = \sum_{i=1}^{\infty} u_i(t)\,\phi_i(x) \tag{15.406}$$

where $u_i(t)$ is the time-varying amplitude component, and $\phi_i(x)$ is the mode-shape function. As before, the semi-inverse method assumes a separation of the variables, or a mode decomposition, from the start.

In practice, a frequent assumption is that a close approximation to the displacement of the beam can be obtained by retaining a finite number n of these modes. The number may sometimes, nevertheless, be quite large. Therefore, let the revised displacement equation be

$$y(x,t) \cong \sum_{i=1}^{n} u_i(t)\,\phi_i(x) \tag{15.407}$$

15.12.1 GENERALIZED BEAM LOADING

The state space approach to beam dynamics allows for a fuller representation of the loading, including especially a collection of concentrated forces at various places along the beam. This is because singularity functions can be invoked, especially the unit impulse function and the unit step function.

[12]Weaver, W., Jr., Timoshenko, S. P., and Young, D. H., *Vibration Problems in Engineering*, Fifth Edition, John Wiley & Sons, New York, 1990, p. 438.

[13]Newland, D. E., *Mechanical Vibration Analysis and Computation*, Longman Scientific & Technical, Essex, England, 1989, p. 392.

[14]Anderson, R. A., "Flexural Vibrations in Uniform Beams According to the Timoshenko Theory," *Transactions of the ASME*, Volume 75, 1953, pp. APM 504-514.

CONCENTRATED FORCES

In this mode, the "distributed" loading can be taken to be

$$w(x,t) = \sum_{j=1}^{k} \delta(x - x_k)\, P_k(t) \tag{15.408}$$

which represents a series of k concentrated forces applied transversely at places named x_k along the beam from the origin of coordinates at, say, one end. Figure 15.19 shows the geometry. The Dirac delta function is eminently suited to the task of representing concentrated forces and other such quantities.

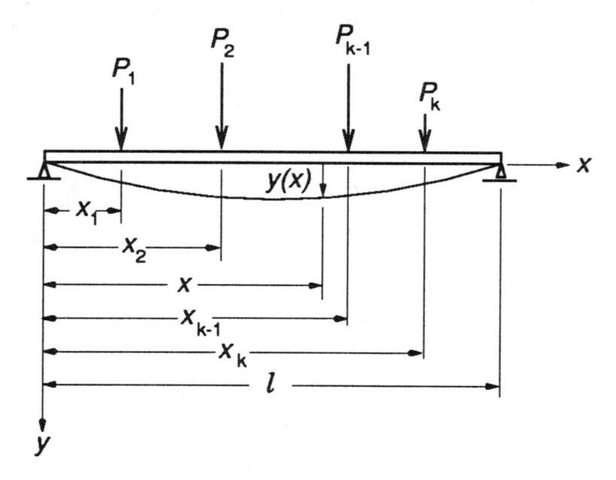

Figure 15.19: Concentrated Forces Loading

GENERAL BEAM LOADINGS

See Figure 15.20 for a more general loading combination which can be expressed in terms of singularity functions. The concentrated force is, as in the previous section, given by its magnitude P multiplied by a Dirac delta operator at the location $x = a$. A concentrated couple (or moment) can also be included in this method of representation by giving its magnitude, C, multiplied by a Dirac operator, $\delta(x - b)$.

A distributed loading of w N/m beginning at the location $x = c$ and ending at $x = d$ is easily acccomodated by employing the unit step function twice, once to activate the loading, a second time to abrogate it. The superposition of these two unit step functions is revealed in Figure 15.21.

Actually, the forces of reaction at the end supports of the beam can also be included by using singularity functions. At the left end, for example, the force is $R_L = -\delta(x - 0)R_L$, and at the right end of the beam it is $R_R = -\delta(x - \ell)R_R$, since these are both concentrated forces applied at the locations $x = 0$, and $x = \ell$, respectively.

For the loads depicted in Figure 15.20, the combined loading distribution is

$$\begin{aligned} w(x,t) &= -\delta(x-0)R_L + \delta(x-a)P + \delta(x-b)C \\ &\quad + w\mathbf{1}(x-c) - w\mathbf{1}(x-d) - \delta(x-\ell)R_R \end{aligned} \tag{15.409}$$

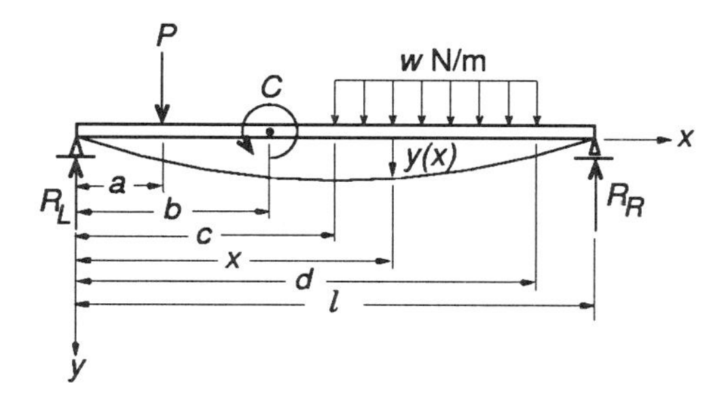

Figure 15.20: General Beam Loading

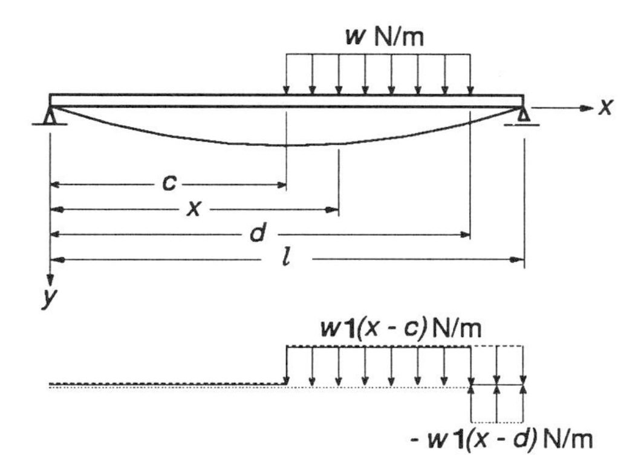

Figure 15.21: Step Function Loadings

SHEARING FORCE AND BENDING MOMENT

To complete the description of generalized beam loadings, it is easy to obtain the shearing force and bending moment diagrams directly from the distributed load equation. For example, in the case illustrated above, Equation 15.409, the shearing force is achieved by integrating the usual relationship

$$\begin{aligned} \frac{dV}{dx} &= w(x,t) \\ &= -\delta(x-0)R_L + P\delta(x-a) + C\delta(x-b) + w\mathbf{1}(x-c) \\ &\qquad -w\mathbf{1}(x-d) - \delta(x-\ell)R_R \end{aligned} \tag{15.410}$$

Therefore,

$$\begin{aligned} V(x) - V(x \to \infty) = &-R_L \int_{-\infty}^{x} \delta(\xi)\,d\xi + P\int_{-\infty}^{x} \delta(\xi - a) \\ &+ C\int_{-\infty}^{x} \delta(x-b)d\xi + w\mathbf{1}(x-c)\int_{c}^{x} d\xi \\ &- w\mathbf{1}(x-d)\int_{d}^{x} d\xi - R_R\int_{-\infty}^{x} \delta(\xi-\ell)d\xi \end{aligned} \tag{15.411}$$

and there are boundary conditions to be adjoined. Since the beam ends at $x = \ell$, there is no shear beyond that, hence $V(x \to \infty) = 0$. Moreover, it is seen that for the concentrated forces the integrals are step functions. Thus

$$\begin{aligned} V(x) = &-R_L\mathbf{1}(x) + P\mathbf{1}(x-b) + C\,\mathbf{1}(x-c) + w\,(x-c)\,\mathbf{1}(x-c) \\ &-w\,(x-d)\,\mathbf{1}(x-d) - R_R\mathbf{1}(x-\ell) \end{aligned} \tag{15.412}$$

This procedure can be carried to the next level with the determination of the bending moment equation.

$$\begin{aligned} \frac{dM}{dx} &= -V(x) \\ &= R_L\mathbf{1}(x) - P\mathbf{1}(x-a) - C\mathbf{1}(x-b) - w\,(x-c)\,\mathbf{1}(x-c) \\ &\qquad +w\,(x-d)\,\mathbf{1}(x-d) + R_R\mathbf{1}(x-\ell) \end{aligned} \tag{15.413}$$

$$\begin{aligned} M(x) - M(x \to \infty) = &R_L\mathbf{1}(x)\int_0^x d\xi - P\,\mathbf{1}(x-a)\int_0^x d\xi - C\,\mathbf{1}(x-b)\int_0^x d\xi \\ &-w\,\mathbf{1}(x-c)\int_c^x (\xi - c)\,d\xi + w\,\mathbf{1}(x-d)\int_d^x (\xi-d)\,d\xi \\ &+R_R\,\mathbf{1}(x-\ell)\int_L^x d\xi \end{aligned} \tag{15.414}$$

where, again, there are boundary conditions. There is no moment at $x \geq \ell$ because it is a free end, so $M(x \to \infty) = 0$. As to the integrals in the fourth and fifth terms, these are elementary forms with the results

$$\int_c^x (\xi - c)\,d\xi = \frac{(x-c)^2}{2} \tag{15.415}$$

$$\int_d^x (\xi - d)\,d\xi = \frac{(x-d)^2}{2} \tag{15.416}$$

For the bending moment equation the outcome is

$$\begin{aligned} M(x) = &R_L\,\mathbf{1}(x) - P\,(x-a)\,\mathbf{1}(x-a) - C\,(x-b)\,\mathbf{1}(x-b) \\ &-w\,\frac{(x-c)^2}{2}\,\mathbf{1}(x-c) + w\,\frac{(x-d)^2}{2}\,\mathbf{1}(x-d) \\ &-R_R\,(x-\ell)\,\mathbf{1}(x-\ell) \end{aligned} \tag{15.417}$$

Graphs showing the shearing force and bending moment over the length of the beam are easily constructed, and are shown in Figure 15.22.

The use of impulse functions and unit step functions greatly facilitates the construction of the shearing force and bending moment diagrams which are so important in structural design activities, as has been seen above. As with most such devices, a few problem-solving practice sessions generally serve to convert even novice designers into accomplished practitioners of the art.

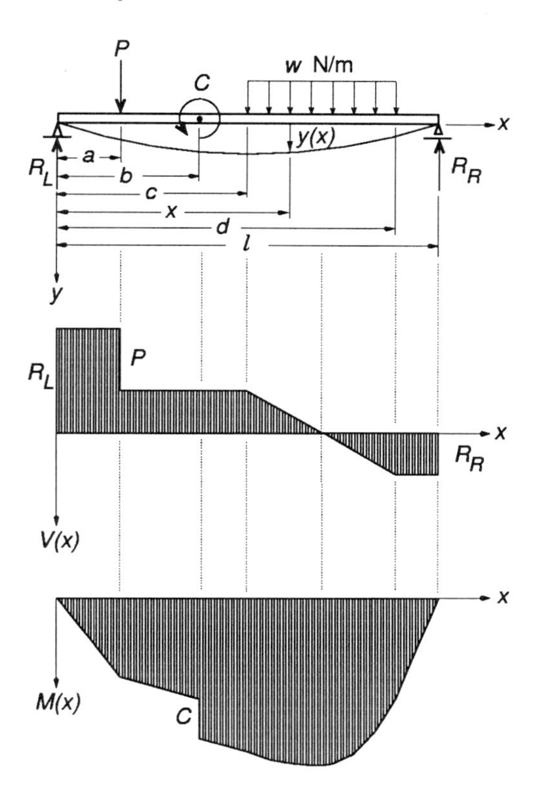

Figure 15.22: Shearing Force and Bending Moment

15.12.2 STATE SPACE EQUATIONS OF MOTION

A state space model of the dynamics of the beam can be constructed by composing the state vector

$$\mathbf{x}(t) = [\ u_1(t) \ \ \dot{u}_1(t) \ \ u_2(t) \ \ \dot{u}_2(t) \ \ \cdots \ \ u_n(t) \ \ \dot{u}_n(t)\ ]^t \tag{15.418}$$

It should be noted especially that this is a $2n$-dimensional vector, formed from the displacements and velocities of each of the modes, as previously revealed in the illustrative examples in Section 15.4.

Another matrix is formed as

$$\mathbf{A} = diag[\mathbf{\Lambda}_1, \mathbf{\Lambda}_2, \ldots, \mathbf{\Lambda}_n] \tag{15.419}$$

with the sub-matrices

$$\mathbf{\Lambda}_i = \begin{bmatrix} 0 & 1 \\ -\omega_i^2 & 0 \end{bmatrix} \tag{15.420}$$

The ω_i in these matrices are the self-same eigenvalues

$$\omega_i^2 = \frac{i^4 \pi^4 EI}{\rho A \ell^4}$$

that were found previously, as in Equation 15.390.

One more matrix, $\mathbf{B}$, is given by

$$\mathbf{B} = [\mathbf{B}_1 \;\; \mathbf{B}_2 \;\; \cdots \;\; \mathbf{B}_n]^t \tag{15.421}$$

where each sub-matrix $\mathbf{B}_i$ is itself a $2 \times k$ matrix

$$\mathbf{B}_i = \frac{2}{\rho A \ell} \begin{bmatrix} 0 & 0 & \cdots & 0 & \cdots & 0 \\ \delta(x - x_1) & \delta(x - x_2) & \cdots & \delta(x - x_j) & \cdots & \delta(x - x_k) \end{bmatrix} \tag{15.422}$$

The system input vector (driving force vector) has the form

$$\mathbf{u}(t) = [P_1(t) \;\; P_2(t) \;\; \cdots \;\; P_i(t) \;\; \cdots \;\; P_k(t)]^t \tag{15.423}$$

in which the $P_i(t)$ are the force magnitudes acting at the several locations, as in Formula (15.408). Using the equipment developed in the previous section, more general loadings can be incorporated at this same juncture.

Now the model can be displayed explicitly as

$$\dot{\mathbf{x}}(t) = \mathbf{A}\,\mathbf{x}(t) + \mathbf{B}\,\mathbf{u}(t) \tag{15.424}$$

$$\mathbf{y}(t) = \mathbf{C}\,\mathbf{x}(t) + \mathbf{D}\,\mathbf{u}(t) \tag{15.425}$$

When the first of these is expanded, the outcome confirms Equation 15.405. The second equation is the one employed to show the displacement of the beam. Then, with

$$\dot{\mathbf{x}}(t) = [\ \dot{u}_1(t) \;\; \ddot{u}_1(t) \;\; \dot{u}_2(t) \;\; \ddot{u}_2(t) \;\; \cdots \;\; \dot{u}_n(t) \;\; \ddot{u}_n(t)\ ]^t \tag{15.426}$$

$$\mathbf{A}\mathbf{x}(t) = \begin{bmatrix} 0 & 1 & 0 & 0 & \cdots & 0 & 0 & \cdots & 0 & 0 \\ -\omega_1^2 & 0 & 0 & 0 & \cdots & 0 & 0 & \cdots & 0 & 0 \\ 0 & 0 & 0 & 1 & \cdots & 0 & 0 & \cdots & 0 & 0 \\ 0 & 0 & -\omega_2^2 & 0 & \cdots & 0 & 0 & \cdots & 0 & 0 \\ \vdots & \vdots & \vdots & \vdots & \ddots & \vdots & \vdots & \ddots & \vdots & \vdots \\ 0 & 0 & 0 & 0 & \cdots & 0 & 1 & \cdots & 0 & 0 \\ 0 & 0 & 0 & 0 & \cdots & -\omega_i^2 & 0 & \cdots & 0 & 0 \\ \vdots & \vdots & \vdots & \vdots & \ddots & \vdots & \vdots & \ddots & \vdots & \vdots \\ 0 & 0 & 0 & 0 & \cdots & 0 & 0 & \cdots & 0 & 1 \\ 0 & 0 & 0 & 0 & \cdots & 0 & 0 & \cdots & -\omega_n^2 & 0 \end{bmatrix} \begin{bmatrix} u_1(t) \\ \dot{u}_1(t) \\ u_2(t) \\ \dot{u}_2(t) \\ \vdots \\ u_i(t) \\ \dot{u}_i(t) \\ \vdots \\ u_n(t) \\ \dot{u}_n(t) \end{bmatrix}$$

$$= \begin{bmatrix} \dot{u}_1(t) \\ -\omega_1^2\, u_1(t) \\ \dot{u}_2(t) \\ -\omega_2^2\, u_2(t) \\ \vdots \\ \dot{u}_i(t) \\ -\omega_i^2\, u_i(t) \\ \vdots \\ \dot{u}_n(t) \\ -\omega_n^2\, u_n(t) \end{bmatrix} \tag{15.427}$$

$$
\mathbf{B}\,\mathbf{u}(t) = \frac{2}{\rho A \ell}
\begin{bmatrix}
0 & 0 & \cdots & 0 & \cdots & 0 \\
\delta(x-x_1) & \delta(x-x_2) & \cdots & \delta(x-x_j) & \cdots & \delta(x-x_k) \\
0 & 0 & \cdots & 0 & \cdots & 0 \\
\delta(x-x_1) & \delta(x-x_2) & \cdots & \delta(x-x_j) & \cdots & \delta(x-x_k) \\
\vdots & \vdots & \ddots & \vdots & \ddots & \vdots \\
0 & 0 & \cdots & 0 & \cdots & 0 \\
\delta(x-x_1) & \delta(x-x_2) & \cdots & \delta(x-x_j) & \cdots & \delta(x-x_k) \\
\vdots & \vdots & \ddots & \vdots & \ddots & \vdots \\
0 & 0 & \cdots & 0 & \cdots & 0 \\
\delta(x-x_1) & \delta(x-x_2) & \cdots & \delta(x-x_j) & \cdots & \delta(x-x_k)
\end{bmatrix}
\begin{bmatrix}
0 \\ P_1(t) \\ 0 \\ P_2(t) \\ \vdots \\ 0 \\ P_i(t) \\ \vdots \\ 0 \\ P_k(t)
\end{bmatrix}
$$

$$
= \frac{2}{\rho A \ell}
\begin{bmatrix}
0 \\
\delta(x-x_1)\,P_1(t) + \cdots + \delta(x-x_k)\,P_k \\
0 \\
\delta(x-x_1)\,P_1(t) + \cdots + \delta(x-x_k)\,P_k \\
\vdots \\
0 \\
\delta(x-x_1)\,P_1(t) + \cdots + \delta(x-x_k)\,P_k \\
\vdots \\
0 \\
\delta(x-x_1)\,P_1(t) + \cdots + \delta(x-x_k)\,P_k
\end{bmatrix}
\tag{15.428}
$$

$$
\begin{bmatrix}
\dot{u}_1(t) \\ \ddot{u}_1(t) \\ \dot{u}_2(t) \\ \ddot{u}_2(t) \\ \vdots \\ \dot{u}_i(t) \\ \ddot{u}_i(t) \\ \vdots \\ \dot{u}_n(t) \\ \ddot{u}_n(t)
\end{bmatrix}
=
\begin{bmatrix}
\dot{u}_1(t) \\ -\omega_1^2\,u_1(t) \\ \dot{u}_2(t) \\ -\omega_2^2\,u_2(t) \\ \vdots \\ \dot{u}_i(t) \\ -\omega_i^2\,u_i(t) \\ \vdots \\ \dot{u}_n(t) \\ -\omega_n^2\,u_n(y)
\end{bmatrix}
+ \frac{2}{\rho A \ell}
\begin{bmatrix}
0 \\
\delta(x-x_1)\,P_1(t) + \cdots + \delta(x-x_k)\,P_k(t) \\
0 \\
\delta(x-x_1)\,P_1(t) + \cdots + \delta(x-x_k)\,P_k(t) \\
\vdots \\
0 \\
\delta(x-x_1)\,P_1(t) + \cdots + \delta(x-x_k)\,P_k(t) \\
\vdots \\
0 \\
\delta(x-x_1)\,P_1(t) + \cdots + \delta(x-x_k)\,P_k(t)
\end{bmatrix}
\tag{15.429}
$$

The state equations come in pairs, each of which gives the motion of one mode, and is uncoupled from the other pairs. On the other hand, the equations of each pair are, in general, coupled. In the present case, one equation of each pair is merely a redundant identity in the modal velocities, therefore adds nothing to the overall solution of the motion.

This set of equations demonstrates the so-called *modal decomposition* associated with a state space vibration.

The state space method evidently allows for a much more comprehensive version of beam vibration models than can usually be had in the conventional manner.

$$\begin{aligned}
\dot{u}_1(t) &= \dot{u}_1(t) \\
\ddot{u}_1(t) &= -\omega_1^2 u_1(t) + \frac{2}{\rho A\ell}[\delta(x-x_1)\,P_1(t) + \cdots + \delta(x-x_k)\,P_k(t)] \\
\dot{u}_2(t) &= \dot{u}_2(t)\,, \\
\ddot{u}_2(t) &= -\omega_2^2 u_2(t) + \frac{2}{\rho A\ell}[\delta(x-x_1)\,P_1(t) + \cdots + \delta(x-x_k)\,P_k(t)] \\
\vdots\ &= \qquad\qquad\qquad\qquad \vdots \\
\dot{u}_i(t) &= \dot{u}_i(t)\,, \\
\ddot{u}_i(t) &= -\omega_i^2 u_i(t) + \frac{2}{\rho A\ell}[\delta(x-x_1)\,P_1(t) + \cdots + \delta(x-x_k)\,P_k(t)] \\
\vdots\ &= \qquad\qquad\qquad\qquad \vdots \\
\dot{u}_n(t) &= \dot{u}_n(t)\,, \\
\ddot{u}_n(t) &= -\omega_n^2 u_n(t) + \frac{2}{\rho A\ell}[\delta(x-x_1)\,P_1(t) + \cdots + \delta(x-x_k)\,P_k(t)]
\end{aligned}$$

or

$$\begin{aligned}
\ddot{u}_1(t) + \omega_1^2 u_1(t) &= \frac{2}{\rho A\ell}[\delta(x-x_1)\,P_1(t) + \cdots + \delta(x-x_k)\,P_k(t)] \\
\ddot{u}_2(t) + \omega_2^2 u_2(t) &= \frac{2}{\rho A\ell}[\delta(x-x_1)\,P_1(t) + \cdots + \delta(x-x_k)\,P_k(t)] \\
\vdots\qquad &= \qquad\qquad\qquad \vdots \\
\ddot{u}_i(t) + \omega_i^2 u_i(t) &= \frac{2}{\rho A\ell}[\delta(x-x_1)\,P_1(t) + \cdots + \delta(x-x_k)\,P_k(t)] \\
\vdots\qquad &= \qquad\qquad\qquad \vdots \\
\ddot{u}_n(t) + \omega_n^2 u_n(t) &= \frac{2}{\rho A\ell}[\delta(x-x_1)\,P_1(t) + \cdots + \delta(x-x_k)\,P_k(t)]
\end{aligned}$$

At this stage, the linear second-order differential equations above are the equivalents of those in Equation 15.387, which were obtained by conventional methods. The terms on the right-hand sides of these equations are the driving forces, expressed in operational terms using the singularity functions.

The equations are solved for their complementary and particular integrals, which correspond to the free vibration modes, and those of the forced vibrations, respectively, as in Equations 15.389 and 15.400. For the free vibrations, there are the initial conditions, including initial displacements and initial velocities; and for the forced vibrations, there are the magnification factors. These determinations are made just as was done previously.

The mode shapes are revealed by substituting first from Equation 15.432 into Equation 15.406, and from thence into Equation 15.405, which is repeated here for ready reference.

$$EI\frac{\partial^4 y(x,t)}{\partial x^4} + \rho A\,\frac{\partial^2 y(x,t)}{\partial t^2} = w(x,t)$$

In preparation for entering these terms, it is well to note that

$$\frac{\partial^2 y}{\partial t^2} = \sum_{i=1}^{n} \ddot{u}_i(t)\,\phi_i(x) \qquad \text{and} \qquad \frac{\partial^4 y}{\partial x^4} = \sum_{i=1}^{n} u_i(t)\,\phi_i''''(x)$$

THE FREE VIBRATION

Then, for the free vibration, the Timoshenko beam equation is

$$EI\sum_{i=1}^{n} u_i(t)\,\phi_i''''(x) + \rho A\sum_{i=1}^{n} \ddot{u}(t)\;\phi_i(x) = 0$$

or, with $u(t) = a_i\, cos(\omega_i t + \lambda_i)$,

$$EI\sum_{i=1}^{n} u_i(t)\,\phi_i''''(x) + \rho A\sum_{i=1}^{n} \Big[-\,\omega_i^2\, u_i(t)\Big]\,\phi_i(x) = 0$$

which is, with Equation 15.390,

$$EI\sum_{i=1}^{n} u_i(t)\,\phi_i''''(x) - \rho A\sum_{i=1}^{n} \frac{i^4\pi^4 EI}{\rho A\ell^4} u_i(t)\,\phi_i(x) = 0\,.$$

Now it can be seen that the solution is $\phi_i(x) = cos\frac{i\pi x}{\ell}$, since

$$\phi_i''''(x) = \frac{\partial^4}{\partial x^4}\Big(cos\frac{i\pi x}{\ell}\Big) = \frac{i^4\pi^4}{\ell^4}\, cos\frac{i\pi x}{\ell}$$

and when this is substituted for $\phi_i''''(x)$ in the previous equation, the result is the identity

$$\sum_{i=1}^{n} u_i(t)\frac{i^4\pi^4}{\ell^4}\, cos\frac{i\pi x}{\ell} - \sum_{i=1}^{n} u_i(t)\frac{i^4\pi^4}{\ell^4}\, cos\frac{i\pi x}{\ell} \equiv 0$$

Therefore, the displacement equation during the free vibration is given by

$$y(x,t) = \sum_{i=1}^{n} a_i\, cos(\omega_i t + \lambda_i)\, cos\frac{i\pi x}{\ell} \tag{15.430}$$

in which a_i is the amplitude coefficient and λ_i is the phase shift for the i-th mode, both to be determined from the initial conditions.

THE FORCED VIBRATION

Forced vibrations are the steady-state components of the beam motion when transients associated with the initiation of the motion and the subsequent free vibration have died down because of damping that is inherent in the system. Accordingly, assume that each of the driving forces is represented by a magnitude P_j which multiplies a temporal function with its own frequency, say $P_j\, cos\,\Omega_j t$. The sustained motions of the beam will then be proportional to these driving forces, and of the same oscillatory form. The resultant time component of the displacement can be represented as

$$u_i(t) = \sum_{j=1}^{k} \mathcal{A}_{ij} \cos \Omega_j t \tag{15.431}$$

in which $\mathcal{A}_{ij}$ is the amplitude coefficient of the i-th mode associated with the j-th excitation, and is yet to be determined.

Now let this be substituted into the modal temporal equation

$$\ddot{u}_i(t) + \omega_i^2 u_i(t) = \frac{2}{\rho A \ell}[\delta(x - x_1) P_1(t) + \cdots + \delta(x - x_k) P_k(t)]$$

to have

$$-\sum_{j=1}^{k} \Omega_{ij}^2 \mathcal{A}_{ij} \cos \Omega_j t + \omega_i^2 \sum_{j=1}^{k} \mathcal{A}_{ij} \cos \Omega_j t = \frac{2}{\rho A \ell} \sum_{j=1}^{k} \delta(x - x_j) P_j \cos \Omega_j t$$

To satisfy this equation, $\mathcal{A}_{ij}$ must be

$$\mathcal{A}_{ij} = \frac{2 P_j}{\rho A \ell (\omega_i^2 - \Omega_j^2)} \delta(x - x_j) \tag{15.432}$$

These are the magnification factors associated with each component of each mode; that is, each $\mathcal{A}_{ij}$ magnifies that particular P_j driving force as it excites the i-th mode.

For the sake of comparison with the previous analysis, assume that there is a single force $P \cos \Omega t$ acting at the center of the beam where $x = \ell/2$; i.e., $k = 1$. Then

$$u_i(t) = \mathcal{A}_i \cos \Omega t = \frac{2 P \cos \Omega t}{\rho A \ell (\omega_i^2 - \Omega^2)} \delta\big(x - \frac{\ell}{2}\big)$$

and

$$y(x,t) = \sum_{i=1}^{n} u_i(t)\, \phi_i(x) = \sum_{i=1}^{n} \frac{2 P \cos \Omega t}{\rho A \ell (\omega_i^2 - \Omega^2)} \delta\big(x - \frac{\ell}{2}\big) \sin \frac{i \pi x}{\ell} \tag{15.433}$$

But since the unit impulse function $\delta(x - \frac{\ell}{2})$ serves to convert $\sin \frac{i\pi x}{\ell}$ to $\sin \frac{i\pi}{2}$, by virtue of its sifting property (i.e., $x \to \ell/2$), the outcome is

$$y\big(\frac{\ell}{2}, t\big) = \sum_{i=1}^{n} \frac{2 P \cos \Omega t}{\rho A \ell (\omega_i^2 - \Omega^2)} \sin\big(\frac{i\pi}{2}\big)$$

and this becomes

$$y\big(\frac{\ell}{2}, t\big) = \sum_{i=1,3,5,\ldots}^{n} (-1)^{\frac{i-1}{2}} \frac{2 P \ell^3}{i^4 \pi^4 E I} \Big(\frac{1}{1 - \frac{\Omega^2}{\omega_i^2}}\Big) \cos \Omega t \tag{15.434}$$

when

$$\omega_i^2 = \frac{i^4 \pi^4 E I}{\rho A \ell^4}$$

is employed.

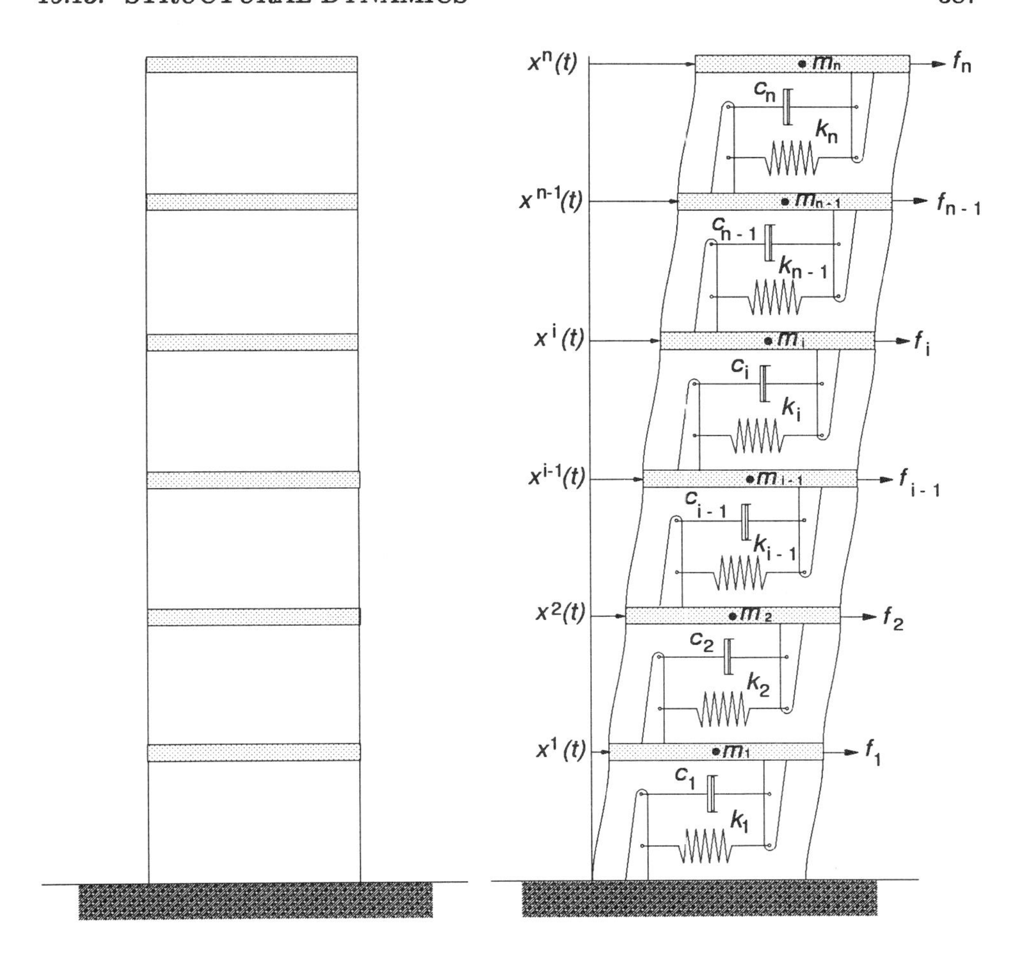

Figure 15.23: High-Rise Structure

15.13 STRUCTURAL DYNAMICS

Figure 15.23 shows the geometry of a multi-story high-rise structure. Each floor has a mass, m_i, which can be considered to be lumped at the centroid of the floor itself. The floors are joined to lower floors by columns at each end, and these columns are modeled as elastic members of negligible mass, and each may be assigned its own stiffness k_i. In addition, provision is made for viscous damping furnished by, say, a dashpot mechanism as shown in the figure. Each floor is allowed its particular damping constant c_i.

When the structure is subjected to forces in the plane of the figure, such as, for instance, wind loads, each story can flex upon its elastic columns in a horizontal motion. The lateral displacement of each story is denoted $x^i(t)$, and the lateral force acting upon it is $f^i(t)$.

The equations of motion for the several storys can be written as follows:

$$m_1\ddot{x}^1 + c_1\,\dot{x}^1 - c_2[\dot{x}^2 - \dot{x}^1] + k_1\,x^1 - k_2[x^2 - x^1] = f^1(t) \tag{15.435}$$

$$m_2\ddot{x}^2 + c_2[\dot{x}^2 - \dot{x}^1] - c_3[\dot{x}^3 - \dot{x}^2] + k_2[x^2 - x^1] - k_3[x^3 - x^2] = f^2(t) \tag{15.436}$$

$$m_i\ddot{x}^i + c_i[\dot{x}^i - \dot{x}^{i-1}] - c_{i+1}[\dot{x}^{i+1} - \dot{x}^i] + k_i[x^i - x^{i-1}] - k_{i+1}[x^{i+1} - x^i] = f^i(t) \tag{15.437}$$

$$m_n\ddot{x}^n + c_n[\dot{x}^n - \dot{x}^{n-1}] + k_n[x^n - x^{n-1}] = f^n(t) \tag{15.438}$$

The state-space form of these equations can be represented as

$$\mathbf{x}(t) = \left[x^1(t)\; x^2(t)\; \cdots\; x^i(t)\; \cdots\; x^n(t)\; y^1(t)\; y^2(t)\; \cdots\; y^i(t)\; \cdots\; y^n(t)\right]^t \tag{15.439}$$

in which $y^i(t) = \dot{x}^i(t)$, etc. Then

$$\dot{x}^1 = y^1 \tag{15.440}$$

$$\dot{x}^2 = y^2 \tag{15.441}$$

$$\cdots = \cdots$$

$$\dot{x}^i = y^i \tag{15.442}$$

$$\cdots = \cdots$$

$$\dot{x}^n = y^n \tag{15.443}$$

$$\dot{y}^1 = -\frac{c_1}{m_1}y^1 + \frac{c_2}{m_1}[y^2 - y^1] - \frac{k_1}{m_1}x^1 + \frac{k_2}{m_1}[x^2 - x^1] + \frac{1}{m_1}f^1(t) \tag{15.444}$$

$$\dot{y}^2 = -\frac{c_2}{m_2}[y^2 - y^1] + \frac{c_3}{m_2}[y^3 - y^2] - \frac{k_2}{m_2}[x^2 - x^1] + \frac{k_3}{m_2}[x^3 - x^2] + \frac{1}{m_2}f^2(t) \tag{15.445}$$

$$\cdots = \cdots$$

$$\dot{y}^i = -\frac{c_i}{m_i}[y^i - y^{i-1}] + \frac{c_{i+1}}{m_i}[y^{i+1} - y^i] - \frac{k_i}{m_i}[x^i - x^{i-1}] + \frac{k_{i+1}}{m_i}[x^{i+1} - x^i] + \frac{f^i(t)}{m_i} \tag{15.446}$$

$$\cdots = \cdots$$

$$\dot{y}^n = -\frac{c_n}{m_n}[y^n - y^{n-1}] - \frac{k_n}{m_n}[x^n - x^{n-1}] + \frac{1}{m_n}f^n(t) \tag{15.447}$$

are the converted forms of the equations of motion.

The state transition equation becomes

$$\dot{\mathbf{x}}(t) = \mathbf{A}\,\mathbf{x}(t) + \mathbf{G}\,\mathbf{f}(t) \tag{15.448}$$

in which the matrices are

$$\mathbf{A} = \begin{bmatrix} \mathbf{A}_{11} & \vdots & \mathbf{A}_{12} \\ \cdots & & \cdots \\ \mathbf{A}_{21} & \vdots & \mathbf{A}_{22} \end{bmatrix} \qquad \mathbf{G} = \begin{bmatrix} \mathbf{O}_n \\ \mathbf{M}^{-1} \end{bmatrix} \qquad \mathbf{f}(t) = \begin{bmatrix} f^1(t) \\ f^2(t) \\ \cdots \\ f^i(t) \\ \cdots \\ f^n(t) \end{bmatrix} \tag{15.449}$$

In these expressions

$$\mathbf{A}_{11} = \mathbf{O}_n = \begin{bmatrix} 0 & 0 & \cdots & 0 \\ 0 & 0 & \cdots & 0 \\ \vdots & \vdots & 0 & \vdots \\ 0 & 0 & \cdots & 0 \end{bmatrix} \qquad \mathbf{A}_{12} = \mathbf{I}_n = \begin{bmatrix} 1 & 0 & \cdots & 0 \\ 0 & 1 & \cdots & 0 \\ \vdots & \vdots & 1 & \vdots \\ 0 & 0 & \cdots & 1 \end{bmatrix} \tag{15.450}$$

are sub-matrices of the partitioned matrix **A** as indicated. Other sub-matrices are

$$\mathbf{A}_{21} = \begin{bmatrix} -\frac{k_1+k_2}{m_1} & \frac{k_2}{m_1} & 0 & 0 & 0 & 0 & 0 & \cdots & 0 & 0 \\ \frac{k_2}{m_2} & -\frac{k_2+k_3}{m_2} & \frac{k_3}{m_2} & 0 & 0 & 0 & 0 & \cdots & 0 & 0 \\ \vdots & \vdots & \vdots & \vdots & \vdots & \vdots & \vdots & \cdots & \vdots & \vdots \\ 0 & 0 & 0 & 0 & \frac{k_i}{m_i} & -\frac{k_i+k_{i+1}}{m_i} & \frac{k_{i+1}}{m_i} & \cdots & 0 & 0 \\ \vdots & \vdots & \vdots & \vdots & \vdots & \vdots & \vdots & \cdots & \vdots & \vdots \\ 0 & 0 & 0 & 0 & 0 & 0 & 0 & \cdots & \frac{k_n}{m_n} & -\frac{k_n}{m_n} \end{bmatrix} \tag{15.451}$$

$$\mathbf{A}_{22} = \begin{bmatrix} -\frac{c_1+c_2}{m_1} & \frac{c_2}{m_1} & 0 & 0 & 0 & 0 & 0 & \cdots & 0 & 0 \\ \frac{c_2}{m_2} & -\frac{c_2+c_3}{m_2} & \frac{c_3}{m_2} & 0 & 0 & 0 & 0 & \cdots & 0 & 0 \\ \vdots & \vdots & \vdots & \vdots & \vdots & \vdots & \vdots & \cdots & \vdots & \vdots \\ 0 & 0 & 0 & 0 & \frac{c_i}{m_i} & -\frac{c_i+c_{i+1}}{m_i} & \frac{c_{i+1}}{m_i} & \cdots & 0 & 0 \\ \vdots & \vdots & \vdots & \vdots & \vdots & \vdots & \vdots & \cdots & \vdots & \vdots \\ 0 & 0 & 0 & 0 & 0 & 0 & 0 & \cdots & \frac{c_n}{m_n} & -\frac{c_n}{m_n} \end{bmatrix} \tag{15.452}$$

$$\mathbf{M} = diag[m_1, m_2, \ldots, m_i, \ldots, m_n] \tag{15.453}$$

$\mathbf{M}^{-1}$ is the inverse of **M**.

15.14 SEISMIC STRUCTURAL ANALYSIS

One of the large areas of engagement among civil engineers is that concerned with the investigation of the response of structures to seismic excitation. A substantial fraction of the world's population is found in regions which can be described as "earthquake prone." From the earliest records of civilization—and even before that from the records of the ruins—the terrible *terremoto* has been a dreaded natural phenomenon.

Extensive earthquakes in Turkey, Greece, the Caucasuses, Mexico, California, Alaska, Chile, and Japan have plagued people the world over in recent years. Events such as the 1994 Northridge (Los Angeles) and the 1995 Kobē (Hyogo-ken Nanbu) quakes have left vivid impressions upon the minds of TV viewers all across the globe, and deep scars upon the survivors at the sites.

Much of the damage inflicted in earthquakes is the fault of poor construction practices, either because of the necessity to erect inexpensive buildings or to utilize land that is known to be bad foundation earth in the event of quakes. Filled land, such as in portions of the San Francisco Bay area and the former lake beds about Mexico City are well-known to contain soils susceptibile to easy transmission of seismic forces. In some of the more highly developed regions, new standards have been incorporated into the building codes, and structures designed under these codes have proved to be substantially more resilient.

Designers and analysts are continually seeking additional ways in which to improve seismic resistance, both for new and existing structures, particularly in the developed areas where the pressures for growth are greatest and where the codes are continually being strengthened.

The so-called "passive control strategies" for reducing damage in earthquakes have been employed for some years in a number of locales. Traditionally, these involve creating redundancies in the building's structural elements, active mass drivers (AMDs), and other means for tuning the structure to remove the effects of forcing vibrations in certain spectra.

Among the more promising methods of improving earthquake response of structures is that known as the "active control strategy." In this technique, actuators and other devices are employed to counteract undesirable dynamic forces when such forces appear on the structure, as in seismic events. The actuators are programmed to react in such an optimal fashion as to reduce the overall stresses and deformations of the building, thus limiting (if not eliminating) the damage.

However, in some areas the trends are to more high-rise types of structures, and it is these that are less suitable to optimal control because the actuators would need to have very long strokes and present very great energy needs. Moreover, these buildings are frequently found in clusters in a densely-developed urban environment. Designers seeking improved seismic resistance under such circumstances have investigated engaging one or more adjoining buildings in a common reduction agenda.

This approach may have been engendered by a knowledge of the proven ability of tuned dampers to reduce vibration susceptibility in other fields. In other words, adjacent buildings may actually be engaged as dynamic absorbers in such a model.

A recent paper[15] gives a review of the pertinent literature and an example of the analysis typical of the current state of the art.

---

[15]Christenson, R. E., Spencer, B. F., Jr., and Johnson, E. A., "Coupled Building Control Using Active and Smart Damping Strategies," ***Optimization and Control in Civil and Structural Engineering***, B. H. V. Topping and B. Kumar, Editors, CIVIL-COMP PRESS, Edinburgh, 1999, pp. 187-195.

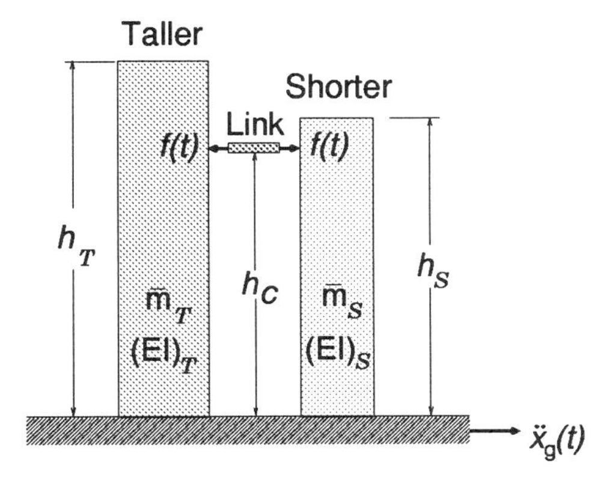

Figure 15.24: Coupled Buildings

15.14.1 A COUPLED BUILDING MODEL

To simplify the problem, the authors chose to consider two adjacent flexible high-rise buildings, with a single linkage between them of a suitable architectural treatment. The taller building has a fixed geometry and properties, while the shorter one is allowed to have variable attributes. The excitation applied to the taller building is also held fixed, while the response reduction is analyzed. Figure 15.24 shows the system.

The dissimilar structures are connected at an elevation h_c by a coupling link with the capacity of producing varied force levels as required to reduce the response of the taller one to seismic (or to wind) loading. The system is assumed to be linearly elastic, and the excitation is a lateral acceleration of the ground. Any effects not in the plane of the figure are neglected in the analysis in question.

The equations of motion of the system are

$$\mathbf{M}\,\ddot{\mathbf{x}}(t) + \mathbf{C}\,\dot{\mathbf{x}}(t) + \mathbf{K}\,\mathbf{x}(t) = -\mathbf{G}\,\ddot{x}_g(t) + \mathbf{P}\,f(t) \tag{15.454}$$

in which

$\mathbf{x}(t) = [\mathbf{x}_{\mathcal{T}}^t(t) \quad \mathbf{x}_{\mathcal{S}}^t(t)]^t$ is the combined lateral displacements vector,

$\mathbf{G} = [(\mathbf{M}_{\mathcal{T}}\,\mathbf{1})^t \quad (\mathbf{M}_{\mathcal{S}}\,\mathbf{1})^t]^t$ is the ground acceleration loading vector,

$\mathbf{P} = [-\mathbf{P}_{\mathcal{T}}^t \quad \mathbf{P}_{\mathcal{S}}^t]^t$ is the vector force in the control link,

$\ddot{x}_g(t)$ is the ground acceleration.

$\mathbf{1}$ is a vector of ones, and $f(t)$ is the control force magnitude.

The motional equations can be presented in state space form using the state vector

$$\mathbf{q}(t) = [\mathbf{x}^t(t) \quad \dot{\mathbf{x}}^t(t)]^t \tag{15.455}$$

Two types of output are considered:

- evaluation/regulated outputs $\mathbf{z}(t)$, which include interstory drift angles, absolute accelerations, and normalized story shears of each building,

- and measured outputs $\mathbf{y}(t)$, including the relative displacement across the coupling link and the absolute accelerations of the buildings at the height of the link.

The state space equations are:

$$\dot{\mathbf{q}}(t) = \mathbf{A}\,\mathbf{q}(t) + \mathbf{B}\,\ddot{x}_g(t) + \mathbf{E}\,f(t) \tag{15.456}$$

$$\mathbf{z}(t) = \mathbf{C}_z\,\mathbf{q}(t) + \mathbf{F}_z\,f(t) \tag{15.457}$$

$$\mathbf{y}(t) = \mathbf{C}_y\,\mathbf{q}(t) + \mathbf{F}_y\,f(t) \tag{15.458}$$

with

$$\mathbf{A} = \begin{bmatrix} \mathbf{0} & \mathbf{I} \\ -\mathbf{M}^{-1}\mathbf{K} & -\mathbf{M}^{-1}\mathbf{C} \end{bmatrix} \qquad \mathbf{B} = \begin{bmatrix} \mathbf{0} \\ -\mathbf{M}^{-1}\mathbf{G} \end{bmatrix} \tag{15.459}$$

$$\mathbf{E} = \begin{bmatrix} \mathbf{0} \\ \mathbf{M}^{-1}\mathbf{P} \end{bmatrix} \tag{15.460}$$

$$\mathbf{C}_z = \begin{bmatrix} \mathbf{\Delta} & \mathbf{0} \\ -\mathbf{M}^{-1}\mathbf{K} & -\mathbf{M}^{-1}\mathbf{C} \\ -\mathbf{\Gamma}_m\mathbf{M}^{-1}\mathbf{K} & -\mathbf{\Gamma}_m\mathbf{M}^{-1}\mathbf{C} \end{bmatrix} \qquad \mathbf{F}_z = \begin{bmatrix} \mathbf{0} \\ \mathbf{M}^{-1}\mathbf{P} \\ \mathbf{\Gamma}_m\mathbf{M}^{-1}\mathbf{P} \end{bmatrix}$$

$$\mathbf{C}_y = \begin{bmatrix} \mathbf{P} & \mathbf{0} \\ -\mathbf{\Lambda}\mathbf{M}^{-1}\mathbf{K} & -\mathbf{\Lambda}\mathbf{M}^{-1}\mathbf{C} \end{bmatrix} \qquad \mathbf{F}_y = \begin{bmatrix} \mathbf{0} \\ \mathbf{\Lambda}\mathbf{M}^{-1}\mathbf{P} \end{bmatrix} \tag{15.461}$$

and

$$\mathbf{\Delta} = \begin{bmatrix} \dfrac{\Delta_{\mathcal{T}}}{h_f} & \mathbf{0} \\ \mathbf{0} & \dfrac{\Delta_{\mathcal{S}}}{h_f} \end{bmatrix} \qquad \mathbf{\Gamma}_m = \begin{bmatrix} \dfrac{\Delta_{\mathcal{T}}}{\overline{m}_{\mathcal{T}}h_{\mathcal{T}}g} & \mathbf{0} \\ \mathbf{0} & \dfrac{\Delta_{\mathcal{S}}}{\overline{m}_{\mathcal{S}}h_{\mathcal{S}}g} \end{bmatrix} \tag{15.462}$$

$$\mathbf{\Lambda} = \begin{bmatrix} \mathbf{P}_{\mathcal{T}} & \mathbf{0} \\ \mathbf{0} & \mathbf{P}_{\mathcal{S}} \end{bmatrix}, \qquad \Delta_i = \begin{bmatrix} 1 & \cdots & \cdots & 0 \\ -1 & 1 & \cdots & 0 \\ \cdots & \cdots & \cdots & \cdots \\ 0 & \cdots & -1 & 1 \end{bmatrix} \tag{15.463}$$

In the equations above, the parameter h_f represents a variable which ranges from zero to the height of each building, taken at each story. The building in question is modeled as a continuous elastic beam, and the response characteristics are then investigated at each location h_f for each building. That is to say, h_f is the test variable in each case.

15.14.2 THE COUPLED BUILDING PARAMETERS

For their investigation, the authors took the following values:

- The taller building is 50 stories high.
- Each story is 4 m, for both structures.
- The mass density of the taller is $\overline{m}_{\mathcal{T}} = 4 \times 10^5$ kg/m; (i.e., mass per unit height).
- The stiffness of the taller is $(EI)_{\mathcal{T}} = 8.18 \times 10^{13}$ N·m^2, in which E is Young's modulus, and I is the second moment of area of the cross-section in bending.

They then employed a Bernoulli-Euler model for each beam-structure as it bends in the plane of the illustration for the coupled system, and utilized the inverse method by which modes of vibration are assumed. Further assumptions made are that there is viscous damping, taken at 2% of the critical damping value for each mode, and that the first five modes will reveal the desired response features. The calculated natural frequencies associated with these modes for the taller structure are 0.20, 1.25, 3.51, 6.89, and 11.39 Hz, respectively. The values for the shorter building depend upon its height, which is a variable in the study.

15.14.3 THE EXCITATION

The ground motion in earthquakes is represented conventionally as a filtered white noise with a spectrum such as that depicted by Kanai-Tajimi.[16,17,18] Kanai and Tajimi investigated the spectral properties of a number of records taken from damaging earthquakes, and concluded that a simple spring-mass-dashpot system subjected to an excitation composed of white noise with spectral density S_0 furnishes a good generic representation of the natural phenomena.

Seismically-generated earth motion is a stochastic variable. This variable exhibits a dependence upon the time parameter subject to probability laws, rather than deterministic considerations. That is, it is a *random* event, not a repetitive or otherwise organized one. It shares common properties with turbulent air and water conditions (winds and waves), random fluctuations in electrical circuits (noise), and surface roughness of terrain in geological formations (hills and valleys.)

The principal features of stochastic variables have been exhibited in Chapter 13 and Appendix I, where the probabilistic nature of such events is treated. Appendix I contains specifically the generation of the linear oscillator model subject to a white noise excitation.

White noise is the sonic or motional analogue of white light in the visible or optical realm. That is, it is like "white light" in that it contains a uniform

---

[16]Soong, T. T., and Grigoriu, M., *Random Vibration of Mechanical and Structural Systems*, Prentice-Hall, Englewood Cliffs, New Jersey, 1993, pp. 64-68.

[17]Kanai, K., "Semi-empirical Formula for the Seismic Characteristics of the Ground," *Bulletin of the Earthquake Research Institute, University of Tokyo*, Number 35, 1957, pp. 309-325.

[18]Tajima, H., "A Statistical Method of Determining the Maximum Response of a Building Structure During an Earthquake," *Proceedings of the Second World Conference on Earthquake Engineering*, Japan, 1960, pp. 781-797.

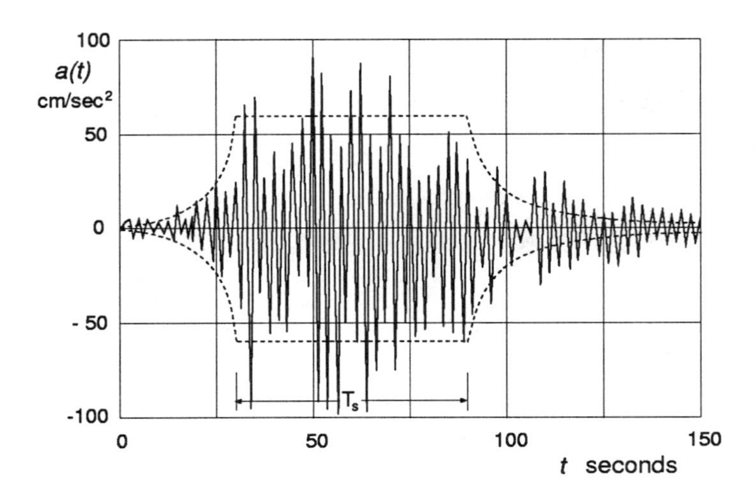

Figure 15.25: Typical Earthquake Ground Acceleration

power spectral density throughout the visible electromagnetic spectrum. In the language often used in elementary schools, "White light contains all the colors of the rainbow," so white noise contains all of the frequencies in the audio range.

An oscillograph record of a typical ground acceleration caused by an earthquake will exhibit features similar to those seen in Figure 15.25. Although it is clear from this illustration that the overall record is not a stationary stochastic process, that portion associated with the strong-motion during which a maximum structural response can be expected is tolerably represented by a simple stationary model. The duration of this effect is shown in the figure to be T_s.

Therefore, the power spectral density of the ground acceleration in the seismic model of a linear oscillator is[19,20,21,22]

$$S_g(\omega) = \frac{\omega_g^4 + 4\,\zeta_g^2\,\omega_g^2\,\omega^2}{(\omega_g^2 - \omega^2)^2 + 4\,\zeta_g^2\,\omega_g^2\,\omega^2}\,S_0 \tag{15.464}$$

The parameters ω_g and ζ_g are the natural frequency and damping ratio for this oscillator model as fitted to the characteristics of the earth's surface stratum at the given location. As shown in Appendix I, the parameters were selected by using moments of the spectral distributions for a large number of strong earthquakes of record.[23] The modeling gives $\omega_g = 15.6$ rad/sec and $\zeta_g = 0.6$, with the magnitude of the root mean square (RMS) acceleration as $1.63\ \text{m/sec}^2$.

[19]Ang, A. H-S., "Probability Concepts in Earthquake Engineering," ***Applied Mechanics in Earthquake Engineering***, American Society of Mechanical Engineers, New York, 1974, p. 227.

[20]Soong, T. T., and Grigoriu, Mircea, ***Random Vibration of Mechanical and Structural Systems***, P T R Prentice-Hall, Inc., Englewood Cliffs, New Jersey, 1993, p. 65.

[21]Kanai, K., loc cit.

[22]Tajimi, H., loc cit.

[23]Lai, S. S. P., "Statistical Characterization of Strong Ground Motions Using Power Spectral Density Function," ***Bulletin of the Seismological Society of America***, Volume 72, 1982, pp. 259-274.

15.14.4 THE TRANSFER FUNCTION

The relationship between the ground acceleration input and the output for evaluation/regulation is the transfer function $H_{z\ddot{x}_g}(i\omega)$. There is also the Laplace transform of the Kanai-Tajimi white noise filter $H_{\ddot{x}_g W}(i\omega)$. The transfer function supplies information as to the frequency characteristics for each building.

The function $H_{zW}(i\omega) = H_{z\ddot{x}_g}(i\omega) \cdot H_{\ddot{x}_g W}(i\omega)$ furnishes the desired information concerning the response the taller building to the ground excitation, and is thus helpful in analyzing the coupled buildings system.

The authors of the paper gave results in the form of graphs showing the interstory drift angle, the absolute acceleration of the roof, and the base shear for the 50-story (200 m) building, as well as for 30-story (120 m) and 20-story (80 m) ones which have the same mass density and stiffness, as individual (i.e., uncoupled) structures. A sample of the results is given in Figure 15.26, below.

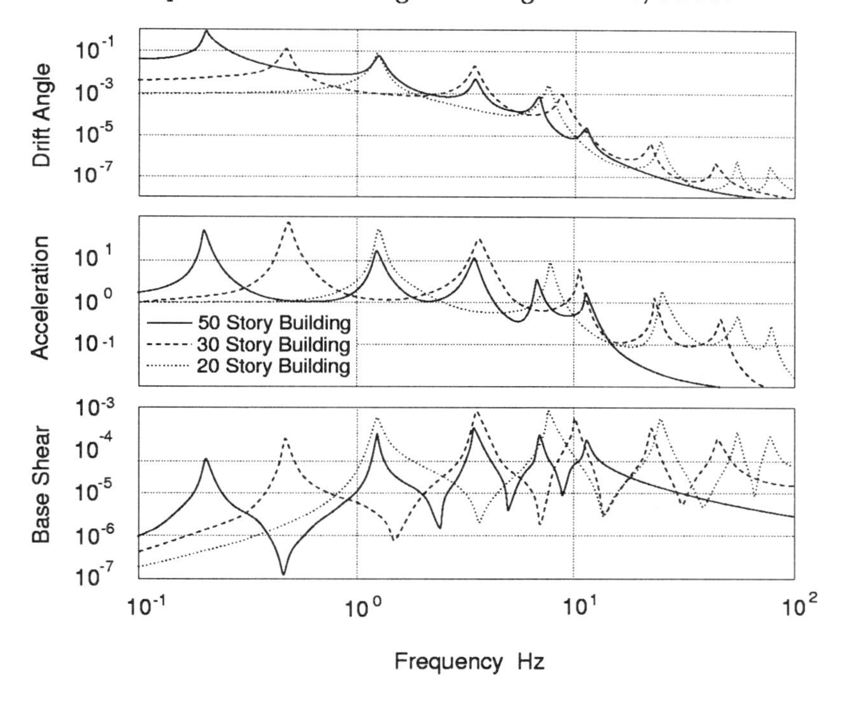

Figure 15.26: Building Responses to Ground Motion

What the graphs illustrate is the manner in which the various modes contribute to the responses for a 50-story, a 30-story, and a 20-story building for the three parameters (1) interstory drift angle (top floor), (2) roof acceleration, and (3) base shear (at the foot of the structure).

The first mode dominates the drift angle. The peak values for the second modes for each of the 50-, 30-, and 20-story buildings are a whole order of magnitude less than are those of the first mode.

The roof acceleration graphs reveal that, although there is attenuation of the higher modes, these participate to a greater degree than is the case with interstory drift. The second mode is still significant in both the 50- and 30-story responses, but is less for the 20-story one.

Higher modes contribute even more to the base shear responses than for accelerations.

For the coupled system, with a taller and a shorter building active, the interstory drifts (roofs) are dominated by the first mode of response of the taller building. On the other hand, the first and second modes of the shorter building are dominant for the accelerations (roofs) and story shears (bases). Therefore, a control strategy for this system should concentrate upon diminishing the magnitudes of these modes.

15.14.5 SOME CONCLUSIONS

In Reference 15, authors Christenson, Spencer, and Johnson formulated a cost function for the model structural system which they described, and then investigated the results of varying the design parameters of the model upon its response to earthquake excitation. Their experiments were displayed in a series of graphs which compared the interstory drift angles, absolute roof accelerations, and story shears at the base for uncoupled buildings, as well as active and smart damping control strategies.

They concluded that coupled building control is an effective method for reducing the hazards of seismic action for neighboring (adjacent) high-rise structures.

A particularly good active strategy, as they found, was to couple 40-(200 m) and 50-(160 m) story buildings at the top of the shorter structure, using mass density and stiffness of the shorter at one-half those of the taller building.

For the smart damping strategy, they found that reductions of 40.9%, 18.5%, and 18.6% were achievable with one 50-story and one 40-story building in the drift angle, acceleration, and story shear, respectively.

The reader is referred to the journal paper (Reference 15) for further details of this important contribution.

15.15 PROJECT #13

1. The stress tensor in an elastic solid is

$$\tau = \begin{bmatrix} \tau^1{}_1 & \tau^1{}_2 & \tau^1{}_3 \\ \tau^2{}_1 & \tau^2{}_2 & \tau^2{}_3 \\ \tau^3{}_1 & \tau^3{}_2 & \tau^3{}_3 \end{bmatrix}$$

in which the τs are individual stress components obtained by measurements or calculations at some place in the solid, all with respect to some particular reference set of axes as shown below.

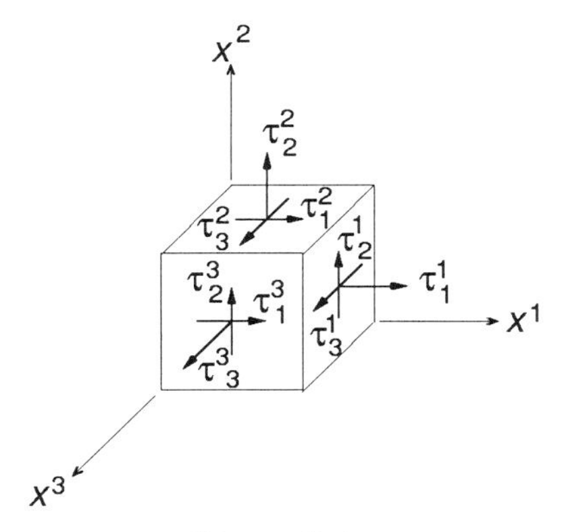

Stress Cube

Given the components

$$\tau = \begin{bmatrix} 5,000 & 2,000 & 3,000 \\ 2,000 & 9,300 & 4,000 \\ 3,000 & 4,000 & 21,000 \end{bmatrix}$$

(a) Compute the invariants I_1, I_2, and I_3 of the tensor.

(b) Write the secular equation $f(\lambda) = 0$; locate the maximum, minimum, and inflection point; and sketch the equation. Solve the cubic equation for the 3 roots (eigenvalues) λ_1, λ_2, and λ_3.

(c) Solve for the direction cosines of the three principal axes of stress for this eigenstate with respect to the original axes.

(d) Sketch the state of stress on a cube similar to the one above, showing its orientation with respect to the original cube.

(e) Confirm the computations by using a computer program of your choice, and include a printout of the results.

2. The strain tensor at a place in an elastic solid body is composed of strains measured with electric resistance strain gages (SR-4s). Its matrix is

$$\epsilon = \begin{bmatrix} 575 & 1,260 & -190 \\ 1,260 & 880 & 220 \\ -190 & 220 & 610 \end{bmatrix}$$

(a) Determine the trace (first invariant) of the strain matrix.

(b) Expand the determinant of the matrix by minors to compute the second and third invariants of the strain tensor.

(c) Write the secular equation, and then solve it for the three roots—the eigenvalues.

(d) Compute the direction cosines of the three principal axes with respect to the original set of axes. Confirm these computations using a computer program of your choice, and include a printout of the results.

3. A spring-mass-dashpot system is disposed as shown in the figure. With the spring grounded, the mass rests in equilibrium at the place named x_0. Now let the ground be given the motion $y = \mathcal{A}\cos\Omega t$.

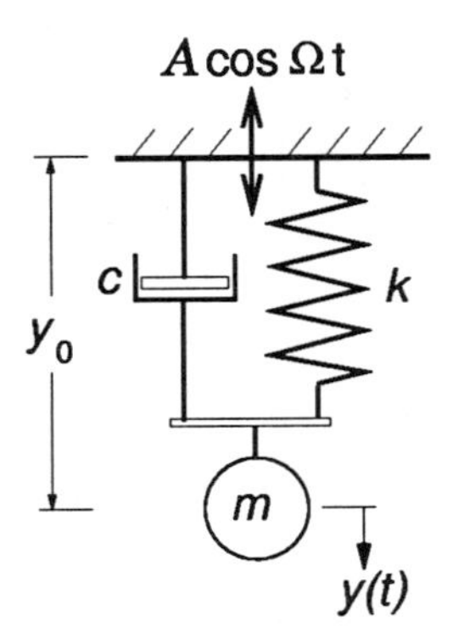

Spring-Mass-Dashpot System

(a) Construct the equation of motion of the mass.

(b) If the initial conditions are $y(0) = 0$, $\dot{y}(0) = 0$, use a Laplace transform method to determine the motion of the mass.

4. The figure shows two masses that are joined by a dashpot, with mass m_1 connected to ground through a spring and dashpot in parallel. Mass m_2 is subjected to an oscillatory force $\mathcal{A}\cos\Omega t$.

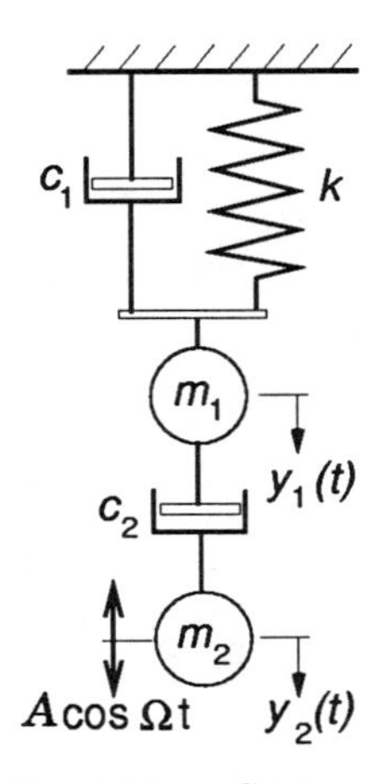

Dual Mass System

(a) Write the equations of motion for both masses.

(b) Select state variables for this system, and display the state vector, as well as its derivative.

(c) Give the input and output vectors.

(d) Write the state equations for the system.

5. The mass in the elementary system shown below is supported by a spring and dashpot.

   (a) Determine the zero-state response of the system.

   (b) Find the impulse response, assuming that k and c are constants.

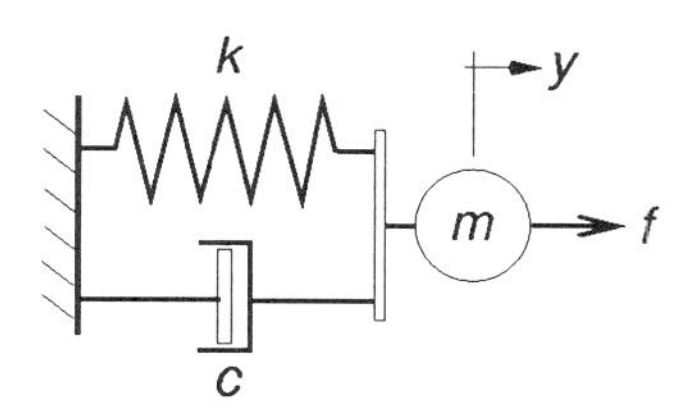

Elementary System

6. For the second Example problem in the Section 15.4, above, let the system parameters have the following values:

$$m_1 = 3\ \text{kg} \qquad m_2 = 2\ \text{kg}$$
$$c_1 = 0.0002\,\text{N}\cdot\text{s/m} \qquad c_2 = 0.00015\,\text{N}\cdot\text{s/m}$$
$$k = 300\ \text{N/m}$$

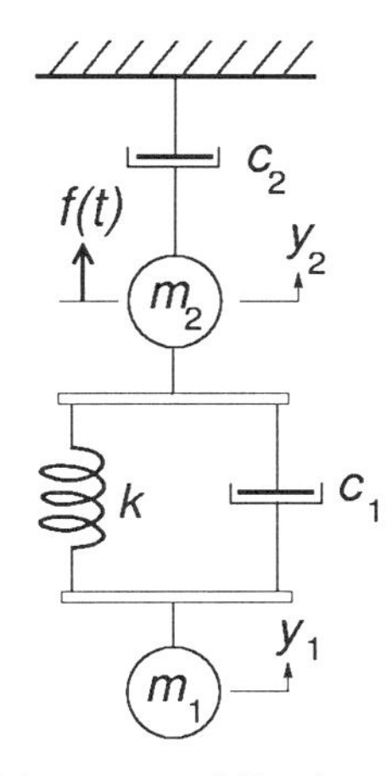

Spring, Masses, and Dashpots System

Determine the free and forced system response if the initial conditions are

$$y_1(0) = 0\,, \quad y_2(0) = 0$$
$$\dot{y}_1(0) = 0 \quad y_2(0) = 0$$

and $f = F\,cos\,\Omega t$, where $F = 30$ N, and $\Omega = 3$ rad/sec.

Use any or all of the Equations 15.169 through 15.179 as needed.

7. Compute the transfer function response for the system shown, where the system parameters are:

$m = 1.5\,\text{kg} \qquad c = 0.00015\,\text{N}\cdot\text{s/m} \quad k = 45\,\text{N/m}$
$\Omega = 4\,\text{rad/sec}$

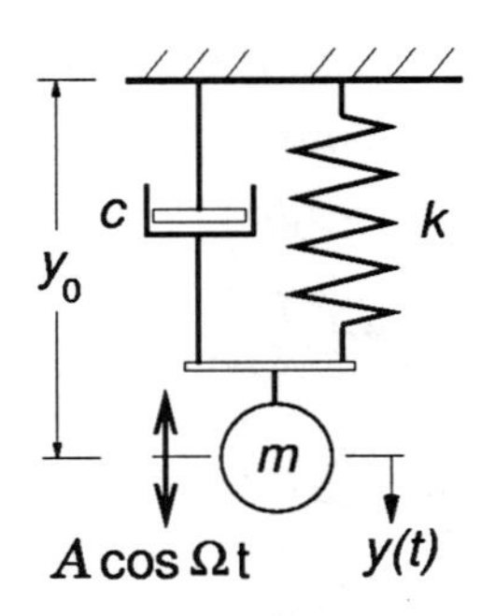

Spring-Mass-Dashpot System

8. A simple spring-mass-dashpot system shown in the figure below is governed by the equation of motion

$$m\,\frac{d^2x}{dt^2} + c\,\frac{dx}{dt} + k\,x = f\,cos\,\omega t$$

or

$$\frac{d^2x}{dt^2} + 2\,\zeta\,\omega_0\,\frac{dx}{dt} + \omega_0^2\,x = \frac{f}{m}\,cos\,\omega t$$

in terms of the natural frequency ω_0 and the critical damping ratio ζ.[24]

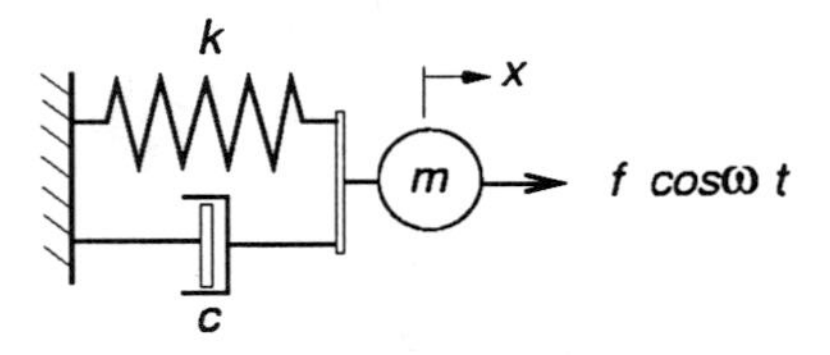

Spring-Mass-Dashpot System

(a) Show that the frequency response function for the system is

$$H(\omega) = \frac{1}{m\,(i\omega)^2 + c\,(i\omega) + k}$$

or

$$H(\omega) = \frac{1}{\omega_0^2 - \omega^2 + 2\,i\,\zeta\,\omega_0\,\omega}$$

(b) Show that the magnification factor for the system is

$$|H(\omega)| = \frac{k}{m\omega_0^2\left[\left(2\,\zeta\,\frac{\omega}{\omega_0}\right)^2 + \left(1 - \frac{\omega^2}{\omega_0^2}\right)^2\right]^{\frac{1}{2}}}$$

[24]See Appendix I, Equation I.9.

(c) Plot the dimensionless magnification factor for critical damping ratios of 0.25, 0.05, 0.1, 0.2, 0.5, and 1.0, for the range of frequency ratio ω/ω_0 from 0 to 3.

(d) Plot the dimensionless magnification factor for a critical damping ratio of 0.1 on a log-log graph for the frequency ratio from 0.1 to 10.

9. The figure illustrates a system of two masses with springs, dashpots, and a driving force.

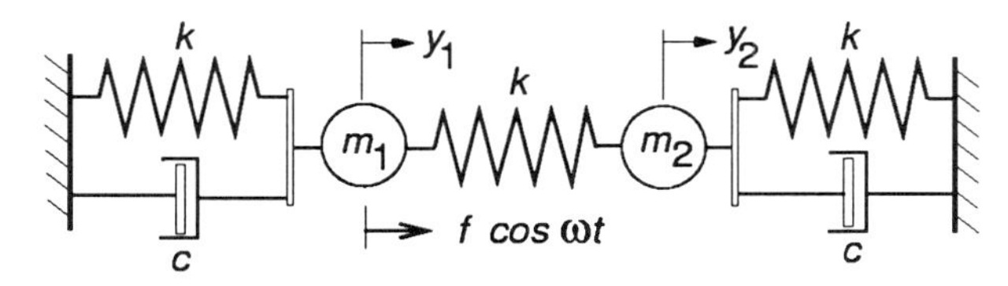

Spring-Mass-Dashpot System

(a) Write the state equations of motion for this system in canonical form.

(b) Find the frequency response function for the mass m_1.

(c) If the masses are each equal to 1.5 kg, both springs have a modulus of 0.8 N/m, and the damping ratio is 0.2 N$\cdot$ s/m, determine the frequency response function $H_1(\omega)$.

(d) Write the secular equation and compute the eigenvalues (the natural frequencies).

(e) Plot the response for both masses on the same graph with $H_1(\omega)$ for the range $\omega = 0$ to $\omega = 3\,\text{rad/s}$.

10. A cantilever beam is illustrated in the figure.

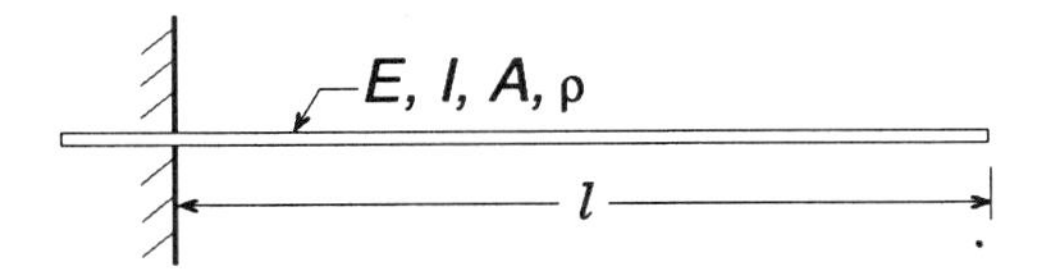

Cantilever Beam

The end conditions are:

$$y(0,t) = 0 \quad \frac{dy(0,t)}{dx} = 0 \quad \frac{d^2y(\ell,t)}{dx^2} = 0 \quad \frac{d^3y(\ell,t)}{dx^3} = 0$$

which correspond to the a built-in end on the left, where the deflection and slope are both zero, and a free end on the right, where there is neither shearing force nor bending moment.

(a) Show that the general form of the modes is

$$\phi(x) = C_1\, sin\, kx + C_2\, cos\, kx + C_3\, Sinh\, kx + C_4\, Cosh\, kx$$

in which C_1, C_2, C_3, C_4, and k are constants to be determined from the end conditions.

(b) Then show that this reduces to

$$\phi(x) = C_2\,(cos\,kx - Cosh\,kx) + C_4\,(sin\,kx - Sinh\,kx)$$

for the built-in end of the cantilever beam.

(c) Now show that the free end conditions reduce this further to

$$cos\,k\ell\,Cosh\,k\ell = -1$$

(d) Compute the first three roots k_1, k_2, and k_3 of the equation above. These are eigenvalues for this beam, and from them find the first three natural frequencies

$$\omega_i = k_i^2\sqrt{\frac{EI}{\rho A}} \qquad i = 1, 2, 3$$

Hint: use the approximation $k_i\,\ell \approx (i - \frac{1}{2})\pi$ as the starting point in a Newton-Raphson procedure.

(e) Sketch the first three mode shapes of the cantilever.

11. The figure shows a three-story structure for which the floors are assumed to be rigid concrete slabs each of mass m, and the columns are elastic members of flexural rigidity EI and negligible mass compared to the other elements of the structure. Internal friction in the system is considered negligible. A lateral thrust S, due to a wind loading, is applied at the top floor.

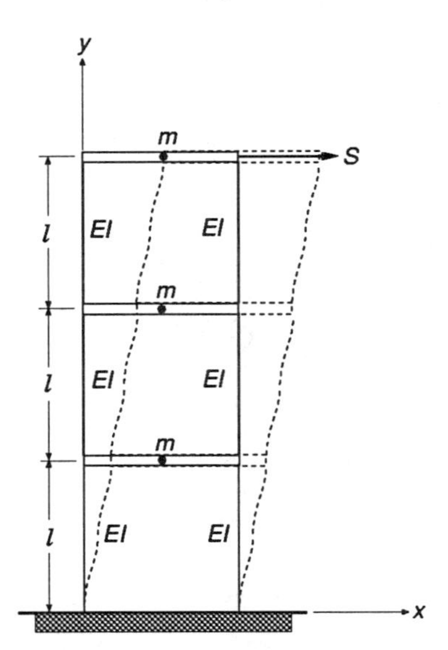

Three-Story Structure

(a) If it is assumed that each column deflects laterally in a symmetrical manner about an inflection point at its center under the action of story shears S_i, compute the deflections at each story of the structure.

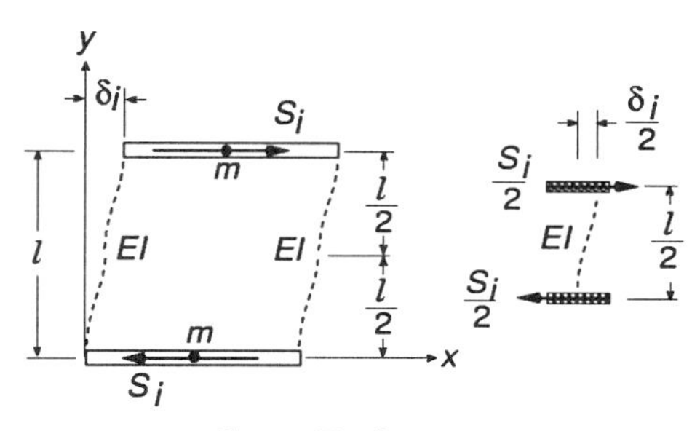

Story Deflection

Hint: Each column behaves as two simple cantilever beams symmetrical about the center, therefore the total deflection of each column from bottom to top is twice that of a cantilever loaded by the horizontal shearing force $S_i/2$; i.e.,

$$\delta_i = 2\left[\frac{(S_i/2)(\ell/2)^3}{3EI}\right] = \frac{S_i\ell^3}{24EI}$$

(b) Write the state-space equations for the motion of the structure in their canonical matrix form, and show explicit forms of the matrices and sub-matrices involved.

(c) Compute the natural frequencies (in terms of m, ℓ, E, and I).

12. For the three-story structure illustrated in 11, above, suppose that an earthquake produces a horizontal ground acceleration $a_g\,\delta(t-\tau)$ m/s^2, in which a_g is a constant.

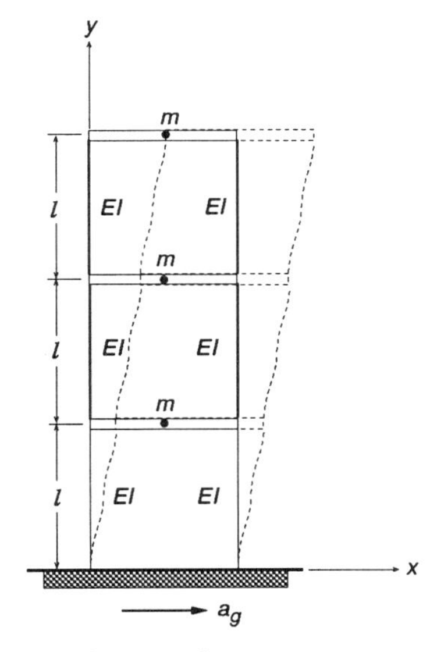

Ground Acceleration

(a) Find the motions of the three floor masses in response to this excitation.

(b) Determine the story shears during the structure's motion subsequent to the quake.

13. The graph shows a histogram of an earthquake record with the abscissæ denoting ground acceleration frequencies. The ordinates are numbers of observations at each frequency.

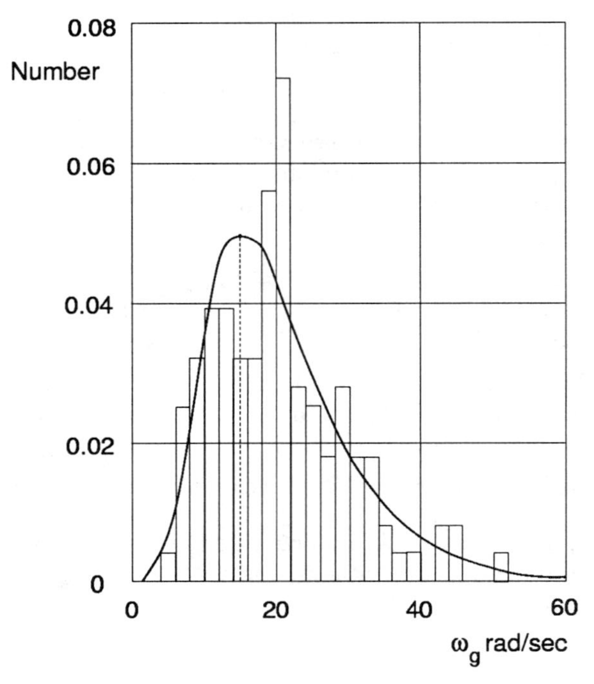

Ground Acceleration Histogram
Frequency

(a) Using the methods illustrated in Chapter 13 and Appendix C, fit the histogram with either a Pearson-III or Pearson-I frequency distribution, as appears most accurate.

(b) Compare the equation for the frequency distribution with that given in Lai, S. S. P., "Statistical Characterization of Strong Ground Motions Using Power Spectral Density Function," *Bulletin of the Seismographic Society of America*, Volume 72, 1982, pp. 259-274, or Soong, T. T., and Grigoriu, Mircea, *Random Vibration of Mechanical and Structural Systems*, P T R Prentice-Hall, Inc., Englewood Cliffs, New Jersey, 1993, p. 66.

(c) Report the mean and variance, and compare with the equivalent parameters for the histogram sample.

14. The graph displays a histogram of an earthquake record with damping coefficient values on the x-axis, and number of observations of each damping constant on the y-axis.

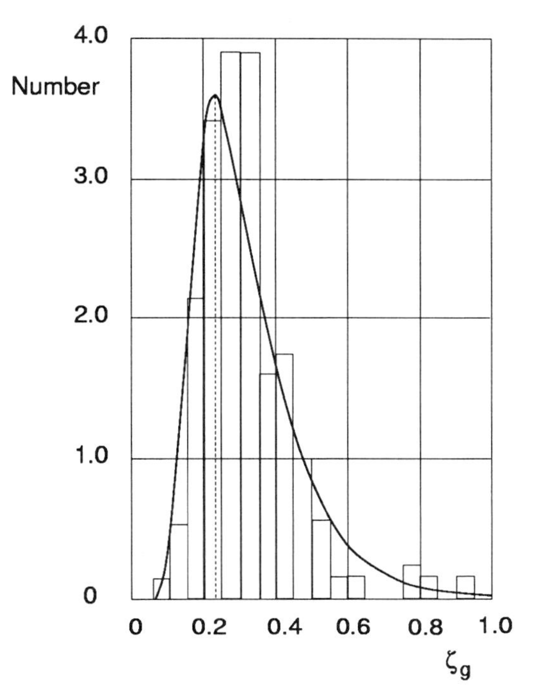

Ground Acceleration Histogram
Damping Coefficient

(a) Fit the histogram with either a Pearson-III or Pearson-I frequency distribution, or both, to obtain the best accuracy.

(b) Compare the frequency distribution equation with that of S. S. P. Lai or T. T. Soong and Mircea Grigoriu cited in 13 (b), above.

(c) Give the mean and variance, and compare these with the same parameters computed for the sample.

15. Make a graph of the Kanai-Tajimi power spectral density ground acceleration function (Equation 15.464)

$$\frac{S_g(\omega)}{S_0} = \frac{\omega_g^4 + 4\,\zeta_g^2\,\omega_g^2\,\omega^2}{(\omega_g^2 - \omega^2)^2 + 4\,\zeta_g^2\,\omega_g^2\,\omega^2}$$

as a function of the frequency ω.

Compare this graph with the power spectral density ground acceleration plots shown in Ravara, A., "Spectral Analysis of Seismic Actions," *Proceedings of the 3rd World Conference on Earthquake Engineering*, Auckland and Wellington, New Zealand, 1965 or Ang, A. H-S., "Probability Concepts in Earthquake Engineering," *Applied Mechanics in Earthquake Engineering*, The American Society of Mechanical Engineeers, 1974, p. 250, for the 1940 El Centro, California earthquake.

CHAPTER 16

AUTONOMOUS SYSTEMS

16.1 INTRODUCTION

Throughout the last two centuries of civil engineering history there have been certain periods when particular techniques have been employed to such an extent as to merit the description "state of the art." One might mention ***moment distribution methods***, the ***Hardy Cross procedure***, ***relaxation***, ***finite element methods***, and a host of computational softwares such as STRUDL, STAAD, etc. Each of these levels of proficiency has had its "day in the sun" and has served well to advance analysis and design in the several branches of practice.

Within the last few decades a number of new methods or techniques have appeared at the forefront of research in civil engineering, and have been rapidly passed into the mainstream of education and practice in the field. These methods utilize the power which computers—main-frames, workstations, and PCs—provide to accomplish new algorithms which transcend those formerly available to researchers, educators, and practitioners. They fly under different colors, or at least under different nomenclatures, but they have in common that they automate or mechanize activities formerly deemed to be the function of engineers. Of course, the new methods have all been created by persons, so it is clear that human mentality is incorporated into their functioning. But the stated (sometimes unstated) object shared by all of them by whatever name is to transfer a maximum amount of analysis, design, and operation to machines, leaving the persons to perform other (presumably higher) tasks.

Some language applied to these schemes is ***Artificial Intelligence***, ***Expert Systems***, ***Self-Designing Systems***, ***CAD [Computer-Aided Design]***, ***CAE [Computer-Aided Engineering]***, ***Fuzzy Systems***, ***Optimal Systems***, ***Genetic Algorithms***, and ***Pareto-Optimal Designs***. Others include ***Control Systems***, ***Parametric Control of Systems***, ***Object-Oriented Design***, and ***IT [Information Technology]***.

Inherent within each of these schemes is an algorithm (either perceived or implied) which begins with some parameters furnished to it as data, and proceeds by its own internalized functioning to ever more refined and optimum outcomes according to the criteria built into its program. There is usually a repertory of stored information which the process may call upon as it advances through its process, and there is often a staged progress which allows an interactive period with the human oversight representation. The more exotic elements of such algorithms even include uncertainties in the data supplied and in the repertory, hence the nomenclature "fuzzy" logic.

Although it is clear that machines have yet to reach the level of human mentality in most arenas (this despite DEEP BLUE's accomplishments at chess against world-class masters), there is a continuing transfer of the thinking-like function to such devices. It is appropriate to refer to many of the current computational resources as *autonomous systems*. That is, they perform on their own, largely without further inputs from their human overlords.

The scope of self-regulating schemes in civil engineering practice is so broad and so complex that several encyclopædic volumes would be required to survey the field. What can be done in a more realistic way is to investigate examples of such, with a view toward illustrating some of the more essential elements which are common to most of them. One of the more important advances in systems theory—an innovation that has been especially fruitful when applied to autonomous systems—is the introduction of fuzzy science.

16.2 FUZZY SETS

Fuzzy sets are sets which have indistinct boundaries or edges. Members of such sets are often described in linguistic rather than in precise mathematical terms or crisp definitions. For example, there is the set of "senior citizens." Some people belong to this set, obviously, but some certainly do not. Various businesses may define "senior citizen" in crisp numbers, say 62 or older, but others may mean the set to include persons 55 or above, and still others might intend it to be for those over 65. To some teen-agers, anyone above the age of 30 is a senior citizen.

In the absence of a precise description, users of fuzzy logic will generally consider the degree of membership in the set by linguistic graduations such as "not at all" senior citizens, "somewhat" senior citizens, "senior" citizens, "quite" senior citizens, and "very" senior citizens. A standard nomenclature creates labels "positive large," "positive medium," "positive small," "zero," "negative small," "negative medium," and "negative large," for a seven-fold norm. In this scheme, the range of membership is from 0 to 1, or from "not member" to "full member." A scale of the vocabulary of these linguistic terms is illustrated in Figure 16.1, below, using five triangles and two trapezoids at the extremes. These are piecewise linear models, of course.

The figure shows clearly that each location on the scale may have a membership grade in more than one linguistic term. Such a multiple-valued condition is characteristic of fuzzy representation, wherein it is seen that there is not mutually exclusive categorization.

A very common membership function is the co-called "plateau" or trapezoidal form shown in Figure 16.2. An especially useful nonlinear membership function for hydraulic systems is that one seen in Figure 16.3. The trapezoidal version is piecewise linear, of course, and the nonlinear one is composed of circular arcs and a line, for the sake of simplicity.

When the time comes to establish rules for the utilization of the membership, the terms are mapped onto the quantized scale, so that each term is identified by

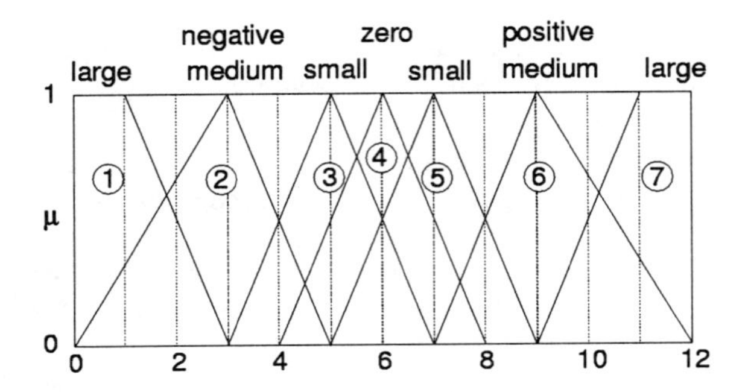

Figure 16.1: Linguistic Vocabulary Terms

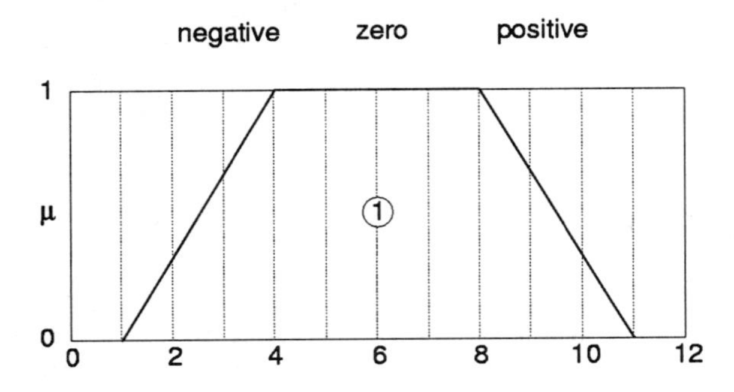

Figure 16.2: "Plateau" Logic

the place where it is centered. Ambiguities are ultimately resolved by a "center-of-gravity" method (or similar one) at the time of defuzzification.

The mapping of linguistic vocabulary terms against a scale of values as revealed in Figure 16.1 is the fundamental trait which allows fuzzy logic to be applied to real technological systems. It is the object of the analysis and design of such systems to be able to describe inter-element relationships with fuzzy language, then metamorphose those fuzzy relationships into quantized form similar to the binary and continuous equivalents. The process is called "defuzzification." There are many different ways to do this, and several of the more common ones will be seen in the sequel.

Comparable graphs can be displayed for crisp (Boolean) logic state transitions as shown in Figure 16.4 and analog (continuous) functions, seen in Figure 16.5. The binary transition of Figure 16.4 displays the abrupt discontinuous translation from state 0 to state 1, with those values assumed. In the second case, Figure 16.5, there is a smooth, continuous movement from 0 to 1 through an infinte universe of values. The latter graph presents what is usually called a *sigmoid* or an S-curve.

16.2.1 ELEMENTS OF FUZZY LOGIC

Fuzzy logic is a relatively recent branch of science, although, as with most other sciences, it has roots reaching back to antiquity. In modern times important con-

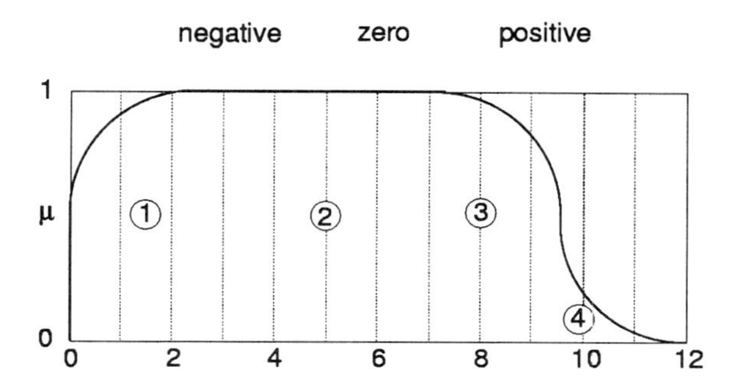

Figure 16.3: Nonlinear Logic

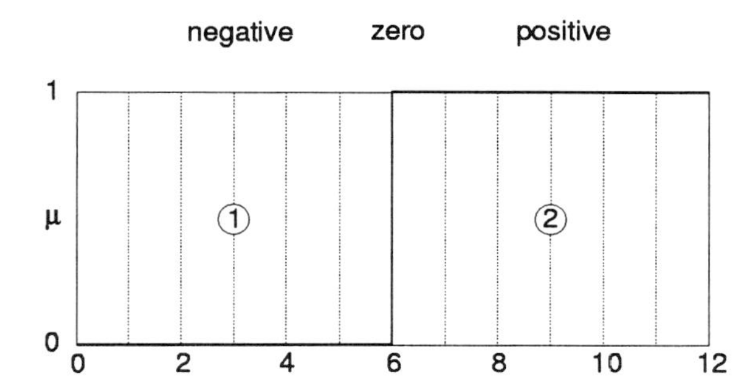

Figure 16.4: Binary Logic Terms

tributions were made by Bertrand Russell[1] at the turn of the Twentieth Century, by Jan Lukasiewicz[2] in the 1920s decade, and by Max Black[3] in 1937.

Jan Lukasiewicz is the inventor of the *max* and *min* operators which are possibly the most common devices employed in fuzzy science applications throughout the realm of technology at present.

The field was given a giant boost by the arrival of Lotfi A. Zadeh, an Iranian immigrant to the United States in the latter days of World War II. By the mid 1960s Zadeh was Chairman of Electrical Engineering at University of California at Berkeley, where he produced his "Fuzzy Sets,"[4] a research paper that not only created the new name (*vagueness* became *fuzziness*), but along with it a new calculus. In the succeeding years he and his associates developed the science from its framework to a full-blown intellectual field.

By the mid-to-late 1980s, technological applications of fuzzy logic were becoming very significant, especially those involving automatic control systems. "The rest," as they say, "is history." A more extensive (but still brief) history is avail-

[1]Russell, B., *The Philosophy of Logical Atomism*, Open Court, 1985.
[2]Lukasiewicz, J., *Selected Works*, Borkowski, Editor, North Holland, 1970.
[3]Black, M., "Vagueness: An Exercise in Logical Analysis," *Philosophy of Science*, Volume 4, 1937, pp. 427-455.
[4]Zadeh, L. A., "Fuzzy Sets," *Information and Control*, Volume 8, 1965, pp. 338-353.

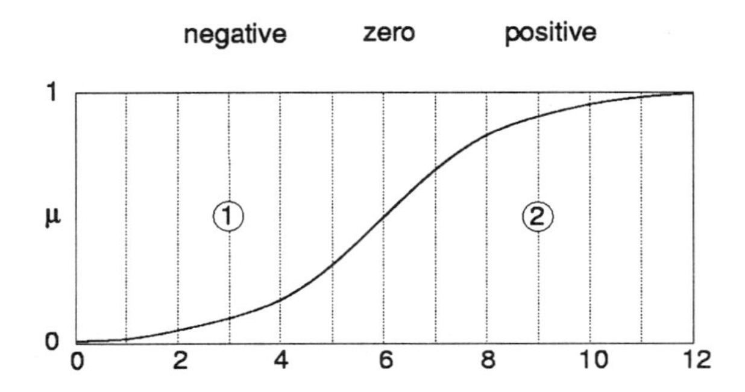

Figure 16.5: Analog Logic Graph

able in the fascinating book written by Zadeh's former student Bart Kosko.[5]

The basic agenda of fuzzy science is to replace the ***functions*** of crisp-based logic with the ***rules*** of fuzzy logic. That is, instead of a "law" in formula format to describe a phenomenon, there are rule-based models of fuzzy-like relationships. These are often referred to as ***fuzzy approximation theorems***, or FATs. The theory is that fuzzy rules can replace equations. It may take a large number of such rules to account for all of the desired stipulations in a given model, but the number is assumed to be always finite. The merit of the representation is that it permits the system of fuzzy rules to be computed (hopefully, by a computer).

The greatest merit of fuzzy theory is that it permits a system having unsure elements to be described by subjective linguistic terms. In this respect, it contrasts sharply with the binary precision of crisp Boolean logic systems. The crisp system is exemplified by the famous Shakespearean quote from Hamlet, "To be or not to be, that is the question." There is nothing in between "being" and "not being" in this scheme. The fuzzy counterpart is found in the modern-day ethical dilemna concerning human life. Does it begin at the moment of conception? Is a fetus "human" (i.e., being) at the end of the first trimester? At the end of the second trimester? Or at birth? Perhaps only at the age of 40!

As with all other science models, fuzzy has its limitations or disadvantages. A clear one is the rapidity with which the number of rules explodes, together with the increased difficulty of computing, but also with increasing degredation of stability or "nervousness," if it's a real system that's being controlled. Ockham's Razor is clearly needed to accompany the analyst at the rule forming stage of fuzzy systems work. The number of rules should be kept to a well-chosen few, and the simpler the better.

Let the universal set be represented by X, and a fuzzy set A in X is defined by its membership function $\mu_A(x)$ that specifies the membership of $x \in X$ to fuzzy set A. The range of membership values is from 0 to 1, as shown above. Thus, the closer the membership value is to 1, the higher is the grade of membership of x to A, and, of course, vice-versa.

This step is probably the most difficult of the modeling process in systems analysis and design, for it involves subjective criteria. On the other hand, its

[5]Kosko, B., ***The Fuzzy Future***, Harmony Books, New York, 1999, 353 + xiv pp.

merit resides in the very fact that it avoids crisp characterizations, which can be difficult or impossible to know.

A few necessary operators have their fuzzy counterparts:

- The *fuzzy union* of the fuzzy sets A and B is defined as the operand

$$\mu_{A\cup B}(x) = u[\mu_A(x), \mu_B(x)] = max[\mu_A(x), \mu_B(x)] \tag{16.1}$$

  whenever the fuzzy union is defined as the *max* operator.

- The *fuzzy intersection* is defined as the operand

$$\mu_{A\cap B}(x) = i[\mu_A(x), \mu_B(x)] = min[\mu_A(x), \mu_B(x)] \tag{16.2}$$

  when the fuzzy intersection is defined as the *min* operator.

The combination of several fuzzy sets which result in a single fuzzy set invokes *aggreagtion operators*. For the combination of A_1, A_2, ..., A_n to the single set A, there is the aggregation operator

$$\mu_A(x) = h[\mu_{A_1}(x), \mu_{A_2}(x), \ldots, \mu_{A_n}(x)] \tag{16.3}$$

Both fuzzy union and fuzzy intersection are aggregation operators, and there is the *generalized mean operator*

$$h_\alpha = \left(\frac{\mu_{A_1}(x)^\alpha + \mu_{A_2}(x)^\alpha + \ldots + \mu_{A_n}(x)^\alpha}{n}\right)^{\frac{1}{\alpha}} \tag{16.4}$$

in which α is a shape parameter to distinguish various means. The arithmetic mean h_1, for example, is that for which α is equal to 1. The geometric mean h_0 has $\alpha \to 0$, and the harmonic mean h_{-1} corresponds to $\alpha = -1$.

In addition, generalized aggregators can be defined by weighted generalized means using weighting factors $w_i \geq 0$, $i = 1, 2, \cdots, n$. The weights reflect the relative importance of each set being aggregated. Thus

$$h_{\alpha,w} = \left(\sum_{i=1}^{n} w_i\, \mu_{A_i}^\alpha\right)^{\frac{1}{\alpha}} \tag{16.5}$$

in which, of course, the total weighting must be

$$\sum_{i=1}^{n} w_i = 1 \tag{16.6}$$

16.3 WATER SUPPLY NETWORKS

Few components of the infrastructure serving contemporary society are more fundamental and important than the systems which furnish adequate water for drinking and all the other things humans wish to do. This is not only a contemporary concern, of course, as other human habitations throughout history have demonstrated. One can cite the excellent delivery systems for water created by the ancient Romans, for example. The aqueducts which reached from Cisterna (site of cisterns) and other locales in the Alban Hills to the citizens of Rome were marvels of hydraulic engineering as of the state of the art in those times.

Reflections of those earlier water supply marvels are present today in a number of Mediterranean countries, where water must be obtained in mountainous regions during times of more abundant rainfall, and moved in regulated conduits to the populous areas throughout the climatological year.

In Greece, for example, Dr. L. S. Vamvakeridou-Lyroudia at the National Technical University in Athens, has been a leader in analyzing and designing water supply systems to become highly regulated (controlled) in an autonomous manner. She and her associates have published a number of recent papers which document the engineering of these water systems.[6,7,8] A study of some features of her work is very instructive.

16.4 PIPE NETWORK WATER SUPPLY

In Chapter 3 an example of a piping network was shown. This elementary system was employed to illustrate the analysis of an important element of infrastructural engineering; namely, a public water supply. The more challenging investigation is that which is associated with the design of such a system—most especially, the ***optimum*** design of a pipe network.

In this context, an optimal design is often characterized as one which satisfies all of the constraints and which furnishes a conservative loading. That is, one which supplies the largest demand for water at each of the network's nodes. In former times this design philosophy might well have been thought to be adequate.

A further level of optimization is that which satisfies all of the constraints, including the rated maxima for discharge at each of the nodes, and which minimizes the cost of the water supply. Cost can include both construction cost and operating cost, as each component can be held to a minimum. Within this framework of design optimization, the problem becomes one of linear programming (for the construction) and dynamic programming (for the operating phase). Accordingly, when the network is analyzed as a linear programming exercise, say, there is some outcome which is declared to be either "satisfactory" or "unsatisfactory," depending upon whether or not the constraints are met. Such a solution of the design process merits the usual description of "crisp logic."

A higher form of design optimization in this arena is that exemplified by the artificial intelligence technique called *fuzzy reasoning*. This device permits the

---

[6]Vamvakeridou-Lyroudia, L. S. "Fuzzy Reasoning in Irrigation Network Design," *Hydrosoft/94, Proceedings, Fifth International Conference on Hydraulic Engineering Software, Chalkidiki, Greece, September 1994*, W. R. Blain and K. L. Katsifarakis, Editors, Computational Mechanics Publications, Southampton Boston, Volume 1, pp. 97-104.

[7]Vamvakeridou-Lyroudia, L. S., "Distribution Network Optimization Using Dynamic Programming," *Hydrosoft/90, Proceedings of the Third International Conference on Hydraulic Engineering Software, Massachusetts*, USA, *April 1990*, W. R. Blain and D. Ouazar, Editors, Computational Mechanics Publications, Southampton Boston, pp. 85-96.

[8]Vamvakeridou-Lyroudia, L. S., and Psarra, E., "Integrated Design Package for Irrigation Networks," *Hydrosoft/90, Proceedings of the Third International Conference on Hydraulic Engineering Software, Massachusetts, USA, April 1990*, W. R. Blain and D. Ouazar, Editors, Computational Mechanics Publications, Southampton Boston, pp. 97-108.

inclusion of constraints which cannot or should not be rendered in crisp formats for a variety of reasons. Utilizing fuzzy reasoning has been shown to yield cost savings of up to 20% in the overall system cost.[9]

Among the considerations which invoke fuzzy logic in piping network design optimizations are the following:

- The demands at the network nodes can be based upon probabilistic models, rather than as fixed, deterministic, loadings.

- Frictional head losses can be permitted to vary from their idealized values as assumed, and flow demands can be allowed to depart from their idealized lumped-node parameters.

- A more realistic model of upper and lower constraint bounds can be permitted without causing failure of the design system.

Certain things are NOT fuzzy in the optimal design of pipe networks. It can be taken as definite that the geometry of the system is furnished as "crisp" information. That is, the source(s) of the water, the locations of the customers (users), and the route-map of the pipe layout are furnished at the outset. Assuming that a new or renovated network is intended, the usual head (energy) considerations at the nodes and continuity requirements of hydraulic science must be maintained. (This is not the case, for example, if the pipes are old and leaky. But that is another problem altogether.)

With these stipulations in place, the designer's task is to determine the diameters of the pipes which will minimize the cost function. The ***intelligent*** model for accomplishing this task mimics the activity of the human mentality as it goes about this project.

16.4.1 SIMPLE PIPEFLOW NETWORK

Consider again the pipeflow network shown in Figure 16.6. There are four nodes, labelled A, B, C, and D, respectively; and there are flowrates Q_A, Q_B, Q_C, and Q_D, all in m^3/s at these nodes. The pipes are named a, b, c, d, e, and f, and their lengths are $\ell_a = 200$ m, $\ell_b = 1200$ m, $\ell_c = 400$ m, $\ell_d = 500$ m, $\ell_e = 1000$ m, and $\ell_f = 300$ m. The pipe diameters are unknown, to be determined.

Also unknown are the pressures and flowrates at each node, but there are requirements placed upon these items as constraints which must be met. If one of the customers at a given node is located in a three-story building, for example, the pressure head will need to be adequate to cause the water to rise to the height of that building's top floor, with sufficient pressure to maintain standard pressure at the taps in addition. If another customer is a small manufacturing enterprise which has a minimum water usage requirement in its operations, that, too, will furnish a constraint which will need to be met.

---

[9]Vamvakeridou-Lyroudia, L. S., "Fuzzy Reasoning in Water Supply Network Design Optimization," *Developments in Computer Aided Design and Modeling for Civil Engineering*, CIVIL-COMP PRESS, B. H. V. Topping, Editor, 1995, pp. 255-262.

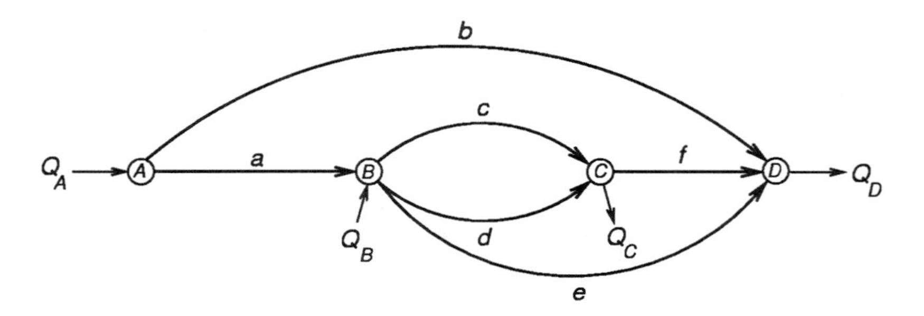

Figure 16.6: Pipeflow Network

The principles of hydraulics which determine the behavior of water in the network are the Darcy-Wiesbach law

$$h = f\frac{\ell}{d}\frac{v^2}{2g} \;\; m \text{ (of fluid flowing)} \tag{16.7}$$

and the continuity equation

$$Q = Av = \frac{\pi}{4} d^2 v \text{ m}^3/\text{s} \tag{16.8}$$

with parameters as defined in Chapter 3, Equations 3.42 and 3.43.

A conventional procedure for optimizing the design of this pipe net by "crisp" logic might follow the flow diagram illustrated in Figure 16.7.

The figure reveals an algorithm which is iterative in the conventional sense. That is, the process can be continued to any desired level of accuracy or precision, but it cuts off sharply in a digital sense when the convergence reaches that pre-set condition. Of course, it can be programmed to pause at each iteration to permit the oversight human to review the progress to that stage and to enter corrections as may be indicated.

The "COMPUTE" boxes in the diagram are based upon the crisp formulas of hydraulics. Appendix J suggests some specific embodiments of these programs.

16.4.2 FUZZY PIPE NET DESIGN

Waterflow in pipe nets is usually optimized by a Dynamic Programming algorithm. This is composed of a staged process of linear programming optimization, in which the outcome of a prior stage is additive with the current stage of the optimization, which then continues throughout the network stage-by-stage until all stages are optimum. A total optimum cost will be the cumulative sum of all stages.

Each stage of the process is joined to each other stage by the equations of hydraulics, which ensure the parameters for optimization of each stage.

As the dynamic programming algorithm is applied to the pipe net, the decision variables will be the diameters of available commercial pipes, since one would not envision making piping of special diameter for just this purpose. Such a pragmatic requirement constrains the process to an "integer"-type of programming, at least insofar as the diameter constraint is concerned. There are also constraints

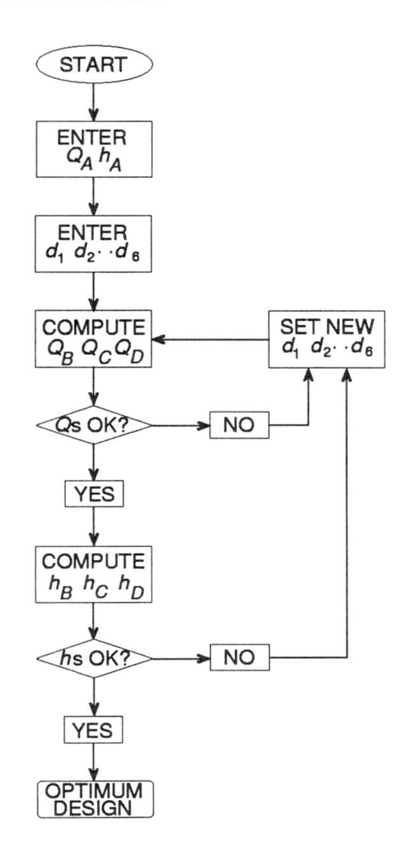

Figure 16.7: Pipe Net Algorithm

established by principles of hydraulics. Each pipe is a "stage" of the algorithm, and it is linked to each other stage through a node in the network. At each such node, there must be continuity of fluid pressure and mass flowrate. Pressure and flowrate are continuous functions (i.e., *not* integer values), and they are computed for each stage by the laws of hydraulics, using algorithms such as those displayed in Appendix J, so that the stages relate properly to each other.

There is then the objective function for minimization of the total cost, which is composed of the sum of the costs of each of the previous stages.

Various amounts of fuzziness can be introduced into the otherwise crisp model of pipe net flow optimization. Certainly, one vague element is that of pipe friction, but there are good algorithms available for computing the friction factor from the Colebrook formula, and, unless the pipes are old, friction factors are probably reliably obtained from calibration experiments. Since the pipe diameters are to be crisp commercial numbers, this variable would be exempt from fuzzification.

Two other elements, however, can profitably be allowed to be vague:

(1) **Pressure constraints** The pipes in a network have maximum rated pressures, and, no doubt, there is also a maximum vacuum rating, both of which may heavily depend upon the joints as well as the strength of the pipewalls

themselves. However, in most real installations, the operating pressures can be held well within those limits, but are otherwise either unknown or very vaguely known. Moreover, the pressure at a given node may be allowed to assume whatever it will be, short of safety considerations, in order to minimize the cost function. It is, of course, directly dependent upon the previous stage, therefore upon all previous stages back to the pressure source (initial head at a reservoir or pumping station).

Assume, therefore, that pressure is a fuzzy variable. At each pipe node it is related to the set of optimum feasible solutions by a membership function μ_P, which can be drawn to include the constraints of rated pressure for that specific section of pipe.

(2) **Discharge constraints** Given a specific flowrate in each link (pipe) of a net, the volume constraint is fuzzy in a similar manner to that of pressure and can be represented by a membership function μ_Q. Discharge, however, depends solely upon the local stage, and there is therefore a membership function for each stage and node.

When pressure and discharge have been admitted to fuzzy status, the design of a pipe net will present a set of fuzzy minimum costs at each stage. The combined effect of fuzzy pressures and flows gives rise to a combined membership function μ_{PQ} at each stage. This function is estimated by an aggregation operator for each stage, which simulates the brokerage between targets and constraints in that stage. This function is then applied *before* the stage optimization, and this allows the objective function to minimize the fuzzy aggregated costs. See Figure 16.8.

The advantage offered by a fuzzy network model above the crisp optimization is that alternativies are not automatically rejected prior to the optimization solely on the basis of constraint violation, since, through its membership function, some alternative may yet be allowed to a grade of membership in the set of acceptable ones. Of course, this acceptance may come at a high price in the local view (at a given node), while minimizing the overall cost function.

In other words, an alternative may be rejected on cost grounds in the fuzzy optimization, whereas it would not automatically be rejected as not feasible.

Furthermore, the application of aggregation operators to render linguistic terms commonly used by systems people into objective mathematical forms to describe feasibilities and qualities of system status is very satisfying.

16.4.3 FUZZY OPTIMIZATION MODEL

Using the rationale outlined in the previous section, a model for optimization of the fuzzy pipe network can be delineated. Of course, pipe networks can be constructed based upon a most tortured topology, therefore it may be prudent to focus upon a rather simple form such as that previously illustrated.

The network is composed of individual pipes and nodes where they are joined to the other pipes. If the number of pipes in the net is NP, the number of nodes is one greater; say, $NN = NP + 1$. There is a one-to-one correspondence between the number of computational stages and the number of pipes, so let the stage

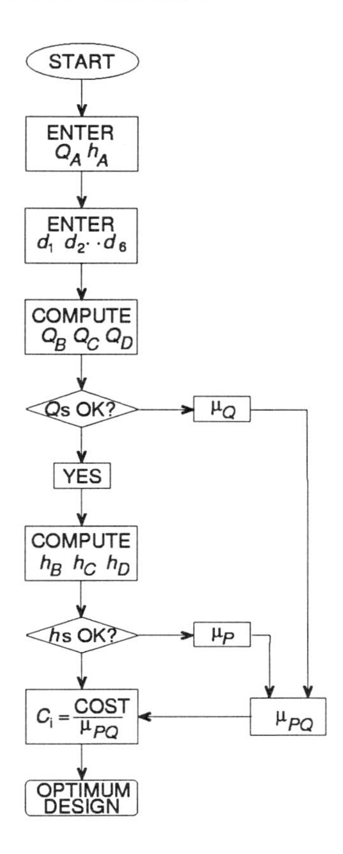

Figure 16.8: Fuzzy Network Algorithm

be designated $i = 1, 2, \ldots, NP$. Let the stage numbering begin at the furthest upstream node, and proceed uniformly to the furthest downstream one.

At a given stage there may be M flows, hence there are

$$Q(i, m) \quad i = 1, 2, \ldots, N \quad m = 1, 2, \ldots, M \tag{16.9}$$

altogether.

As for the diameters, there is the crisp commercially available set

$$DPOS(i, k_i) \quad i = 1, 2, \ldots, N \quad k_i = 1, 2, \ldots, NK(i) \tag{16.10}$$

where the smallest diameter is $k_i = 1$, and the largest is $NK(i)$, for that stage.

Therefore, the set of potential flowrates is

$$QPOS(i, k_i, m) \tag{16.11}$$

$$i = 1, 2, \ldots, N \quad k_i = 1, 2, \ldots, NK(i) \quad m = 1, 2, \ldots, M$$

This permits the following constraint equations to be formed:

$$Q_{min}(D) \leq QPOS(i, k_i, m) \leq Q_{max}(D) \tag{16.12}$$

In the fuzzy optimization model, each $QPOS(i, k_i, m)$ holds membership in the set of feasible alternatives according to the nonlinear membership function $\mu_{QM}(i, i_k, m)$, which takes on the value 1 if the flowrate is within the constraint limits, and declines to 0 as the flowrate exceeds either the upper or lower bound. A typical picture of this behavior is that shown in Figure 16.9, below.

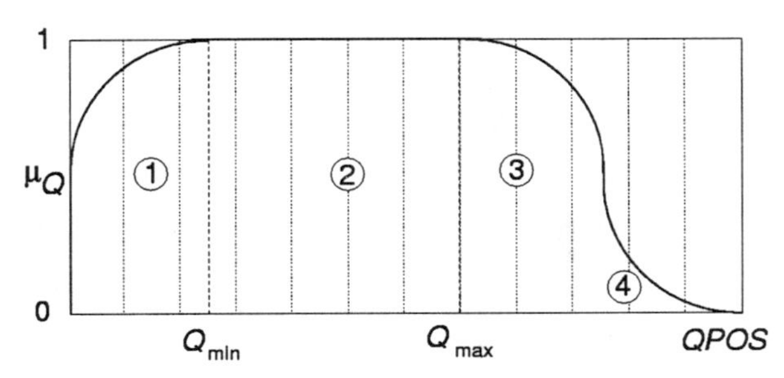

Figure 16.9: Flowrate Constraint Membership Function

The constraint membership function is rendered in mathematical form as:

$$\mu_{QM}(i, k_i, m) = \begin{cases} 1, & Q_{\min}(D) \le QPOS(i, k_i, m) \le Q_{\max} \\ (1 + X1^{p1})^{-1}, & Q_{\min}(D) > QPOS(i, k_i, m) \\ (1 + X2^{p2})^{-1}, & Q_{\max}(D) < QPOS(i, k_i, m) \end{cases} \tag{16.13}$$

wherein

$$X1 = \frac{|QPOS(i, k_i, m) - Q_{\min}(D)|}{Q_{\min}(D)} \tag{16.14}$$

$$X2 = \frac{|QPOS(i, k_i, m) - Q_{\max}(D)|}{Q_{\max}(D)} \tag{16.15}$$

$X1$ is required to satisfy $0 \le X1 \le 1$, but, practically speaking, $X2$ has no upper bound, because, although theoretically the range is $0 \le X2 \le \alpha$, say, the membership value approaches 0 as $QPOS$ goes large. Thus it is not likely to be selected in the optimization process.

The nonlinear flowrate constraint membership function illustrated above emphasizes the reality that small departures from fulfilment of a constraint are tolerable, whereas increasingly larger ones are increasingly intolerable. This accounts for the $p1$ and $p2$ exponents in Equation 16.13, which allows these arcs to incorporate variable degrees of degredation in the fulfillment of the constraint. A different degree of tolerance for the lower bound and upper bound is also permitted.

At this juncture, it becomes possible to display the membership functions

$$\mu_{QM}(i, k_i, m) \qquad m = 1, 2, \ldots, M \tag{16.16}$$

for each stage which correspond to the flows M at the nodes. For the total membership function of the feasible flowrates set, this gives

$$\mu_Q(i, k_i) = h_Q[\mu_{QM}(i, k_i, 1), \ldots, \mu_{QM}(i, k_i, M)] \tag{16.17}$$

$$i = 1, 2, \ldots, N; \qquad k_i = 1, 2, \ldots, KN(i)$$

In this equation h_Q is the flowrate aggregation operator. This operator can vary from the *min* operator of Equation 16.2 to some generalized weighted mean, such as Equation 16.5 gives, by apportioning differing weights to the node flows M.

The flowrate membership functions $\mu_Q(i, k_i)$ are computed only once in advance of the dynamic programming process, since they are unaffected by the stage variables.

Two more arrays are established. These are the head losses

$$HPOS(i, k_i, m) \tag{16.18}$$

$$i = 1, 2, \ldots, N \quad k_i = 1, 2, \ldots, NK(i) \quad m = 1, 2, \ldots, M$$

and the array of possible costs for all of the alternatives

$$CPOS(i, k_i) \quad i = 1, 2, \ldots, N \quad k_i = 1, 2, \ldots, NK(i) \tag{16.19}$$

This latter incorporates the cost at each pipe for each commercially available diameter.

The decision variables in the optimization are the pipe diameters $D(i, k_i)$. Actually, the k_i constitute the range for each stage.

The piezometric head at the node immediately upstream of the stage under computation is

$$(H_i, k_i)_m \quad i = 1, 2, \ldots, N \quad m = 1, 2, \ldots, M \tag{16.20}$$

These values must satisfy the following pressure constraints:

$$H_i - Z(i) \geq P_{\min}(i) \tag{16.21}$$

with $Z(i)$ the elevation at the node (in meters), and $P_{\min}(i)$ the minimum desired pressure at the node (in meters of fluid flowing).

Membership functions for the pressure parameters can be treated in a manner similar to those for the flowrate constraints. For the diameter alternative k_i, node flow m, and stage i, the feasible solutions set is $\mu_{HM}(H_i, k_i)$. A model for the pressure constraint membership function can be drawn as in Figure 16.10, where the two arcs reflect the more acceptable small variations and the less-acceptable larger ones by their convex and concave shapes.

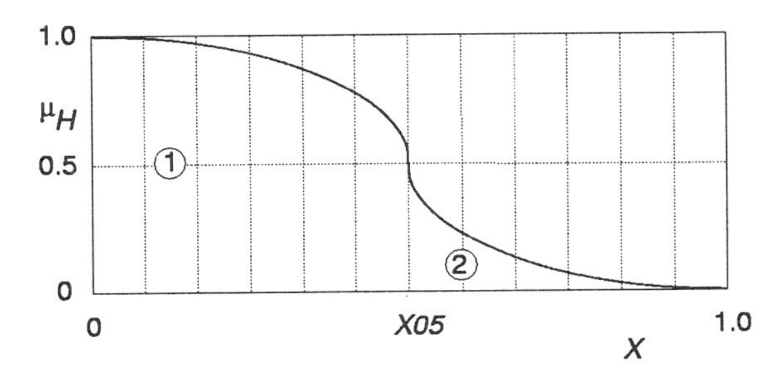

Figure 16.10: Pressure Constraint Membership Function

In Figure 16.10, there are three geometric particulars in the nonlinear relationship illustrated: to wit, a convex shape, a concave shape, and a point of inflection which marks their juncture. The abscissa is a variable X, which is defined as

$$X = \frac{|P(i) - P_{\min}(i)|}{P_{\min}(i)} \tag{16.22}$$

where $P(i)$ is the pressure at node i.

These features are rendered in mathematical terms as:

$$\mu_{HM}(H_i, k_i) = \begin{cases} 0.5 + 0.5\left[1 - \left(\frac{X}{X05}\right)^{p3}\right]^{\frac{1}{p3}} & 0 \leq X \leq X05 \\ 0.5\left[1 - \left(\frac{X - X05}{1 - X05}\right)^{p4}\right]^{\frac{1}{p4}} & X05 < X \leq 1 \\ 1 & P_{\min}(i) \leq Z(i) \end{cases} \tag{16.23}$$

However, the piezometric head at a stage is a function of that of the previous stage. Because the heads are state variables in this model, two additional partial aggregators are constructed for the pressure membership function: one of these is an aggregator for the stage h_p, and a second is for the downstream place h_D.

Suppose there is the possible state value (H_i, k_i) at a node i. There are then $k_i = 1, \ldots, KN(i)$ individual diameters to be considered, each of which comprises an alternative. Also at the stage is the flowrate membership function value $\mu_Q(i, k_i)$ from Equation 16.17, as well as a stage membership function $\mu_P(H_i, k_i)$. The latter can be formulated as

$$(H_{i+1}, k_i) = (H_i, k_i) - DHPOS(i, k_i, m) \tag{16.24}$$

$$\mu_P(H_i, k_i) = h_P[\mu_{PM}(H_i, k_i), \mu_P(H_{i+1}, k_i)_{m=1} \cdots \mu_P(H_{i+1}, k_i)_{m=M}] \tag{16.25}$$

in which h_P is the stage pressure aggregation operator.

In Equation 16.25 the element $\mu_{PM}(H_i, k_i)$ represents the membership function of H_i at node i. Should the succeeding node $i+1$ be a downstream end node, there are no further conditions to incorporate. In this case the downstream conditions may be represented as

$$\begin{aligned} \mu_P(H_{i+1}, k_i)_m = h_D[&\mu_{PM}(H_{i+1}, k_i)_m \\ &\mu_P(H_{i+1}, k_{i+1})_{m,s1}, \ldots, \mu_P(H_{i+1}, k_{i+1})_{m,sn}] \end{aligned} \tag{16.26}$$

with H_D as the downstream aggregation operator, and $s1, \ldots, sn$ are the downstream branches.

For both h_P and h_D the form may take either the *min* operator or some generalized weighted mean operator, just as was the case with the flowrate aggregator.

At last the combined membership function is at hand in the form

$$\mu_{PQ}(H_i, k_i) = h_{PQ}[\mu_P(H_i, k_i), \mu_Q(i, k_i)] \tag{16.27}$$

The total stage aggregator is the quantity h_{PQ}. It yields the overall total value of acceptability for each of the decision variables—the pipe diameters. Hence, the stage cost function for each diameter is

$$C(i, k_i) = \frac{CPOS(i, k_i)}{\mu_{PQ}(H_i, k_i)} \tag{16.28}$$

With this result, the total cumulative objective function can be expressed as

$$F_i(H_i, k_i) = min[C(i, k_i) + \sum_{s=1, t \in s}^{NS} F_{i+t}(H_{i+t}, k_{i+t})] \tag{16.29}$$

$$i = 1, \ldots, N \quad t = i + 1, \ldots, N$$

In the formula, s takes on the values for each node. The summation in the formula stands for the partial optimal cost functions, which are joined cumulatively.

In application, the programming starts with the most downstream stage, and moves up the network to the farthest stage upstream, and when that stage is reached, the optimal diameter values are brought back by a forward searching algorithm, which touches all stages.

16.5 CONTROL OF A WATER SUPPLY SYSTEM

Another interesting and informative model is furnished in the recent paper presented by Dr. L. S. Vamvakeridou-Lyroudia.[10] The Athens water supply is operated by Athens Water Company (EYDAP), which decided during the 1994-95 year that the existing software for controlling this system was an archaic technology that should be replaced. The company selected 20 Programmable Logic Controllers (PLCs), a Supervisory Control and Data Acquisition (SCADA) system, and a real-time expert system model named DANAIS, an acronym for Dynamic Analysis Aqueduct Intelligent System, which was then assigned to National Technical University of Athens as a research project.

Dr. Vamvakeridou-Lyroudia was the principal researcher, and was assisted by her graduate students of the Faculty of Civil Engineering.

16.5.1 THE AQUEDUCT

The main feature of the Athens water supply system is an aqueduct which runs from its source at the Mornos/Evinos dam site in mountainous terrain down some 95 kilometers to a plain, then 50 km further into the city and its environs.

An additional 43-odd km extend the conduit to the last large user. The total length is about 190 km. Two additional water sources add to the flow along the aqueduct. Many outlets take flows from the system: among these are the Marathon Dam reservoir (used to supply Athens, the city itself), a large distillery (Menidi) near the downstream end, a second (new) distillery, and a number of small towns and villages along the way.

A general schematic diagram of the system is illustrated in Figure 16.11.

Marathon Lake is an artificially created reservoir constructed by the oldest water supply company operating in Greece from 1929. It is located about 4 to 5

[10]Vamvakeridou-Lyroudia, L. S., and Giovanopoulos, H., "An Original Real Time Expert Model for the Operation and Control of the Water Supply Aqueduct of the Greater Athens Area, Using Fuzzy Reasoning and Computer Assisted Learning," ***Novel Design and Information Technology Applications for Civil and Structural Engineering***, CIVIL-COMP PRESS, B. Kumar and B. H. V. Topping, Editors, 1999, pp. 205-213.

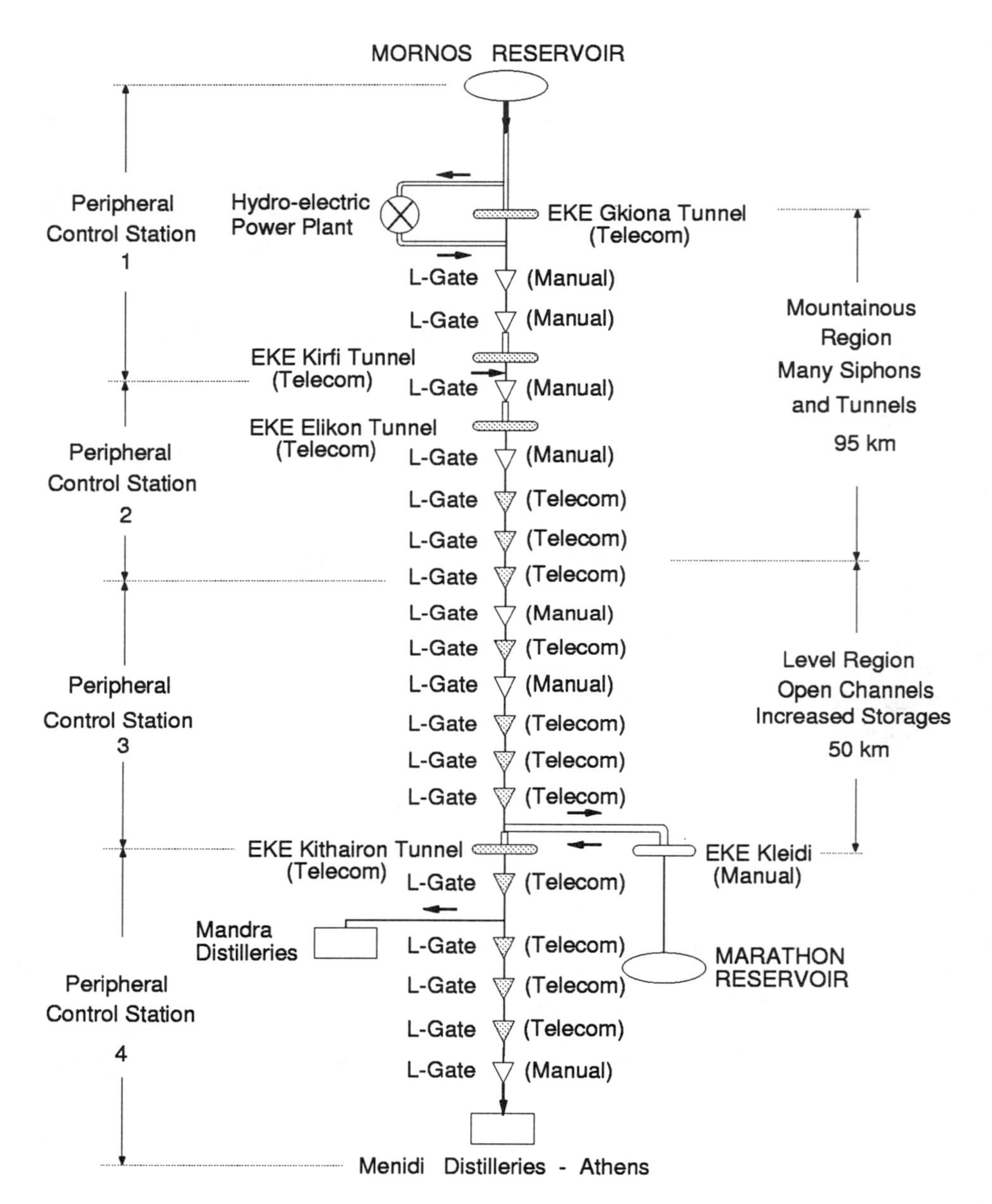

Figure 16.11: The Mornos/Evinos Aqueduct

kilometers upstream from the famous Plain of Marathon, the famous historic site of ancient Greek history. It was there that the Greeks turned back the invading Persians in a battle formative in the history of the Mediterranean world and western culture.

At the time of its construction in 1929, Athens was a much smaller city, and the marathon Dam served as its sole water source, but nowadays its capacity is sufficient for only 30 to 40 days supply, so it serves as a large "water tank" (of 40 million cubic meters capacity).

The Mornos/Evinos aqueduct is gravity driven. About 112 km of the total length is open channel flow, and some 71 km is in 15 tunnels. Another 7 km is devoted to 10 siphons, and there is one hydro-electric power plant. The water flow is regulated by some 23 structures distributed over the aqueduct's length. There are 18 gates of nominal size (called "L-gates"), and 5 larger ones incorporating stilling basins for energy dissipation purposes (called "E-gates" or "EKE" as the Greek acronym). The gates are controlled either manually or by telecommand. Water takes from 18 to 25 hours to traverse the length of the aqueduct, depending upon the conditions.

Oversight of the aqueduct is allocated to SCADA system, which receives inputs from many sensors along the route. There are both analog and digital inputs from 20 PLCs located at the various control sites, and these are submitted every 30 seconds to 4 Peripheral Control Stations, as well as to a Central Control Station in Athens where a graphical status is displayed and records are kept.

Every 15 minutes the SCADA main operational module activates a real-time simulation algorithm—DANAIS. The inputs, historical data bank, and intelligent information from the algorithm are employed to process the signals, validate hydraulic computations, and estimate new gate positions. Then new orders are sent back to SCADA and through it to the PLCs to set the new gate openings.

Inasmuch as the aqueduct is essentially one-dimensional, the location of the several control gates and other hydraulic structures is specified by giving one number, a distance from the upstream node in kilometers. The various reaches of the system are defined by nodes, which thereby furnish beginning and ending locations for all of the hydraulic computations. The simulation allocates the following properties to each reach:

- Both an upstream node and a downstream node as defined by distances from the origin of the aqueduct.
- Mode of flow within the reach. Three types are modeled: free surface flow, pressure driven flow, and chute flow (free surface flow with steep controls).
- Free surface flows are trapezoidal channels with uniform geometry: bottom width B, sides-slope z, maximum allowable water depth $y_{\max}$, and bottom slope S_0. Pressure flow takes place in single or double circular pipes or tunnels of diameter D.
- Uniform Manning's coefficient n for channel flows, uniform surface roughness k for pipe flows.
- Inflows and outflows are allocated to the nearest nodes.

- Control gates are designated at nodes only.
- Flow meters and level guages are permitted at any place along the reach.

These stipulations create a model of 61 reaches, which are the computational elements of the system.

In common with other metropolitan areas throughout the world, the demand for water grows continually as new residential areas and industrial installations are being added to the region. This fact is the principal agent motivating the re-design of the Mornos/Evinos aqueduct in 1994-95. It is also a large factor in planning for the future of this system.

16.5.2 THE DANAIS ALGORITHM

The heart of the water supply control for Athens is the real-time expert computational scheme DANAIS. This algorithm is, in turn, composed of four separate modules for performing its computations. The schematic diagram of this arrangement is shown in Figure 16.12.

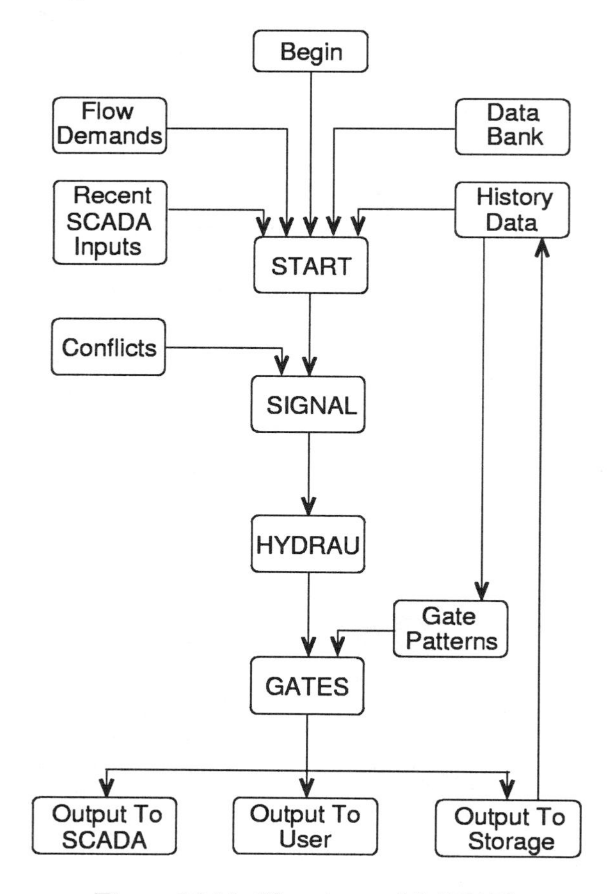

Figure 16.12: Flowchart of DANAIS

The component modules are:

- START This block in the flowchart reads, classifies, and arranges data. The three types of data involved are:
  - *Stationary data*; i.e., those which need not be expected to change every 15 minutes, such as geometric measurements of the aqueduct, etc.
  - *Variable data*; that is, items which *are* expected to alter every fifteen minutes, such as flow rates, sensor inputs, etc.
  - *Data bank items*; these are data which are built, stored, and updated each time the program is run, such as the recent history (retained for the 8 most recent runs), and a real-time auto-corrective memory which allows the algorithm to learn from its own mistakes and correct these during run-time.
- SIGNAL The purpose of this block is to process and evaluate signal inputs. The analog and digital inputs from all sensors are examined, checked, and cross-checked for validity. Where there are inputs missing, an attempt is made to furnish these. Where conflicting information is found, it is tested against stored patterns of such conflict as determined during calibration of the model, so that on-screen warnings can be issued to users. Most of the conflicts that had been anticipated during the initial operation of the system have been resolved satisfactorily, so that in recent operation no new ones have appeared.
- HYDRAU This module is at the heart of the physical operation of the system. Herein are applied computations of the hydraulic flow-rates in the several reaches of the aqueduct. The Muskingum-Cunge algorithm is the one employed for this purpose. The results are the targeted rates of flow at each control gate every 15 minutes.
- GATES Gates module determines and issues the necessary gate control orders to the local PLCs. From the desired flow rates at the gates, the height of the gate is determined to produce this flow, and the gate then moves automatically to assume this position.

  However, there are many reasons why the actual position may not be in accord with the desired one. There are also unexpected emergencies to be dealt with. For these reasons, the complex possibilities of gate behavior were reduced to establishing an heirarchy pattern of gate movements, then fuzzy logic linguistic terms were added, and auto-corrective data banks created.

16.5.3 KNOWLEDGE FRAMEWORK

A number of factors contributed to the selection of fuzzy and Boolean logic patterns for the DANAIS expert system model. In the first place, the aqueduct is sufficiently long that direct visual inspection or confirmation of gate movements and other physical variables during operation can not be reliably utilized, even though there are personnel on duty round-the-clock at each of the four peripheral control stations to perform maintenance and security functions. Secondly, it was determined that simpler signal processing modes—such as IF···THEN rules—were unreliable and ineffective in detecting malfunctions in the communications network for this widespread system.

In fact, it was realized that the major weakness to effective operation of the aqueduct system was unreliable ("strange" or conflicting) signals, either analog or digital, or both. So the decision was made early that the knowledge based model would be designed to permit the resolution of such fuzzy data in an auto-corrective manner to the maximum extent possible.

Specifically, the conditions concerning gate openings were a cause of major concern. So the model was prepared with the potential to test each input signal for veracity against not only its programmed or expected value, but also against the pattern of inputs from the remainder of the system. In this way, an anomalous reading can be called into question, remedied (within limits) by the program itself, and/or made the basis for an alarm to the oversight staff. Such intelligent behavior is difficult to achieve with even a large number of IF···THEN statements, therefore the alternate approach was taken.

Stored in the algorithm's data bank as simple ASCII files are a number of "conflict" patterns, based upon the types and styles of input signals associated with previous incidences of such conflicts. Each time the program runs, it scans the patterns for congruence with the one being processed to see if there are similarities, and if such are found, an attempt is made to rectify the conflict or call for assistance. When the user oversight is activated, it may add the current conflict pattern to the data bank as a novel storage for future reference. The program itself, however, is prohibited from automatically adding these patterns to the knowledge base on its own, as certification by humans is considered the higher form.

There is a heirarchy of gate movements prescribed for this system. That is, an agenda is prescribed for which gates shall move first and by how much, then which gates shall follow afterward in what order. Considerations which dominate this protocol include several gates moving in parallel; operational parameters such as time from most recent motion, inter-movement time minimum (30 minutes); maximization of hydropower energy; and emergency motions. For this scheme, three-valued fuzzy logic is utilized in the form of a wild character that may indicate either ***any value*** or ***unknown but existing value***, both distinct from a ***not existing*** state. The latter is represented by a Boolean *false*, signified by a negative value. "0" stands for *closed gate*, and "1" designates *open gate*.

A library of pre-defined functions which define gate movements is included within the code. Some are: *same* (order the same gate position), *move* (permission to move the gate, amount not designated), *m1*, *m2*, *m3*, (for priority of movements), > or < (for comparative magnitudes), *old* (to repeat an order), etc.

Some special notations are involved in the codes. In Greece, the public electric company is widely known by the acronym formed of its three initials *D. E. H.* Therefore the variable named *deh* actually refers to the power plant, and thus *mvdeh* means "MoVement of the power plant's turbine to a new stage." Similarly, *iodeh* translates to "I order the power plant to ...."

A representative portion of the fuzzy gate pattern for controlling the farthest upstream gate movement pattern at the Gkiona EKE is shown in Table 16.1.

Table 16.1: Part of Fuzzy Gate Movement Patterns for Gkiona EKE

PRESENT STATUS									ORDER		
ideh	iopen gate1	iopen gate2	iopen gate3	qgate	mvbig	mvdeh	sos	iodeh	iopen gate1	iopen gate2	iopen gate3
0	1	1	*	*	*	*	*	0	same	0	same
0	-9	-9	*	*	*	*	*	0	pvo8	0	same
0	0	-9	*	*	*	*	*	0	same	old	move
0	-9	0	*	*	*	*	*	0	old	same	move
0	0	0	*	<7.5	no	*	*	0	same	same	move
0	1	0	*	<7.5	yes	*	*	0	m2	same	m1
0	0	1	*	<7.5	yes	*	*	0	same	m2	m1
0	0	0	*	<7.5	yes	*	*	0	m2	m3	m1
0	1	0	*	>7.5	no	no	*	0	same	same	move
0	0	0	*	>7.5	no	yes	yes	0	same	same	move
0	0	0	*	>7.5	no	yes	no	Qon	same	same	move
0	1	0	*	>7.5	yes	no	yes	0	same	same	move
0	0	1	*	>7.5	yes	no	yes	0	same	same	move
0	0	1	*	>7.5	yes	yes	no	Qon	same	0	move
0	0	0	*	>7.5	yes	yes	no	Qon	same	same	move
1	1	1	*	<7.5	*	*	*	Qoff	same	0	same
1	1	-9	*	<7.5	*	*	*	Qoff	same	0	same
1	0	0	*	>7.5	*	yes	no	m1	same	same	m2

16.5.4 FUZZY REGULATORS

As was previously mentioned, the regulation of the aqueduct is accomplished by the movements of some 23 system gates. The two types of gates are common sluice gates and E-gates (EKEs). The latter, which are operated under pressure, are sited at the end of long tunnels which connect upstream reservoirs to downstream channels, and are provided with stilling basins incorporated into their exit structures.

The E-gates also contain surge towers, to mitigate waterhammer effects. There are in all 4 EKE that are driven by telecommands, and one which is manually operated. Four of the structures have double gates which operate in parallel. The largest EKE, which has 3 parallel gates, is located at the exit of the Gkiona tunnel.

Figure 16.13 shows some details of a typical EKE, including the upstream reservoir, the tunnel, the control gate, the stilling basin, and the downstream channel.

The gate opening X, is related to the flow rate Q, and the head drop through the gate H, by the formula

$$X = \frac{aQ^b}{H^c} \tag{16.30}$$

where a, b, and c are parameters determined by calibration measurements under steady conditions of flow. The calibration parameters were carefully determined for each gate individually, since they are the principal flow controls for the aqueduct.

The control system logic is designed to drive each gate to some position X, by a command *move*, but these commands do not contain the magnitude of the movement. In practice, the operating circumstances may vary from the idealized

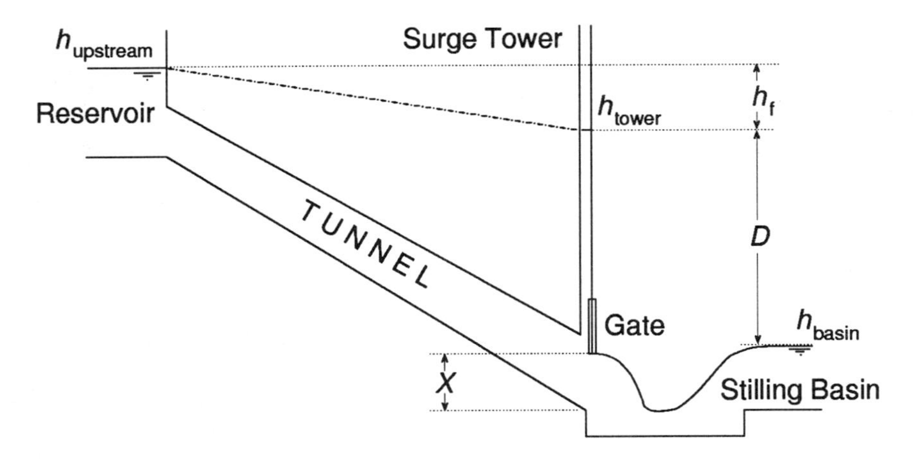

Figure 16.13: Typical Regulatory Structure

command motion. The surge tower level meters may be giving erroneous readings, friction losses in the pipe-tunnel must be estimated, not measured, and except for the Gkiona EKE there is no flow meter (except during calibration) near the structure, so flow rates must also be estimated. Also, level meters at places such as the surge tower and at the tunnel intake may be inoperative.

Considerations of overall operation in a satisfactory manner dictate that there should be a fuzzy expert rule for gate openings, since the gates are large, and errors in their opening positions could be dangerous. In this scheme, flow rates, level meter readings, and friction losses are treated as trapezoidal fuzzy numbers.

The fuzzy gate movement X as output from Equation 16.30 is entered into an overall fuzzy aggregator to combine the membership functions. Conventional Yager, Sugeno, and Dubois-Prade *and* aggregators were applied, as well as a simple harmonic aggregator because of its proven effectiveness in a similar application.

When the aggregators were employed with real data, it was seen that the total membership function, even when using an α-cut of 0.80, did not produce a smooth or quasi-smooth defuzzification plateau, as illustrated in Figure 16.14. This suggested that a modified highest plateau method might be employed. In this method, max_{near} and max_{far} positions were automatically determined to represent the "far" maximum X and "near" maximum X places for an optimal imaginary plateau. These, in turn, are joined to language terms in the following manner: max_{far} gives a somewhat more open gate position that signifies a shortage of water in the channel, and max_{near} stands for a more closed gate indicating "much" water in the conduit.

Thereafter, in the overall operating mode of the aqueduct, when the linguistic terms "full," "medium," or "low" are encountered, the corresponding defuzzified value is chosen automatically. After the fact, as seen in Figure 16.14, the Sugeno aggregator was seen to be inadequate, the Yager and Dubois-Prade aggregators yielded quasi-similar values, and the harmonic mean aggregator proved to be the best.

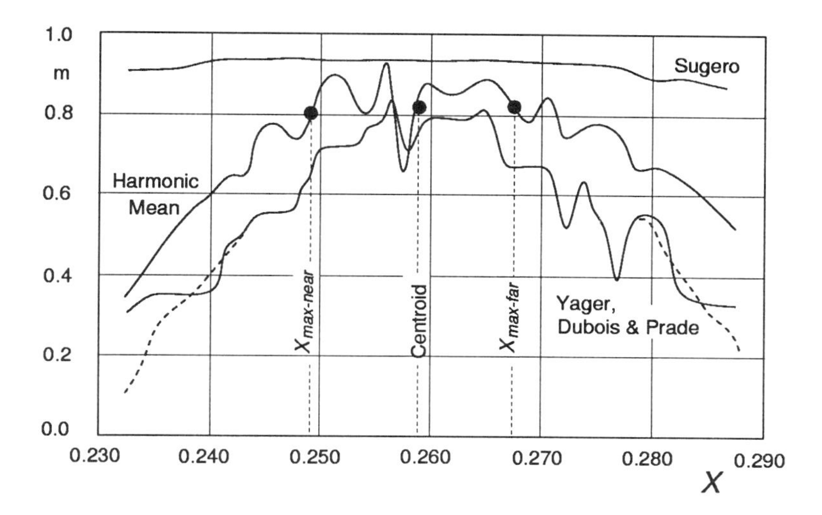

Figure 16.14: Fuzzy Aggregators EKE Elikon

16.5.5 AUTOCORRECTIVE OPERATIONS

In spite of the fact that calibrations are done carefully and other measurements are made with good precision, certain quantities escape from exact definition. Friction factors are notorious for changing with time and other conditions. Shape factors for gates and weirs also vary with passage of time, and may be poorly calibrated. All such inaccuracies can lead to faulty operation of the system.

An autocorrective process is incorporated into the expert regulatory model of the system, whereby deviations and aberrations from targeted quantities are corrected. By this mechanism the expert model "learns" on the job while operating in a real-time mode, and its learnings are stored in its repertory. The feature was originally intended for use during the "run-in" period of the system, but emerged as a valuable permanent artifact.

Therefore the real-time autocorrective data bank contains a file called *autolearn*, which permits the model to learn from its mistakes, and then to incorporate its new knowledge into the data bank for future use whenever it is enabled by a toggled switch engaged by the superuser human.

Two general classes of corrective mistakes, miscues, and mischiefs are seen: (1) wrong flow rates to the large distillery users due to malfunctioning in meeting demands and (2) water levels either too high or too low in the channel itself, particularly in the upstream reaches from gates and tunnels. Provision is therefore made for the user to choose between these two types of mistakes to be corrected.

AUTOLEARN is scanned each time the system model runs, so that the previous errors are corrected. Actually, the expert model may overcorrect in order to relieve a former mistake, whence a gate position may be such that a water level becomes too low after some hours of operation. This results in a new mistake, and further correction procedures. These changes are duly recorded in the data bank for each of the control gates.

In general, the overall operation stabilizes after a day or two of operation under continual autocorrection or slowly-varying hydraulic conditions. When the flow conditions change considerably, however, the autocorrective function engages again.

Since it is possible for automatic correction of hydraulic quantities to take place only when steady conditions near the gates prevail, DANAIS must first estimate whether or not this is the case. The steady condition is often not difficult to obtain, especially since the distilleries often run at or near a desired flow rate for long periods of time. Therefore, unless there have been emergency gate movements somewhere, most of the aqueduct will be operating under steady conditions, and this will automatically be detected by DANAIS through its historical data file. Thereafter, the model senses water levels and flow rates, and makes corrections accordingly.

Naturally, if the project manager should initiate changes in targeted flows, a wave propagation will occur along the aqueduct and unsteady flows will be the consequence. The autocorrective feature is automatically disabled under this scenario.

In the case of the smaller—L-type—gates, the automatic correction program for hydraulic parameters has been satisfactory. These structures are conventional sluice gates, either single or double, with long weirs on each side. This configuration is sometimes called *bec de canard*, or "duck's nose," and in Greek is known as Λ-shape: that is, L-gates. Of the 18 spaced along the aqueduct, 11 are operated by telecom signals, and 7 are manually directed. The model treats all in a similar manner, but telecom signals are directed only at the automatic gates.

The gate configuration is illustrated in Figure 16.15.

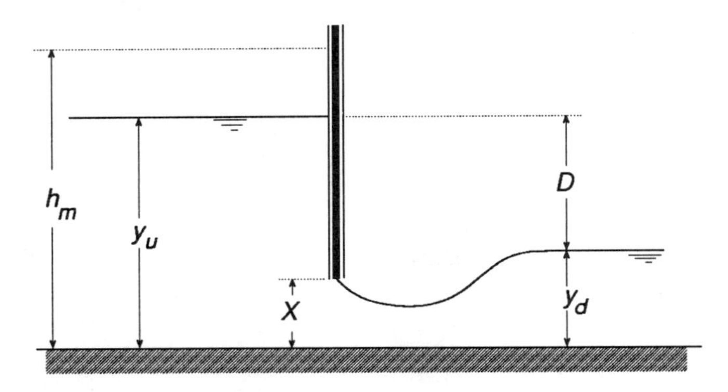

Figure 16.15: L-Gates Geometry

Three possible flow conditions pertain to these gates:

- The gate is partially open, as seen in Figure 16.15, and the water level behind the sluice is less than the level of the weir crest, h_m, therefore the flow-rate passing under the gate is controlled by the depths $D = y_u - y_d$, in which y_u is the upstream depth, and y_d is the depth downstream.
- If the water level upstream exceeds that of the weir crest, some flow passes

over the weir. Whether or not the gate is closed or partially open, it cannot control the flow. In this circumstance, with maximum water depth, the volume of water held behind the weir is greatest, and propagation of a flow wave is attenuated by the gate. If the gate is fully open, water flows under it without touching the gate bottom, hence the gate exercises no control. Consequently, the water level and water volume held upstream is a minimum, and wave propagation is faster than for other conditions. However, the gate cannot regulate the flow.

The nature of the aqueduct is such that there is no requirement to maintain a prescribed water level in reaches upstream from any given L-gate, as with irrigation canals. Accordingly, the best state of operation is that of the first category listed above; that is, the gate is partially open, there is neither overflow of the weir nor free flow under the gate. Thus the flow is regulated by movements of the gate.

For this case, the gate opening X, the discharge through the gate Q, and the head difference $D = y_u - y_d$, are related by Equation 16.1, as with the E-gates. A common set of values for L-gates is $b \approx 1.00$, $c \approx 0.50$. In theory, the constants a, b, and c for a given gate may be determined by calibration measurements, but in the case of the Mornos L-gates, acceptable accuracy was not realized in this way.

Since the importance of the E-gates to the regulation of the system is much greater than those for the L-gates, the decision was reached to perform L-gate calibrations at run time, using the autocorrective data bank. The process was as follows:

1. A first estimate of the shape factor a was entered into the program data, then the autocorrect mode was turned on.

2. Steady or quasi-steady state conditions were allowed to obtain at the gate in question. This condition was confirmed by the program itself.

3. If the upstream water level was either too high or too low, a correction factor for the shape factor constant was used as:

$$a_{\text{new}} = a_{\text{old}} \left(\frac{X + STEP}{X} \right) \tag{16.31}$$

   in which a_{new} is the new value of a, a_{old} is the former value, X is the present gate position, and $STEP$ is an increment in the gate opening. For water too high, $STEP$ was taken as 0.02, while for low water, $STEP$ was -0.02.

Comparable correction factors were employed for the discharge in Formula 16.7 for all gates, except that such corrections were not allowed at the same time as for the shape factor ones. In the event, such flow rate corrections proved insignificant except for the EKE close to Athens (the Kitharon tunnel).

Water level corrections were made for all gates, both L- and E-types except for the uppermost EKE at Gkiona, for which the autocorrective mode is never employed.

16.5.6 HIERARCHY EFFECTS

Some facts dominate the Mornos aqueduct performance: the length—190 km; the time required for water to travel from top to bottom—about 1 day; the flow rate—variable from 8 to 16 m^3/s; the volume of water in the canal—between 1,500,000 to 2,000,000 m^3 at any time, only a small portion of which is engaged. The tunnels must be kept full, and the amount of water which can be stored or retrieved upon short notice in a reach between two gates is determined by hydraulic conditions and weir levels at sluice gates.

Given the gross features enumerated above, a time scale for the operation of the system allows the description "short" (a fuzzy linguistic variable), which means "a time brief enough that flow-rate changes at the entering end of the conduit at its most upstream gate (EKE Gkiona)" has not yet "affected" (also a fuzzy logic term) a given gate under discussion. Also, note that water storage capabilities are unequal along the channel length.

The three general regions of the aqueduct are:

- The first 95 km across a mountainous region where there are many siphons and tunnels (with smaller storages).
- The second portion, about 50 km, built across a plains region, comprised mainly of open channel reaches and gates, having fairly significant storages.
- The third region, of 43 km length, nearest Athens, beginning at the EKE Kithairon, where the channel cross sections are narrower and storages are smaller. Also, damage due to overflows will be greater. Exit from this section is solely to the distilleries.

During the earliest phases of the Mornos aqueduct project it became clear that a critical role is played by the EKE Kithairon gate. The movements of this gate are the real determinant of the system's performance, therefore it became the true "expert" which decides upon priorities and perils in the system. Thus EKE Kithairon's gate opening position is set by a complex sequence of IF···THEN rules and criteria. These take into account:

- The distillery flow demands.
- The hydraulic conditions.
- The conditions upstream and downstream of Kithairon.
- The historical data.
- The autolearn data.
- The time of day (because of differing danger factors).

When EKE Kithairon has set its operating conditions, the expert then begins to apply corrective functions at all other downstream L-gates and to the upstream gates within its sphere of influence; i.e., in the second region of the aqueduct. These gates are thereby slaved to EKE Kithairon. The storages along this region are thus managed by a single expert with an overview of the whole sub-system.

In practice, adding this extra weighting to the functioning of EKE Kithairon has produced very satisfactory operations. This reflects the fact that calibration of this region has been effective, and the result is an accuracy of about 99% in the determination of the flow rate by Equation 16.1.

This contrasts with the experience in the upper region near EKE Gkiona. Flow meters and level meters in the gates and at the hydroplant have resisted accurate calibration. There are also inaccuracies in flow rates associated with additional constraints on the possible gate motions.

Recently, a new problem has entered the scenario: live fresh water mussels in large amounts have become trapped in the stilling basin at EKE Gkiona. Since they come from the reservoir above the gate, they delay the response and affect the sensitivity of level meters immediately downstream from the gate. The result is that EKE Gkiona has become the expert for the uppermost region of the aqueduct, and altered rules for the gate positions have been established. These emphasize safety in the upper 93 km of the system, and allow additional gate movements, as required, to conduct the water in this region without emptying or overflowing tunnels and siphons.

Options were created in the *autolearn* data bank for the two master stations at EKE Gkiona and EKE Kithairon to "talk" to each other about conditions and problems within their respective areas. This feature is under a toggle control at the disposal of the system administrator.

A trial period was established during which this feature was enabled. It did not perform well, as both stations became "nervous" when they learned about the other's problems. The contradictory decisions they issued soon provoked the following comment:[11]

> "... in artificial intelligence a system should not become too informed or too intelligent."

16.6 AUTOMATIC SYSTEMS

In almost every contemporary technological field there are urgent motivations to increase the utilization of automatic machines and automatic control techniques. These tendencies are now nearing one century old. They were pioneered by assembly-line methods in auto manufacture; continuous and batch process production in chemical and related industries; and the design and production of navigation, stabilization, and detection apparatus for air and space vehicles as well as land and sea craft.

Nowadays the achievements in this area have made of it a science in its own right, and are global and penetrate every technical discipline and most commercial and business activities as well.

The reasons driving this explosive growth are self-evident. In logistics, the science of automatics permits operations on a scale many times more complex than would otherwise be possible. On the economic side, there is attractive reduction in operating costs (in the face of rising labor problems), thereby making automation a natural choice for manufacturing. To engineers, confronted with the space age and

[11] Vamvakeridou-Lyroudia, L. S., and Giovanopoulos, H., op. cit., p. 213.

its requirements for exceedingly close control, remote operations, and optimum analysis, the automatic system offers the only hope of meeting the transcendent demands of design needed to move nature.

Short of writing a whole treatise on the subject, it will suffice to note that the central concept of all automatic apparatus is the capability of such systems to exhibit other than a purely mechanistic or deterministic response.

An ordinary machine does only what it is designed to do. A lever is actuated, a button is pushed, and one and only one response is forthcoming. The ordinary machine is expected to proceed according to the law of mechanics, once the initial conditions are established.

Yet automatic machinery is intended to do more than this. In a sense, it is required to "think"—to perform its design function whatever may be the additional stimuli which it may receive during the process. Thereby it is "liberated" from the purely mechanical response required by natural laws, and is thus autonomous.

This "thinking" function of automatic systems can be expressed in more precise terms by noting the following properties:

- The autonomous machine must sense the result of its actions and use its findings to take corrective measures.
- It must operate without regard to external disturbances.
- An automatic machine must have the capacity to correct errors or minimize the difference between desired result and obtained result, all with great accuracy and high speed.

16.6.1 THE CONCEPT OF FEEDBACK

The generalized descriptions itemized above are a linguistic way of introducing the concept known as feedback in automatic systems. This idea can be elaborated by using a simple system described by a first-order linear differential equation.

Take first the case where there is not any feedback, just a mechanistic response. Extract from some piece of machinery a rotating "shaft" with its entrained mass moment of inertia I, viscous damping constant c, and nominal rotational speed Ω, Figure 16.16. Suppose further that all the "torques" which are acting upon the "shaft" are regulated as well as possible. Nevertheless, there will always exist some small error between the torque applied to the shaft and that absorbed by the load.

The system is represented by

$$I\frac{d\omega}{dt} + c\omega = T(t) \tag{16.32}$$

where ω is the variation in speed about the steady value Ω, and $T(t)$ is the error torque.

The solution to Equation 16.32 is composed of two parts, the complementary integral corresponding to

$$I\frac{d\omega_c}{dt} + c\omega_c = 0 \tag{16.33}$$

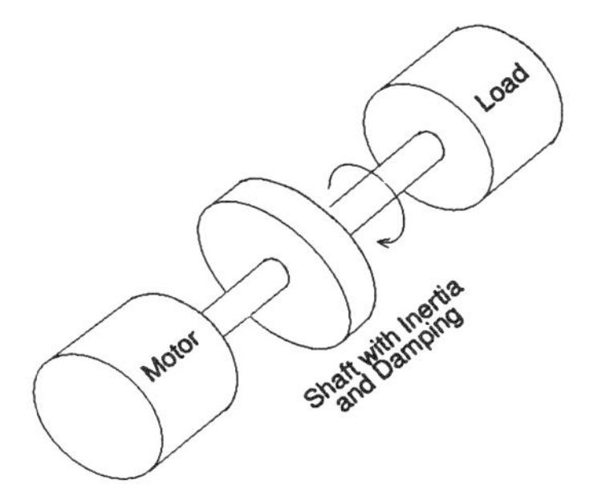

Figure 16.16: Rotary System

and a particular integral which is obtained from

$$I\frac{d\omega_p}{dt} + c\,\omega_p = T(t) \tag{16.34}$$

so that $\omega = \omega_c + \omega_p$.

The complementary part is

$$\omega_c = \omega_0\, e^{-t/\tau} \qquad \tau = \frac{I}{c} \tag{16.35}$$

In this part, ω_0 is the constant of integration, found from the initial condition $t = 0,\ \omega = \omega_0$.

As for the particular integral, its form depends upon the nature of the exciting torque $T(t)$. However, much insight can be obtained into the nature of the response by noting what happens when a constant torque T_0 is suddenly applied at time zero. (In the language of the previous chapter, this is a ***step torque*** of magnitude T_0). Subject to this excitation the particular integral is

$$\omega_p = \frac{T_0}{c}(1 - e^{-t/\tau}) \tag{16.36}$$

A graph showing these two elementary responses is given in Figure 16.17. An inspection of this figure shows that the response of the shaft is governed by a characteristic response time τ. In other words, a certain period of time is required for the shaft to come to the asymptotic value of its new speed. This "time constant" is about 3τ.

Now the entrained inertia of a shaft is nominally a substantial quantity, and the mechanical damping is generally a very small quantity (because the designers have designed it away to the maximum extent), so that $\tau = I/c$ is evidently a long period of time. That is to say, any errors in speed of the shaft will require a long time to correct, and furthermore, the error torque will need to be small, since, from Equation 16.36, it is magnified by the ratio I/c.

Consequently, the purely mechanistic device is not a very satisfactory solution in any design which needs a rapid and accurate response.

By contrast, the same system with the addition of a feedback element presents a very different picture. Let the shaft speed be sensed by a velocity transducer

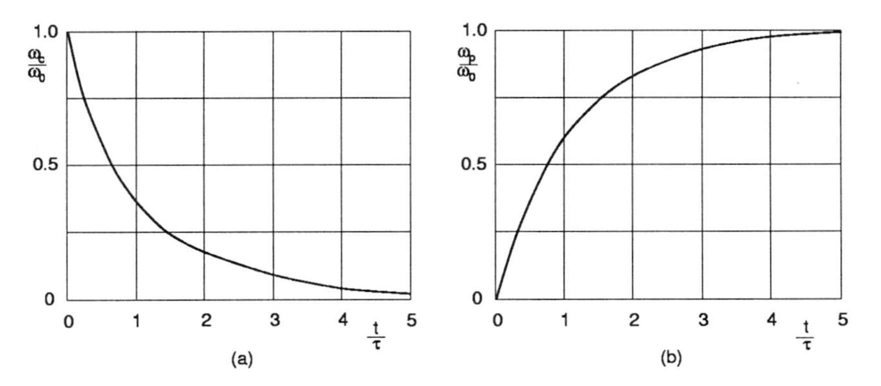

Figure 16.17: Complementary and Particular Integrals

which then actuates a torque reduction in the driving motor that is proportional to the speed variation ω. The setup is depicted in Figure 16.18. (Incidentally, this ratio of the torque reduction to the speed variation is the source of the language "proportional control" in systems theory.)

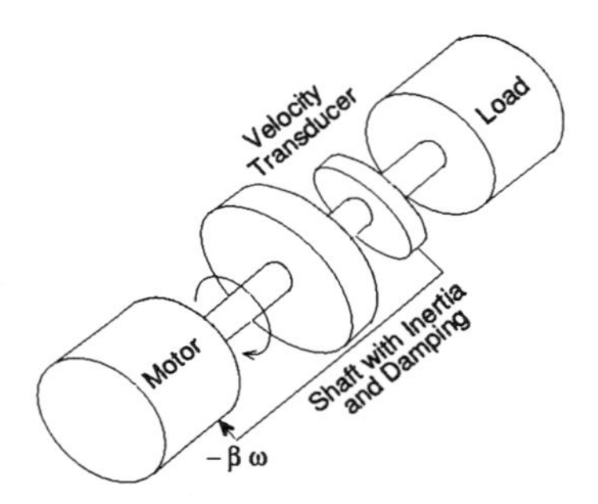

Figure 16.18: Rotary System with Feedback

Now the equation comparable to Equation 16.32 is

$$I\frac{d\omega}{dt} + c\omega = T(t) - \beta\,\omega \tag{16.37}$$

or

$$I\frac{d\omega}{dt} + (c+\beta)\omega = T(t) \tag{16.38}$$

and it is seen that $c+\beta$ stands in the place of c in the former solution.

The response time is now $\tau = I/(c+\beta)$, and the magnitude of the speed variation is proportional to $T_0/(c+\beta)$. It is very easy to make β large compared to c, and therefore it is easy to improve the speed of response and to reduce the magnitude of the undesirable speed variation.

So a little feedback makes a dramatic improvement in the performance of this simple machine, and it is also clear that this improvement is accomplished within the machine itself. It has in fact begun to "think," or it has become automatic.

16.6.2 EXTENSION OF THE CONCEPT OF FEEDBACK

The fundamental rôle played by feedback as demonstrated in the previous elementary example can be extended to more complex systems. Other exciting force systems can be modeled, and higher order systems such as spring-mass-dashpot devices governed by equations of motion of the form

$$m\left(\frac{d^2 s}{dt^2}\right) + c\left(\frac{ds}{dt}\right) + k\, s = F(t) \tag{16.39}$$

can be considered, and nonlinear mechanisms can be investigated as well.

An example of the latter is supplied by the double-sided hydraulic actuator shown schematically in Figure 16.19, the equation of motion of which turns out to be

$$\frac{dv}{dt} = c_1 - c_2\, v^{7/4} \tag{16.40}$$

The constants in this equation are

$$c_1 = \frac{g}{W}(P_0 A_1 - P_3 A_2 - W - f) \tag{16.41}$$

and

$$c_2 = \frac{0.241\, g\, \rho\, \nu^{1/4}}{W}\left(\frac{\ell_1\, A_1^{11/4}}{d_1^{19/4}} + \frac{\ell_2\, A_2^{11/4}}{d_2^{19/4}}\right) \tag{16.42}$$

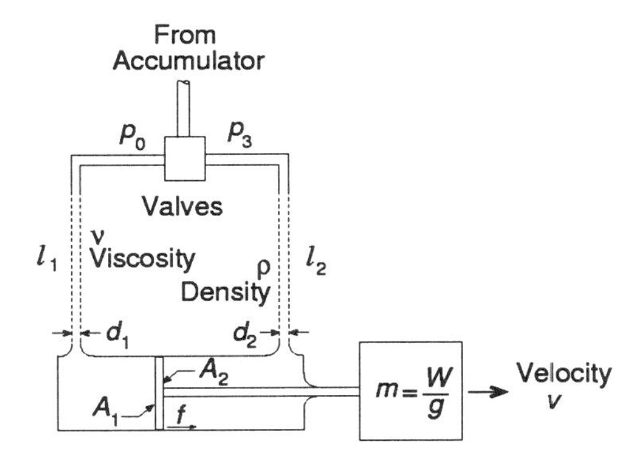

Figure 16.19: Hydraulic Actuator System

Once again the overall system performance can readily be improved by the addition of feedback. In this case a velocity transducer is employed to sense the motion of the mass, and a fraction of the variation from desired speed is used to reduce the applied pressure at the control valves.

However, in the case of nonlinear systems, the solutions to the equations of motion of the system, both with and without feedback, are ordinarily effected by numerical (or even graphical) means, since closed-form analytical answers are seldom known.

An even further degree of complexity in the application of feedback to automatic machinery comes about by regarding the multi-degree-of-freedom system.

In such a system it is possible to include feedback between variables, as well as just that for a specific variable.

For those systems which have variable coefficients in the equations of motion, or which have elements with time-varying components, additional complications are presented, especially with respect to the stability of the response.

16.6.3 THE ULTRASTABLE AUTONOMOUS SYSTEM

It is axiomatic that the extension of the feedback concept to increasingly more complex systems as suggested above leads to increasingly more complex analyses and designs. The increases are of a nonlinear order in most cases, so that an explosive magnitude of difficulty looms in a variety of examples. Systematic development of criteria for analysis, prediction of response, stability, and synthesis become increasingly more difficult to effect, and there is considrably more doubt concerning reliability.

Engineers involved with the selection of components and optimizing the performance of complex automatic machines are presented with the dilemma of error, when accuracy is required, and instability, when the need is reliability.

Happily, there is still the possibility of an automatic device, using feedback principles, which "thinks" for itself well enough that it can even compensate for some of its designer's mistakes. This machine is called the ***ultrastable autonomous system.***

To illustrate an elementary version of such a system, consider the two variables

$$\frac{dx^1}{dt} = a\,x^1 - b\,x^2 \tag{16.43}$$

$$\frac{dx^2}{dt} = c\,x^1 - a\,x^2 \tag{16.44}$$

in which a, b, c are constants that in turn depend upon an auxiliary parameter λ. The equations could represent a mechanical system composed of two shafts with their entrained inertias and dampings, and some cross-feedback between the two.

To advance to the analysis of the system, note first that, by eliminating the time variable between the two Equations 16.43 and 16.44, there is the differential relationship

$$2\,c\,dx^1 - 2\,a\,(x^1\,dx^2 + x^2\,dx^1) + 2\,b\,x^2\,dx^2 = 0 \tag{16.45}$$

A first integral of this is evident immediately as

$$c\,(x^1)^2 - 2\,a\,x^1\,x^2 + b\,(x^2)^2 = K \tag{16.46}$$

in which K is an integration constant that depends upon the initial conditions.

Upon differentiating each of Equations 16.43 and 16.44 and using Equation 16.45, it it found that

$$\frac{d^2x^1}{dt^2} + \omega^2\,x^1 = 0 \qquad \omega^2 = b\,c - a^2 \tag{16.47}$$

$$\frac{d^2x^2}{dt^2} + \omega^2\,x^2 = 0 \qquad \omega^2 = b\,c - a^2 \tag{16.48}$$

so that if $\omega^2 > 0$,

$$x^1 = A\,cos(\omega t + p) \tag{16.49}$$
$$x^2 = B\,cos(\omega t + q) \tag{16.50}$$

and if $\omega^2 < 0$,

$$x^1 = C\,Cosh(\omega t + r) \tag{16.51}$$
$$x^2 = D\,Cosh(\omega t + s) \tag{16.52}$$

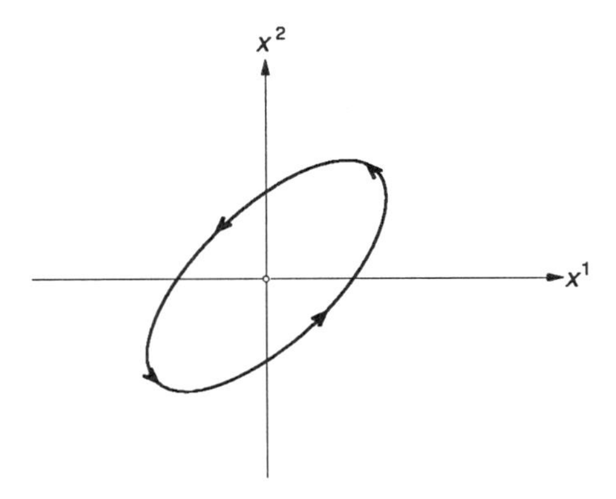

Figure 16.20: Stable Limit Cycle

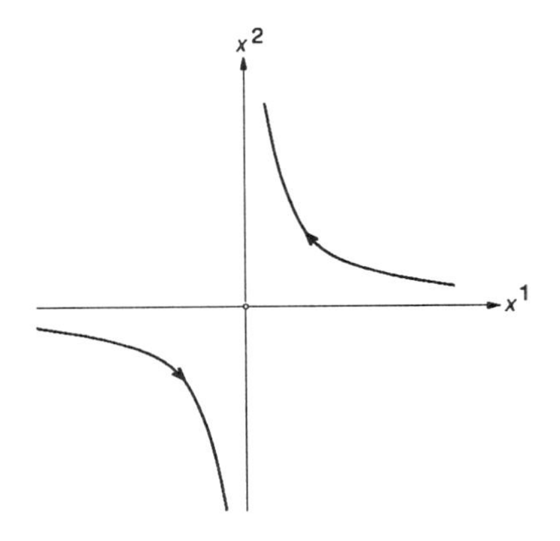

Figure 16.21: Unstable Limit Cycle

In the first case ($\omega^2 > 0$), the motion is a stable oscillation of both variables; in the latter case ($\omega^2 < 0$), the motions of both equantities diverge, and the system is unstable. Equation 16.46, for the stable case, is that of an ellipse, so its plot in phase space shows the limit cycle depicted in Figure 16.20, whereas for the unstable case it is an hyperbola, Figure 16.21. The exact features of the curves are established by the particular values of the constants a, b, c, and it may be recalled that these are in turn functions of a parameter λ; i.e., $a = a(\lambda)$, $b = b(\lambda)$, $c = c(\lambda)$.

Now imagine that an automatic machine is required that will give a stable operation irrespective of the values of a, b, c. Assume that there are extreme values of the variables x^1 and x^2 beyond which a satisfactory performance does not exist. These then constitute constraints, in the usual systems parlance. The constraints establish a fixed boundary in phase space. The additional equipment needed to transform the system into an ultrastable autonomous machine is a set of limit switches and feedback in the parameter λ.

The operation is as follows (See Figures 16.22, 16.23, and 16.24):

- Starting with a parameter λ_1, which gives constants a_1, b_1, c_1, a designer's choice, and under specified initial conditions, the system point moves along an unstable path from P to Q (Figure 16.22), at which time a limit switch on constraint $x^2 \leq C_2$ operates (because the machine has reached its extreme value beyond which it must not work).

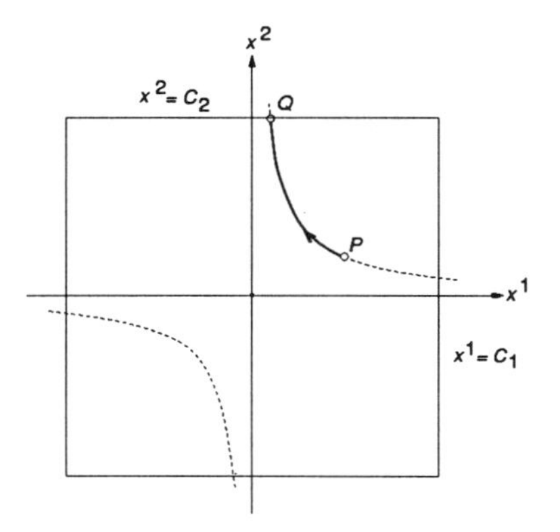

Figure 16.22: Unstable Phase Trajectory

- A new choice of the system parameter λ_2 is made (by feedback) to give new values to the constants a_2, b_2, c_2, and this time the machine selects a stable path Q to R, (Figure 16.23). But the path selected, though stable, still requires the system point to go outside the boundary, which it may not do, so the limit switch on x^1 variable operates. (That is, $x^1 \leq -C_1$.)

- A third choice of λ, namely λ_3, is obtained. For this selection the choice gives rise to constants a_3, b_3, c_3, for which a stable limit cycle exists totally within the acceptable boundary. The system point moves from R to S (Figure 16.24), and thence around the limit cycle ellipse continuously.

In this fashion the machine has "thought" its way forward to a happy function, even to the extent of redesigning itself to fulfill the intended role.

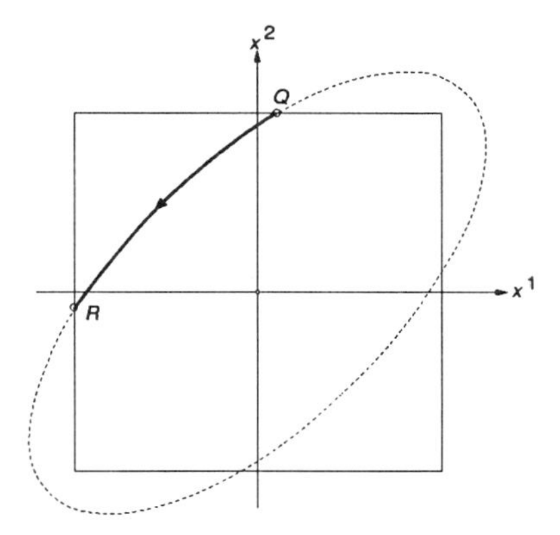

Figure 16.23: Stable But Unsatisfactory Path

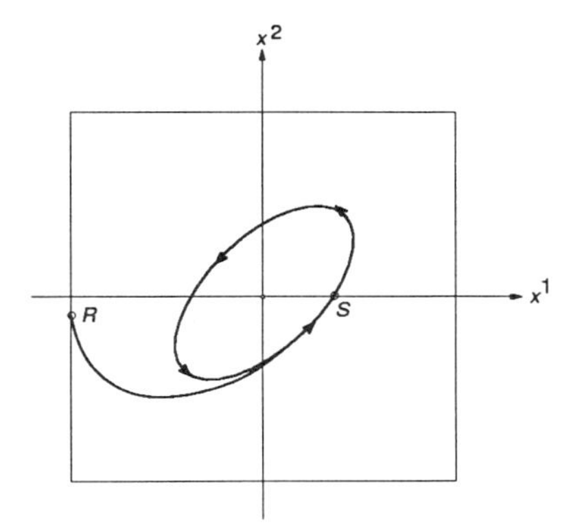

Figure 16.24: Acceptable Stable Limit Cycle

16.6.4 A COMMENT

The role of feedback in systems is to make the machinery automatic, and the role of feedback in automatic machinery is to bring automatic machines a step closer to the human mind.

16.7 PROJECT #14

1. Listed below are several topics which fall in the general category of autonomous systems. Select one of these topics of interest and do the following for the topic selected, or some major branch of it:

   Artificial Intelligence, Self-Designing Systems, Computer-Aided Design, Computer-Aided Engineering, Fuzzy Systems, Optimal Systems, Genetic Algorithms, Pareto-Optimal Systems, Object-Oriented Design, Control Systems, Information Technology.

(a) Find two or three books (textbooks, research monographs, or other reference works) which focus upon the topic.

(b) Locate ten or fifteen major journal articles concerning the topic within the last, say, five years.

(c) Determine from the sources thus collected, who are the originators of scholarship in the topic in question. Then locate the formative papers or books produced by these authors, whatever may be the dates of their production.

(d) Construct a project report which contains the following information:

   i. A section which describes the origins of the topic, including the circumstances of its generation, who were the people most responsible for it, and the scope of its reach.

   ii. A section relating the development of the topic from its inception until very recent times. Follow any branches in the subject as it developed and matured.

   iii. A section which comments upon the applicability of the topic to civil engineering systems and practice. Give the merits and limitations appropriate to it for the purposes at hand.

   iv. Write a summary section which tells the most interesting and valuable aspects of the topic.

   v. Create a bibliography of the subject, including reference books, journal articles, and other reference matter, all of which would permit someone unfamiliar with it to quickly find the prime source material about it.

2. Refer to the EWASA example in Chapter 4, especially to Figures 4.4 and 4.5, which show the Eno River in Orange and Durham Counties.

   (a) Conduct your own personal research concerning the more important features of this river basin from its upper reaches in Northern Orange County to Falls of the Neuse Lake, to find the data necessary to model the stream as a system. You may wish to contact state, regional, county, and municipal agencies to assist you in this task.

   Don't forget the possibilities inherent in newspaper and magazine accounts, as sources of data. You might visit the morgue of *The NEWS & OBSERVER*, Raleigh's newspaper, for feature articles on this topic.

   (b) Using the information you have collected, classify the inputs and outputs of the system model as to their varying degrees of fuzziness. Make similar assessments of the many constraints upon the river basin as a system.

   (c) Imagine that you are the system administrator of EWASA, and you wish to control and optimize the operation of the several components of the system. Try your hand at writing a rule-based algorithm for this purpose.

   (d) Describe how you perceive the measures needed to carry out the program of control and optimization you have modeled.

3. The Neuse River is sometimes characterized as North Carolina's most polluted stream. Its basin traverses a region which incorporates several of the area's fastest growing cities and towns, some of the most fertile agricultural and foretry lands, wetlands and estuaries that are vital to fisheries, and prime recreational and tourism sites.

   (a) Conduct your personal research project to determine the main point sources and distributed origins of pollutants along the river. To do this, you may wish to consult state, regional, county, and municipal sources of information on this topic. The Wilmington office of the U. S. Corps of Engineers will also be a useful source of data. (There is a Raleigh telephone number for the Corps Office.)

       Also, remember the potential of newspapers and other periodicals as prime sources of information through feature articles and series.

   (b) Try your hand at modeling the Neuse basin as a system which can be controlled and optimized. Some inputs and outputs may need to be treated as fuzzy parameters, as well as some constraints. Write a rule-based algorithm for the main elements of the system.

   (c) Describe what you perceive to be the principal problems that must be solved to improve the overall quality of this river basin.

   (d) Comment upon the management policies at the Falls of Neuse Lake, with special reference to the effect of water release in times of (a) excessive flooding during heavy rains (such as those which accompany hurricanes), and (b) in times of drought when the stream flows may need to be augmented to maintain water quality and volume for downstream municipal uses.

4. An historic river in eastern North Carolina is the Tar. From its upper reaches in the lower Piedmont region to its confluence with the sounds near the coast, it passes through a number of towns and cities, vast areas of productive farmlands and forests, and a prime region for tourism and recreation. In 1999 it was the central axis of the most severe flooding on record, in which some communities were essentially destroyed.

   (a) Initiate a personal research project which will enable you to model this river basin as a system which, hopefully, can be controlled and optimized. You will probably wish to consult state, regional, county, and local government sources to obtain data on the flows, pollutant loads, inputs, outflows, and other relevant elements of the basin.

   (b) Try to model the important elements, using fuzzy logic methods and other devices as needed. Then make an attempt to write a rule-based algorithm for the management of the basin.

   (c) Devote a special feature of your report to the Princeville-Tarboro area (and others like it), with special regard to the economics and politics of rebuilding in the wake of the great flood. Comment upon the feasibility of analysis, design, and construction for another potential 500-year event in this region.

5. You are invited to try your hand at creating an ultrastable autonomous system after the fashion illustrated in the last section of this chapter. That is, there are two protagonists engaged in an interactive (feedback) mode as in Equations 16.43 and 16.44. Let the system start from an initial condition that is along an hyperbolic path in phase space, and let the purpose of the system be to advance to a stable operation within prescribed boundaries.

   (a) Design an algorithm that will alter the system coefficients a, b, and c at the moment the system phase point reaches the limiting boundary to new values which will produce a new trajectory that is an ellipse in phase space. No doubt you will initiate this altering event somewhat previous to the system point's actually attaining the limit boundary. You will also wish to have a built-in test of whether the new system's parameters are those of an ellipse, and if not, to take further actions.

       Hint: This could be a ratchet mechanism, or perhaps a rule-based fuzzy logic statement.

   (b) Now design another such algorithm to reduce an oversized ellipse to one which is entirely within the limit boundaries. Actually, it may be necessary to perform several such reductions before coming to a stable envelope. Built-in tests at each level are probably indicated.

   (c) Perhaps you would wish to introduce a dedicated computer for each of the major steps outlined above. Possibly you could design microprocessors for this function.

   (d) Can you suggest an anticipatory mechanism which could possibly predict the proximity of a limiting boundary so that advance preparation could be made for altering the system parameters?

       Hint: Try a series representation of the trajectory of Equation 16.46, in which the present position and present rates of change can be entered to project a new position at a new time increment.

   (e) Can you think of a hardware embodiment of the systems you are designing?

APPENDIX A

COMPUTING THE DISCOUNT RATE

In Chapter 5 it was shown that the Present Worth Uniform Series factor (P-WUS) relates the annual return A on an investment P as $P = A(PWUS)$ where

$$PWUS = \frac{P}{A} = \frac{(1+i)^n - 1}{i(1+i)^n} \tag{A.1}$$

Thus, with

$$f(i) = \frac{P}{A} i\,(1+i)^n - (1+i)^n + 1 \tag{A.2}$$

and

$$f'(i) = \frac{P}{A} n\,i\,(1+i)^{n-1} + \frac{P}{A}(1+i)^n - n\,(1+i)^{n-1} \tag{A.3}$$

the Newton-Raphson algorithm permits the computation of a next (better) value of i_1 from a starting value i_0 as

$$i_1 = i_0 - \frac{f(i_0)}{f'(i_0)} \tag{A.4}$$

The process starts from an initial choice of i_0 and proceeds by iteration until there is a convergence in value between i_{m+1} and i_m to any preset desired degree of accuracy. It is important to start with a value i_0 which is substantially larger than the expected final value of i, so that convergence is thereby guaranteed.

Equation A.2 is of the degree $n + 1$, where n, the number of years, is a variable parameter that depends upon the particular problem. It might normally be expected, therefore, that there would be $n + 1$ roots to be found. However, the nature of the equation is such that there are two fixed singular points, one at $i = 0$, where the value of $f(i)$ is also 0, and one at $i = -1$, where the function equals 1. The generic shape of the graph of the function is shown in Figure A.1.

The shape reveals that there is one negative root, which of course has no significance in money computations, and one positive root, which is the desired one. Sometimes there is a double root at $i = 0$, which indicates that there is an infeasible solution.

In the theory of equations it is shown that for the fourth degree and higher, there is no solution for the roots of such an equation which consists of a finite number of finite operations. In other words, numerical procedures such as the Newton-Raphson algorithm are indicated.

As previously mentioned, it is best to start at a value i_0 such that convergence to the root is assured. It can be seen in Figure A.2 that i_0 should be larger than the root, so that a tangent to the curve at i_0, when extended, will intersect the axis at a place closer to the root than was the starting point. If this rule is followed, the desired convergence is obtained.

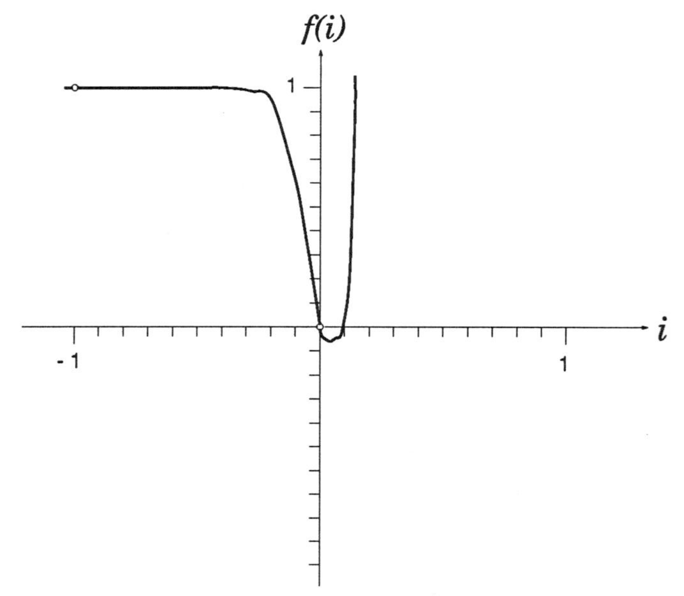

Figure A.1: Discount Factor Equation

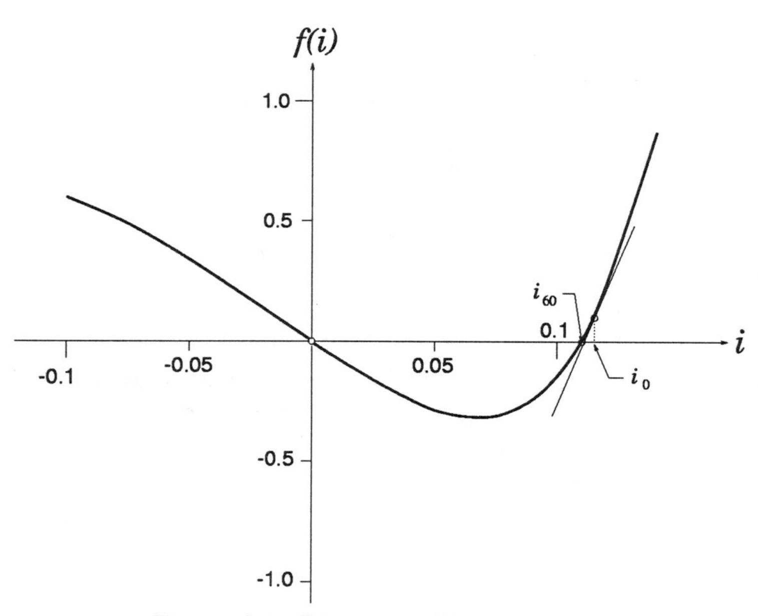

Figure A.2: Selecting a Starting Point

Newton-Raphson Algorithm for Discount Factor

The algorithm can be rendered in all standard computing languages, of course, and is easy to program for hand-held machines as well. Shown below are the keystrokes needed to perform this exercise when the device employs straight logic. There are only minor alterations for those which employ RPN[1] logic.

1 f9 s/s	3	a^x rcl 7 =	3	rcl × 8	2
Enter n		rcl × 0	2	sto chs	2
sto 7	2	rcl × 8	2	rcl 9	2
2 f9 s/s	3	sto ·	2	a^x rcl 7 =	3
Enter P		rcl 9	2	rcl × 8	2
sto 8	2	a^x rcl 7 =	3	sto + chs	2
3 f9 s/s	3	st − ·	2	rcl 9	2
Enter A		1	1	a^x rcl 4 =	3
sto ÷ 8	2	st + ·	2	sto − chs	1
4 f9 s/s	3	rcl 7	2	rcl ·	2
Enter i_0		− 1 =	3	rcl ÷ chs	2
sto 0	2	sto 4	2	chs	3
f lab 1	1	rcl 9	2	rcl + 0 s/s	3
rcl 0	2	a^x rcl 4 =	3	rcl · s/s	3
+ 1 =	3	rcl × 0	2	rcl chs s/s	3
sto 9	2	rcl × 7	2	jmp 1	1
					98

Most of the commands are self-interpreting; a few which may not be are the following.

sto 7	store the contents of the entry register in register 7
rcl 4	recall the contents of register 4 to the entry register
rcl × 8	recall the contents of register 8 and multiply by the entry register amount and leave the result in the entry register
a^x	raise the entry register amount to the power following a^x
chs	change the sign of the contents of the entry register
s/s	start/stop (run/stop)
f lab 1	internal flag — place named "1"
jmp 1	jump to the place named "1"
f 9	external flag — round up entry register amount to integer, and leave result at far left of entry register

The machine should have a capacity of 98 keystrokes to perform this routine.

The entry register contents are noted at each passage of the *s/s* (at the 94th keystroke) and recorded. These data furnish the values of the discount rate to convergence. The procedure is terminated when two successive values differ by less than a pre-set amount. This test can easily be built-in to the program by a conditional if-then-else command sequence if desired and if sufficient capacity is available. A round-off command should be programmed just prior to the test, of course.

[1]Reverse Polish Notation.

Computing the Discount Rate Example

To illustrate the use of this program in computing the discount rate for a uniform series of annual repayments of a principal sum, let the principal be \$60, 000, the annual amounts be \$8, 600, and the time be 14 years. The computation proceeds as follows.

$i_0 = 0.115000000$	$f(i_0) = 0.092600337$	$f'(i_0) = 74.152575973$
$i_1 = 0.113751218$	$f(i_1) = 0.067354272$	$f'(i_1) = 72.549910811$
$i_2 = 0.112822833$	$f(i_2) = 0.049249602$	$f'(i_2) = 71.377702254$
$i_3 = 0.112132847$	$f(i_3) = 0.036154315$	$f'(i_3) = 70.517014937$
$i_4 = 0.111620144$	$f(i_4) = 0.026619971$	$f'(i_4) = 69.883219312$
$i_5 = 0.111239223$	$f(i_5) = 0.019643559$	$f'(i_5) = 69.415483387$
$i_6 = 0.110956238$	$f(i_6) = 0.014519562$	$f'(i_6) = 69.069734103$
$i_7 = 0.110746022$	$f(i_7) = 0.010745444$	$f'(i_7) = 68.813845229$
$i_8 = 0.110589869$	$f(i_8) = 0.007959680$	$f'(i_8) = 68.624289707$
$i_9 = 0.110473880$	$f(i_9) = 0.005900175$	$f'(i_9) = 68.483776969$
$i_{10} = 0.110387726$	$f(i_{10}) = 0.004375785$	$f'(i_{10}) = 68.379565872$
$i_{11} = 0.110323733$	$f(i_{11}) = 0.003246473$	$f'(i_{11}) = 68.302248838$
$i_{12} = 0.110276202$	$f(i_{12}) = 0.002409297$	$f'(i_{12}) = 68.244869230$
$i_{13} = 0.110240898$	$f(i_{13}) = 0.001788381$	$f'(i_{13}) = 68.202277032$
$i_{14} = 0.110214677$	$f(i_{14}) = 0.001327692$	$f'(i_{14}) = 68.170656477$
$i_{15} = 0.110195201$	$f(i_{15}) = 0.000985791$	$f'(i_{15}) = 68.147178608$
$i_{16} = 0.110180735$	$f(i_{16}) = 0.000731998$	$f'(i_{16}) = 68.129745099$
$i_{17} = 0.110169991$	$f(i_{17}) = 0.000543579$	$f'(i_{17}) = 68.116799014$
$i_{18} = 0.110162011$	$f(i_{18}) = 0.000403679$	$f'(i_{18}) = 68.107184829$
$i_{19} = 0.110156084$	$f(i_{19}) = 0.000299795$	$f'(i_{19}) = 68.100044773$
$i_{20} = 0.110151681$	$f(i_{20}) = 0.000222651$	$f'(i_{20}) = 68.094742011$
$i_{21} = 0.110148412$	$f(i_{21}) = 0.000165361$	$f'(i_{21}) = 68.090803692$
$i_{22} = 0.110145983$	$f(i_{22}) = 0.000122814$	$f'(i_{22}) = 68.087878692$
$i_{23} = 0.110144179$	$f(i_{23}) = 0.000091215$	$f'(i_{23}) = 68.085706263$
$i_{24} = 0.110142840$	$f(i_{24}) = 0.000067747$	$f'(i_{24}) = 68.084092764$
$i_{25} = 0.110141845$	$f(i_{25}) = 0.000050317$	$f'(i_{25}) = 68.082894385$
$i_{26} = 0.110141106$	$f(i_{26}) = 0.000037371$	$f'(i_{26}) = 68.082004321$
$i_{27} = 0.110140557$	$f(i_{27}) = 0.000027757$	$f'(i_{27}) = 68.081343246$
$i_{28} = 0.110140149$	$f(i_{28}) = 0.000020615$	$f'(i_{28}) = 68.080852246$
$i_{29} = 0.110139846$	$f(i_{29}) = 0.000015312$	$f'(i_{29}) = 68.080487567$
$i_{30} = 0.110139621$	$f(i_{30}) = 0.000011372$	$f'(i_{30}) = 68.080216709$
$i_{31} = 0.110139454$	$f(i_{31}) = 0.000008446$	$f'(i_{31}) = 68.080015533$
$i_{32} = 0.110139330$	$f(i_{32}) = 0.000006273$	$f'(i_{32}) = 68.079866114$
$i_{33} = 0.110139238$	$f(i_{33}) = 0.000004659$	$f'(i_{33}) = 68.079755136$
$i_{34} = 0.110139169$	$f(i_{34}) = 0.000003460$	$f'(i_{34}) = 68.079672708$
$i_{35} = 0.110139119$	$f(i_{35}) = 0.000002570$	$f'(i_{35}) = 68.079611487$
$i_{36} = 0.110139081$	$f(i_{36}) = 0.000001909$	$f'(i_{36}) = 68.079566016$
$i_{37} = 0.110139053$	$f(i_{37}) = 0.000001418$	$f'(i_{37}) = 68.079532243$
$i_{38} = 0.110139032$	$f(i_{38}) = 0.000001053$	$f'(i_{38}) = 68.079507159$
$i_{39} = 0.110139017$	$f(i_{39}) = 0.000000782$	$f'(i_{39}) = 68.079488528$

$$
\begin{array}{lll}
i_{40} = 0.110139005 & f(i_{40}) = 0.000000581 & f'(i_{40}) = 68.079474690 \\
i_{41} = 0.110138997 & f(i_{41}) = 0.000000431 & f'(i_{41}) = 68.079464412 \\
i_{42} = 0.110138990 & f(i_{42}) = 0.000000324 & f'(i_{42}) = 68.079456779 \\
i_{43} = 0.110138985 & f(i_{43}) = 0.000000238 & f'(i_{43}) = 68.079451109 \\
i_{44} = 0.110138982 & f(i_{44}) = 0.000000176 & f'(i_{44}) = 68.079446897 \\
i_{45} = 0.110138979 & f(i_{45}) = 0.000000131 & f'(i_{45}) = 68.079443770 \\
i_{46} = 0.110138977 & f(i_{46}) = 0.000000097 & f'(i_{46}) = 68.079441447 \\
i_{47} = 0.110138976 & f(i_{47}) = 0.000000072 & f'(i_{47}) = 68.079439721 \\
i_{48} = 0.110138975 & f(i_{48}) = 0.000000053 & f'(i_{48}) = 68.079438440 \\
i_{49} = 0.110138974 & f(i_{49}) = 0.000000039 & f'(i_{49}) = 68.079437488 \\
i_{50} = 0.110138974 & f(i_{50}) = 0.000000029 & f'(i_{50}) = 68.079436781 \\
i_{51} = 0.110138973 & f(i_{51}) = 0.000000022 & f'(i_{51}) = 68.079436256 \\
i_{52} = 0.110138973 & f(i_{52}) = 0.000000016 & f'(i_{52}) = 68.079435866 \\
i_{53} = 0.110138972 & f(i_{53}) = 0.000000012 & f'(i_{53}) = 68.079435576 \\
i_{54} = 0.110138972 & f(i_{54}) = 0.000000009 & f'(i_{54}) = 68.079435361 \\
i_{55} = 0.110138972 & f(i_{55}) = 0.000000006 & f'(i_{55}) = 68.079435201 \\
i_{56} = 0.110138972 & f(i_{56}) = 0.000000004 & f'(i_{56}) = 68.079435082 \\
i_{57} = 0.110138972 & f(i_{57}) = 0.000000003 & f'(i_{57}) = 68.079434994 \\
i_{58} = 0.110138972 & f(i_{58}) = 0.000000002 & f'(i_{58}) = 68.079434929 \\
i_{59} = 0.110138972 & f(i_{59}) = 0.000000002 & f'(i_{59}) = 68.079434880 \\
i_{60} = 0.110138972 & f(i_{60}) = 0.000000001 & f'(i_{60}) = 68.079434817
\end{array}
$$

As can easily be seen from the numbers above, the convergence to a discount rate with 9 decimal accuracy is somewhat slow. The nature of the equation for i is largely responsible for this feature, because of the very steep slope of the curve $f(i)$ as it crosses the i-axis.

Nevertheless, the computation should be carried out to the limit of accuracy of the computer in question, because the discount rate usually operates upon large dollar amounts, at least in systems work, so the last significant figures represent real money.

In any case, with contemporary machines, the computation takes place at speeds which mask the tedium of many large numbers.

APPENDIX B

TREATMENT OF DATA

B.1 INTRODUCTION

The world of science and technology is laden with a plethora of data to be used in analysis, modeling, and design. Contemporary data-gathering methods permit the acquisition of great volumes of information with little effort, so the principal need is to reduce the raw numbers into a significant set of parameters which have meaning for the purposes at hand. The winnowing of the data field into nuggets of golden kernels is the task facing all those who stand at the crucial place in the visceral stream of technological life.

Generic methods employed to accomplish this transformation are subsumed under the description "treatment of data," although the current jargon is more like "data massage."

B.2 LEAST-SQUARES FITTING

One of the venerable methods of reducing data to model form is the technique called "least-squares fitting." There are several aspects of the process which have contributed to its popularity and longevity. For one thing, it is intellectually satisfying in that its basic premise is that an equation or curve can be passed through a data distribution in such a way that the total error of representation is minimimized. In this regard it is attractive to systems analysts because of its optimizing potential.

It is assumed that data are acquired under random error conditions; i.e., that any systematic bias has been suppressed, so that a given data point is as likely to be in error in a positive sense as in a negative way. In addition, the "error" or difference between a data value and the "correct" or model value corresponding to it is squared, to remove the compensating potential of positive plus negative amounts. Thus, only positive (squared) quantities enter the sum of error which is to be minimized.

The process is the mathematical analog of that invoked by a graphical fitting of a data set by eye, when a template is so placed that "as many data points are above it as are below it" and the curve drawn along the edge of the template "fits" the whole field of the data. Actually, the human eye is rather good at

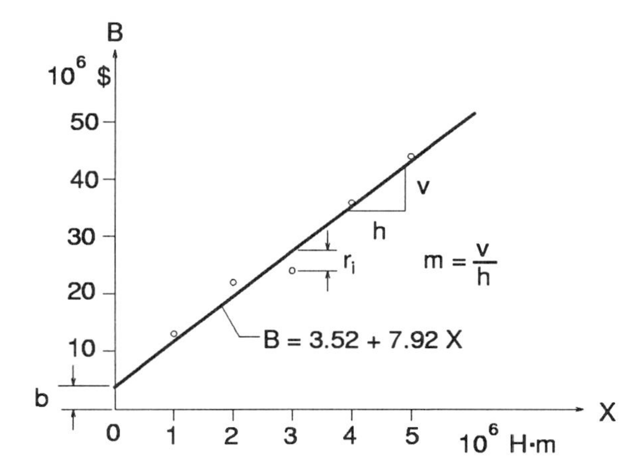

Figure B.1: Benefit versus Reservoir Size

discriminating an optimum graphical fit of this kind, and the fundamental thesis of the least-squares method is reinforced because of it.

B.2.1 THE LINEAR MODEL

Let the analysis begin with fitting a linear model by the least-squares method, with the data as shown in Figure B.1. Each data point consists of a pair of numbers x_i, y_i giving the abscissa and ordinate values. There are n data points in the set. Following the basic assumption of the process, there is a "best-fit" straight line which passes through the set of points that minimizes the error sum of the actual data points as compared to the model straight line.

Thus, if the equation of the line is $y = m\,x + b$ in the customary slope-intercept form, it may be said to be a "two-parameter model." The two parameters are m and b, obviously. The entire purpose of the exercise is to determine m and b, and thus replace the many specific pairs of data with one equation which best represents all of it.

A particular pair of data, the ith, should have the representation

$$y_i = m\,x_i + b \tag{B.1}$$

if it perfectly derives from the linear, two-parameter model. In fact, it will not match the line exactly, and will have some overage or underage or residue that can be given by

$$r_i = y_i - m\,x_i - b \tag{B.2}$$

Now let the residue be squared to make all positive quantities, and add to obtain a measure of the global residue

$$R = \sum_{i=1}^{n} r_i^2 = \sum_{i=1}^{n} (y_i - m\,x_i - b)^2 \tag{B.3}$$

In the calculus the minimum of such a two-parameter system is obtained by differentiating the quantity R with respect to each parameter, then setting the result equal to zero and solving simultaneously for the specific values of m and b. Hence

$$\frac{\partial R}{\partial m} = 2\sum_{i=1}^{n}(y_i - m\,x_i - b)\frac{\partial}{\partial m}(y_i - m\,x_i - b)$$

$$= 2\sum_{i=1}^{n}(y_i - m\,x_i - b)(x_i) = 2\sum_{i=1}^{n}(x_i y_i - m\,x_i^2 - b\,x_i) = 0 \qquad \text{(B.4)}$$

$$\frac{\partial R}{\partial b} = 2\sum_{i=1}^{n}(y_i - m\,x_i - b)\frac{\partial}{\partial b}(y_i - m\,x_i - b)$$

$$= 2\sum_{i=1}^{n}(y_i - m\,x_i - b)(-1) = 2\sum_{i=1}^{n}(-y_i + m\,x_i + b) = 0 \qquad \text{(B.5)}$$

or

$$m\sum_{i=1}^{n}x_i^2 + b\sum_{i=1}^{n}x_i = \sum_{i=1}^{n}x_i\,y_i \qquad \text{(B.6)}$$

$$m\sum_{i=1}^{n}x_i + b\sum_{i=1}^{n}1 = \sum_{i=1}^{n}y_i \qquad \text{(B.7)}$$

These are now two linear simultaneous equations which, when solved, yield

$$m = \frac{\sum_{i=1}^{n}(1)\sum_{i=1}^{n}x_i\,y_i - \sum_{i=1}^{n}x_i\sum_{i=1}^{n}x_i\,y_i}{n\sum_{i=1}^{n}x_i^2 - (\sum_{i=1}^{n}x_i)^2} \qquad \text{(B.8)}$$

$$b = \frac{\sum_{i=1}^{n}x_i^2\sum_{i=1}^{n}y_i - \sum_{i=1}^{n}x_i\sum_{i=1}^{n}x_i\,y_i}{n\sum_{i=1}^{n}x_i^2 - (\sum_{i=1}^{n}x_i)^2} \qquad \text{(B.9)}$$

The denominators of these expressions are identical, and $\sum_{i=1}^{n}(1) = n$ has been used.

B.2.2 RESERVOIR BENEFITS EXAMPLE

The professional staff of the Black Ankle Authority have gathered from other locations where such reservoirs have been built the following information concerning the benefits to be derived from a reservoir of the proposed type:

$$x_1 = 1.01 \quad y_1 = 11.9 \quad x_2 = 2.10 \quad y_2 = 21.3 \quad x_3 = 3.05 \quad y_3 = 24.8$$

$$x_4 = 4.04 \quad y_4 = 36.4 \quad x_5 = 4.98 \quad y_5 = 43.5 \qquad \text{(B.10)}$$

Determine the least-squares straight-line which fits this data set.

Perform the indicated sums below:

$$\sum_{i=1}^{5} x_i = 1.01 + 2.10 + 3.05 + 4.04 + 4.98 = 15.18 \tag{B.11}$$

$$\sum_{i=1}^{5} y_i = 11.9 + 21.3 + 24.8 + 36.4 + 43.5 = 137.9 \tag{B.12}$$

$$\begin{aligned}\sum_{i=1}^{5} x_i^2 &= (1.01)^2 + (2.10)^2 + (3.05)^2 + (4.04)^2 + (4.98)^2 \\ &= 1.02 + 4.41 + 9.30 + 16.32 + 24.80 = 55.85\end{aligned} \tag{B.13}$$

$$\begin{aligned}\sum_{i=1}^{5} x_i\, y_i &= (1.01)(11.9) + (2.10)(21.3) + (3.05)(24.8) + (4.40)(36.4) + (4.98)(43.5) \\ &= 12.02 + 44.73 + 75.64 + 147.06 + 216.63) = 496.03\end{aligned} \tag{B.14}$$

The slope and intercept are:

$$m = \frac{5\,(496.03) - (15.18)(137.9)}{5\,(55.85) - (15.18)^2} = 7.92 \tag{B.15}$$

$$b = \frac{(55.85)(137.9) - (15.18)(496.03)}{5\,(55.85) - (15.18)^2} = 3.52 \tag{B.16}$$

The equation for benefits as a function of reservoir size is

$$B = 3.52 + 7.92\,X \tag{B.17}$$

B.2.3 A SECOND-DEGREE MODEL

Consider next the formula

$$y = a + b\,x + c\,x^2 \tag{B.18}$$

as the proposed three-parameter model for a data field that is assumed to be fitted with an equation of this degree. The a, b, and c are the three parameters which are to be determined by a least-squares technique so that the global error is minimized.

Following the same procedure as that employed previously, the residue at each data point is

$$r_i = y_i - a - b\,x_i - c\,x_i^2 \tag{B.19}$$

so that the total squared residue is

$$R = \sum_{i=1}^{n} r_i^2 = \sum_{i=1}^{n} (y_i - a - b\,x_i - c\,x_i^2)^2 \tag{B.20}$$

The requirement for a global minimum is given by

$$\frac{\partial R}{\partial a} = 2\sum_{i=1}^{n}(y_i - a - b\,x_i - c\,x_i^2)\frac{\partial}{\partial a}(y_i - a - b\,x_i - c\,x_i^2)$$
$$= 2\sum_{i=1}^{n}(y_i - a - b\,x_i - c\,x_i^2)(-1) = 0 \tag{B.21}$$

$$\frac{\partial R}{\partial b} = 2\sum_{i=1}^{n}(y_i - a - b\,x_i - c\,x_i^2)\frac{\partial}{\partial b}(y_i - a - b\,x_i - c\,x_i^2)$$
$$= 2\sum_{i=1}^{n}(y_i - a - b\,x_i - c\,x_i^2)(-x_i) = 0 \tag{B.22}$$

$$\frac{\partial R}{\partial c} = 2\sum_{i=1}^{n}(y_i - a - b\,x_i - c\,x_i^2)\frac{\partial}{\partial c}(y_i - a - b\,x_i - c\,x_i^2)$$
$$= 2\sum_{i=1}^{n}(y_i - a - b\,x_i - c\,x_i^2)(-x_i^2) = 0 \tag{B.23}$$

or

$$a\sum_{i=1}^{n}(1) + b\sum_{i=1}^{n}x_i + c\sum_{i=1}^{n}x_i^2 = \sum_{i=1}^{n}y_i \tag{B.24}$$
$$a\sum_{i=1}^{n}x_i + b\sum_{i=1}^{n}x_i^2 + c\sum_{i=1}^{n}x_i^3 = \sum_{i=1}^{n}x_i\,y_i \tag{B.25}$$
$$a\sum_{i=1}^{n}x_i^2 + b\sum_{i=1}^{n}x_i^3 + c\sum_{i=1}^{n}x_i^4 = \sum_{i=1}^{n}x_i^2\,y_i \tag{B.26}$$

These three equations, simultaneous in a, b, and c, are solved to yield

$$a = \frac{\mathcal{CBG} + \mathcal{FEB} + \mathcal{HAE} - \mathcal{HB}^2 - \mathcal{AFG} - \mathcal{CE}^2}{n\mathcal{BG} + \mathcal{AEB} + \mathcal{BAE} - \mathcal{B}^3 - \mathcal{A}^2\mathcal{G} - n\mathcal{E}^2} \tag{B.27}$$
$$b = \frac{n\mathcal{FG} + \mathcal{AHB} + \mathcal{BEC} - \mathcal{FB}^2 - n\mathcal{HE} - \mathcal{ACG}}{n\mathcal{BG} + \mathcal{AEB} + \mathcal{BAE} - \mathcal{B}^3 - \mathcal{A}^2\mathcal{G} - n\mathcal{E}^2} \tag{B.28}$$
$$c = \frac{n\mathcal{BH} + \mathcal{AEC} + \mathcal{BFA} - \mathcal{B}^2\mathcal{C} - n\mathcal{EF} - \mathcal{A}^2\mathcal{H}}{n\mathcal{BG} + \mathcal{AEB} + \mathcal{BAE} - \mathcal{B}^3 - \mathcal{A}^2\mathcal{G} - n\mathcal{E}^2} \tag{B.29}$$

in which

$$\mathcal{A} = \sum_{i=1}^{n}x_i \tag{B.30}$$
$$\mathcal{B} = \sum_{i=1}^{n}x_i^2 \tag{B.31}$$

$$\mathcal{C} = \sum_{i=1}^{n} y_i \tag{B.32}$$

$$\mathcal{E} = \sum_{i=1}^{n} x_i^3 \tag{B.33}$$

$$\mathcal{F} = \sum_{i=1}^{n} x_i\, y_i \tag{B.34}$$

$$\mathcal{G} = \sum_{i=1}^{n} x_i^4 \tag{B.35}$$

$$\mathcal{H} = \sum_{i=1}^{n} x_i^2\, y_i \tag{B.36}$$

Note that, again, the denominators in the three expressions above are equal.

B.2.4 RESERVOIR COSTS EXAMPLE

The Black Ankle Authority have acquired the following data on the costs of constructing reservoirs as follows:

$$x_1 = 0.97 \quad y_1 = 6.1 \quad x_2 = 2.05 \quad y_2 = 15.9 \quad x_3 = 3.10 \quad y_3 = 21.2$$

$$x_4 = 3.98 \quad y_4 = 40.2 \quad x_5 = 5.06 \quad y_5 = 53.3 \tag{B.37}$$

The task is to determine the equation $y = a + b\,x + c\,x^2$ which best fits this data.

The necessary sums are computed below.

$$\mathcal{A} = \sum_{i=1}^{5} x_i = 15.16 \quad \mathcal{C} = \sum_{i=1}^{5} y_i = 136.7 \quad \mathcal{B} = \sum_{i=1}^{5} x_i^2 = 56.20 \tag{B.38}$$

$$\mathcal{E} = \sum_{i=1}^{5} x_i^3 = 231.92 \quad \mathcal{F} = \sum_{i=1}^{5} x_i\, y_i = 533.93 \tag{B.39}$$

$$\mathcal{G} = \sum_{i=1}^{5} x_i^4 = 1017.36 \quad \mathcal{H} = \sum_{i=1}^{5} x_i^2\, y_i = 2602.88 \tag{B.40}$$

When these are inserted into Formulas B.27, B.28, and B.29, the results are:

$$a = 2.16, \quad b = 2.89, \quad c = 1.46 \tag{B.41}$$

The desired cost equation as a function of reservoir size can then be presented as

$$C = 2.16 + 2.89\,X + 1.46\,X^2 \tag{B.42}$$

B.2.5 UTILITY FUNCTION EXAMPLE

A contractor's risk records have produced the following data from past decisions:

$$\begin{aligned} x_1 &= -\$20,000 \quad y_1 = -100U \quad x_2 = -\$16,750 \quad y_2 = -70U \\ x_3 &= -\$15,000 \quad y_3 = -56U \quad x_4 = -\$11,600 \quad y_4 = -33.6U \\ x_5 &= -\$8,200 \quad y_5 = -16.8U \quad x_6 = -\$5,800 \quad y_6 = -8.4U \end{aligned} \tag{B.43}$$

The origin $x_1 = \$0$, $y_1 = 0U$ is also to be a data point. Again, the task is to determine the equation $y = a + b\,x + c\,x^2$ which best fits this data.

The required sums are displayed below.

$$\mathcal{A} = \sum_{i=1}^{7} x_i = -77,350 \quad \mathcal{B} = \sum_{i=1}^{7} x_i^2 = 1,141,002,500 \quad \mathcal{C} = \sum_{i=1}^{7} y_i = -284.8 \tag{B.44}$$

$$\mathcal{E} = \sum_{i=1}^{7} x_i^3 = -1.838179787 \times 10^{13} \quad \mathcal{F} = \sum_{i=1}^{7} x_i\, y_i = 4,588,740 \tag{B.45}$$

$$\mathcal{G} = \sum_{i=1}^{7} x_i^4 = 3.130995772 \times 10^{17} \quad \mathcal{H} = \sum_{i=1}^{7} x_i^2\, y_i = -78,172,799,000 \tag{B.46}$$

When these are inserted into Formulas B.27, B.28, and B.29, the results are:

$$a = -0.037441096 \quad b = -0.000017849 \quad c = -0.000000250 \tag{B.47}$$

The desired utility function can then be presented as

$$C = -0.037441096 - 0.000017849\,X - 0.000000250\,X^2 \tag{B.48}$$

B.3 THE LOGISTIC CURVE

Fitting the logistic equation to data begins with the formula

$$y = \frac{b}{e^{-at} + c} \tag{B.49}$$

in which a, b, and c are the parameters to be determined. The process is continued by re-writing it in the form

$$y\,e^{-at} + y\,c = b \tag{B.50}$$

and then expanding the exponential factor in a Taylor's series

$$e^{-at} = e^{-a_o t} - \tfrac{1}{1!} t e^{-a_o t}(\delta a) + \tfrac{1}{2!} t^2 e^{-a_o t}(\delta a)^2 - \tfrac{1}{3!} t^3 e^{-a_o t}(\delta a)^3 + \cdots \tag{B.51}$$

about an initial place a_o. That is, $a = a_o + \delta a$, in which δa is a small parameter. Therefore, the starting place value of a_o needs to be a reasonably good approximation to the final value of a.

Then

$$y[e^{-a_o t} - \tfrac{1}{1!} t e^{-a_o t}(\delta a) + \tfrac{1}{2!} t^2 e^{-a_o t}(\delta a)^2 - \tfrac{1}{3!} t^3 e^{-a_o t}(\delta a)^3 + \cdots] + y\,c = b \tag{B.52}$$

and because δa is small, higher powers than the linear term will be neglected. The truncated series is

$$ye^{-a_o t} - y\frac{1}{1!}te^{-a_o t}(\delta a) + y\,c = b \tag{B.53}$$

a form in which a_o is a constant, and the optimizing parameter is δa.

For a given data point Equation B.53 above should be perfectly satisfied, but in fact will not be. Call the residual r_i, so that

$$r_i = y_i e^{-a_o t_i} - y_i t_i e^{-a_o t_i}(\delta a) + y_i\,c - b \tag{B.54}$$

and the global sum of squared residuals is

$$R = \sum_{i=1}^{n} r_i^2 = \sum_{i=1}^{n} \{y_i e^{-a_o t_i} - y_i t_i e^{-a_o t_i}(\delta a) + y_i\,c - b\}^2 \tag{B.55}$$

As with the previous models, the optimum set of parameters to minimize R is obtained by solving simultaneously from

$$\begin{aligned}\frac{\partial R}{\partial \delta a} &= 2\sum_{i=1}^{n}\{y_i e^{-a_o t_i} - y_i t_i e^{-a_o t_i}(\delta a) + y_i\,c - b\}\frac{\partial r_i}{\partial \delta a}\\ &= 2\sum_{i=1}^{n}\{y_i e^{-a_o t_i} - y_i t_i e^{-a_o t_i}(\delta a) + y_i\,c - b\}\{-y_i t_i e^{-a_o t_i}\}\\ &= 2\sum_{i=1}^{n}\{-y_i^2 t_i e^{-2a_o t_i} + y_i^2 t_i^2 e^{-2a_o t_i}(\delta a)\\ &\qquad -y_i^2 t_i c(e^{-a_o t_i}) + b y_i t_i (e^{-a_o t_i})\} = 0\end{aligned} \tag{B.56}$$

$$\begin{aligned}\frac{\partial R}{\partial b} &= 2\sum_{i=1}^{n}\{y_i e^{-a_o t_i} - y_i t_i e^{-a_o t_i}(\delta a) + y_i\,c - b\}\frac{\partial r_i}{\partial b}\\ &= 2\sum_{i=1}^{n}\{y_i e^{-a_o t_i} - y_i t_i e^{-a_o t_i}(\delta a) + y_i c - b\}\{-1\}\\ &= 2\sum_{i=1}^{n}\{-y_i e^{-a_o t_i} + y_i t_i e^{-a_o t_i}(\delta a) - y_i\,c + b\} = 0\end{aligned} \tag{B.57}$$

$$\begin{aligned}\frac{\partial R}{\partial c} &= 2\sum_{i=1}^{n}\{y_i e^{-a_o t_i} - y_i t_i e^{-a_o t_i}(\delta a) + y_i\,c - b\}\frac{\partial r_i}{\partial c}\\ &= 2\sum_{i=1}^{n}\{y_i e^{-a_o t_i} - y_i t_i e^{-a_o t_i}(\delta a) + y_i\,c - b\}\{y_i\}\\ &= 2\sum_{i=1}^{n}\{y_i^2 e^{-a_o t_i} - y_i^2 t_i e^{-a_o t_i}(\delta a) + y_i^2\,c - b y_i\} = 0\end{aligned} \tag{B.58}$$

or

$$(\delta a)\sum_{i=1}^{n} y_i^2 t_i^2 e^{-2a_o t_i} + b\sum_{i=1}^{n} y_i t_i e^{-a_o t_i} - c\sum_{i=1}^{n} t_i y_i^2 = \sum_{i=1}^{n} y_i^2 t_i e^{-2a_o t_i} \quad \text{(B.59)}$$

$$(\delta a)\sum_{i=1}^{n} y_i t_i e^{-a_o t_i} + b\sum_{i=1}^{n}(1) - c\sum_{i=1}^{n} y_i = \sum_{i=1}^{n} y_i e^{-a_o t_i} \quad \text{(B.60)}$$

$$(\delta a)\sum_{i=1}^{n} y_i^2 t_i e^{-a_o t_i} + b\sum_{i=1}^{n} y_i - c\sum_{i=1}^{n} y_i^2 = \sum_{i=1}^{n} y_i^2 e^{-a_o t_i} \quad \text{(B.61)}$$

For convenience, the set of Equations B.59, B.60, B.61 can be written

$$\mathcal{A}\,(\delta a) + \mathcal{B}\,b - \mathcal{C}\,c = \mathcal{D} \quad \text{(B.62)}$$

$$\mathcal{B}\,(\delta a) + n\,b - \mathcal{E}\,c = \mathcal{F} \quad \text{(B.63)}$$

$$\mathcal{C}\,(\delta a) + \mathcal{E}\,b - \mathcal{G}\,c = \mathcal{H} \quad \text{(B.64)}$$

in which

$$\mathcal{A} = \sum_{i=1}^{n} y_i^2\, t_i^2\, e^{-2a_o t_i} \quad \text{(B.65)}$$

$$\mathcal{B} = \sum_{i=1}^{n} y_i\, t_i\, e^{-a_o t_i} \quad \text{(B.66)}$$

$$\mathcal{C} = \sum_{i=1}^{n} y_i^2\, t_i \quad \text{(B.67)}$$

$$\mathcal{D} = \sum_{i=1}^{n} y_i^2\, t_i\, e^{-2a_o t_i} \quad \text{(B.68)}$$

$$\mathcal{E} = \sum_{i=1}^{n} y_i \quad \text{(B.69)}$$

$$\mathcal{F} = \sum_{i=1}^{n} y_i\, e^{-a_o t_i} \quad \text{(B.70)}$$

$$\mathcal{G} = \sum_{i=1}^{n} y_i^2 \quad \text{(B.71)}$$

$$\mathcal{H} = \sum_{i=1}^{n} y_i^2\, e^{-a_o t_i} \quad \text{(B.72)}$$

The simultaneous solution of Equations B.62, B.63, and B.64 produces

$$\delta a = \frac{n(\mathcal{C}\mathcal{H} - \mathcal{D}\mathcal{G}) + \mathcal{B}\mathcal{F}\mathcal{G} + \mathcal{D}\mathcal{E}^2 - \mathcal{C}\mathcal{E}\mathcal{F} - \mathcal{B}\mathcal{E}\mathcal{H}}{n(\mathcal{C}^2 - \mathcal{A}\mathcal{G}) + \mathcal{B}^2\mathcal{G} + \mathcal{A}\mathcal{E}^2 - 2\mathcal{B}\mathcal{C}\mathcal{E}} \quad \text{(B.73)}$$

$$b = \frac{\mathcal{C}^2\mathcal{F} + \mathcal{B}\mathcal{D}\mathcal{G} + \mathcal{A}\mathcal{E}\mathcal{H} - \mathcal{A}\mathcal{F}\mathcal{B} - \mathcal{B}\mathcal{H}\mathcal{C} - \mathcal{C}\mathcal{D}\mathcal{E}}{n(\mathcal{C}^2 - \mathcal{A}\mathcal{G}) + \mathcal{B}^2\mathcal{G} + \mathcal{A}\mathcal{E}^2 - 2\mathcal{B}\mathcal{C}\mathcal{E}} \quad \text{(B.74)}$$

$$c = \frac{n(\mathcal{C}\mathcal{D} - \mathcal{A}\mathcal{H}) + \mathcal{B}^2\mathcal{H} + \mathcal{A}\mathcal{E}\mathcal{F} - \mathcal{B}\mathcal{D}\mathcal{E} - \mathcal{B}\mathcal{C}\mathcal{F}}{n(\mathcal{C}^2 - \mathcal{A}\mathcal{G}) + \mathcal{B}^2\mathcal{G} + \mathcal{A}\mathcal{E}^2 - 2\mathcal{B}\mathcal{C}\mathcal{E}} \quad \text{(B.75)}$$

B.3.1 LOGISTIC EXAMPLE

In Chapter 13 the population of England and Wales is fitted with the logistic equation for the years 1911 through 1971. The three-point fit procedure is employed to obtain the results

$$a = 0.00484651 \tag{B.76}$$
$$b = 35,314,334 \tag{B.77}$$
$$c = -0.02095000 \tag{B.78}$$

The a in this parameter set can be regarded as the first approximation to the better value which may be obtained using the least squares fitting of the logistic curve. In other words, it is a_o to be used in Equations B.65 through B.72. The other data are:

$$t_1 = 0 \qquad y_1 = 36,070,000 \tag{B.79}$$
$$t_2 = 10 \qquad y_2 = 37,887,000 \tag{B.80}$$
$$t_3 = 20 \qquad y_1 = 39,952,000 \tag{B.81}$$
$$t_4 = 40 \qquad y_4 = 43,758,000 \tag{B.82}$$
$$t_5 = 50 \qquad y_5 = 46,072,000 \tag{B.83}$$
$$t_6 = 60 \qquad y_6 = 48,594,000 \tag{B.84}$$

Use these numbers to calculate $\mathcal{A}$, $\mathcal{B}$, $\mathcal{C}$, $\mathcal{D}$, $\mathcal{E}$, $\mathcal{F}$, $\mathcal{G}$, and $\mathcal{H}$. Then enter the results in Equations B.73, B.74, and B.75 to get

$$a = 0.0033173321 \tag{B.85}$$
$$b = 26,007,657 \tag{B.86}$$
$$c = -0.280899108 \tag{B.87}$$

These results are employed in Chapter 14, Equation 14.38 and Figure 14.3, to display the population growth model for England and Wales.

APPENDIX C

THE PEARSON DIFFERENTIAL EQUATION

C.1 THE GENERIC FORM

The frequency distribution illustrated in Figure 13.1 is a generic shape common to many such distributions found in nature. The figure is reproduced below, for convenience. Specifically, the curve starts out on the left with a tangent to the x-axis, rises to a maximum at the mode where it has a horizontal tangent and descends to approach the x-axis asymptotically for large values of x. There are thus three places where it has a horizontal tangent.

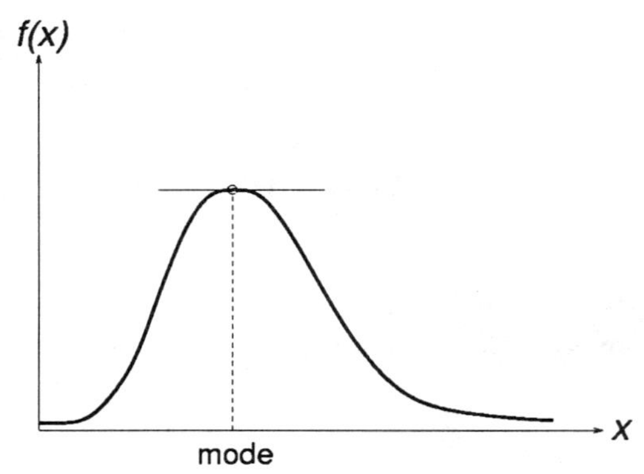

Figure C.1: Generic Frequency Distribution

Pearson devised a generic form of differential equation which could be made to represent the behavior above with the potential of being adjustable to fit a number of different particular circumstances. The fundamental form of this differential equation is

$$\frac{dy}{dx} = \frac{y(x-a)}{F(x)} \tag{C.1}$$

in which

$$F(x) = b_1 + b_1\,x + b_2\,x^2 \cdots \tag{C.2}$$

It is easy to see that dy/dx, the slope, is zero where y is zero (at the far left, the far right), where $x = a$, and where $F(x)$ is very large (where x is large).

The function $F(x)$ is shown as a power series in x, with a sufficient number of adjustable constants (the bs) so that it can be caused to fit a particular data set. In practice, only a few terms are retained because the adaptation to a data set becomes increasingly more difficult and inherently more unstable with increasing numbers of terms. Furthermore, it will be seen that only a few are needed to furnish a good representation.

To determine the coefficients in $F(x)$, first cross-multiply and separate the variables in the differential equation to get

$$\int x^n (b_0 + b_1 x + b_2 x^2 + \cdots)\, dy = \int x^n y (x - a)\, dx \tag{C.3}$$

Now integrate by parts

$$x^n (b_0 - b_1 x + b_2 x^2 + \cdots)\, y| - \int \{n b_0 x^{n-1} + (n+1) b_1 x^n + (n+2) b_2 x^{n+1} + \cdots\}\, y\, dy$$
$$= \int y x^{n+1}\, dy - \int a y x^n\, dx \tag{C.4}$$

and note that the first term vanishes at both the upper and lower limits because y is zero there. The remainder yields the following recurrence relation

$$-n b_0 \mu_{n-1} - (n+1) b_1 \mu_n - (n+2) b_2 \mu_{n+1} - \cdots = \mu_{n+1} - a \mu_n \tag{C.5}$$

in which the n-th moment is

$$\mu_n = \int y x^n\, dx \tag{C.6}$$

Now put $n = 0, 1, 2, 3$, respectively and get four equations

$$a \mu_0 + 0 b_0 + b_1 \mu_0 + 2 b_2 \mu_1 + \cdots = -\mu_1 \tag{C.7}$$
$$a \mu_1 + b_0 \mu_0 + 2 b_1 \mu_1 + 3 b_2 \mu_2 + \cdots = -\mu_2 \tag{C.8}$$
$$a \mu_2 + 2 b_0 \mu_1 + 3 b_1 \mu_2 + 4 b_2 \mu_3 + \cdots = -\mu_3 \tag{C.9}$$
$$a \mu_3 + 3 b_0 \mu_2 + 4 b_1 \mu_3 + 5 b_2 \mu_4 + \cdots = -\mu_4 \tag{C.10}$$

Set $\mu_1 = 0$ (this puts the origin of coordinates at the mode) and keep the three parameters b_0, b_1, and b_2. Solve the three independent simultaneous equations which remain to get

$$a + b_1 = 0 \qquad b_0 + 3 b_2 \mu_2 = -\mu_2 \qquad a \mu_2 + 3 b_1 \mu_2 + 4 b_2 \mu_3 = -\mu_3 \tag{C.11}$$

$$a \mu_3 + 3 b_0 \mu_2 + 4 b_1 \mu_3 + 5 b_2 \mu_4 = -\mu_4 \tag{C.12}$$

With

$$\beta_1 = \frac{\mu_3}{\mu_2^{3/2}} \qquad \beta_2 = \frac{\mu_4}{\mu_2^2} \tag{C.13}$$

and using the solution of the above simultaneous equations for the bs in the function $F(x)$ yields

$$\frac{1}{y}\frac{dy}{dx} = -\frac{x + \frac{\sqrt{\mu_2}\sqrt{\beta_1}(\beta_2+3)}{2(5\beta_2-6\beta_1-9)}}{\frac{\mu_2(4\beta_2-3\beta_1)+\sqrt{\mu_2}\sqrt{\beta_1}(\beta_2+3)\,x+(2\beta_2-3\beta_1-6)\,x^2}{2(5\beta_2-6\beta_1-9)}} \tag{C.14}$$

which can be simplified to

$$\frac{d\log y}{dx} = \frac{x + a}{b_0 + b_1\, x + b_2\, x^2} \tag{C.15}$$

C.2 PEARSON-III STATISTICS

For the case when $b_2 = 0$, the differential equation is integrated to give

$$\log y = \int \frac{x + a}{b_0 + b_1\, x}\, dx = \frac{x}{b_1} + \frac{1}{b_1}(a - \frac{b_0}{b_1}) \log(b_1\, x + b_0) + \text{constant} \tag{C.16}$$

Upon taking the anti-log of both sides, the result is

$$y = y_1 e^{x/b_1}\, (b_1\, x + b_0)^{(a - b_0/b_1)b_1} \tag{C.17}$$

Translate the origin of the coordinate system to the mode to obtain

$$y = y_0\, e^{-\gamma x}\, (1 + \frac{x}{a})^{\gamma a} \qquad x \geq -a \tag{C.18}$$

or, using the standard variable

$$t = \frac{x - \overline{x}}{s_x} \qquad x \geq -A \tag{C.19}$$

there is

$$y = K\, (A + t)^{A^2 - 1}\, e^{-At} \tag{C.20}$$

This is the Pearson Type III frequency distribution.

In the equation above for the Pearson Type III frequency distribution, the parameters $\overline{x}$, s_x, and A are determined from the first, second, and third moments of the data, respectively. There is therefore one remaining parameter—namely, K—to be ascertained. Its value is set by the normalizing condition that the area under the frequency distribution curve must be equal to unity, so that the certainty criterion is fulfilled. In other words, it is certain that any given possible value of T will be found somewhere under the curve between the lower and upper limits of an integral of y over the range.

To compute K, make a change of variable $t = u - A$ so that

$$\int_{-A}^{\infty} y(t)\, dt = 1 = K \int_0^{\infty} u^{A^2-1}\, e^{-Au+A^2}\, du = K e^{A^2} \int_0^{\infty} u^{A^2-1}\, e^{-Au}\, du \tag{C.21}$$

Now integrate by parts with $u = v/A$, $du = dv/A$ to obtain

$$1 = K\, e^{A^2} \int_0^{\infty} \frac{v^{A^2-1}}{A^{A^2-1}}\, e^{-v}\, \frac{dv}{A} = K\, \frac{e^A}{A^{A^2}} \int_0^{\infty} v^{A-1}\, e^{-v}\, dv \tag{C.22}$$

The integral on the right is recognized as that of the Gamma function with the argument A^2, so K can be determined as

$$K = \frac{A^{A^2}}{e^A\, \Gamma(A^2)} \tag{C.23}$$

C.3 CUMULATIVE PROBABILITY

The Pearson-III frequency distribution derived above is valid on the range

$$-x/s_x \leq t \leq x \tag{C.24}$$

Therefore the cumulative probability up to an indefinite upper limit x is given by the integral

$$P(x) = \int_{-x/s_x}^{x} y\, dt = \int_{\beta}^{x} K(A+t)^{A^2-1}\, e^{-At}\, dt \tag{C.25}$$

in which $\beta = -x/s_x$ has been substituted for convenience in lieu of the lower limit. It may be expected that this integral is a form of the indefinite Gamma function with the argument x, and this is indeed the case. Up to a point, a procedure similar to that invoked in the case of the Gamma function of argument A^2 can be followed. Thus, make the change of variable

$$t = u - A \quad dt = du \quad t = x \quad u = x + A \quad t = \beta \quad u = \beta + A \tag{C.26}$$

so that the integral becomes

$$P(x) = K\, e^{A^2} \int_{\beta A+A^2}^{xA+A^2} u^{A^2-1}\, e^{-Au}\, du \tag{C.27}$$

Perform a transformation as in the former case to get

$$P(x) = K\, e^{A}/A^{A^2} \int_{\beta A+A^2}^{xA+A^2} v^{A^2-1}\, e^{-v}\, dv = \frac{1}{\Gamma(A^2)} \int_{\beta A+A^2}^{xA+A^2} v^{A^2-1}\, e^{-v}\, dv \tag{C.28}$$

Unfortunately, there seems to be no closed-form version of this integral, since the upper limit is indefinite. Instead, the preferred procedure appears to be a direct numerical computation based upon a series expansion of the exponential function under the integral sign. Accordingly, the series

$$e^{-v} = 1 - v + \tfrac{1}{2!}\, v^2 - \tfrac{1}{3!}\, v^3 + \cdots \tag{C.29}$$

is substituted and the integration is performed term-by-term to yield

$$\begin{aligned}
P(x) = \left\{ \frac{e^{A^2}\, a^2}{\sqrt{2\pi}\, Q(A)} \right\} \Bigg\{ & \frac{1}{A^2} \left[\left(1 + \frac{x}{A}\right)^{A^2} - \left(1 + \frac{\beta}{A}\right)^{A^2} \right] \\
& - \frac{A^2}{1!(A^2+1)} \left[\left(1 + \frac{x}{A}\right)^{A^2+1} - \left(1 + \frac{\beta}{A}\right)^{A^2+1} \right] \\
& + \frac{A^4}{2!(A^2+2)} \left[\left(1 + \frac{x}{A}\right)^{A^2+2} - \left(1 + \frac{\beta}{A}\right)^{A^2+2} \right] \\
& - \frac{A^6}{3!(A^2+3)} \left[\left(1 + \frac{x}{A}\right)^{A^2+3} - \left(1 + \frac{\beta}{A}\right)^{A^2+3} \right] + \cdots \Bigg\}
\end{aligned} \tag{C.30}$$

In this form it can easily be seen that each term of the expansion can be generated from the preceding one by operating upon it in an iterative manner, so that large exponents and factorials are avoided. However, a large number of terms is still required for convergence.

APPENDIX D

THE PERT FORMULA

The authors of the original PERT Summary Report chose to represent the frequency distribution associated with the expected activity time as that of the Beta function. In terms of the variable t, this is

$$f(t) = k\,(t-a)^{\alpha}(b-t)^{\gamma} \tag{D.1}$$

where a is the optimistic time, b is the pessimistic time, and k, α, and γ are functions of a, b, and m, the most frequent time. Figure D.1 shows the generic form of this function.

This distribution is transformed into the canonical Beta function form by the relationship

$$x = \frac{t-a}{b-a} \tag{D.2}$$

so that

$$f(x) = \frac{x^{\alpha}(1-x)^{\gamma}}{B(\alpha+1,\gamma+1)} \tag{D.3}$$

is that form.

The most frequent x (the mode) is denoted x_{max}, so that

$$x_{\text{max}} = \frac{m-a}{b-a} \tag{D.4}$$

Then

$$x_{max} = \frac{\alpha}{\alpha+\gamma} \tag{D.5}$$

$$e_x = \frac{\alpha+1}{\alpha+\gamma+2} \tag{D.6}$$

$$s_x = \frac{(\alpha+1)(\gamma+1)}{(\alpha+\gamma+2)^2(\alpha+\gamma+3)} \tag{D.7}$$

where e_x (actually, $\overline{x}$), is the expected value of x and s_x is the variance.

The PERT authors also named the variance of t as

$$s_t = \frac{(b-a)^2}{36} \tag{D.8}$$

or

$$\sigma_t = \sqrt{s_t} = \frac{b-a}{6} \tag{D.9}$$

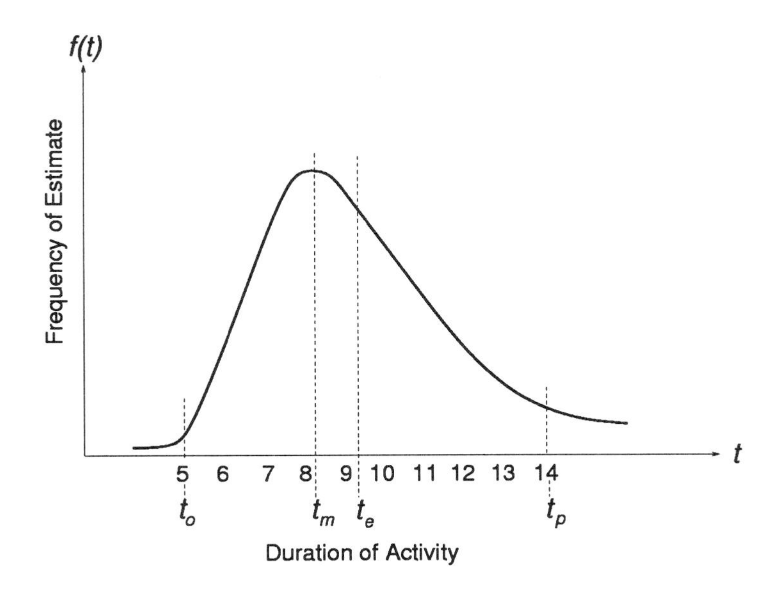

Figure D.1: Beta Function Frequency Distribution

hence

$$s_x = \tfrac{1}{36} \tag{D.10}$$

Upon solving Equation D.5 for γ and inserting this into Equation D.7, there is the result

$$\alpha^3 + (36\, x_{\max}^3 - 36\, x_{\max}^2 + 7\, x_{\max})\, \alpha^2 - 20\, x_{\max}^2\, \alpha - 24\, x_{\max}^3 = 0 \tag{D.11}$$

Therefore, when m, a, and b are furnished, $x_{\max}$, α, γ, e_x, and e_t can be computed, in that order. Note that

$$e_t = a + (b - a)\, e_x \tag{D.12}$$

Note first that γ can be solved for in terms of α and $x_{\max}$ from Equation D.5 as

$$\gamma = \alpha \left(\frac{1 - x_{\max}}{x_{\max}} \right) \tag{D.13}$$

so that

$$e_x = \frac{(\alpha + 1)\, x_{\max}}{x_{\max}\, (\alpha + 2) + \alpha\, (1 - x_{\max})} \tag{D.14}$$

by using Equation D.6.

For example, if $x_{\max} = 1/3$, say, then

$$\alpha^3 + [36(1/3)^3 - 36(1/3)^2 + 7(1/3)]\, \alpha^2 - 20\, (1/2)^2\, \alpha - 24\, (1/3)^3 = 0 \tag{D.15}$$

$$\alpha^3 - 0.3333\, \alpha^2 - 2.2222\, \alpha - 0.8888 = 0 = f(\alpha) \tag{D.16}$$

$$3\, \alpha^2 - 0.6666\, \alpha - 2.2222 = f'(\alpha) \tag{D.17}$$

$$6\, \alpha_i - 0.6666 = 0 \tag{D.18}$$

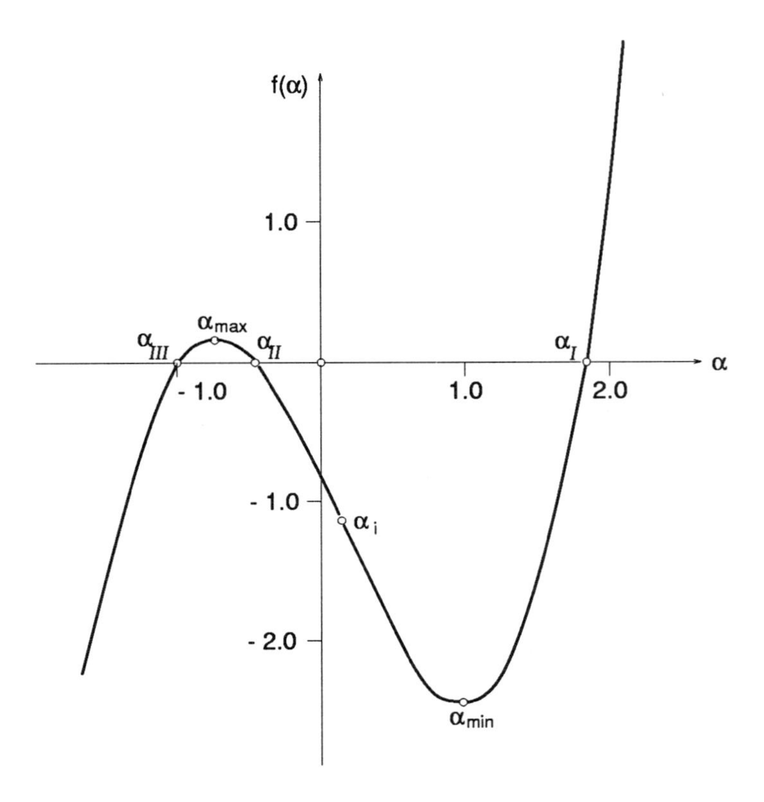

Figure D.2: $f(\alpha)$ versus α

The positive root of this cubic equation can be found using a Newton-Raphson algorithm. Figure D.2 shows a graph of the function $f(\alpha)$, with the maximum, minimum, and inflection point obtained from Equations D.17 and D.18. With the starting value $\alpha_0 = 3$, the iteration proceeds as follows:

$$
\begin{aligned}
\alpha_1 &= \alpha_o - \frac{f(\alpha_0)}{f'(\alpha_0)} = 3 - \frac{16.4444}{22.7777} = 3 - 0.721951219 = 2.278048780 \\
\alpha_2 &= 2.278048780 - \frac{4.140893448}{11.827597329} = 2.278048780 - 0.350104364 = 1.927944415 \\
\alpha_3 &= 1.927944415 - \frac{0.753911217}{7.643390512} = 1.927944415 - 0.098635705 = 1.829308710 \\
\alpha_4 &= 1.829308710 - \frac{0.052068299}{6.597349708} = 1.829308710 - 0.007892305 = 1.821416404 \\
\alpha_5 &= 1.821416404 - \frac{0.000320580}{6.516173331} = 1.821416404 - 0.000049197 = 1.821367206 \\
\alpha_6 &= 1.821367206 - \frac{0.000000012}{6.515668480} = 1.821367206 - 0.000000001 = 1.821367205 \\
\alpha_I &\to 1.821367205
\end{aligned}
\tag{D.19}
$$

The other two roots can be found by dividing Equation D.15 by $\alpha - 1.821367205$

and using the quadratic formula on the result.

$$\begin{array}{r l}
 & \alpha^2 + 1.488033871\,\alpha + 0.488033871 \\
\alpha - 1.821367205 & \overline{\big)\,\alpha^3 - 0.3333\,\alpha^2 - 2.2222\,\alpha - 0.8888} \\
 & \underline{\alpha^3 - 1.821367205\,\alpha^2} \\
 & \quad 1.488033871\,\alpha^2 - 2.2222\,\alpha \\
 & \quad \underline{1.488033871\,\alpha^2 - 2.710256093\,\alpha} \\
 & \qquad\qquad 0.488033871\,\alpha - 0.8888 \\
 & \qquad\qquad \underline{0.488033871\,\alpha - 0.8888}
\end{array}$$

$$\alpha^2 + 1.488033871\,\alpha + 0.488033871 = 0$$

$$\begin{aligned}
\alpha_{II,III} &= -\tfrac{1}{2}(1.488033871) \pm \tfrac{1}{2}\sqrt{(1.488033871)^2 - 4(1)(0.488033871)} \\
&= -0.744016935 \pm \tfrac{1}{2}\sqrt{2.214244803 - 1.952135486} \\
&= -0.744016935 \pm \tfrac{1}{2}\sqrt{0.262109316} \\
&= -0.744016935 \pm \tfrac{1}{2}(0.511966128) \\
&= -0.744016935 \pm 0.255983064 \\
&= -0.488033871, -1.0000000
\end{aligned}$$

Next,

$$e_x = \frac{(1.8214 + 1)(1/3)}{(1/3)(1.8214 + 2) + 1.8214(1 - 1/3)} = 0.377991532 \tag{D.20}$$

$$x_{\max} = \frac{\alpha}{\alpha + \gamma} = 1/3 \qquad \gamma = 2\,\alpha = 3.642734410 \tag{D.21}$$

Other values of $x_{\max}$ can be chosen in the range 0 to 1/2, and the corresponding αs and e_xs computed in the same fashion.

An elementary relationship between $x_{\max}$ and e_x is displayed in Table D.1:

Table D.1

$x_{\max}$	α	e_x	$\frac{4\,x_{\max}+1}{6}$
0	0.0000	0.2053	0.1667
1/8	0.4774	0.2539	0.2500
1/4	1.2174	0.3228	0.3333
1/3	1.8214	0.3780	0.3889
1/2	3.0000	0.5000	0.5000

The function $(4\,x_{\max} + 1)/6$ is included for comparison with e_x, and the comparison in Table D.1 and in Figure D.3 shows that the function $(4\,x_{\max} + 1)/6$ is approximately equal to e_x for $x_{\max} \leq 1/2$. The original PERT authors were seeking such a simplified function, which would allow practical computations of e_x without the necessity of solving the cubic equation in each instance.

Hence,

$$e_x \cong \tfrac{1}{6}\,(4\,x_{\max} + 1) \tag{D.22}$$

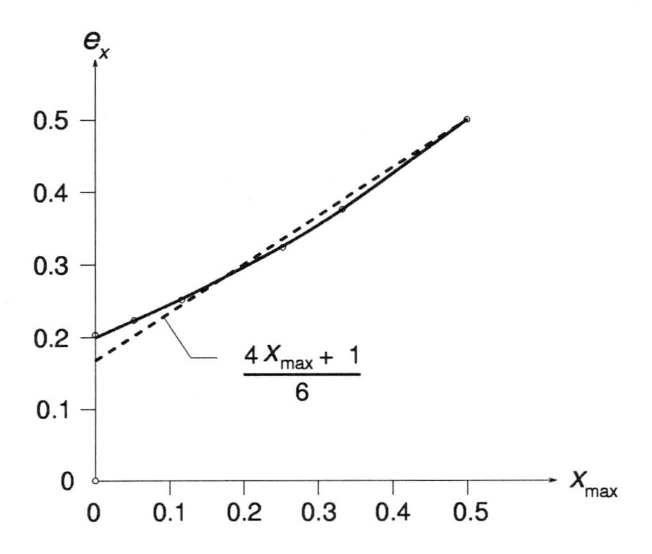

Figure D.3: e_x versus $x_{\max}$

and by substituting from Equation D.4 and Equation D.12 into Equation D.13, there is

$$e_t = \tfrac{1}{6}\,(a + 4\,m + b) \tag{D.23}$$

This is the famous PERT formula for the expected time of an activity.

The Beta function, or Euler's integral of the first kind, is a higher transcendental of the same family as the Gamma function or generalized factorial function. Possibly the most available references are Karl Pearson, Editor, ***Tables of the Incomplete Beta Function***, Cambridge University Press, 1945; and E. S. Pearson and H. O. Hartley, Editors, ***Biometrika Tables for Statisticians***, Cambridge University Press, 1954, Volume I, Table 16, "Tables of Percentage Points of the Incomplete Beta Function."

Specifically, the ***Incomplete Beta Function*** is

$$B_r(p,q) = \int_0^r x^{p-1}\,(1-x)^{q-1}\,dx \qquad 0 \le r \le 1 \tag{D.24}$$

and the ***Complete Beta Function*** is

$$B(p,q) = \int_0^1 x^{p-1}\,(1-x)^{q-1}\,dx \tag{D.25}$$

APPENDIX E

MATRICES

E.1 MATRICES

A matrix is an array of numbers such as that which is shown in Equation E.1, below.

$$\mathbf{A} = \begin{bmatrix} a_{11} & a_{12} & \cdot & \cdot & \cdot & a_{1n} \\ a_{21} & a_{22} & \cdot & \cdot & \cdot & a_{2n} \\ \cdot & \cdot & \cdot & \cdot & \cdot & \cdot \\ \cdot & \cdot & \cdot & \cdot & \cdot & \cdot \\ \cdot & \cdot & \cdot & \cdot & \cdot & \cdot \\ a_{m1} & a_{m2} & \cdot & \cdot & \cdot & a_{mn} \end{bmatrix} \tag{E.1}$$

It has m rows and n columns, and is usually described as an $m \times n$ (m by n) matrix. The product $m \times n$ is called the ***dimension*** of the matrix, and if $m = n$, it is called a ***square matrix***.

The indices i and j indicate the location of the element a_{ij} in the ith row and jth column of the array. The matrix itself is often given as $\mathbf{A} = [a_{ij}]$, $i = 1, 2, \ldots, m$, $j = 1, 2, \ldots, n$. Other matrices, those such as $\mathbf{B}$ and $\mathbf{C}$, would customarily be written $\mathbf{B} = [b_{ij}]$, $i = 1, 2, \ldots, m$, $j = 1, 2, \ldots, n$ and $\mathbf{C} = [c_{ij}]$, $i = 1, 2, \ldots, m$, $j = 1, 2, \ldots, n$.

If a matrix has only one row, then it is usually called a ***row matrix***, as with $\mathbf{R} = [r_1\ r_2 \cdots r_n]$, and if it has but a single column, it is named a ***column matrix***,

$$\mathbf{C} = \begin{bmatrix} c_1 \\ c_2 \\ \cdot \\ \cdot \\ \cdot \\ c_m \end{bmatrix}$$

Row matrices and column matrices are especially useful in describing vectors, since the elements of the rows or columns can be taken to be the components of the vectors. It is most important, however, to distinguish between matrices and vectors. While vectors can, for convenience, be written in a matrix form, either row or column, the matrix must not be identified as the vector itself, since a matrix is only an array of numbers, while a vector usually carries some physical dimensions as well. Moreover, not all matrices are vectors by any means, as the constitutive formulas, incidence matrices, continuity laws, in Chapter 3, and the link-node incidence matrices in Chapter 6 illustrate.

Associated with a matrix is another matrix called the *transpose*, which is obtained by interchanging (transposing) the rows and columns of the matrix. Thus

$$\mathbf{A}^t = \begin{bmatrix} a_{11} & a_{21} & \cdot & \cdot & \cdot & a_{m1} \\ a_{12} & a_{22} & \cdot & \cdot & \cdot & a_{m2} \\ \cdot & \cdot & \cdot & \cdot & \cdot & \cdot \\ \cdot & \cdot & \cdot & \cdot & \cdot & \cdot \\ \cdot & \cdot & \cdot & \cdot & \cdot & \cdot \\ a_{1n} & a_{2n} & \cdot & \cdot & \cdot & a_{mn} \end{bmatrix} \tag{E.2}$$

There is the *zero* or *null* matrix

$$\mathbf{0} = \begin{bmatrix} 0 & 0 & \cdot & \cdot & \cdot & 0 \\ 0 & 0 & \cdot & \cdot & \cdot & 0 \\ \cdot & \cdot & \cdot & \cdot & \cdot & \cdot \\ \cdot & \cdot & \cdot & \cdot & \cdot & \cdot \\ \cdot & \cdot & \cdot & \cdot & \cdot & \cdot \\ 0 & 0 & \cdot & \cdot & \cdot & 0 \end{bmatrix} \tag{E.3}$$

and the *unit* or *identity* matrix

$$\mathbf{I} \equiv \mathbf{1} = \begin{bmatrix} 1 & 0 & \cdot & \cdot & \cdot & 0 \\ 0 & 1 & \cdot & \cdot & \cdot & 0 \\ \cdot & \cdot & \cdot & \cdot & \cdot & \cdot \\ \cdot & \cdot & \cdot & \cdot & \cdot & \cdot \\ \cdot & \cdot & \cdot & \cdot & \cdot & \cdot \\ 0 & 0 & \cdot & \cdot & \cdot & 1 \end{bmatrix} \tag{E.4}$$

E.1.1 MATRIX ARITHMETIC

Two matrices are equal, $\mathbf{A} = \mathbf{B}$, if they are of the same dimension and if each element of $\mathbf{A}$ equals its corresponding element of $\mathbf{B}$. That is, if $a_{ij} = b_{ij}$.

The sum of two matrices is obtained as $\mathbf{C} = \mathbf{A} + \mathbf{B}$ or

$$\begin{bmatrix} c_{11} & c_{12} \\ c_{21} & c_{22} \\ c_{31} & c_{32} \end{bmatrix} = \begin{bmatrix} a_{11} & a_{12} \\ a_{21} & a_{22} \\ a_{31} & a_{32} \end{bmatrix} + \begin{bmatrix} b_{11} & b_{12} \\ b_{21} & b_{22} \\ b_{31} & b_{32} \end{bmatrix}$$

$$= \begin{bmatrix} a_{11} + b_{11} & a_{12} + b_{12} \\ a_{21} + b_{21} & a_{22} + b_{22} \\ a_{31} + b_{31} & a_{32} + b_{32} \end{bmatrix} \tag{E.5}$$

if $\mathbf{A}$ and $\mathbf{B}$ are of the same dimension, and then

$$c_{11} = a_{11} + b_{11} \tag{E.6}$$

$$c_{12} = a_{12} + b_{12} \tag{E.7}$$

$$c_{21} = a_{21} + b_{21} \quad \text{(E.8)}$$
$$c_{22} = a_{22} + b_{22} \quad \text{(E.9)}$$
$$c_{31} = a_{31} + b_{31} \quad \text{(E.10)}$$
$$c_{32} = a_{32} + b_{32} \quad \text{(E.11)}$$

Subtraction of matrices is accomplished as negative addition, so $\mathbf{C} = \mathbf{A} - \mathbf{B}$ with

$$c_{11} = a_{11} - b_{11} \quad \text{(E.12)}$$
$$c_{12} = a_{12} - b_{12} \quad \text{(E.13)}$$
$$c_{21} = a_{21} - b_{21} \quad \text{(E.14)}$$
$$c_{22} = a_{22} - b_{22} \quad \text{(E.15)}$$
$$c_{31} = a_{31} - b_{31} \quad \text{(E.16)}$$
$$c_{32} = a_{32} - b_{32} \quad \text{(E.17)}$$

The sum of k matrices which are identical is

$$\mathbf{B} = \mathbf{A} + \mathbf{A} + \cdots + \mathbf{A} = k\,\mathbf{A} \quad \text{(E.18)}$$

since

$$b_{11} = a_{11} + a_{11} + \cdots + a_{11} = k\,a_{11} \quad \text{(E.19)}$$
$$b_{12} = a_{12} + a_{12} + \cdots + a_{12} = k\,a_{12} \quad \text{(E.20)}$$
$$b_{21} = a_{21} + a_{21} + \cdots + a_{21} = k\,a_{21} \quad \text{(E.21)}$$
$$b_{22} = a_{22} + a_{22} + \cdots + a_{22} = k\,a_{22} \quad \text{(E.22)}$$
$$b_{31} = a_{31} + a_{31} + \cdots + a_{31} = k\,a_{31} \quad \text{(E.23)}$$
$$b_{32} = a_{32} + a_{32} + \cdots + a_{32} = k\,a_{32} \quad \text{(E.24)}$$

Therefore, multiplying a *scalar* times a matrix $\mathbf{A}$ gives a new matrix each of whose elements is k times those of $\mathbf{A}$. If $k = -1$, there is $-\mathbf{A} = [-a_{ij}]$.

Multiplying matrix $\mathbf{A}$ by matrix $\mathbf{B}$ requires that the two matrices be *commensurate*; which is, *the number of rows in* $\mathbf{A}$ *must equal the number of columns in* $\mathbf{B}$. Then, for example,

$$\mathbf{C} = \mathbf{AB} = \begin{bmatrix} a_{11} & a_{12} & a_{13} \\ a_{21} & a_{22} & a_{23} \end{bmatrix} \begin{bmatrix} b_{11} & b_{12} \\ b_{21} & b_{22} \\ b_{31} & b_{32} \end{bmatrix} \quad \text{(E.25)}$$

in which

$$c_{11} = a_{11}b_{11} + a_{12}b_{21} + a_{13}b_{31} \quad \text{(E.26)}$$
$$c_{12} = a_{11}b_{12} + a_{12}b_{22} + a_{13}b_{32} \quad \text{(E.27)}$$
$$c_{21} = a_{21}b_{11} + a_{22}b_{21} + a_{23}b_{31} \quad \text{(E.28)}$$
$$c_{22} = a_{21}b_{12} + a_{22}b_{22} + a_{23}b_{32} \quad \text{(E.29)}$$

The addition, subtraction, and multiplication procedures described above can be summarized in the following:

1. Distributive operations

$$(\mathbf{a}+\mathbf{B})\mathbf{C} = \mathbf{AC}+\mathbf{BC} \tag{E.30}$$
$$\mathbf{A}(\mathbf{B}+\mathbf{C}) = \mathbf{AB}+\mathbf{AC} \tag{E.31}$$
$$(a+b)\mathbf{B} = a\mathbf{B}+b\mathbf{B} \tag{E.32}$$
$$a(\mathbf{A}+\mathbf{B}) = a\mathbf{A}+a\mathbf{B} \tag{E.33}$$

2. Associative operations

$$(k\mathbf{A})\mathbf{B} = k(\mathbf{AB}) = \mathbf{A}(k\mathbf{B}) \tag{E.34}$$
$$\mathbf{A}(\mathbf{BC}) = (\mathbf{AB})\mathbf{C} \tag{E.35}$$
$$(\mathbf{A}+\mathbf{B})+\mathbf{C} = \mathbf{A}+(\mathbf{B}+\mathbf{C}) \tag{E.36}$$

3. Commutative operations

$$\mathbf{A}+\mathbf{B} = \mathbf{B}+\mathbf{A} \tag{E.37}$$

In general, matrix multipliers do not commute.

4. Transpose operations

$$(\mathbf{A}+\mathbf{B})^t = \mathbf{A}^t+\mathbf{B}^t \tag{E.38}$$
$$(\mathbf{AB})^t = \mathbf{B}^t\mathbf{A}^t \tag{E.39}$$

E.1.2 MATRIX INVERSION

Matrix division is not defined. In a number of cases the practical effect of division can be obtained by using the *inverse* of a matrix, if the inverse is available. That is, if

$$\mathbf{A}^{-1}\mathbf{A} = \mathbf{I} = \mathbf{A}\mathbf{A}^{-1} \tag{E.40}$$

where $\mathbf{I}$ is the identity matrix. In order to be invertible, a matrix must be *square*; i.e., have the same number of rows as columns.

There exists a variety of methods for computing a matrix inverse, given the matrix. One such formula is

$$\mathbf{M}^{-1} = \frac{\overset{*}{\mathbf{M}}}{|\mathbf{M}|} \tag{E.41}$$

in which $|\mathbf{M}|$ is the determinant of the matrix

$$det\,\mathbf{M} = \begin{vmatrix} m_{11} & m_{12} & \cdot & \cdot & \cdot & m_{1n} \\ m_{21} & m_{22} & \cdot & \cdot & \cdot & m_{2n} \\ \cdot & \cdot & \cdot & \cdot & \cdot & \cdot \\ \cdot & \cdot & \cdot & \cdot & \cdot & \cdot \\ \cdot & \cdot & \cdot & \cdot & \cdot & \cdot \\ m_{m1} & m_{m2} & \cdot & \cdot & \cdot & m_{mn} \end{vmatrix} \tag{E.42}$$

The matrix $\overset{*}{\mathbf{M}}$ has elements which are formed from the *cofactors* of $\mathbf{M}$. Thus, if cof_{ij}, $(i,j = 1,2,\cdots,n)$, are the cofactors, then

$$\overset{*}{\mathbf{M}} = [cof_{ij}]^t \tag{E.43}$$

Each cofactor cof_{ij} is created as the determinant formed from the determinant of $\mathbf{M}$ by striking the row i and column j, and multiplying the result by $(-1)^{(i+j)}$.

Suppose, for instance, the matrix is

$$\mathbf{M} = \begin{bmatrix} 8 & 3 & 1 \\ 5 & 2 & 7 \\ 4 & 6 & 9 \end{bmatrix} \tag{E.44}$$

The determinant is

$$|\mathbf{M}| = 8\times2\times9+5\times6\times1+4\times3\times7-1\times2\times4-3\times5\times9-8\times6\times7 = -221 \tag{E.45}$$

The cofactors are

$$cof_{11} = (-1)^{(1+1)} \begin{vmatrix} 2 & 7 \\ 6 & 9 \end{vmatrix} = 2\times9-6\times7 = -24 \tag{E.46}$$

$$cof_{12} = (-1)^{(1+2)} \begin{vmatrix} 5 & 7 \\ 4 & 9 \end{vmatrix} = -(5\times9-4\times7) = -17 \tag{E.47}$$

$$cof_{13} = (-1)^{(1+3)} \begin{vmatrix} 5 & 2 \\ 4 & 6 \end{vmatrix} = 5\times6-4\times2 = 22 \tag{E.48}$$

$$cof_{21} = (-1)^{(2+1)} \begin{vmatrix} 3 & 1 \\ 6 & 9 \end{vmatrix} = -(3\times9-6\times1) = -21 \tag{E.49}$$

$$cof_{22} = (-1)^{(2+2)} \begin{vmatrix} 8 & 1 \\ 4 & 9 \end{vmatrix} = 8\times9-4\times1 = 68 \tag{E.50}$$

$$cof_{23} = (-1)^{(2+3)} \begin{vmatrix} 8 & 3 \\ 4 & 6 \end{vmatrix} = -(8\times6-3\times4) = -36 \tag{E.51}$$

$$cof_{31} = (-1)^{(3+1)} \begin{vmatrix} 3 & 1 \\ 2 & 7 \end{vmatrix} = 3\times7-2\times1 = 19 \tag{E.52}$$

$$cof_{32} = (-1)^{(3+2)} \begin{vmatrix} 8 & 1 \\ 5 & 7 \end{vmatrix} = -(8\times7-5\times1) = -51 \tag{E.53}$$

$$cof_{33} = (-1)^{(3+3)} \begin{vmatrix} 8 & 3 \\ 5 & 2 \end{vmatrix} = 8\times2-5\times3 = 1 \tag{E.54}$$

Then, using Equation E.43,

$$\overset{*}{\mathbf{M}} = \begin{bmatrix} -24 & -21 & 19 \\ -17 & 68 & -51 \\ 22 & -36 & 1 \end{bmatrix} \tag{E.55}$$

Now the inverse matrix is gotten from Equations E.41 and E.43 as

$$\mathbf{M}^{-1} = \begin{bmatrix} \dfrac{-24}{-221} & \dfrac{-21}{-221} & \dfrac{19}{-221} \\ \dfrac{-17}{-221} & \dfrac{68}{-221} & \dfrac{-51}{-221} \\ \dfrac{22}{-221} & \dfrac{-36}{-221} & \dfrac{1}{-221} \end{bmatrix} \tag{E.56}$$

This result may be confirmed by using

$$\mathbf{MM}^{-1} = \mathbf{I}$$

The method illustrated above is one approach to computing the inverse matrix from the given matrix. There are a number of other methods exposed in the literature. Unfortunately, the difficulty of carrying out the indicated operations increases rapidly with the dimension of the matrix in question. However, there are also a number of good programs for accomplishing this by using computers.

E.1.3 GABLE ARCH MATRIX INVERSION

As another example of matrix inversion, consider the determination of $\mathbf{A}^{-1}$ for the gable arch of Chapter 3. There it was shown that

$$\mathbf{A} = \begin{bmatrix} 0 & 0 & 8 & 4 & 0 & 0 \\ 1 & 0 & 1 & 0 & 0 & 0 \\ 0 & 1 & 0 & 1 & 0 & 0 \\ 0 & 0 & 0 & 0 & 8 & -4 \\ 0 & 0 & -1 & 0 & 1 & 0 \\ 0 & 0 & 0 & 1 & 0 & -1 \end{bmatrix} \tag{E.57}$$

and since $\mathbf{AA}^{-1} = \mathbf{I}$, there is

$$\begin{bmatrix} 0 & 0 & 8 & 4 & 0 & 0 \\ 1 & 0 & 1 & 0 & 0 & 0 \\ 0 & 1 & 0 & 1 & 0 & 0 \\ 0 & 0 & 0 & 0 & 8 & -4 \\ 0 & 0 & -1 & 0 & 1 & 0 \\ 0 & 0 & 0 & 1 & 0 & -1 \end{bmatrix} \begin{bmatrix} a_{11} & a_{12} & a_{13} & a_{14} & a_{15} & a_{16} \\ a_{21} & a_{22} & a_{23} & a_{24} & a_{25} & a_{26} \\ a_{31} & a_{32} & a_{33} & a_{34} & a_{35} & a_{36} \\ a_{41} & a_{42} & a_{43} & a_{44} & a_{45} & a_{46} \\ a_{51} & a_{52} & a_{53} & a_{54} & a_{55} & a_{56} \\ a_{61} & a_{62} & a_{63} & a_{64} & a_{65} & a_{66} \end{bmatrix}$$

$$= \begin{bmatrix} 1 & 0 & 0 & 0 & 0 & 0 \\ 0 & 1 & 0 & 0 & 0 & 0 \\ 0 & 0 & 1 & 0 & 0 & 0 \\ 0 & 0 & 0 & 1 & 0 & 0 \\ 0 & 0 & 0 & 0 & 1 & 0 \\ 0 & 0 & 0 & 0 & 0 & 1 \end{bmatrix} \tag{E.58}$$

Upon invoking the usual matrix multiplication algorithm, Equation E.25, it is seen that there are the following scalar equations involving the elements a_{ij} of the matrix $\mathbf{A}^{-1}$.

$$8\,a_{31} + 4\,a_{41} = 1 \tag{E.59}$$
$$a_{11} + a_{31} = 0 \tag{E.60}$$
$$a_{21} + a_{41} = 0 \tag{E.61}$$
$$8\,a_{51} - 4\,a_{61} = 0 \tag{E.62}$$
$$-a_{31} + a_{51} = 0 \tag{E.63}$$
$$a_{41} - a_{61} = 0 \tag{E.64}$$

$$8\,a_{32} + 4\,a_{42} = 0 \tag{E.65}$$
$$a_{12} + a_{32} = 1 \tag{E.66}$$
$$a_{22} + a_{42} = 0 \tag{E.67}$$
$$8\,a_{52} - 4\,a_{62} = 0 \tag{E.68}$$
$$-a_{32} + a_{52} = 0 \tag{E.69}$$
$$a_{42} - a_{62} = 0 \tag{E.70}$$

$$8\,a_{33} + 4\,a_{43} = 0 \tag{E.71}$$
$$a_{13} + a_{33} = 0 \tag{E.72}$$
$$a_{23} + a_{43} = 1 \tag{E.73}$$
$$8\,a_{53} - 4\,a_{63} = 0 \tag{E.74}$$
$$-a_{33} + a_{53} = 0 \tag{E.75}$$
$$a_{43} - a_{63} = 0 \tag{E.76}$$

$$8\,a_{34} + 4\,a_{44} = 0 \tag{E.77}$$
$$a_{14} + a_{34} = 0 \tag{E.78}$$
$$a_{24} + a_{44} = 0 \tag{E.79}$$
$$8\,a_{54} - 4\,a_{64} = 1 \tag{E.80}$$
$$-a_{34} + a_{54} = 0 \tag{E.81}$$
$$a_{44} - a_{64} = 0 \tag{E.82}$$

$$8\,a_{35} + 4\,a_{45} = 0 \tag{E.83}$$
$$a_{15} + a_{35} = 0 \tag{E.84}$$
$$a_{25} + a_{45} = 0 \tag{E.85}$$
$$8\,a_{55} - 4\,a_{65} = 0 \tag{E.86}$$
$$-a_{35} + a_{55} = 1 \tag{E.87}$$
$$a_{45} - a_{65} = 0 \tag{E.88}$$

$$8\,a_{36} + 4\,a_{46} = 0 \tag{E.89}$$
$$a_{16} + a_{36} = 0 \tag{E.90}$$
$$a_{26} + a_{46} = 0 \tag{E.91}$$
$$8\,a_{56} - 4\,a_{66} = 0 \tag{E.92}$$
$$-a_{36} + a_{56} = 0 \tag{E.93}$$
$$a_{46} - a_{66} = 1 \tag{E.94}$$

From Equation E.64 of the first set,

$$a_{61} = a_{41} \tag{E.95}$$

and when this is inserted into the fourth equation of the set, Equation E.62,

$$8\,a_{51} - 4\,a_{41} = 0 \tag{E.96}$$

Then from the fifth equation of the set, Equation E.63,

$$a_{51} = a_{31} \tag{E.97}$$

so that the fourth equation, Equation E.62, becomes

$$8\,a_{31} - 4\,a_{41} = 0 \tag{E.98}$$

Use this and the first equation of the set, Equation E.59, to solve simultaneously and get

$$a_{31} = \tfrac{1}{16} \tag{E.99}$$

Then

$$a_{41} = 2\,a_{31} = \tfrac{1}{8} \tag{E.100}$$
$$a_{51} = a_{31} = \tfrac{1}{16} \tag{E.101}$$
$$a_{21} = -a_{41} = -\tfrac{1}{8} \tag{E.102}$$
$$a_{61} = a_{41} = \tfrac{1}{8} \tag{E.103}$$
$$a_{11} = -a_{31} = -\tfrac{1}{16} \tag{E.104}$$

Similar operations are carried out for each of the other five groups of equations. In this way each coefficient is determined, and the desired matrix can be assembled as

$$\mathbf{A}^{-1} = \begin{bmatrix} -\frac{1}{16} & 1 & 0 & -\frac{1}{16} & \frac{1}{2} & \frac{1}{4} \\ -\frac{1}{8} & 0 & 1 & \frac{1}{8} & -1 & -\frac{1}{2} \\ \frac{1}{16} & 0 & 0 & \frac{1}{16} & -\frac{1}{2} & -\frac{1}{4} \\ \frac{1}{8} & 0 & 0 & -\frac{1}{8} & 1 & \frac{1}{2} \\ \frac{1}{16} & 0 & 0 & \frac{1}{16} & \frac{1}{2} & -\frac{1}{4} \\ \frac{1}{8} & 0 & 0 & -\frac{1}{8} & 1 & -\frac{1}{2} \end{bmatrix} \tag{E.105}$$

APPENDIX F

TABLEAU ALGORITHMS

- **Step One Algorithm** Place a scanner on the Z-line and determine the ***largest negative number***. The variable above this column is declared the ***entering variable***. The column containing this number is highlighted to indicate its premier status.

- **Step Two Algorithm** Construct the upper bound of the entering variable column entries. The Z-row entry in this column is vacant, of course. For the first variable row, take the coefficient of the entering variable in this row and divide it into the RHS value at this row. Enter this in the upper bound column in this row. Continue with the other variable rows until each upper bound column element is obtained.

- **Step Three Algorithm** Put a scanner on the Upper-Bound Entering Variable Column and find the ***smallest positive number*** there. That variable now becomes the ***leaving variable***, and its row is highlighted to emphasize its exalted role.

  At the intersection of the entering variable column and the leaving variable row is the cell with the ***pivot*** in it, and the number in the cell is the ***pivot element***. The location and the value are both critical to the next steps in the solution. All entries in this column will need to be 0s, except the pivot.

  Next is the mechanization of the procedure employed in the Simplex Method. A new tableau—Second Tableau—will be created from the Initial Tableau and the identification of the pivot.

- **Step Four Algorithm** Divide the pivot element by itself to produce the required number 1 at that place. Divide each of the entries in the old entering variable column by the pivot element, change signs, and add to the original elements in the row in question. This makes the new numbers there 0. Now divide each of the elements of the leaving variable row by the pivot element and record the results in the entering variable row of the Second Tableau. Do all of the other elements in this row the same (except, of course, the upper-bound on the entering variable).

- **Step Five Algorithm** Create the new Second Tableau entries which are not on the pivot row or in the pivot column. For a given row, divide the pivot column element in that row by the pivot element, change the sign, then go to the pivot row element under the cell where the computation is being done and multiply by the element found there and add the result to the existing element in the cell where the computation is being done. Record this as the element of that cell in the Second Tableau.

Figures F.1, F.2, F.3, F.4, F.5, and F.6 display the operations involved in progressing from the initial tableau to the final tableau for the example of Section 11.4.

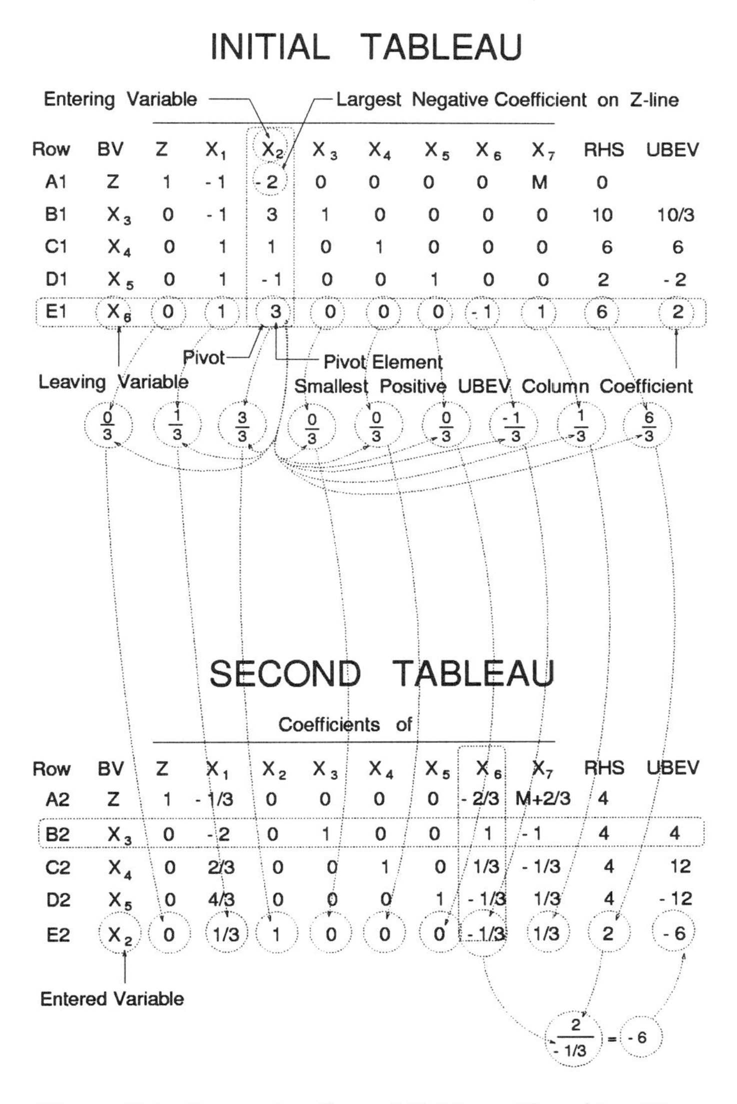

INITIAL TABLEAU

Row	BV	Z	X_1	X_2	X_3	X_4	X_5	X_6	X_7	RHS	UBEV
A1	Z	1	-1	-2	0	0	0	0	M	0	
B1	X_3	0	-1	3	1	0	0	0	0	10	10/3
C1	X_4	0	1	1	0	1	0	0	0	6	6
D1	X_5	0	1	-1	0	0	1	0	0	2	-2
E1	X_6	0	1	3	0	0	0	-1	1	6	2

$\frac{0}{3}$ $\frac{1}{3}$ $\frac{3}{3}$ $\frac{0}{3}$ $\frac{0}{3}$ $\frac{0}{3}$ $\frac{-1}{3}$ $\frac{1}{3}$ $\frac{6}{3}$

SECOND TABLEAU

Row	BV	Z	X_1	X_2	X_3	X_4	X_5	X_6	X_7	RHS	UBEV
A2	Z	1	-1/3	0	0	0	0	-2/3	M+2/3	4	
B2	X_3	0	-2	0	1	0	0	1	-1	4	4
C2	X_4	0	2/3	0	0	1	0	1/3	-1/3	4	12
D2	X_5	0	4/3	0	0	0	1	-1/3	1/3	4	-12
E2	X_2	0	1/3	1	0	0	0	-1/3	1/3	2	-6

Figure F.1: Generating Second Tableau, Two Algorithms

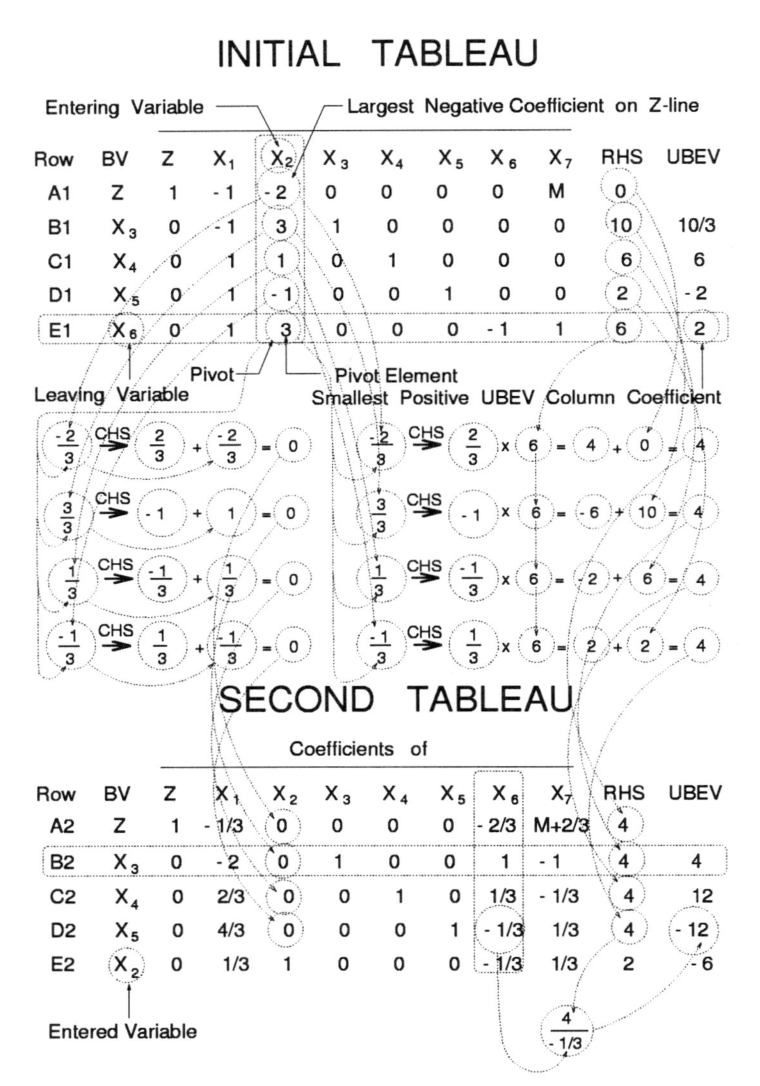

Figure F.2: Generating Second Tableau, Three Algorithms

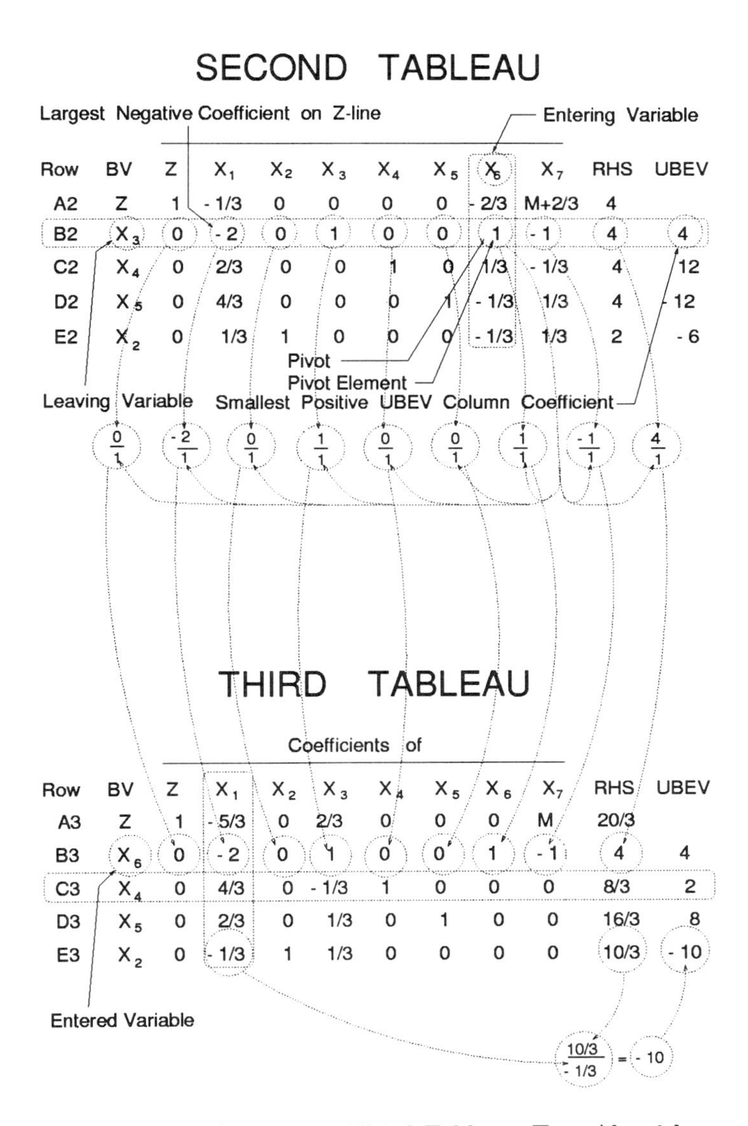

Figure F.3: Generating Third Tableau, Two Algorithms

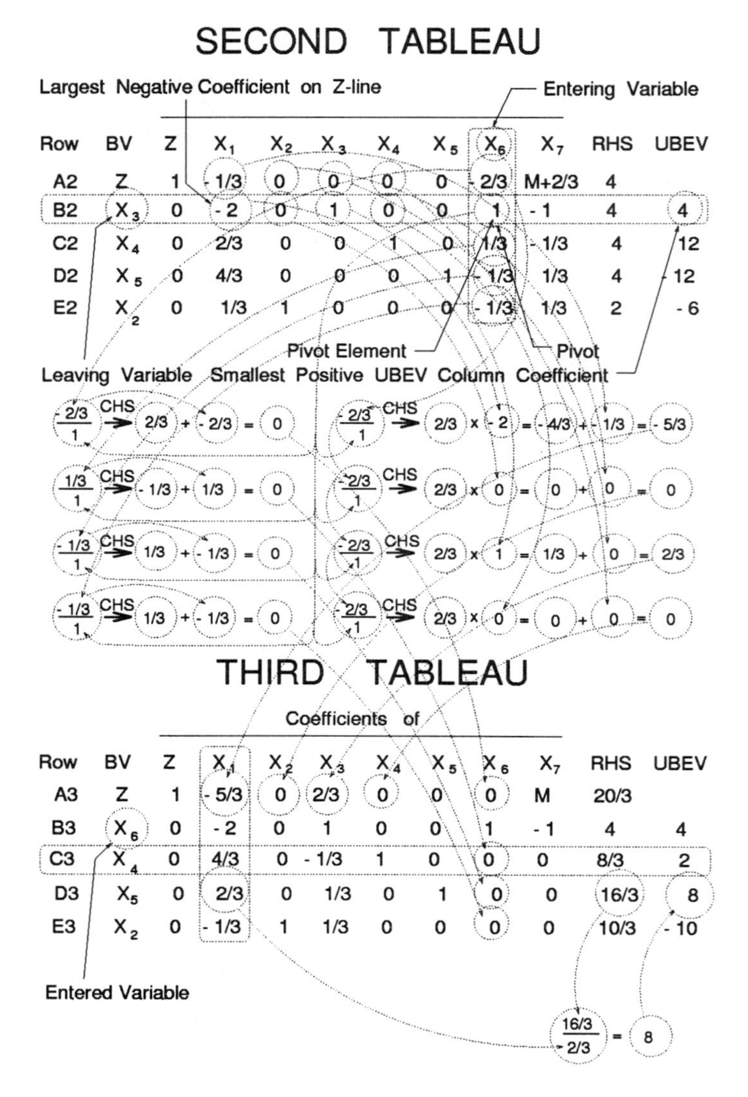

Figure F.4: Generating Third Tableau, Three Algorithms

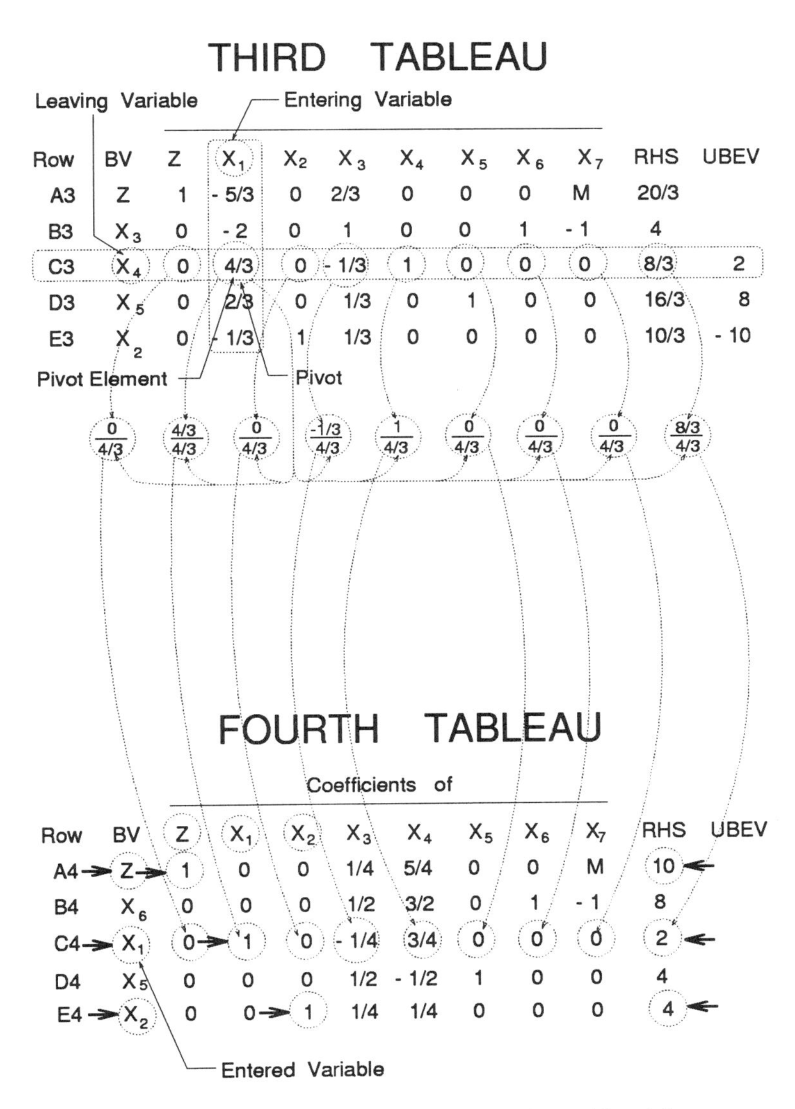

Figure F.5: Generating Fourth Tableau, Two Algorithms

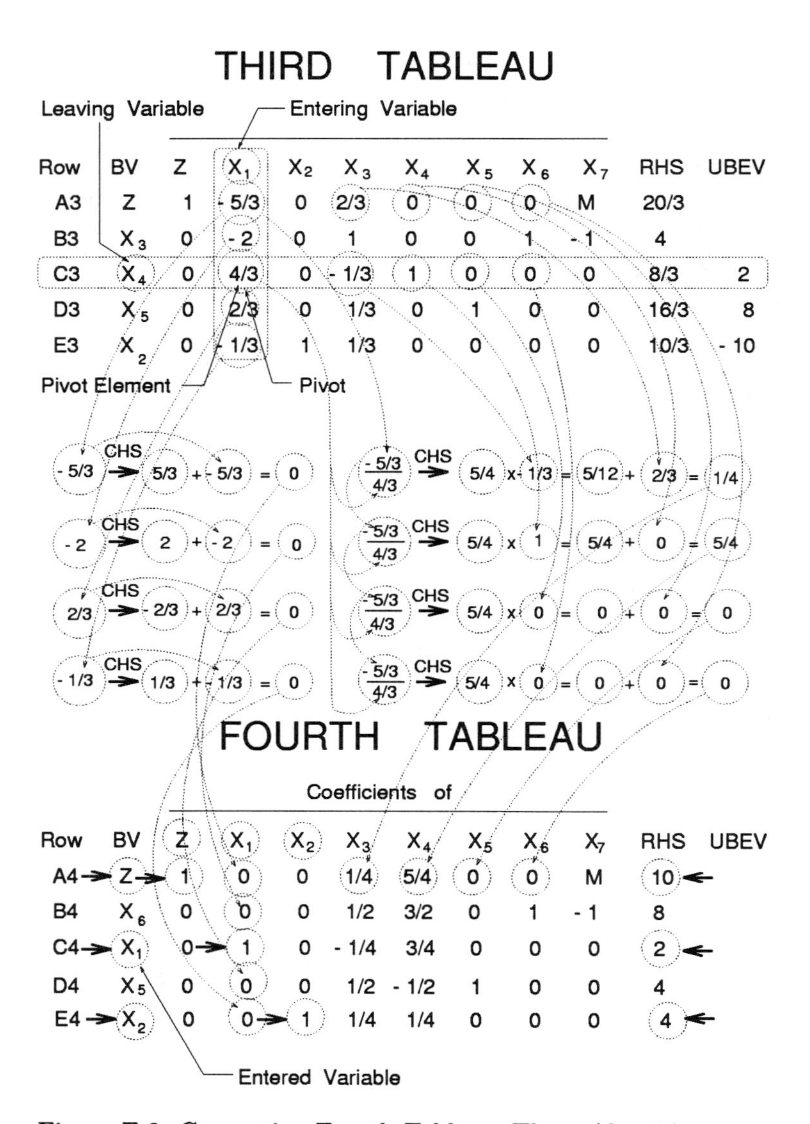

Figure F.6: Generating Fourth Tableau, Three Algorithms

APPENDIX G

NORMAL FREQUENCY DISTRIBUTION CUMULATIVE PROBABILITY

G.1 CUMULATIVE PROBABILITY

In Chapter 13 it was shown that the cumulative probability associated with the normal frequency distribution is

$$P(x) = \int_0^x f(x)\,dx = \int_0^x \frac{1}{\sigma\sqrt{2\pi}} e^{-\frac{1}{2}\left(\frac{x-\overline{x}}{\sigma}\right)^2} dx = \frac{1}{\sqrt{2\pi}} \int_{-\frac{\overline{x}}{\sigma}}^{\frac{x-\overline{x}}{\sigma}} e^{-\frac{1}{2}t^2}\,dt \qquad \text{(G.1)}$$

in which $\overline{x}$ and σ are the two statistics obtained from sample data. There is not known any closed form expression of this integral, so it is evaluated by replacing the integrand with its series expansion obtained from the generic formula

$$e^{-v} = 1 - v + \tfrac{1}{2!}\,v^2 - \tfrac{1}{3!}\,v^3 + \cdots \qquad \text{(G.2)}$$

Thus

$$\begin{aligned}
P(x) &= \frac{1}{\sqrt{2\pi}} \int_{-\frac{\overline{x}}{\sigma}}^{\frac{x-\overline{x}}{\sigma}} e^{-\frac{1}{2}t^2}\,dt \\
&= \frac{1}{\sqrt{2\pi}} \int_{-\frac{\overline{x}}{\sigma}}^{\frac{x-\overline{x}}{\sigma}} \left[1 - \tfrac{1}{2}\,t^2 + \tfrac{1}{2!}\left(\tfrac{1}{2}\,t^2\right)^2 - \tfrac{1}{3!}\left(\tfrac{1}{2}\,t^2\right)^3 + \cdots\right] dt \\
&= \frac{1}{\sqrt{2\pi}} \int_{-\frac{\overline{x}}{\sigma}}^{\frac{x-\overline{x}}{\sigma}} \left[1 - \tfrac{1}{2}\,t^2 + \tfrac{1}{4\cdot 2!}\,t^4 - \tfrac{1}{8\cdot 3!}\,t^6 + \cdots\right] dt \\
&= \frac{1}{\sqrt{2\pi}} \left[t - \tfrac{1}{3\cdot 2}\,t^3 + \tfrac{1}{5\cdot 4\cdot 2!}\,t^5 - \tfrac{1}{8\cdot 7\cdot 3!}\,t^7 + \cdots\right]_{-\frac{\overline{x}}{\sigma}}^{\frac{x-\overline{x}}{\sigma}} \\
&= \frac{1}{\sqrt{2\pi}} \Bigg\{\left[\left(\tfrac{x-\overline{x}}{\sigma}\right) - \tfrac{1}{3\cdot 2}\left(\tfrac{x-\overline{x}}{\sigma}\right)^3 + \tfrac{1}{5\cdot 4\cdot 2\cdot 1}\left(\tfrac{x-\overline{x}}{\sigma}\right)^5 - \tfrac{1}{8\cdot 7\cdot 3\cdot 2\cdot 1}\left(\tfrac{x-\overline{x}}{\sigma}\right)^7 + \cdots\right] \\
&\qquad - \left[\left(\tfrac{-\overline{x}}{\sigma}\right) - \tfrac{1}{3\cdot 2}\left(\tfrac{-\overline{x}}{\sigma}\right)^3 + \tfrac{1}{5\cdot 4\cdot 2\cdot 1}\left(\tfrac{-\overline{x}}{\sigma}\right)^5 - \tfrac{1}{8\cdot 7\cdot 3\cdot 2\cdot 1}\left(\tfrac{-\overline{x}}{\sigma}\right)^7 + \cdots\right]\Bigg\} \\
&= \frac{1}{\sqrt{2\pi}} \Bigg\{\left[\left(\tfrac{x-\overline{x}}{\sigma}\right)^1 - \left(\tfrac{-\overline{x}}{\sigma}\right)^1\right] - \tfrac{1}{3\cdot 2}\left[\left(\tfrac{x-\overline{x}}{\sigma}\right)^3 - \left(\tfrac{-\overline{x}}{\sigma}\right)^3\right] \\
&\qquad + \tfrac{1}{5\cdot 4\cdot 2\cdot 1}\left[\left(\tfrac{x-\overline{x}}{\sigma}\right)^5 - \left(\tfrac{-\overline{x}}{\sigma}\right)^5\right] - \tfrac{1}{8\cdot 7\cdot 3\cdot 2\cdot 1}\left[\left(\tfrac{x-\overline{x}}{\sigma}\right)^7 - \left(\tfrac{-\overline{x}}{\sigma}\right)^7\right] + \cdots\Bigg\} \\
&= \frac{1}{\sqrt{2\pi}} \sum_{n=1}^{\infty} \frac{(-1)^{n+1}}{2^{n-1}(2n-1)(n-1)!} \left[\left(\frac{x-\overline{x}}{\sigma}\right)^{2n-1} - \left(\frac{-\overline{x}}{\sigma}\right)^{2n-1}\right] \qquad \text{(G.3)}
\end{aligned}$$

G.2 COMPUTER PROGRAM

```
program NormalProbability(input,output,lstNp);

 { This program computes the probability associated with a Normal }
 { frequency distribution with parameters xbar and sigma which are }
 { previously determined from sample data.                          }

label
  10;

var
  a, b, c, d, i, j, k, p, s, t, x, xbar, sigma, term: double;
  ans: char; lstNp: text;

begin
  rewrite(lstNp,'normalp.dat');
  write('Enter value of xbar '); readln(xbar); writeln(' ');
  write('Enter value of sigma '); readln(sigma); writeln(' ');
  writeln(lstNp,'Cumulative Probability for Normal Frequency Distribution');
  writeln(lstNp,' ');
  writeln(lstNp,'xbar = ',xbar:8:6,' sigma = ',sigma:8:6,'');
  writeln(lstNp,' ');
  writeln(' '); writeln(' ');

  10:

  write('Enter value of x '); readln(x);
  writeln(' '); writeln(' ');
  i:=1; j:=1; s:=1; t:=1;
  a:=(x-xbar)/sigma; b:=(-xbar/sigma); c:=a; d:=b;
  term:=a-b; p:=term;

  while abs(term)≥1.0E-9 do
    begin
      i:= i+1.0; s:=(-1.0)*s; j:=j*(i-1.0);
      k:=2.0*i-1.0; t:=2.0*t; c:=a*(a*c); d:=b*(b*d);
      term:=(s*(c-d))/(j*k*t); p:=p+term;
    end;

  p:=p/sqrt(2*3.1415926536);
  writeln(' x = ',x:8:6,' P = ',p:8:6,'');
  writeln(lstNp,'x = ',x:8:6,' P = ',p:8:6,'');
  writeln('Do you wish to compute another P? '); readln(ans);
  if (ans = 'y') or (ans = 'Y') then
    begin
      writeln(lstNp,' '); goto 10;
    end
  else
end.
```

G.3 EXAMPLE COMPUTATION

The data are from the Parsons & Whitted example of Chapter 13.

Cumulative Probability for Normal Distribution

xbar = 478.600000 sigma = 153.499000

x	P
x = 100.000000	P = 0.005912
x = 200.000000	P = 0.033852
x = 300.000000	P = 0.121398
x = 400.000000	P = 0.303395
x = 500.000000	P = 0.554528
x = 600.000000	P = 0.784584
x = 700.000000	P = 0.924488
x = 800.000000	P = 0.980952
x = 900.000000	P = 0.996067
x = 1000.000000	P = 0.998748
x = 1100.000000	P = 0.999064
x = 1200.000000	P = 0.999088

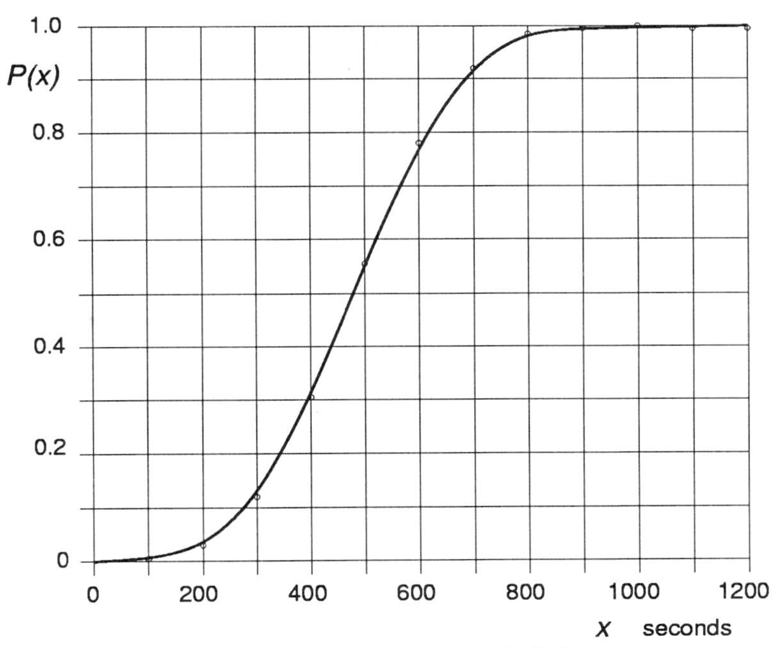

Figure G.1: Cumulative Probability

Figure G.1 displays the cumulative probability $P(x)$ versus x for this example.

APPENDIX H

DRAWING RANDOM NUMBERS

H.1 INTRODUCTION

It was shown in Chapter 13 that randomly drawn numbers are the mechanism by which systems can be simulated. It was also seen that the larger the sample, the more accurate is the representation of the simulation. For this reason it will be desirable in many such simulations to draw large numbers of random numbers.

It is convenient to write a program for producing large numbers of random numbers, as well as making the identification of these with the classes which have been established for the model and its simulation. The class limits will have been determined by obtaining the statistics of the distribution, and then computing the cumulative probabilities for this distribution at the class limits.

Such probabilities can then be designated as "class marks," which represent the class boundaries in probability space. When a random number is drawn, it will lie within the class limits of some class, so it is only necessary to test whether or not a given number is in which one of these ranges.

A desirable program will write the random number when it is drawn, for reference, and will then add one unit to the frequency of that class within which the number belongs. At the end of the drawing, the totals for each class will be presented.

A program written in Pascal source code for the purposes mentioned above is shown below, together with an example of its application.

H.2 COMPUTER PROGRAM

```
program RandomNumbers(input,output,lstRN);

{ This program draws random numbers and places each in one of }
{ five classes according to pre-defined class limits, then totals the }
{ number in each class upon completion of the drawing. }

label
  10;
```

```
var
  x1, x2, x3, x4, x5, x: real;
  i, j, n, f1, f2, f3, f4, f5, sum: integer;
  ans: char; lstRN: text;



begin
  rewrite(lstRN,'random.dat');
  write(' Enter class mark x1 '); readln(x1);
  writeln(' ');
  write(' Enter class mark x2 '); readln(x2);
  writeln(' ');
  write(' Enter class mark x3 '); readln(x3);
  writeln(' ');
  write(' Enter class mark x4 '); readln(x4);
  writeln(' ');
  write('Enter number of Random Numbers to be drawn ');
  readln(n);
  writeln(' ');
  f1:=0; f2:=0; f3:=0; f4:=0; f5:=0; i:=0; j:=0;
  writeln(' ');
  while i ¡ n do
    begin
      x:=random;
      j:=j + 1;
      if j = 4 then
        begin
          writeln(lstRN,' x =',x,'');
          j:=0;
        end
        else
      write(lstRN,' x =',x,'');
      if x ≤ x1 then
        begin
          f1:=f1+1;
        end
        else
      if x ≤ x2 then
        begin
          f2:=f2 + 1;
        end
        else
      if x ≤ x3 then
        begin
          f3:=f3 + 1;
        end
        else
```

```
        if x ≤ x4 then
          begin
            f4:= f4 + 1;
          end
          else

        if x ≥ x5 then
          begin
            f5:= f5 + 1;
          end;
        i:=i + 1;
      end;
    sum:=f1 + f2 + f3 + f4 + f5;
    writeln(' f1 = ',f1,'');
    writeln(' f2 = ',f2,'');
    writeln(' f3 = ',f3,'');
    writeln(' f4 = ',f4,'');
    writeln(' f5 = ',f5,'');
    writeln(' sum = ',sum,''); writeln(' ');
    writeln(lstRN,' x1 =',x1,'');
    writeln(lstRN,' x2 =',x2,'');
    writeln(lstRN,' x3 =',x3,'');
    writeln(lstRN,' x4 =',x4,'');
    writeln(lstRN,' f1 = ',f1,'');
    writeln(lstRN,' f2 = ',f2,'');
    writeln(lstRN,' f3 = ',f3,'');
    writeln(lstRN,' f4 = ',f4,'');
    writeln(lstRN,' f5 = ',f5,'');
    writeln(lstRN,' sum = ',sum,'');
    write('Do you wish to draw another set? ');
    readln(ans);
    writeln(' ');
    if (ans = 'y') or (ans = 'Y') then
      begin
        writeln(lstRN,' ');
        goto 10;
      end
      else
  end.
```

H.3 EXAMPLE COMPUTATION

The data are $f_1 = 4, f_2 = 17, f_3 = 9, f_4 = 2, f_5 = 1$. Assuming that it is a Pearson-III distribution, the statistics are

$$\overline{x} = 2.3636, \qquad s_x = 0.8814, \qquad A = 2.4499.$$

Class marks which correspond to the class limits are

$$P_{1.5} = 0.155882, \quad P_{2.5} = 0.613129, \quad P_{3.5} = 0.893582,$$

$$P_{4.5} = 0.978847, \quad P_{5.5} = 0.996603.$$

100 random numbers are to be drawn.

The printout of the results is given below.

```
x = 6.67309612e-02 x = 9.09533739e-01 x = 9.32533920e-01 x = 2.10703164e-02
x = 2.02018976e-01 x = 2.04082459e-01 x = 7.93113887e-01 x = 2.67157227e-01
x = 1.28373310e-01 x = 7.37413764e-01 x = 9.15231586e-01 x = 7.62057543e-01
x = 4.71062392e-01 x = 6.12559915e-01 x = 4.98185664e-01 x = 4.14520353e-02
x = 6.97868645e-01 x = 9.16608810e-01 x = 6.05422318e-01 x = 1.42607182e-01
x = 8.77655864e-01 x = 1.86270401e-01 x = 6.34889901e-01 x = 2.75923982e-02
x = 1.56724811e-01 x = 5.65784812e-01 x = 2.76387244e-01 x = 5.28337181e-01
x = 6.27392173e-01 x = 8.20829928e-01 x = 3.60483706e-01 x = 6.90494478e-01
x = 8.13269138e-01 x = 6.88754380e-01 x = 5.44780910e-01 x = 2.26131842e-01
x = 1.78051203e-01 x = 9.32061225e-02 x = 8.65828931e-01 x = 5.76204836e-01
x = 8.85803223e-01 x = 9.45082381e-02 x = 4.69831556e-01 x = 9.09641087e-01
x = 7.63741553e-01 x = 2.22802505e-01 x = 5.92753470e-01 x = 1.82578996e-01
x = 5.20398796e-01 x = 8.24713171e-01 x = 7.02694476e-01 x = 4.87616092e-01
x = 4.63247716e-01 x = 6.93686783e-01 x = 7.01619804e-01 x = 3.51630330e-01
x = 4.11880761e-01 x = 7.49615431e-01 x = 6.89858794e-02 x = 9.78937924e-01
x = 3.89727980e-01 x = 9.56953168e-01 x = 1.15036480e-01 x = 3.88504565e-01
x = 1.08640837e-02 x = 7.00613141e-01 x = 2.64064163e-01 x = 5.94303250e-01
x = 7.69051433e-01 x = 5.63880391e-02 x = 5.57881594e-01 x = 5.80942184e-02
x = 8.54157925e-01 x = 8.51476550e-01 x = 4.66967076e-01 x = 7.55870283e-01
x = 3.21540162e-02 x = 8.39177489e-01 x = 8.52354586e-01 x = 7.82624364e-01
x = 4.40901697e-01 x = 5.76777220e-01 x = 6.71468139e-01 x = 4.32276338e-01
x = 4.59071368e-01 x = 9.11189318e-01 x = 8.96683693e-01 x = 9.86716270e-01
x = 7.76995540e-01 x = 6.73984945e-01 x = 6.68605924e-01 x = 1.00657672e-01
x = 1.42679542e-01 x = 8.47950459e-01 x = 1.67518139e-01 x = 8.58253300e-01
x = 1.13813691e-01 x = 3.13384265e-01 x = 6.06110930e-01 x = 2.25800872e-01

x1 = 1.55882001e-01
x2 = 6.13129020e-01
x3 = 8.93581986e-01
x4 = 9.78847027e-01
x5 = 9.96603012e-01

f1 = 17
f2 = 40
f3 = 33
f4 =  8
f5 =  2

sum = 100
```

APPENDIX I

THE LINEAR OSCILLATOR

I.1 UNDAMPED OSCILLATOR

Suppose the system is composed of a mass m attached to a linear spring of modulus k. The mass is subjected to an arbitrary force $f(t)$, which is the input to the system. The system output is $x(t)$, the displacement of the mass. These features are displayed in Figure I.1.

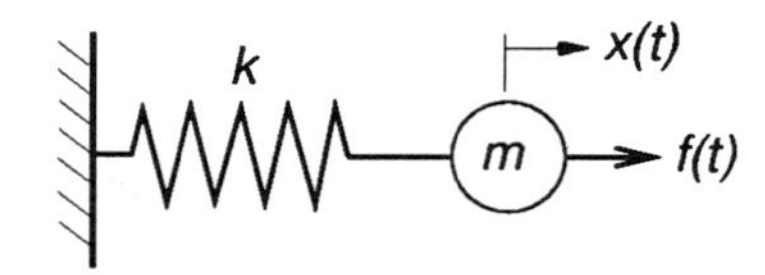

Figure I.1: Undamped Linear Oscillator

The equation of motion of the mass is

$$m\,\ddot{x}(t) + k\,x(t) = f(t) \tag{I.1}$$

with the initial conditions $x(t_0) = x_0$, $\dot{x}(t_0) = \dot{x}_0$, which are the initial displacement and the initial velocity, respectively.

As was shown in Chapter 15, the state-space model of this system is subsumed in the equations

$$\mathbf{x}(t) = \begin{bmatrix} x(t) \\ \dot{x}(t) \end{bmatrix} \qquad \mathbf{x}_0 = \begin{bmatrix} x(t_0) \\ \dot{x}(t_0) \end{bmatrix} = \begin{bmatrix} x_0 \\ \dot{x}_0 \end{bmatrix}$$

in which $\mathbf{x}$ is the state vector and $\mathbf{x}_0$ is the initial state. The equation of motion (state equation), in canonical form, is

$$\dot{\mathbf{x}}(t) = \mathbf{A}\,\mathbf{x}(t) + \mathbf{B}\,f(t) \tag{I.2}$$

with

$$\mathbf{A} = \begin{bmatrix} 0 & 1 \\ -k/m & 0 \end{bmatrix} \qquad \mathbf{B} = \begin{bmatrix} 0 \\ -1/m \end{bmatrix} \tag{I.3}$$

The system response is

$$x(t) = x_0\,cos\omega(t-t_0) + \frac{\dot{x}_0}{\omega} sin\omega(t-t_0) + \frac{1}{m\omega}\int_{t_0}^{t} f(\tau)\,sin\omega(t-\tau)\,d\tau \tag{I.4}$$

where $\omega = \sqrt{k/m}$ is the natural frequency of the spring-mass combination. This response holds, of course, for any time greater than $t = t_0$. The first two terms on the right-hand side of the response equation (the ones containing x_0 and $\dot{x}_0$) are the free vibration (in the absence of a forcing term $f(t)$), and the last term is that component of the motion associated with the particular integral or forced response.

I.2 LINEAR OSCILLATOR WITH DAMPING

The previous model contains no element that reflects the losses associated with the "real" world. That is, there is no term which accounts for the frictional losses experienced in almost every natural system. The model can be altered to incorporate such a feature by including a term proportional to the velocity of the mass, which represents viscous damping. The usual mechanism is a dashpot—a close-fitting piston in a cylinder filled with a viscous fluid such as oil, and containing a small orifice in the piston that allows the fluid to flow out, accompanied by the viscous frictional losses found in such flow. The new equation of motion is written

$$m\,\ddot{x}(t) + c\,\dot{x} + k\,x = f(t) \tag{I.5}$$

wherein c is the damping coefficient, which is assumed to be a constant. The illustration is Figure I.2.

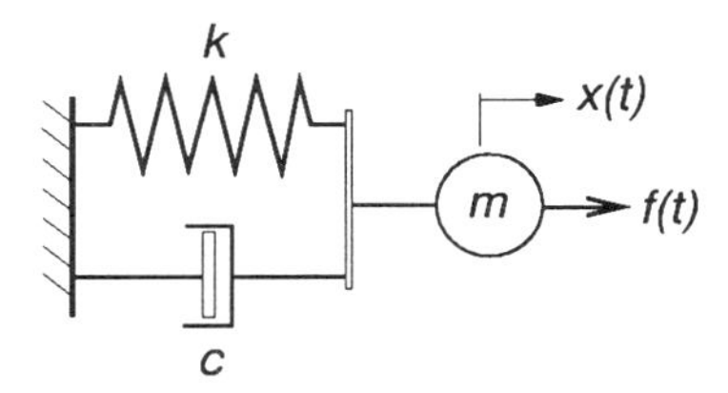

Figure I.2: Linear Oscillator with Damping

In anticipation of commonplace applications of this model, it is convenient to define a quantity called the *critical damping ratio*, ζ. This is the ratio of the actual damping constant in a given system to that which would cause the free motion to become a damped subsidence instead of a damped oscillation. In elementary mechanics textbooks[1] this is shown to be

$$c_{cr} = 2\sqrt{km} \tag{I.6}$$

thus

$$\zeta = \frac{c}{c_{cr}} = \frac{c}{2\sqrt{km}} \tag{I.7}$$

When this is put into the equation of motion it becomes

$$m\,\ddot{x}(t) + 2\sqrt{km}\,\dot{x} + k\,x = f(t) \tag{I.8}$$

[1]Higdon, A., and Stiles, W. B., *Engineering Mechanics*, Vector Edition, *Prentice-Hall, Inc.*, Englewood Cliffs, New Jersey, 1962, p. 718.

and if it is afterwards divided through by m, there is

$$\ddot{x}(t) + 2\,\zeta\,\omega_0\,\dot{x} + \omega_0^2\,x(t) = \frac{1}{m}\,f(t) = F(t) \tag{I.9}$$

with $\omega_0^2 = k/m$, the undamped natural frequency. Again, the initial conditions are adjoined. Let these be $x(0) = 0$, $\dot{x}(0) = 0$.

The state equation is

$$\dot{\mathbf{x}}(t) = \mathbf{A}\,\mathbf{x}(t) + \mathbf{B}\,F(t) \tag{I.10}$$

with

$$0\mathbf{x}(t) = \begin{bmatrix} x(t) \\ \dot{x}(t) \end{bmatrix} \quad \mathbf{A} = \begin{bmatrix} 0 & 1 \\ -\omega_0^2 & -2\zeta\omega_0 \end{bmatrix} \quad \mathbf{B} = \begin{bmatrix} 0 \\ \frac{1}{m} \end{bmatrix} \tag{I.11}$$

The motion in this case becomes

$$\mathbf{x}(t) = \int_0^t \mathbf{\Phi}(t-\tau)\,\mathbf{B}\,F(\tau)\,d\tau \tag{I.12}$$

In the above, the kernel of the integral is given by

$$\mathbf{\Phi}(t) = e^{-\zeta\omega_0 t} \begin{bmatrix} cos\beta t + \frac{\zeta\omega_0}{\beta} sin\beta t & \frac{1}{\beta} sin\beta t \\ -\frac{\omega_0^2}{\beta} sin\beta t & cos\beta t - \frac{\zeta\omega_0}{\beta} sin\beta t \end{bmatrix} \tag{I.13}$$

with $\beta = \omega_0\sqrt{1-\zeta^2}$.

I.3 STOCHASTIC PROCESSES

A stochastic process is composed of two components: one is a function of the deterministic variable t, which represents the mean or average motion; and a second, ω, which is random in nature, and is therefore governed by probability laws. A scalar stochastic process can furnish each component of a vector stochastic process.

The formal representations of stochastic processes are largely based upon the seminal works of Kolmogorov.[2] One of the more fundamental and useful definitions is

$$\mathbf{X}(t) = f(t; Y^1, Y^2, \ldots, Y^k) \tag{I.14}$$

in which the dependence of $\mathbf{X}$ upon t is deterministic (i.e., made specific), and its dependence upon $Y^1, Y^2, \ldots, Y^k$ is thus a function of those numerable random variables. It is then said that $\mathbf{X}$ has k *degrees of randomness*.

Under this definition, the function $\mathbf{X}(t)$ is fully specified by giving the joint probability distribution of $\mathbf{Y}$, where $\mathbf{Y}^t = [Y^1\ Y^2\ \cdots\ Y^k]$.

Moreover, it is assumed that when samples of $\mathbf{Y}$ are available, standard statistical methods can be applied to compute the statistics of the Y^k, and thereby

[2] Kolmogorov, A. N., "Über die Analytischen Methoden in der Wahrscheinlichkeitsrechnung," *Mathematische Annalen*, Bande 104, 1931, pp. 415-458.

furnish a sample of $\mathbf{X}(t)$. In other words, the usual requirements of continuity, differentiabilty, etc., are met. This form of representation also permits the application of computers to the task of simulating the stochastic process in question.

Figure I.3 illustrates a random variable that is a function of time, and is thus called a random process. Such a graph may have been taken from a sensor as it records some physical variable such as ground acceleration in an earthquake. After the record is obtained, it is no longer random, but specific, so that coordinates $X^i(t)$ can be matched with times t_i. However, in some other recording, a different (random) trace may be obtained. Figure I.4 shows several such and the probabilistic nature of each as it contributes to the distribution function of the random process.

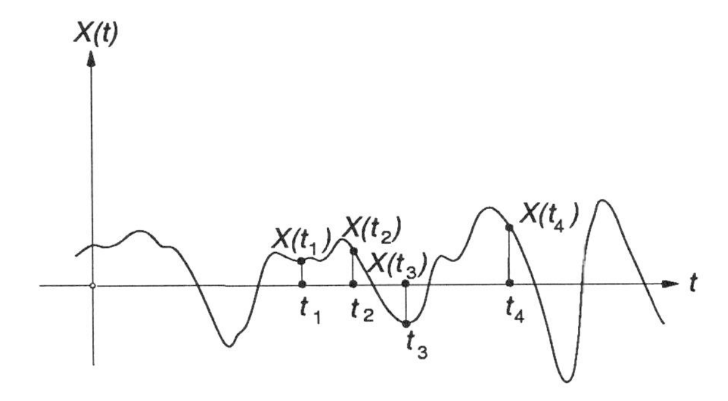

Figure I.3: A Random Process

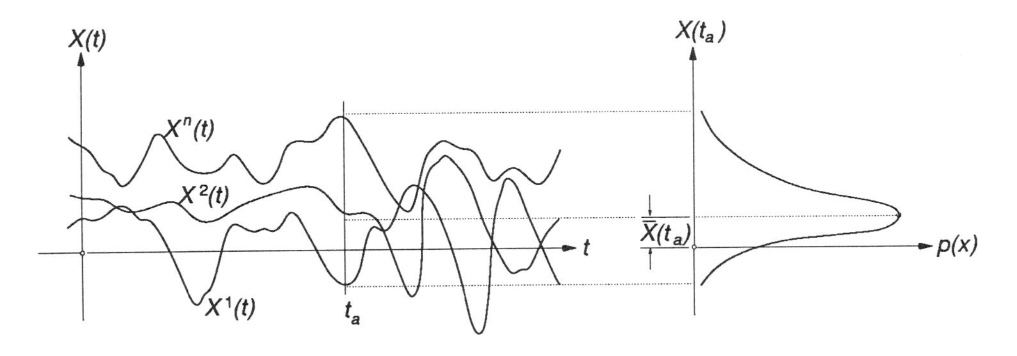

Figure I.4: Random Process Distribution

At time $t = t_a$, the probability is that $X(t)$ will conform to the probability distribution shown, which is obtained statistically from the sample of values $X^1(t_a)$, $X^2(t_a)$, ... , $X^n(t_a)$. The distribution will possess a mean, $\overline{X}(t)$, and a variance, σ, calculated as

$$\mu(t) = \int x\, p(x,t)\, dx \tag{I.15}$$

and

$$\sigma^2(t) = \int x^2\, p(x,t)\, dx - \mu^2(t) \tag{I.16}$$

There may also well be higher statistics if the distribution is significantly skewed.

But this is not the end of the story. Consider another random process—one which has the self-same statistics (mean, variance, etc.) as the former $X(t)$. Call it $Y(t)$. A graph comparing these two processes, Figure I.5, reveals at once that the frequency components of $Y(t)$ are very much higher than those of $X(t)$. Therefore, a probability distribution such as that shown in Figure I.4 for $X(t)$ is inadequate to represent a random process since it does not account for the correlation of $X(t_1)$ with $X(t_2)$, $X(t_3)$, ... , $X(t_n)$, and so forth. The distribution of Figure I.4 is a one-dimensional distribution, appropriate only to $X(t)$, as a "static" random variable.

In other words, $X(t_1)$ and the other $X(t_i)$ are ***strongly*** dependent upon each other, and this interdependence must be accounted for in settling the probability distribution of the random process.

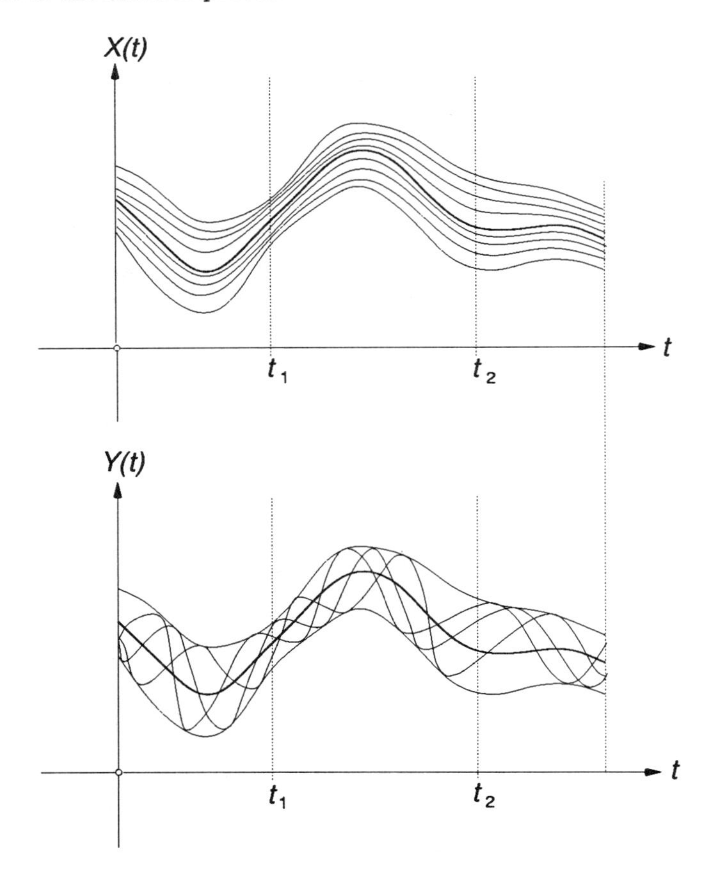

Figure I.5: Dissimilar Random Processes
with Same Temporal Distributions

At the same time, $Y(t_1)$ and its other $Y(t_i)$ are practically independent, at least for the times t_1 and t_2 illustrated in Figure I.5. They become more dependent as the times move closer to coincidence, obviously.

To resolve the dependence of $X(t_i)$ and $X(t_j)$, the process should be represented, as in Figure I.6, by a two-dimensional probability density distribution:

$$p(x^i, x^j; t_i, t_j)\, dx^i\, dx^j = P[x^i < X(t_i) \le x^i + dx^i, x^j < X(t_j) \le x^j + dx^j]$$

The probability density for this case is shown as a surface over the x^1–x^2 plane, and the probability is an element of volume $p(x, y)\, dx\, dy$. The total volume under the surface is, of course,

$$\int_{-\infty}^{+\infty} \int_{-\infty}^{+\infty} p(x^1, x^2)\, dx^1\, dx^2 = 1 \tag{I.17}$$

which represents certainty. To account for the dependence of many statistics upon each other, the probability density should be

$$p(x^1, x^2, \ldots, x^k; t_1, t_2, \ldots, t_k) \tag{I.18}$$

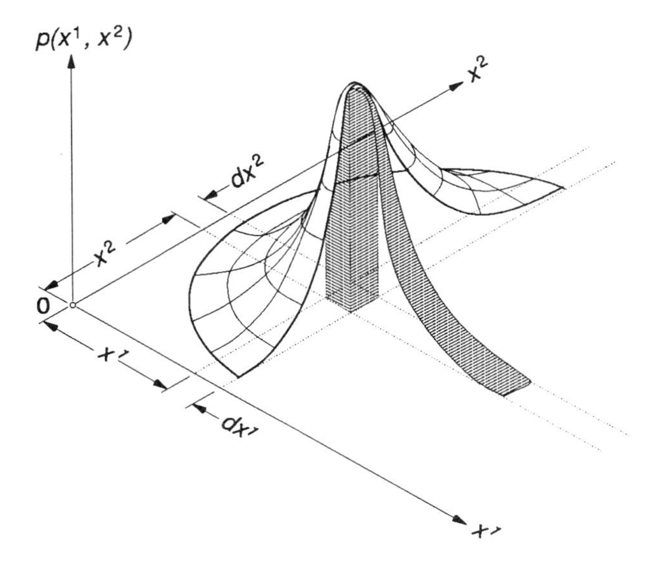

Figure I.6: Two-Dimensional Probability Density

The procedure can be carried to its logical conclusion by invoking the characteristic function associated with the probability distribution of the immediately preceding formula

$$\chi(v^1, v^2, \ldots, v^k; t_1, t_2, \ldots, t_k) = E\{exp\,[i \sum_{j=1}^{k} v^j\, X(t_k)]\} \tag{I.19}$$

Finally, upon taking the statistical moments at t_j infinitesimally close to each other and passing to the limit, there is the most complete representation for the random process

$$\chi[v(t)] = E\{exp\,[i \int v(t)\, X(t)\, dt]\} \qquad i = \sqrt{-1} \tag{I.20}$$

In the two equations immediately above, the notation $E\{\ \}$ is employed to signify the expected (mean) value of the quantity enclosed by the $E\{\ \}$ marks.

I.3.1 STOCHASTIC PROCESS MOMENTS

In many cases, the two-dimensional probability density distribution will be found to adequately represent random processes in systems analyses, and the mean and variance are then given as in Formulas I.15 and I.16. These are, effectively, joint statistical moments of two random variables. Since the density distribution is two-dimensional, there exists the possibility of higher moments not present in the case of one-dimensional distributions. The so-called ij^{th} *joint moment* is

$$m_{ij}(t_1, t_2) = E\{X^i(t_1)X^j(t_2)\} = \int_{-\infty}^{+\infty} \int_{-\infty}^{+\infty} x_i^1\, x_j^2\, p(x^1, x^2; t_1, t_2)\, dx^1\, dx^2 \tag{I.21}$$

In this formula, the symbol $E\{X^i(t_1)X^j(t_2)\}$ represents the mean of the product $X^i(t_1)$ and $X^j(t_2)$. In other words, this moment is like the mixed second moments of area, volume, and mass (the so-called "products of inertia") that are so familiar from elementary mechanics.

Two particular joint moments are of special importance in stochastic process analysis: $m_{11}(t_1, t_2) = E\{X(t_1)X(t_2)\}$ is known as the *auto-correlation function*, and $m_{12}(t_1, t_2) = E\{X(t_1)Y(t_2)\}$ bears the name *cross-correlation function.* As the names imply, these functions correlate two random processes from the same stochastic process (auto-correlation function), and from differing stochastic processes (cross-correlation function.) In fact, the utility of these functions has produced the special designations R_{XX} for the auto-correlation one and R_{XY} for the cross-correlation one.

The auto-correlation function for the two-dimensional case is therefore

$$R_{XX}(t_1, t_2) = E\{X(t_1)X(t_2)\} = \int_{-\infty}^{+\infty} \int_{-\infty}^{+\infty} x^1\, x^2\, p(x^1, x^2; t_1, t_2)\, dx^1\, dx^2 \tag{I.22}$$

I.3.2 STATIONARY RANDOM PROCESSES

A large fraction of the random processes observed in systems analysis possess a behavior that can be called "stationary." This property is very desirable from the point of view of computing the correlation function in those cases where it occurs. The stationary property means that the statistics of the process (mean, variance, etc.,) are not functions of time.

The processes $X(t)$ and $Y(t)$ displayed in Figure I.5 are not stationary, because their distributions and their statistics change with time, as shown.

Basically, a stationary process is one for which the distribution and the statistics are not functions of the time origin or of a shift of the origin in time. This can be formulated mathematically as

$$p(x^1, x^2, \ldots, x^k; t_1, t_2, \ldots, t_k) = p(x^1, x^2, \ldots, x^k; t_1 + \tau, t_2 + \tau, \ldots, t_k + \tau) \quad \text{(I.23)}$$

for any k and τ. If this condition holds for a random process, then (for $k = 1$) its one-dimensional distribution does not depend upon time.

$$p(x; t) = p(x; t + \tau) = p(x) \qquad \text{(I.24)}$$

for any t and τ.

Likewise, put $k = 2$ or $t_2 = t_1 + \tau$ so that for a two-dimensional stationary process the distribution depends on the difference $\tau = t_2 - t_1$, but not on t. Hence,

$$p(x^1, x^2; t_1, t_2) = p(x^1, x^2; t_1, t_1 + \tau) = p(x^1, x^2; t_2 - t_1) = p(x^1, x^2; \tau) \qquad \text{(I.25)}$$

Therefore, for a stationary process, the correlation function is

$$R_{XX}(t_1, t_2) = \iint x^1\, x^2\, p(x^1, x^2; \tau)\, dx^1\, dx^2 = R_{XX}(\tau) \qquad \text{(I.26)}$$

This means that the mutual dependence of $X(t_1)$ and $X(t_2)$ is the same at any place in the process, so long as $t_2 - t_1 = \tau$ is the same.

It is still possible for the random process in question to be time dependent, since the k-dimensional distribution may be time dependent for the higher terms $k \geq 3$. But for many applications in systems work, the less stringent requirement based upon the one-dimensional stationary distribution is considered adequate. For this, the "weakly stationary process," the correlation function is

$$p(x; t) = p(x) \qquad R(t_1, t_2) = R(\tau) \qquad \text{(I.27)}$$

I.3.3 CORRELATION FUNCTION

In the previous section the stationary random process was explored, and the interdependence of two representations of the random process was shown to be defined by the correlation function $R(t_1, t_2)$. It is further seen that the correlation function is only dependent upon the time difference $\tau = t_2 - t_1$ or

$$R(\tau) = E\{X(t)X(t + \tau)\} \qquad \text{(I.28)}$$

There are several properties affiliated with the correlation function.

- $R(\tau)$ is an even function of τ; that is, $R(\tau) = R(-\tau)$
- $R(\tau)$ exists, is finite, and $|R(\tau)| \leq R(0)$
- An $R(\tau)$ continuous at the origin is uniformly continuous over τ. This is exhibited by employing the increment

$$R(\tau + \epsilon) - R(\tau) = E\{[X(t - \epsilon) - X(t)]\, X(t + \tau)\}$$

then using the Schwartz inequality[3] $E\{XY\}^2 = |E\{XY\}|^2 \le E\{X^2\}E\{Y^2\}$ to achieve

$$|R(\tau+\epsilon) - R(\tau)|^2 \le 2R(0)[R(0) - R(\epsilon)]$$

Because $R(\tau)$ is continuous at the origin, $R(0) - R(\epsilon) \to 0$ as $\epsilon \to 0$. Therefore, $R(\tau+\epsilon) - R(\tau) \to 0$ as $\epsilon \to 0$ uniformly on τ.

- $R(\tau)$ is nonnegative definite. That is, for every $t_1, t_2, \ldots, t_k$, and for an arbitrary function $f(t)$, all on the range T,

$$\sum_{i=1}^{n}\sum_{j=1}^{n} R(t_j - t_i)\, f(t_i)\, f(t_j) \ge 0$$

- For $R(\tau)$ continuous at the origin, there is the representation

$$R(\tau) = \int_{-\infty}^{+\infty} e^{i\omega\tau}\, d\Phi(\omega)$$

and for $\Phi(\omega)$ absolutely continuous, there is also

$$S(\omega) = \frac{d\Phi(\omega)}{d\omega}$$

and then

$$R(\tau) = \int_{-\infty}^{+\infty} e^{j\omega\tau}\, S(\omega)\, d\omega$$

in which $S(\omega)$ is real, nonnegative. This is the first of a pair of Fourier transforms,[4] with the inverse transform given by

$$S(\omega) = \frac{1}{2\pi}\int_{-\infty}^{+\infty} e^{-i\omega\tau}\, R(\tau)\, d\tau$$

Fourier transforms are, like their Laplace counterparts, a method of removing a relationship from one domain (where they are usually difficult to treat) into a different realm where they are considered more amenable to manipulation. They occur in pairs, so that the inverse transformation (mapping) can be rendered, and the interpretation into the original language made evident. There are tables of Fourier transforms, just as there are Laplace transforms, and, indeed, many other transforms by name, each suitable to some special purposes. A good source which treats some of the more important transforms, together with tables of some common pairs, is the Sneddon book previously cited.[5]

---

[3] Zadeh, L., and Desoer, C. A., *Linear System Theory*, McGraw-Hill Book Company, Inc., New York, 1963, p. 553.

[4] Sneddon, I. H., *The Use of Integral Transforms*, McGraw-Hill Book Company, New York, 1972, pp. 36,37.

[5] Sneddon, I. H., op. cit.

I.3.4 POWER SPECTRAL DENSITY FUNCTION

An entity known as the *power spectral density function* is defined by the inverse Fourier transform of the correlation function for the weakly stationary stochastic process. The word "spectral" here refers to the correlation *qua function* of the frequency ω rather than the time τ. (Actually, there is a factor of $1/\sqrt{2\pi}$ difference in the definition of the power spectral density function and that of the usual classical version of the Fourier transforms, but this factor is incorporated in the versions given above.) Thus,

$$S(\omega) = \frac{1}{2\pi} \int_{-\infty}^{+\infty} e^{-i\omega\tau} R(\tau)\, d\tau \tag{I.29}$$

defines the function.

The definition given above is the so-called *exponential Fourier transform* of $R(\tau)$, in harmony with the exponential form of the kernel in the defining integral. In practice, it is common to see the *Fourier cosine (and sine) transforms* instead of the exponential form.[6] In these terms, the cosine transform pair is

$$S(\omega) = \frac{1}{\pi} \int_{0}^{\infty} R(\tau)\, cos\omega\tau\, d\tau \qquad R(\tau) = 2 \int_{0}^{\infty} S(\omega)\, cos\omega\tau\, d\omega \tag{I.30}$$

The latter version accounts for the facts that $S(\omega)$ is both real and nonnegative, and that $R(\tau)$ is an even function (of τ).

Further, a customary practice in applications is to employ the *one-sided power spectral density function*

$$G(\omega) = 2\, S(\omega) \qquad \omega > 0 \tag{I.31}$$

$$= 0 \qquad \omega \neq 0 \tag{I.32}$$

With these adjustments, there are

$$G(\omega) = \frac{2}{\pi} \int_{0}^{\infty} R(\tau)\, cos\omega\tau\, d\tau \qquad \omega > 0 \qquad R(\tau) = \int_{0}^{\infty} G(\omega)\, cos\omega\tau\, d\omega \tag{I.33}$$

In addition, when $\tau = 0$, the correlation becomes

$$R(0) = E\{X^2(t)\} = 2 \int_{0}^{\infty} S(\omega)\, d\omega = \int_{0}^{\infty} G(\omega)\, d\omega \tag{I.34}$$

The power spectral density function plays, for the weakly stationary stochastic process, a role similar to that of a Fourier analysis in the nonrandom dynamic system case. It is thus a most potent instrument for both analysis and design of such systems. One might even say, it is the preferred approach to such tasks.

I.3.5 WHITE NOISE POWER SPECTRAL DENSITY

The definition of a white noise is that its power spectral density function is constant over the entire spectrum. In other words,

$$G(\omega) = G_0 \qquad 0 < \omega < \infty \tag{I.35}$$

---

[6] Sneddon, I. H., ibid, p. 7.

This means that the correlation function is

$$R(\tau) = \pi G_o \delta(\tau) \tag{I.36}$$

where $\delta(\tau)$ is the Dirac delta.

Figure I.7 shows the correlation function and the spectral density function for white noise.

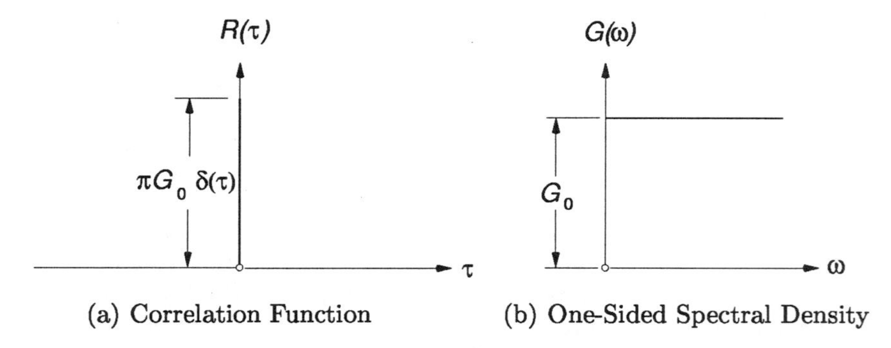

(a) Correlation Function (b) One-Sided Spectral Density

Figure I.7: White Noise Process

White noise is, of course, a fiction: In the first place, its power would need to be infinitely great to produce a spectral density that extends uniformly over the range of frequency from 0 to ∞. But it is nevertheless a useful idealization. For instance, in a comparison of a real noise with the white noise ideal model, it is seen (as in Figure I.8) that, although the real noise is neither uniform over the range nor of infinite extent, it is a suitable approximation, over a band which is, although finite, sufficiently wide to cover most of the effects in a given natural phenomenon.

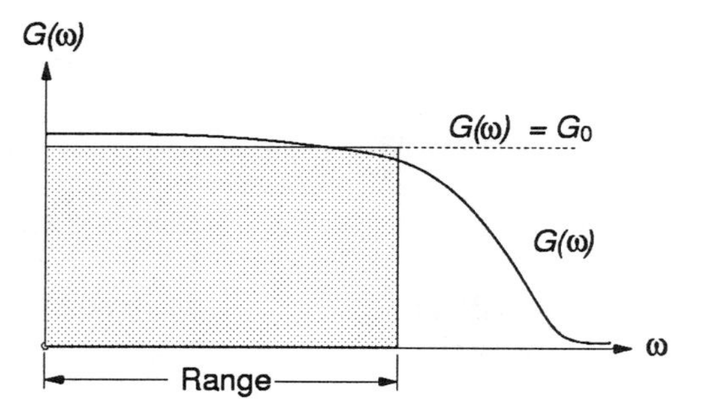

Figure I.8: White Noise Approximation

I.3.6 LINEAR OSCILLATOR WITH WHITE NOISE EXCITATION

Return to the case of the linear oscillator with damping, and let the excitation be a white noise $\mathbf{g}\,W(t)$ which has a mean value of zero and a one-sided spectral density G_0. The correlation function can be taken to be $R_{FF}(\tau) = \pi\, G_0\, \delta(\tau)$, and the initial state may be $\mathbf{X}(0) = \mathbf{0}$.

The state equation as previously given in Equations (I.10), (I.11), (I.12), and (I.13) is

$$\dot{\mathbf{x}}(t) = \mathbf{A}\,\mathbf{x}(t) + \mathbf{g}\,W(t) \tag{I.37}$$

wherein

$$\mathbf{x}(t) = \begin{bmatrix} x(t) \\ \dot{x}(t) \end{bmatrix} \qquad \mathbf{A} = \begin{bmatrix} 0 & 1 \\ -\omega^2 & -2\zeta\omega_0 \end{bmatrix} \qquad \mathbf{g} = \begin{bmatrix} 0 \\ \frac{1}{m} \end{bmatrix} \tag{I.38}$$

and the state transition matrix is given by

$$\mathbf{x}(t) = \int_0^t \mathbf{\Phi}(t-\tau)\,\mathbf{g}\,W(\tau)\,d\tau \tag{I.39}$$

In the integral above, the function $\mathbf{\Phi}$ is given by

$$\mathbf{\Phi}(t) = e^{-\zeta\omega_0 t} \begin{bmatrix} \cos\beta t + \dfrac{\zeta\omega_0}{\beta}\sin\beta t & \dfrac{1}{\beta}\sin\beta t \\ -\dfrac{\omega_0^2}{\beta}\sin\beta t & \cos\beta t - \dfrac{\zeta\omega_0}{\beta}\sin\beta t \end{bmatrix} \tag{I.40}$$

and, as before, $\beta = \omega_0\sqrt{1-\zeta^2}$.

The mean value of $\mathbf{x}(t)$ is zero for this model, thus the next statistic of interest is the variance $x^1(t) = x(t)$. The appropriate component of the response variance $\Gamma(t)$ is

$$\sigma_x^2(t) = \int_0^t \int_0^t \phi_{12}(t-u)\phi_{12}(t-v)\,R_{FF}(u-v)\,du\,dv \tag{I.41}$$

in which $\phi_{12}(t)$ is the element of $\mathbf{\Phi}(\mathbf{t})$ that is appropriate to this correlation; namely,

$$\phi_{12}(t) = \frac{1}{\beta}e^{-\zeta\omega_0 t}\sin\beta t \tag{I.42}$$

Carrying out the multiple integration involved in Equation I.41 is a very formidable task. It is usually not possible to produce an explicit closed-form version, and the best tactic seems to be to retreat to the power spectral density function. Using the one-sided version shown in Equation I.33,

$$R_{FF}(\tau) = \int_0^\infty G_{FF}(\omega)\,\cos\omega\tau\,d\omega \tag{I.43}$$

and entering this into Equation I.41, the new result is

$$\sigma_x^2(t) = \int_0^\infty \int_0^t \int_0^t \phi_{12}(t-u)\,\phi_{12}(t-v)\,G_{FF}(\omega)\,\cos\omega(u-v)\,du\,dv\,d\omega \tag{I.44}$$

Upon injecting Equation I.42, the full expression for the variance is

$$\begin{aligned} \sigma_x^2(t) = \int_0^\infty |h(\omega)|^2\; G_{FF}(\omega) \Bigg\{ 1 + e^{-2\zeta\omega_o t} \Bigg[1 + \frac{2\zeta\omega_o}{\beta}\sin\beta t\cos\beta t \\ - e^{\zeta\omega_o t}\left(2\cos\beta t + \frac{2\zeta\omega_o}{\beta}\sin\beta t\right)\cos\omega t \\ - \frac{2\omega}{\beta}e^{\zeta\omega_o t}\sin\beta t\,\sin\omega t + \frac{1}{\beta^2}\left(\zeta^2\omega_o^2 - \beta^2 + \omega^2\right)\sin^2\beta t \Bigg] \Bigg\}\,d\omega \end{aligned} \tag{I.45}$$

with

$$h(\omega) = \int_{-\infty}^{\infty} \phi_{12}(\tau)\, e^{-i\omega\tau}\, d\tau = \frac{1}{\omega_0^2 - \omega^2 + 2i\zeta\omega_0\omega} \tag{I.46}$$

which is the response function appropriate to $x(t)$.

When G_{FF} is a smooth function (i.e., without sharp peaks) and if ζ is small (that is, small damping), the function $G_{FF}(\omega)$ can be approximated by $G_{FF}(\omega_0)$, since $|\,h(\omega)\,|^2$ has a very sharp peak at $\omega = \omega_0$, (resonance condition), therefore its main importance will be at that frequency. These adjustments allow the integral in Equation I.45 to be carried out by contour integration.[7] The outcome is

$$\sigma_x^2(t) \cong \frac{\pi G_{FF}(\omega_0)}{4\zeta\omega_0^3}\left[1 - \frac{1}{\beta^2}e^{-2\zeta\omega_0 t}\left(\beta^2 + 2\zeta^2\omega_0^2 sin^2\beta t + \zeta\omega_0\beta sin2\beta t\right)\right] \tag{I.47}$$

The result above is seen to be sensitive to the damping ratio, ζ, even if ζ is a small quantity. In practical applications, ζ ranges from $\zeta = 0$ to $\zeta = 0.1$, or so. When $\zeta = 0$, (the undamped case), the variance comes down to

$$\sigma_x^2(t) = \frac{\pi G_{FF}(\omega_0)}{4\omega_0^3}\left(2\omega_0 t - sin2\omega_0 t\right) \tag{I.48}$$

This is the same result which is obtained for the comparable component of the covariance matrix of a model without damping under white noise excitation, as it should be.

A graph showing the manner in which $\sigma_x^2(t)$ varies with several parameters of ζ is displayed in Figure I.9. The nondimensional ordinate is

$$S_x(t) = \frac{2\sigma_x^2(t)\omega_0^3}{\pi G_{FF}(\omega_0)} \tag{I.49}$$

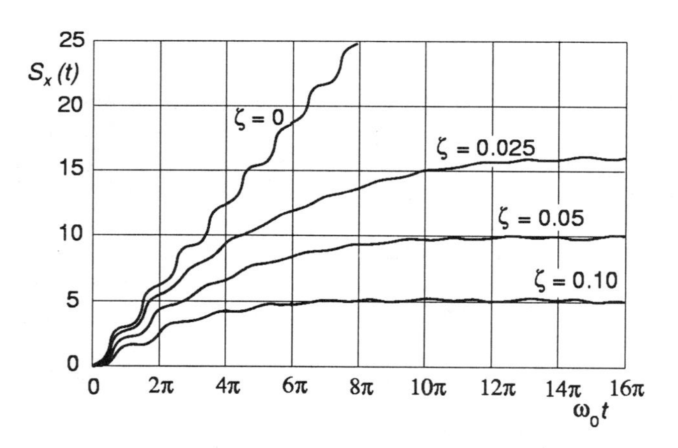

Figure I.9: Variance as a Function of Damping

[7] Caughey, T. K., and Stumpf, H. J., "Transient Response of a Dynamics System Under Random Excitation," *Journal of the Applied Mechanics Division*, *ASME*, Volume 18, 1961, pp. 563-566.

APPENDIX J

HYDRAULIC COMPUTATIONS

J.1 FLOWS IN PIPES

The flow of fluids in pipes is a major applications area of fluid mechanics (hydraulics) in Civil Engineering practice. The fundamental equation which governs the phenomenon is the Darcy-Wiesbach formula

$$h_f = f\,\frac{L}{D}\,\frac{v^2}{2g} \tag{J.1}$$

in which

h_f	is the head loss in *meters* (of fluid flowing),
f	is the friction factor, dimensionless,
L	is the pipe length in *meters*,
D	is the pipe diameter, also in *meters*,
v	is the mean velocity in *meters/second*,
g	is local gravity, m/s^2.

The formula is established by applying principles of dimensional analysis, and is taught in all good undergraduate colleges of engineering.

The friction factor (or resistance coefficient) in the Darcy-Wiesbach law is a function of the Reynolds number and the pipe roughness. The accepted relationship in the turbulent regime is the well-known Colebrook formula [1][1]

$$\frac{1}{\sqrt{f}} = -0.86\,ln\left(\frac{\epsilon/D}{3.7} + \frac{2.51}{\mathcal{R}\sqrt{f}}\right) \tag{J.2}$$

wherein $\mathcal{R} = vD/\nu$ is the Reynolds number and ϵ is the pipe roughness in *meters*. ν is the kinematic viscosity of the fluid in m$^2/s$.

The Colebrook formula is a very good representation of the friction factor, but because of the transcendental embedment of f in it, practical applications have usually resorted to graphical or approximate computation of f for subsequent use in the Darcy-Wiesbach law. The Moody diagram[2] is the classical example of a graphical design chart, and the Wood[3] and Jain[4] formulas are representative of the approximations in recent times.

[1]Numbers in brackets refer to the list of references at the end of Section J.1.

However, it is not necessary to resort to such approximate solutions of the Colebrook formula. A recent paper[5] displays a Newton-Raphson algorithm applied to the task of excerpting f from the formula. By employing the algorithm in a subprogram, the analysis and design problems of pipeflow can be facilitated through direct computation with the Darcy-Wiesbach law.

J.1.1 THE ALGORITHM

The Colebrook equation can be written

$$F(f) = \sqrt{f}\left(\exp^{-\left(\frac{1}{0.86\sqrt{f}}\right)} - \frac{\epsilon/D}{3.7}\right) + \frac{2.51}{\mathcal{R}\sqrt{f}} = 0 \tag{J.3}$$

Thereafter, a starting value f_o is selected, following which some better numbers

$$f_1 = f_o - \frac{F(f_o)}{F'(f_o)} \quad \cdots \quad f_{n+1} = f_n - \frac{F(f_n)}{F'(f_n)} \tag{J.4}$$

are obtained by iteration. The process continues until whatever pre-set degree of accuracy is reached at convergence. Convergence is guaranteed, if a suitable starting point (such as is illustrated in Figure J.1) is taken. Three or four iterations are usually adequate for engineering purposes.

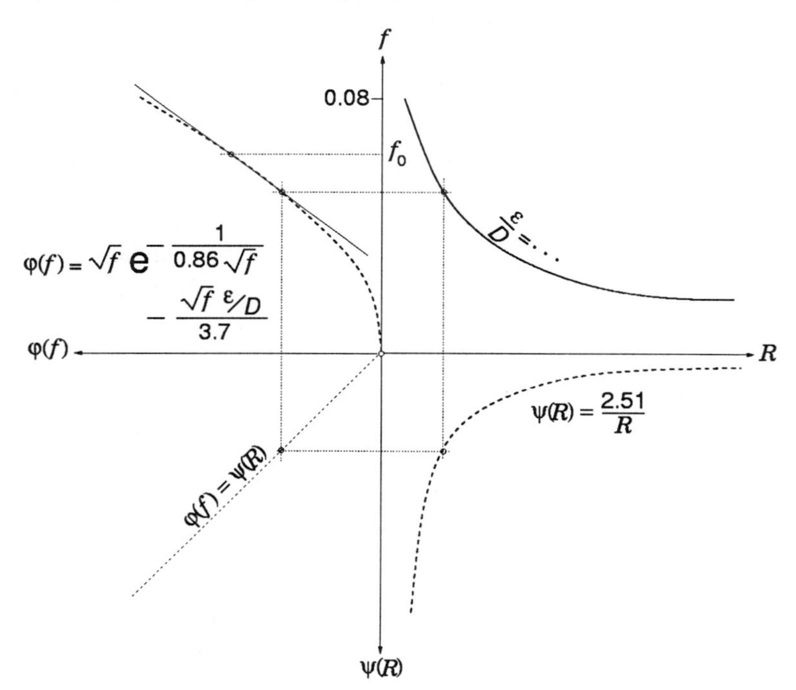

Figure J.1: Solution Algorithm for Colebrook Formula

In practice, the friction factor does not exceed 0.08, so this value can be built into the iterative procedure as a starting point. Normally this costs, at most, one additional iteration, which is unnoticed since it occurs at machine speed for present computers.

Figure J.1 displays the understructure of the solution algorithm. From the equation for $F(f)$, above, it is seen that there are two functions $f_2(f)$ and $f_1(\mathcal{R})$ in which the variables f and $\mathcal{R}$ have been separated. A graph of f versus $\mathcal{R}$ is constructed, and then the third quadrant of the space is utilized to plot $f_2(f)$ versus $f_1(\mathcal{R})$ so that the solution for f occurs along the line $f_2(f) = f_1(\mathcal{R})$.

Then, beginning with the initial value of f (0.08) on the curve $f_2(f)$ in the upper reaches of the second quadrant of the space, the iterative process advances to its conclusion at a value of $f \equiv f_{n+1}$ which represents a fulfilment of the Colebrook formula. Note that the first quadrant of the f versus $\mathcal{R}$ space is the same as that of the Moody diagram, in which ϵ/D has a particular value. That is, ϵ/D plays the role of a parameter in the entire process, one which was selected at the outset as an item of data.

J.1.2 THE TYPES OF PIPEFLOW PROBLEMS

Pipeflow problems can be classified into three types: unknown head loss, unknown discharge, and unknown diameter. Design engineers may encounter any of these types in practice, depending upon what is known under the particular circumstances. For example, if the pipe diameter, the roughness, the length, the temperature and the discharge are known, the computation is a straightforward matter of substituting in the Darcy-Weisbach law so as to produce the head loss directly.

A second type of computation—determining the discharge—is not much more difficult if one starts with the head loss, the pipe diameter, the roughness, the length and the temperature. In this case however, an iterative process ensues because both the Reynolds number and the mean velocity are unknown.

The third category of problems is that of sizing the pipe; *i.e.*, finding the diameter of the pipe which is required to carry a specified discharge at a certain headloss. This requires iterating through the Darcy-Wiesbach, the Reynolds number and the discharge.

To achieve a maximum utility, a computational program for pipeflow would incorporate the capacity to treat all three of these types of problems.

J.1.3 THE PROGRAM

Based upon the desired features mentioned above, the following program is written in Pascal source code with the filename ***pipeflow.p***, and the compiled program has the filename ***pipeflow.go***.

```
program PipeFlow(input,output,lst);
{ This program computes the head loss, the discharge, or the   }
{ diameter when appropriate pipe flow parameters are entered. }

var
   A, D, E, F, F1, F2, H, J, K, L, M, N, Q, R, V, NEWF: real;
   reply: char; lst: text;
   name: string{22};
   date: string{18};
   thetime: string{10};

label
   10;

procedure FrictionFactor;
   begin
      R:=V*D/N;
      J:= −1/(0.86*sqrt(F));
      K:= E/(3.7*D);
      M: (sqrt(F)*(2.51/R));
      F1:= 2*(F*(exp(J)−K)−M);
      F2:= ((1−J)*exp(J))-K;
      NEWF:= F − (F1/F2);
   end;

begin
   for i:=1 to 25
      begin
         writeln(' ');
      end;
   write(' Please enter your name. '); readln(name); writeln(' ');
   write(' Give the date. '); readln(date); writeln(' ');
   write(' What time is it? '); readln(thetime);
   for i:=1 to 20
      begin
         writeln(' ');
      end;
   10:
   rewrite(lst,'pipeflow.dat');
   for i:=1 to 25
      begin
         writeln(' ');
      end;
   writeln(' This program computes the head loss, the discharge or the');
   writeln(' diameter when appropriate pipe flow parameters are entered.');
   writeln(' '); writeln(' '); writeln(' ');
   writeln(' If H, Q or D is unknown, enter 0 for it');
```

```
for i:=1 to 10
  begin
    writeln(' ');
  end;
write(' Please give the pipe diameter in meters '); readln(D); writeln(' ');
write(' Please enter the roughness in meters '); readln(E); writeln(' ');
write(' List the head loss in meters '); readln(H); writeln(' ');
write(' Give the pipe length in meters '); readln(L); writeln(' ');
write(' What is the kinematic viscosity in square meters/s '); readln(N); ');
write(' writeln(' ');
write(' Please furnish the discharge in cubic meters/s '); readln(Q);

for i:=1 to 25
  begin
    writeln(' ');
  end:
NEWF:= 0.08;

if H = 0 then
  begin
    repeat
      A:=3.1415926536*D*D/4;
      V:=Q/A;
      FrictionFactor;
    until abs(NEWF - F)≤ 0.000001;
    writeln(' The friction factor is ',F:9:5,''); writeln(' ');
    H:=NEWF*(L/D)*(V*V)/19.612;
    writeln(' The head loss is ',H:6:2,' m'); writeln(' ');
  end;

if Q = 0 then
  begin
    repeat
      F:=NEWF;
      V:=sqrt(19.612*H*D/(F*L));
      A:=3.1415926536*D*D/4;
      FrictionFactor;
    until abs(NEWF - F)≤ 0.000001;
    writeln(' The friction factor is ',F:9:5,''); writeln(' ');
    Q:=A*V;
    write(' The discharge is ',Q:6:2,' cubic meters/s'); writeln(' ');
  end;

if D = 0 then
  begin
    repeat
      F:=NEWF;
```

```
         D:=exp(0.2*ln(8*L*Q*Q*F/(H*9.806*3.1415926536*3.1415926536)));
         V:=4*Q/(3.1415926536*D*D);
         FrictionFactor;
      until abs(NEWF − F)≤ 0.000001;
      writeln(' The friction factor is ',F:9:5,''); writeln(' ');
      writeln(' The diameter is ',D:7:3,' m');
   end;


for i:=1 to 15
   begin
      writeln(' ');
   end;
write(' Do you wish to have a printout of this? '); readln(reply);
if (reply = 'y') or (reply = 'Y') then
   begin
      writeln(' '); writeln(lst,'username: ',name:22,' ',date:18,' ',thetime:10,' ');
      writeln(lst,' '); writeln(lst,' '); writeln(lst,' ');
      writeln(lst,' PIPEFLOW'); writeln(lst,' ');
      write(lst,' The pipe diameter is ',D:8:6,' m'); writeln(lst,' ');
      write(lst,' The roughness is ',E:10:8,' m'); writeln(lst,' ');
      write(lst,' The head loss is ',H:8:6,' m'); writeln(lst,' ');
      write(lst,' The pipe length is 'L:8:6,' m'); writeln(lst,' ');
      write(lst,' The kinematic viscosity is 'N:10:8,' sq m/s'); writeln(lst,' ');
      writeln(lst,' The discharge is 'Q:8:6.' cms'); writeln(lst,' ');
      write(lst,' The friction factor is 'F:9:5,' ');
      writeln(' '); writeln(' '); writeln(lst,' '); writeln(lst,' ');
   end;


write(' Do you wish to do another problem? '); readln(reply);
if (reply = 'y') or (reply = 'Y') then
   begin
      writeln(' '); writeln(' '); goto 10;
   end;
   else
   begin
      for i:=1 to 25
         begin
            writeln(' ');
         end;
      writeln(' Thank you for using this program.'); writeln(' '); writeln(' ');
      writeln(' Have a nice day!');
      for i:=1 to 10
         begin
            writeln(' ');
         end;
   end;
end.
```

The program writes to an output file named ***pipeflow.dat*** which must be printed to see the hard copy of the results. The file is re-written each time the program is run. The results are also displayed on the monitor of the workstation.

As can easily be seen from the code, the program is user-friendly and interactive. When the execute command ***pipeflow.go*** is given, the program runs and displays prompts on the monitor screen for the entry of data. As the last item of data is entered, the results are shown and the user is queried to learn if a hard copy is wanted. The last inquiry is whether or not a new problem is to be solved.

J.1.4 EXAMPLE PROBLEMS

The following are examples of the types of problems which are solved using the program ***pipeflow.go***.

1. Water flows at a rate of 0.056 m^3/s in a 20 cm diameter cast-iron pipe. What is the head loss per kilometer of pipe?

   Water (taken under ambient nominal conditions) has a kinematic viscosity of $\nu = 1.12 \times 10^6$ m^2/s, and cast-iron pipe has a roughness of 0.00026 m.

   In this example, the data above were entered, along with 0 for the head loss. The machine produced the following output:

   ```
   PIPEFLOW

   The pipe diameter is 0.200000 m
   The roughness is 0.00026000 m
   The head loss is 17.941660 m
   The pipe length is 1000.000000 m
   The kinematic viscosity is 0.00000112 sq m/s
   The discharge is 0.056000 cms
   The friction factor is 0.02215
   ```

2. The head loss per kilometer of 20 cm diameter cast-iron pipe is 20 m. What is the flow-rate of the water in the pipe?

   Using these data and the roughness and viscosity from the previous problem, the following output was obtained when 0 was entered for Q:

   ```
   PIPEFLOW

   The pipe diameter is 0.200000 m
   The roughness is 0.00026000 m
   The head loss is 20.000000 m
   The pipe length is 1000.000000 m
   The kinematic viscosity is 0.00000112 sq m/s
   The discharge is 0.059178 cms
   The friction factor is 0.02211
   ```

3. What size cast-iron pipe is required to carry water at a discharge of 0.01 cms with a head loss of 1.8 m per 300 m of pipe?

   The stated data and previous values of roughness and viscosity were entered, along with 0 for the diameter, to produce the following:

PIPEFLOW

```
The pipe diameter is 0.308044 m
The roughness is 0.00026000 m
The head loss is 1.800000 m
The pipe length is 300.000000 m
The kinematic viscosity is 0.00000112 sq m/s
The discharge is 0.100000 cms
The friction factor is 0.02013
```

These results are easily confirmed using the Moody diagram.

J.1.5 REFERENCES

1. Colebrook, C. F., "Turbulent Flow in Pipes, with Particular Reference to the Transition Between the Smooth and Rough Pipe Laws," *Journal of the Institution of Civil Engineers*, Volume 11, pp. 133-156, 1938.

2. Moody, L. F., "Friction Factors for Pipe Flow," *Transactions, American Society of Mechanical Engineers*, Volume 66, pp. 671-684, 1944; "An Approximate Formula for Pipe Friction Factors," *Mechanical Engineering*, Volume 69, p. 1005, 1947.

3. Wood, D. J., "An Explicit Friction Factor Relationship," *Civil Engineering*, Volume 36, Number 12, pp. 60-61, 1966.

4. Jain, A. K., "Accurate Explicit Equation for Friction Factor," *Journal of the Hydraulics Division, American Society of Civil Engineers*, Volume HY5, pp. 674-677, 1976.

5. McDonald, P. H., "Computing the Friction Factor from the Colebrook Formula," *Proceedings, I Pan American Congress of Applied Mechanics*, Rio de Janiero, Brasil, pp. 356-359, 1989.

J.2 CHANNEL FLOW COMPUTATIONS

The flow of water and other liquids in open channels—often called *free surface flows*—is a topic of considerable importance in hydraulics and hydrology, and is a major feature of Civil Engineering practice in the so-called "water area." Canals, ditches, rivers, streams, and culverts are among the natural or man-made works which exhibit channel flows.

There are three fundamental governing equations which dominate the computational agenda in analysis and design with open channel flow. These are:[2]

1. The Chézy-Manning law

$$Q = \frac{1}{n} \; A \; R_H^{2/3} \; S_o^{1/2} \tag{J.5}$$

in which

Q	is the discharge in *cubic meters per second*,
A	is the channel cross-sectional flow area in *sq meters*,
R_H	is the hydraulic radius in *meters*,
S_o	is the bedslope of the channel (dimensionless)
n	is Manning's roughness coefficient for the channel lining.

2. The Critical Depth Formula

$$\frac{Q^2 b}{g A^3} = 1 \tag{J.6}$$

wherein

Q	is the discharge in *cubic meters per second*,
b	is the top width of the liquid in *meters*,
A	is the cross-sectional area of the liquid in *square meters*,
g	*islocalgravity*, $\mathrm{m/s^2}$.

3. The Free-Surface Differential Equation

$$dx = \frac{1 - Q^2 b / g A^3}{S_o - S} \; dy \tag{J.7}$$

with

x	the channel length measure, *meters*,
y	the depth of liquid at place x in *meters*,
S	the slope of the energy line (dimensionless),

and the other parameters as before.

The Chézy-Manning law applies to *uniform flow*, a condition in which the water depth is constant (uniform) along the *reach* (length) of the channel. The value of this constant depth, y_o, is called *the normal depth*, and is obtained from

[2]See reference at the end of this section.

Chézy-Manning by solving the fifth-degree algebraic equation which results from the substitution of Q, S_o, and n, the channel's properties. In other words, there is associated with each channel a certain depth—the normal depth—at which water will flow at constant depth and while carrying a certain discharge Q.

The ***critical depth***, y_c, is a measure of whether the flow of water in a channel is *supercritical* (*rapid, shooting*) or *subcritical* (*slow, tranquil*). The particular shape of the free-surface profile is a function of the relationship between y_o and y_c, so both must be determined in order that the profile can be made definite. y_c is obtained as the solution of the critical depth formula when Q, b, g and the parameters of A are inserted. The resulting equation is (for, say, a trapezoidal cross-section) of the sixth degree in y_c.

The differential equation for the free surface is a quadrature. Ideally, one would wish to know the depth as a function of the length (y as a function of x), but instead one gets x as a function of y. It is easy to see that when the variables are separated, dx integrates to x, evaluated between definite limits. That is,

$$\int_{x_1}^{x_2} dx = |x|_{x_1}^{x_2} = x_2 - x_1 = L = \int_{y_1}^{y_2} \frac{1 - Q^2 b/gA^3}{S_o - S} dy = \int_{y_1}^{y_2} f(y)dy \tag{J.8}$$

The customary assumption is that for a differential reach dx the slope of the energy line is given by the Chézy-Manning formula, so that

$$L = \int_{y_1}^{y_2} \frac{1 - Q^2 b/gA^3}{S_o - \left(nQ/AR_H^{2/3}\right)^2} dy = \int_{y_1}^{y_2} f(y)dy \tag{J.9}$$

and the most efficaceous procedure is usually taken to be a numerical integration to obtain number pairs (L, y_2), when y_1 is given.

J.2.1 THE PROGRAM

A desired program for machine computation would incorporate all of the three elements delineated above. As input data, it would be expected that Q, n, and the shape of the cross-section would be furnished. As output, the program should render y_o, y_c, S_c and a selection of pairs (L, y_2) for a chosen y_1.

Possibly the most common cross-section is that of the trapezoid. For this shape, there would be given a bottom width and a sides slope parameter. So the program needs to be capable of computing A and R_H internally, given values of bottom width and sides slope.

The trapezoidal section is taken as the shape for the purpose of displaying a computational program. Accordingly, the program is named ***trapchan.p*** with a compiled or execute name ***trapchan.go***. The source code language is Pascal.

A Newton-Raphson algorithm is selected for computing the normal depth, and a bisection or half-interval method is chosen for the critical depth algorithm. The latter is as crude and unenriched, computationally speaking, as the former is elegant and efficient. However, the bisection method seems ideally suited to the critical depth task because the nature of the function whose root is to be determined is so favorable (*i.e.*, monotonic and slowly-varying). For both algorithms a starting value of y = 100 m is selected. In other words, it is assumed that all channels to be solved have a maximum depth of 100 m of water. This is a very safe assumption, unless one may be working with Loch Ness or Lake Superior, for instance.

The trapezoidal formula [not related to the cross-section shape] is selected for the numerical integration to determine the free-surface profile. It is

$$L = \sum_{i=1}^{m} \tfrac{1}{2} \left[f\left(y_{i+1}\right) + f\left(y_i\right) \right] \Delta y_i \tag{J.10}$$

hence the program user must decide upon the number of steps m that are to be taken and the size of increment Δy. Because the free-surface profile is of gently-changing depth (indeed, it is often called ***slowly-varying flow***), only a few intervals are generally employed in hand computations of the profile. However, when using a computer, a very large number can easily be accomodated. The program user must keep in mind that the intervals are with respect to ***depth***, so it is necessary to know the upper limit (as well as the starting value) of the depth, or else let the value of y_2 run indefinitely and afterwards interpolate between steps to obtain the depth at a given reach.

It is considered preferable to leave control of these latter parameters in the hands of the program user, rather than to build in a specific number of steps in view of the differing utilities associated with the options mentioned above.

Given below is the code for the program.

```
program TrapChan(input, output, lst);
{ This program computes the normal depth, the critical depth }
{ and the free surface profile for trapezoidal channel flow.   }

var
   b, c, g, h, l, n, q, s, z, a1, a2, a3, c1, c2, f1, f2, f3,
   p2, p3, r2, sc, so, r3, t1, t2, t3, y1, yn, den, num, ycr,
   delx, dely, fony, funy, newyn, ycrmax, ycrmin, y2, l1: real;
   k, m, m1, i: integer; reply: char; lst: text;

label
   10, 20, 30;

begin

   10:

   rewrite(lst,'trapchan.dat');

   for i:=1 to 25 do
      begin
         writeln(' ');
      end;
```

```
write('Please give the discharge in cubic meters/s '); readln(q); writeln(' ');
write('Please say what the bedslope is '); readln(so); writeln(' ');
write('List the Mannings n '); readln(n); writeln(' ');
write('What is the sides slope parameter? '); readln(z); writeln;
write('Please furnish the initial depth in meters '); readln(y1); writeln;
write('What is the depth increment in meters? '); readln(dely); writeln(' ');
write('How many steps do you wish to have? '); readln(m);
y2:=y1;
m1:=m;

writeln(' '); writeln(' ');

if so = 0 then
   begin
      writeln(' The normal depth is infinite');
      goto 20;
   end
else
if so ≤ 0 then
   begin
      writeln(' The normal depth is imaginary');
      goto 20;
   end
else
begin
   c:=q*n/sqrt(so);
   c1:=-2*exp(1.5*ln(c))*sqrt(1+z*z);
   c2:=-b*exp(1.5*ln(c));
   newyn:=100;
   repeat
   yn:=newyn;
   f1:=exp(2.5*ln(b*yn+z*yn*yn))+c1*yn+c2;
   f2:=2.5*exp(1.5*ln(b*yn+z*yn*yn))*(b+2*z*yn)+c1;
   newyn:=yn-(f1/f2);
   until abs(newyn-yn) ≤ 0.000001;
   writeln(' The normal depth is ',yn:6:2,' m');
end;

20:

writeln(' ');
ycrmax:=100;
ycrmin:=0;
repeat
   ycr:=0.5*(ycrmax+ycrmin);
   t1:=b+2*z*ycr;
   a1:=b*ycr+z*exp(2*ln(ycr));
```

```
      f3:=1-((q*q*t1)/(9.806*a1*a1*a1));
      if f3 ≥ 0 then
         ycrmax:=ycr
      else
         ycrmin:=ycr;
   until abs(ycrmax-ycrmin) ≤ 0.000001;
   writeln(' The critical depth is ',ycr:6:2,' m'); writeln(' ');

   p2:=2*sqrt(1+z*z)*ycr+b;
   a2:=z*(ycr*ycr)+b*ycr;
   r2:=a2/p2;
   t2:=b+2*z*ycr;
   sc:=(9.806*(n*n)*a2)/(t2*exp(1.3333333333*ln(r2)));
   writeln(' The critical slope is ',sc,''); writeln(' ');

   k:=m;
   l:=0;

   writeln(' L = ',l:6:2,' m       Y = ',y1:6:2,' m');
   while m ≥ = 0 do

      begin
         a3:=b*y1+z*(y1*y1);
         p3:=b+2*(sqrt(z*z+1)*y1);
         r3:=a3/p3;
         t3:=b+2*z*y1;
         g:=(q*q)*t3;
         s:=exp(2*ln(n*q/(a3*exp(0.6666666667*ln(r3)))));
         h:=9.806*a3*a3*a3;
         den:=so-s;
         num:=1-(g/h);
         fony:=num/den;
         if k ≤≥ m then

            begin
               delx:=0.5*(fony+funy)*dely;
               l:=l+delx;
               writeln(' L = ',l:6:2,' m       Y = ',y1:6:2,' m');
            end;

         m:=m-1;
         y1:=y1+dely;
         funy:=fony;
      end;

   writeln(' ');
```

```
write(' Do you wish to have a printout of this? '); readln(reply);
if (reply = 'y') or (reply = 'Y') then

  begin
    writeln(lst,' TRAPEZOIDAL CHANNEL'); writeln(lst,' ');
    writeln(lst,' The discharge is ',q:8:6,' cubic meters/s'); writeln(lst,' ');
    writeln(lst,' The bedslope is ',so:10:8,'  '); writeln(lst,' ');
    writeln(lst,'The Mannings n is ',n,'  '); writeln(' ');
    writeln(lst,'The bottom width is ',b:6:4,' m'); writeln(lst,' ');
    writeln(lst,'The sides slope parameter is ',z:6:4,'  '); writeln(lst,' ');
    writeln(lst,'The initial depth is ',y2,' m'); writeln(lst,' ');

    if so = 0 then
      begin
        writeln(lst,'The normal depth is infinite');
        goto 30;
      end
    else

    if so ≤ 0 then
      begin
        writeln(lst,'The normal depth is imaginary');
        goto 30;
      end
    else

    begin
      writeln(lst,'The normal depth is ',yn:10:8,' m');          writeln(lst,' ');
    end;

    30:

    writeln(lst,'The critical depth is ',ycr:10:8,' m'  '); writeln(lst,' ');
    writeln(lst,'The critical slope is ',sc,'  '); writeln(lst,' ');

    k:=m1;
    l1:=0;
    writeln(lst,'L = ',l1:6:4,' m'       Y = ',y2:6:4,' m');

    while m1 ≥ = 0 do
      begin
        a3:=b*y2+z*(y2*y2);
        p3:=b+2(sqrt(z*z+1)*y2);
        r3:=a3/p3;
        t3:=b+2*z*y2;
        g:=(q*q)*t3;
        s:=exp(2*(ln(n*q/(a3*exp(0.6666666667*ln(r3))))));
```

```
                h:=9.806*a3*a3*a3;
                den:=so−s;
                num:=1-(g/h);
                if k <> m1 then
                   begin
                      delx:=0.5*(fony+funy)*dely;
                      l1:=l1+delx;
                      writeln(lst,'L = ',l1:6:4,' m       Y = ',y2:6:4,' m');

                   end;
                m1:=m1−1;
                y2:=y2+dely;
                funy:=fony;
             end;
          end
       else
       writeln(' ');
       writeln(' Do you wish to do another problem? ');
       readln(reply);
       if (reply = 'y') or (reply = 'Y') then
          begin
             goto 10;
          end
          else
       for i:=1 to 25 do
          begin
             writeln(' ');
          end;
       writeln(' Thank you for using this program');
       writeln(' ');
       writeln(' Have a nice day! ');
end.
```

The program writes to an output file named *trapchan.dat*, which is re-written each time the program is run. This facilitates the task of importing the data into a graphics display software, should that be desired.

J.2.2 EXAMPLE PROBLEM

A trapezoidal channel with sides slope $z = 1$, Manning's $n = 0.012$, bedslope $S_o = 0.001$ and bottom width 4 m carries a discharge of 35 m^3/s. Compute the normal depth, the critical depth, the critical slope and the free-surface profile downstream from the place where the depth is 3.0 m to where its is 3.6 m.

These data were entered in response to program prompts, along with 6 steps of 0.01 m increment each, to produce the following machine output:

TRAPEZOIDAL CHANNEL

The discharge is 35.000000 cubic meters/s
The bedslope is 0.00100000
The Mannings n is 1.20000001e−02
The bottom width is 4.0000 m
The sides slope parameter is 1.0000
The initial depth is 3.00000000e+00 m
The normal depth is 1.95315480 m
The critical depth is 1.70788383 m
The critical slope is 1.62695546e−03

L = 0.0000 m Y = 3.0000 m
L = 107.5056 Y = 3.1000
L = 213.8784 Y = 3.2000
L = 319.3300 Y = 3.3000
L = 424.0261 Y = 3.4000
L = 528.0953 Y = 3.5000
L = 631.6401 Y = 3.6000

J.2.3 REFERENCE

Vennard, J. K., and Street, R. L., *Elementary Fluid Mechanics*, Sixth Edition, John Wiley & Sons, New York, 1982, pp. 452, 466, 476.

APPENDIX K

EIGENVALUE COMPUTATIONS

K.1 THE ALGORITHM

A great many eigenvalue problems associated with vectors and tensors of problems in engineering and science are formulated in terms of three-dimensional Cartesian matrices composed of real elements. For such circumstances, the eigenvalue problem is amenable to a special treatment with a straightforward procedure in the computation. Let the matrix in question be represented as

$$\mathbf{a} = \begin{bmatrix} a^1_1 & a^2_1 & a^3_1 \\ a^1_2 & a^2_2 & a^3_2 \\ a^1_3 & a^2_3 & a^3_3 \end{bmatrix} \tag{K.1}$$

In three dimensions, with real elements, adjoined to the other input data are the relationships incident to the geometry of the space itself; specifically, the orthonormality of the coordinate frame of reference. The eigenvectors lie along principal axes that possess the ortho-normal property, and this adds an additional equation to the set of three equations for determining each eigenvalue. As a consequence, the sets of equations are rendered independent, so that the eigenvalues themselves can be computed.

These equations are displayed in the section entitled EIGENVALUES in Chapter 15, Section 15.2.11.

It can be assumed that the elements of the 3×3 matrix are furnished as a matter of data at the outset, therefore the three invariants can be computed in a straightforward manner, so that the secular equation

$$f(\lambda) = -\lambda^3 + I_1\,\lambda^2 - I_2\,\lambda + I_3 = 0 \tag{K.2}$$

is explicitly available.

Because it is a cubic equation, the easiest way to obtain the roots will normally be a numerical procedure, and the most efficient one for use with contemporary computers is probably a Newton-Raphson algorithm. This procedure is very easy to program and to incorporate within a larger shell for the remainder of the determination. The three roots of the secular equation are the eigenvalues, and are named in accordance with the usual convention $\lambda_1 > \lambda_2 > \lambda_3$.

The second major component of the overall algorithm is the determination of the direction cosines of the eigenvectors. This requires the selection of any two of the first three equations in each set, plus the addition of the normalizing equation for the direction cosines of each principal axis. The latter is an equation of the second degree, therefore the set of three independent equations consists of two linear ones and one nonlinear one, simultaneous in the three unknown ℓs associated with the axis in question.

A favored solution procedure for this combination is the *Gauss elimination algorithm*, which is normally carried out numerically. For the three dimensional problem, however, it can actually be done in an algebraic format without too much tedium. The end result is a set of formulas for getting the numerical answers by direct substitution in the formulas. The formulas are shown in Equations K.3, K.4, and K.5, below.

$$\left\{\left[\left(\frac{a_2^1}{a_1^1-\lambda_i}\right)\left(\frac{a_3^2(a_1^1-\lambda_i)-a_1^2a_3^1}{(a_2^2-\lambda_i)(a_1^1-\lambda_i)-(a_2^1)^2}\right)-\frac{a_3^1}{a_1^1-\lambda_i}\right]\right\}^2$$

$$+\left\{\frac{a_3^2(a_1^1-\lambda_i)-a_1^2a_3^1}{(a_2^2-\lambda_i)(a_1^1-\lambda_i)-(a_2^1)^2}\right\}^2+1=\frac{1}{(\ell_i^3)^2} \tag{K.3}$$

$$\ell_i^2=-\left(\frac{a_3^2(a_1^1-\lambda_i)-a_1^2a_3^1}{(a_1^1-\lambda_i)(a_2^2-\lambda_i)-(a_2^1)^2}\right)\ell_i^3 \tag{K.4}$$

$$\ell_i^1=-\frac{a_2^1}{a_1^1-\lambda_i}\ell_i^2-\frac{a_3^1}{a_2^1-\lambda_i}\ell_i^3 \tag{K.5}$$

K.2 THE PROGRAM

The source code for the program is written in Turbo-Pascal. The inputs are the a^i_j elements of the matrix. Outputs are to the screen of the monitor and to a printer for hard copy.

The outputs consist of the three eigenvalues (principal values), and the three direction cosines for each of the three principal axes of the eigenstate.

```
program Eigenvalue(input,output,lst);
{ This program computes the eigenvalues and eigenvectors  }
{ of a 3x3 matrix with real elements. }

label
   10;
var
   a11, a12, a13, a21, a22, a23, a31, a32, a33, l1, l2, l3,
   C, B, E, D, G, H,
   f, fprime, lambda, newlambda, lambda1, lambda2, lambda3,
   L131, L132, L121, L122, L111, L112, L231, L232, L221, L222,
   L211, L212, L331, L332, L321, L322, L311, L312, temp: real;
   J, start, max, tomp; integer;
   a: array[1..9] of real; reply: char;
   control: array[1..9] of integer;
```

```
begin
   10:  clrscr;
   writeln(' '); writeln(' ');
   write(' Enter a11 '); readln(a11);
   writeln(' '); writeln(' ');
   write(' Enter a12 '); readln(a12);
   writeln(' '); writeln(' ');
   write(' Enter a13 '); readln(a13);
   writeln(' '); writeln(' ');
   write(' Enter a21 '); readln(a21);
   writeln(' '); writeln(' ');
   write(' Enter a22 '); readln(a22);
   writeln(' '); writeln(' ');
   write(' Enter a23 '); readln(a23);
   writeln(' '); writeln(' ');
   write(' Enter a31 '); readln(a31);
   writeln(' '); writeln(' ');
   write(' Enter a32 '); readln(a32);
   writeln(' '); writeln(' ');
   write(' Enter a33 '); readln(a33);
   writeln(' '); writeln(' ');

   I1:=a11+a22+a33;
   I2:=a11*a22-a21*a12+a11*a33-a31*a13+a22*a33-a32*a23;
   I3:=a11*a22*a33+a21*a32*a13+a31*a12*a32-a31*a22*a13
         -a12*a21*a33-a11*a32*a23;

   a[1]:=a11;  a[2]:=a12;  a[3]:=a13;
   a[4]:=a21;  a[5]:=a22;  a[6]:=a23;
   a[7]:=a31;  a[8]:=a32;  a[9]:=a33;

   for start:= 1 to 9 do

       begin
          max:=start;
          for J:=start to 9 do
             if a[J] > a[max] then
             max:=J;
             temp:=a[start];
             tomp:=control[start];
             a[start]:=a[max];
             control[start]:=control[max];
             a[max]:=temp;
             control[max]:=tomp;
          end;

   newlambda:=10*a[1];
```

```
repeat
   lambda:=newlambda;
   f:=lambda*lambda*lambda-I1*lambda*lambda+I12*lambda-I3;
   fprime:=3*lambda*lambda-2*I1*lambda+I2;
   newlambda:=lambda-f/fprime;
until abs(newlambda-lambda) < 0.00001*abs(a[1]);

lambda1:=newlambda;
lambda2:=-((lambda1-I1)/2)+sqrt(((lambda1-I1)/2)+((lambda1-I1)/2)
            -(lambda1*(lambda1-I1)+I2));
lambda3:=-((lambda1-I1)/2)-sqrt(((lambda1-I1)/2)*((lambda1-I1)/2)
            -(lambda1*(lambda1-I1)+I2));

C:=(a23*(a11-lambda1)-a21*a13)/((a22-lambda1)*(a11-lambda1)-sqr(a12));
B:=1+sqr(C)+sqr((a12/(a11-lambda1))*C-(a13/(a11-lambda1)));

L131:=sqrt(1/B); L132:=-L131;
L121:=((a21*a13-a23*(a11-lambda1))/((a11-lambda1)*(a22-lambda1)
        -sqr(a12)))*L131;
L122:=((a21*a13-a23*(a11-lambda1))/((a11-lambda1)*(a22-lambda1)
        -sqr(a12)))*L132;
L111:=((-a12)/(a11-lambda1))*L121-((a13)/(a11-lambda1))*L131;
L112:=((-a12)/(a11-lambda1))*L122-((a13)/(a11-lambda1))*L132;

E:=(a23*(a11-lambda2)-a21*a13)/((a22-lambda2)*(a11-lambda2)-sqr(a12));
D:=1+sqr(E)+sqr((a12/(a11-lambda2))*E-(a13/(a11-lambda2)));

L231:=sqrt(1/D); L232:=-L231;
L221:=((a21*a13-a23*(a11-lambda2))/((a11-lambda2)*(a22-lambda2)
        -sqr(a12)))*L231;
L222:=((a21*a13-a23*(a11-lambda2))/((a11-lambda2)*(a22-lambda2)
        -sqr(a12)))*L232;
L211:=((-a12)/(a11-lambda2))*L221-((a13)/(a11-lambda2))*L231;
L212:=((-a12)/(a11-lambda2))*L222-((a13)/(a11-lambda2))*L231;

G:=(a23*(a11-lambda3)-a21*a13)/((a22-lambda3)*(a11-lambda3)-sqr(a12));
H:=1+sqr(G)+sqr((a12/(a11-lambda3))*G-(a13/(a11-lambda3)));

L331:=sqrt(1/H); L332:=-L331;
L321:=((a21*a13-a23*(a11-lambda3))/((a11-lambda3)*(a22-lambda3)
        -sqr(a12)))*L331;
L322:=((a21*a13-a23*(a11-lambda3))/((a11-lambda3)*(a22-lambda3)
        -sqr(a12)))*L332;
L311:=((-a12)/(a11-lambda3))*L321-((a13)/(a11-lambda3))*L331;
L312:=((-a12)/(a11-lambda3))*L322-((a13)/(a11-lambda3))*L332;

writeln('a11 = ',a11:6:2,'  a12 = ',a12:6:2,'  a13 = ',a13:6:2,'');
```

```
writeln('a21 = ',a21:6:2,'  a22 = ',a22:6:2,'  a23 = ',a23:6:2,'');
writeln('a31 = ',a31:6:2,'  a32 = ',a32:6:2,'  a33 = ',a33:6:2,'');
writeln(' ');  writeln(' ');
writeln('I1 = ',I1:6:2,'  I2 = ',I2:6:2,'  I3 = ',I3:6:2,'');
writeln(' ');  writeln(' ');

writeln('lambda0 = ',10*a[1]:6:2,'  lambda1 = ',lambda1:6:2,'');
writeln(' ');  writeln(' ');
writeln('lambda2 = ',lambda2:6:2,'  lambda3 = ',lambda3:6:2,'');
writeln(' '); writeln(' ');

writeln('L111 = ',L111:6:2,'  L121 = ',L121:6:2,'  L131 = ',L131:6:2,'');
writeln('L112 = ',L112:6:2,'  L122 = ',L122:6:2,'  L132 = ',L132:6:2,'');
writeln(' '); writeln(' ');
writeln('L211 = ',L211:6:2,'  L221 = ',L221:6:2,'  L231 = ',L231:6:2,'');
writeln('L212 = ',L212:6:2,'  L222 = ',L222:6:2,'  L232 = ',L232:6:2,'');
writeln(' '); writeln(' ');
writeln('L311 = ',L311:6:2,'  L321 = ',L321:6:2,'  L331 = ',L331:6:2,'');
writeln('L321 = ',L312:6:2,'  L322 = ',L322:6:2,'  L332 = ',L332:6:2,'');

write(' Do you wish to have a printout of this? ');  readln(reply);
if (reply = 'y') or (reply = 'Y') then
  begin
    writeln(lst,' ');
    write(lst,' a11 = ',a11:6:2,'  a12 = ',a12:6:2,' ');
    writeln(lst,'a13 = ',a13:6:2,'');
    writeln(lst,' ');
    write(lst,' a21 = ',a21:6:2,'  a22 = ',a22:6:2,' ');
    writeln(lst,'a23 = ',a23:6:2,'');
    write(lst,' a31 = ',a31:6:2,'  a32 = ',a32:6:2,' ');
    writeln(lst,'a33 = ',a33:6:2,'');
    writeln(lst,' ');
    write(lst,' I1 = ',I1:6:2,'  I2 = ',I2:6:2,'');
    writeln(lst,'I3 = ',I3:6:2,'');
    writeln(lst,' ');
    write(lst,' lambda1 = ',lambda1:6:2,' ');
    write(lst,'lambda2 = ',lambda2:6:2,' ');
    writeln(lst,'lambda3 = ',lambda3:6:2,' ');
    writeln(lst,' ');
    write(lst,' L111 = ',L111:6:8,'  L121 = ',L121:6:8,'');
    writeln(lst,'L131 = ',L131:6:8,'');
    writeln(lst,' ');
    write(lst,' L112 = ',L112:6:8,'  L122 = ',L122:6:8,'');
    writeln(lst,'L132 = ',L132:6:8,'');
    writeln(lst,' ');
    write(lst,' L211 = ',L211:6:8,'  L221 = ',L221:6:8,'');
    writeln(lst,'L231 = ',L231:6:8,'');
```

```
        writeln(lst,' ');
        write(lst,' L212 = ',L212:6:8,'  L222 = ',L222:6:8,'');
        writeln(lst,'L232 = ',L232:6:8,'');
        writeln(lst,' ');
        write(lst,' L311 = ',L311:6:8,'  L321 = ',L321:6:8,'');
        writeln(lst,'L331 = ',L331:6:8,'');
        writeln(lst,' ');
        write(lst,' L312 = ',L312:6:8,'  L322 = ',L322:6:8,'');
        writeln(lst,'L332 = ',L332:6:8,'');
        writeln(lst,' ');
        writeln(' '); writeln(' '); writeln(lst,' '); writeln(lst,' ');
      end;

    write(' Do you wish to do another problem? '); readln(reply);
    if (reply = 'y') or (reply = 'Y') then
      begin
        writeln(' '); writeln(' '); goto 10;
      end;
      else
      begin
        for i:=1 to 25
          begin
            writeln(' ');
          end;
        writeln(' Thank you for using this program.'); writeln(' '); writeln(' ');
        writeln(' Have a nice day!');
        for i:=1 to 10
          begin
            writeln(' ');
          end;
      end;
end.
```

If one is accustomed to programming with other contemporary languages, such as C, C^{++}, etc., the code above can be called, converted, and run by appropriate commands. If one is running LINUX, say, the commands are *p2c*, followed by *gcc* or g^{++}, with pertinent options.

K.3 EXAMPLE PROBLEM

Let the matrix be the symmetrical one

$$\mathbf{a} = \begin{bmatrix} 50.0 & 27.0 & 33.0 \\ 27.0 & 60.0 & 40.0 \\ 33.0 & 40.0 & 40.0 \end{bmatrix} \tag{K.6}$$

The program ***eigen*** is run, and the printout is as follows.

a11 = 50.00   a12 = 27.00   a13 = 33.00

a21 = 27.00   a22 = 60.00   a23 = 40.00

a31 = 33.00   a32 = 40.00   a33 = 40.00

I1 = 150.00   I2 = 3982.00   I3 = 16780.00

lambda1 = 117.262253   lambda2 = 27.542143   lambda3 = 5.195603

L111 = 0.52998729   L121 = 0.63918019   L131 = 0.55728103
L112 = -0.52998729   L122 = -0.63918019   L132 = -0.55728103

L211 = -0.77223774   L221 = 0.63530772   L231 = 0.00574231
L212 = 0.77223774   L222 = -0.63530772   L232 = -0.00574231

L311 = -0.35037457   L321 = -0.43339680   L331 = 0.83030409
L312 = 0.35037457   L322 = 0.43339680   L332 = -0.83030409

It may be noted that the first invariant, I_1, has the value 150, and that

$$I_1 \equiv a_1^1 + a_2^2 + a_3^3 \equiv \lambda_1 + \lambda_2 + \lambda_3 \equiv 150 \tag{K.7}$$

Index